ECONOMICS, ECONOMETRICS AND THE LINK
Essays in Honor of Lawrence R. Klein

CONTRIBUTIONS
TO
ECONOMIC ANALYSIS

226

Honorary Editor:
J. TINBERGEN†

Editors:
D. W. JORGENSON
J. -J. LAFFONT
T. PERSSON
H. K. VAN DIJK

ELSEVIER
Amsterdam – Lausanne – New York – Oxford – Shannon – Tokyo

ECONOMICS, ECONOMETRICS AND THE LINK
Essays in Honor of Lawrence R. Klein

Executive Editor:

M. DUTTA
Rutgers University
New Brunswick, N.J., U.S.A.

Board of Editors:
Sir James Ball, Ronald G. Bodkin, Arthur S. Goldberger, Shinichi Ichimura,
Richard F. Kosobud, John M. Letiche, Roberto S. Mariano,
Joel Popkin, Paul Taubman

1995

ELSEVIER
Amsterdam – Lausanne – New York – Oxford – Shannon – Tokyo

HB
139
E343
1995

ELSEVIER SCIENCE B.V.
Sara Burgerhartstraat 25
P.O. Box 211, 1000 AE Amsterdam, The Netherlands

Library of Congress Cataloging-in-Publication Data

Economics, econometrics and the link : essays in honor of Lawrence R.
Klein / editor M. Dutta.
        p.   cm. -- (Contributions to economic analysis ; 226)
    Includes bibliographical references and index.
    ISBN 0-444-81787-5 (acid-free paper)
    1. Econometrics. 2. Economics. 3. Klein, Lawrence Robert.
4. Klein, Lawrence Robert--Bibliography.   I. Dutta, M. II. Series.
HB139.E343   1994
330'.01'5195--dc20
                                                            94-34470
                                                               CIP

ISBN: 0 444 81787 5

This book is printed on acid-free paper.

PRINTED IN THE NETHERLANDS

INTRODUCTION TO THE SERIES

This series consists of a number of hitherto unpublished studies, which are introduced by the editors in the belief that they represent fresh contributions to economic science.

The term "economic analysis" as used in the title of the series has been adopted because it covers both the activities of the theoretical economist and the research worker.

Although the analytical methods used by the various contributors are not the same, they are nevertheless conditioned by the common origin of their studies, namely theoretical problems encountered in practical research. Since for this reason, business cycle research and national accounting, research work on behalf of economic policy, and problems of planning are the main sources of the subjects dealt with, they necessarily determine the manner of approach adopted by the authors. Their methods tend to be "practical" in the sense of not being too far remote from application to actual economic conditions. In additon they are quantitative.

It is the hope of the editors that the publication of these studies will help to stimulate the exchange of scientific information and to reinforce international cooperation in the field of economics.

The Editors

Lawrence R. Klein

PREFACE

On Friday September 20, 1991, more than 150 friends of Lawrence R. Klein spent an evening with Sonia and Lawrence Klein at the magnificent Museum of the University of Pennsylvania in Philadelphia. These friends came from all continents. They proposed to prepare and publish a *festschrift* in honor of Lawrence R. Klein. This volume, *Economics, Econometrics and The LINK* is the end product from all his friends, including those who have contributed the twenty-nine chapters to this volume and the many others who were unable to join in this project. This volume is a tribute to the scholarly contributions of Lawrence R. Klein.

In the McCarthy era, Lawrence Klein left home "for the peace and academic freedom of Oxford," England. After a productive engagement in econometric model building of the United Kingdom at the Institute of Statistics at Oxford University, in 1958 he returned to the United States as a professor of economics at the University of Pennsylvania, where he has been engaged in many productive years. Beginning in 1968, he held the distinguished Chair - Benjamin Franklin Professor of Economics and Finance - at the University of Pennsylvania. In 1991, the title of emeritus of the same chair was accorded to him.

Friends of Lawrence Klein became members of a global family by sharing a common agenda: to relate the study of economics to the "studies of mankind." The core of the agenda has been conducting economic research on the broadest and deepest interlinkages of the world economies, indeed, of humankind. We believe that the best preface to this festschrift is to chronicle the works of Lawrence Klein. Although a complete bibliography of Klein's writings is provided, it seems most fitting to chronicle his major works.

The Keynesian Revolution (1947), a term Lawrence Klein coined, was based on his doctoral dissertation at the Massachusetts Institute of Technology (1944), and it won an instantaneous global readership. Professor Paul Samuelson cites him as "Columbus" in the voyage of exploration of macroeconomics.

Lawrence Klein brought to his studies of macroeconomics the theory of probability and mathematical specifications of behavioral equations of macroeconomic actors. He persistently searched for the basic structure with stable behavioral parameters of a macroeconomic system, thereby immensely enriching the literature of economics and econometrics. In 1950, Klein published *Economic Fluctuations in the United States, 1921-1941*, followed in 1955 by *An Econometric Model of the United States, 1929-1951* (with A.S.Goldberger). Klein was soon recognized as a leading authority on the art and science of econometric modeling of macroeconomies. Reference was made to him as America's Jan Tinbergen.

In 1959 Klein was awarded the John Bates Clark Medal, granted by the American Economic Association to the most distinguished American economist under the age of 40. In 1960, he was elected President of the Econometric Society and in 1977 Klein was elected President of the American Economic Association.

Lawrence Klein broadened his research in two directions - first, construction of econometric models of other macroeconomies; and second, construction of macroeconomic model of the United States economy on a much larger scale, so that many more economic variables could be treated as endogenous, enabling the model to answer many more economic questions. The two volumes were published: *An Econometric Model of the United Kingdom* (1961), prepared in collaboration with Ball, Hazlewood, and Vandome and *The Brookings Quarterly Econometric Model of the United States* (1965), edited with Duesenberry, Fromm, and Kuh. Successive generations of the Wharton Econometric Models of the United States and the Wharton Econometric Forecasting Unit's quarterly forecasts have been a rich addition to the treasure of our knowledge in economics.

Klein's leadership in constructing econometric models of additional macroeconomies has extended to many countries in Western Europe, Eastern Europe, Asia, Russia and the other members of the Commonwealth of Independent States, as well as the Americas. As early as 1968, Klein initiated "a new project to model the international transmission mechanism." His was a landmark recognition of the fact "that the U.S. economy is no longer *closed*". Indeed, the U.S. economy had to be treated analytically as an open economy and the specification that followed is the now famous LINK Model which Klein directs in collaboration with Bert G. Hickman. It is one of the first systematic approaches to interlink many sovereign macro-economies in a global scheme. The LINK Model, by endogenizing international trade volume and prices, represents an extension of the modeling exercise to specify the macro core of the world economy.

While Lawrence Klein is an econometrician, he has been an economist first. His contributions to econometrics have invariably been based on the teachings of great "masters" in economics; and his message has been never to compromise core economic theory with the "jungle-land of facts." Lawrence Klein reminded us that "there is a misconception, in general, as to what constitutes macroeconomics, because there are two dimensions to aggregation - over commodities and services, and over economic units (firms and households)". He rejected the doctrinaire views of *monetarism*, and forcefully argued that "money matters" in economic analysis but that its role should not be exaggerated.

His presidential address at the American Economic Association meetings (1977) on "The Supply Side" invites our attention to "a full supply-side *and* demand-side economics", with broad-based interdisciplinary relationships. He cautioned his audience not to confuse the legitimate aspects of the "supply side" with "the simplistic populist approaches through tax cuts". In 1983, he elaborated his analysis in the Royer lectures on *The Economics of Supply and Demand* and in *Lectures in Econometrics.* These were followed by his *Economics in Theory and Practice: An Eclectic Approach* (1989) with Jaime Marquez.

Economic studies, Klein emphasized, must not be based on doctrinaire precepts. He argued for "a new approach" for macroeconomic studies, *"neutral as far as doctrine is concerned".* Characteristically, he called for a new look at *The Keynesian Revolution.*

Highest professional honors have been bestowed on Lawrence R. Klein for his many contributions to economics and econometrics. He was elected a Distinguished Fellow of the American Economic Association in 1978, and was awarded the Alfred Nobel Memorial Prize in Economic Sciences in 1980. As his friends joined him and his wife, Sonia, in celebrating these events, it is hoped that the reader of this volume will have occasion to join them in paying homage to Lawrence R. Klein for his truly significant influence on our profession.

As Executive Editor of this volume, I would be remiss in not expressing my sincerest gratitude to the contributing authors and to my fellow editors, all of whom have given me their best cooperation. I owe particular editorial indebtedness to Arthur S. Goldberger and Ronald G. Bodkin. My sincerest acknowledgement is due to Barbara Ruth Campbell, who has given me invaluable technical assistance by way of "normalizing" the 29 diskettes, written in many different computer languages. Thanks are also due to Michael Ahearn for the photograph of Lawrence R. Klein.

M. Dutta, Executive Editor

Table of Contents

Macroeconometric Modeling

Econometric Analysis
- The Balance of Payments and The Exchange Rate

Price Competitiveness: The Home Price and The World Price

Sectoral Econometric Modeling

Country/Area Specific Econometric Modeling

The Works of Lawrence R. Klein

BOOKS

The Keynesian Revolution (New York: Macmillan, 1947), 2nd edition 1966.

Economic Fluctuations in the United States, 1921-1941 (New York: Wiley, 1950).

A Textbook of Econometrics (Evanston: Row, Peterson, and Co., 1953), 2nd edition (Englewood Cliffs, New Jersey: Prentice-Hall, Inc., 1974).

Contributions of Survey Methods to Economics (with G. Katona, J. Lansing and J. Morgan), (New York: Columbia University Press, 1954).

An Econometric Model of the United States, 1929-1951 (with A. S. Goldberger), (Amsterdam: North-Holland, 1955).

An Econometric Model of the United Kingdom (with R.J. Ball, A. Hazlewood, and P. Vandome), (Oxford: Basil Blackwell, 1961).

An Introduction to Econometrics (Englewood Cliffs, New Jersey: Prentice-Hall, 1962).

Readings in Business Cycles, ed. (with R.A. Gordon), for the American Economic Association (Homewood: Richard D. Irwin, 1965).

The Brookings Quarterly Econometric Model of the United States, ed. (with J. Duesenberry, G. Fromm, and E. Kuh), (Chicago: Rand McNally, 1965).

The Wharton Index of Capacity Utilization (with R. Summers) (Philadelphia: Wharton School of Finance and Commerce, 1967).

The Wharton Econometric Forecasting Model (with M.K. Evans) (Philadelphia: Wharton School of Finance and Commerce, 1967) 2nd enlarged edition, 1968.

Economic Growth: The Japanese Experience Since the Meiji Era, ed. (with K. Ohkawa), (Homewood: Irwin, 1968).

The Brookings Model: Some Further Results, ed. (with J. Duesenberry, G. Fromm, and E. Kuh), (Chicago: Rand McNally, 1969).

An Essay on the Theory of Economic Prediction (Helsinki: Yrjo Jahnsson Foundation, 1969), 2nd enlarged edition (Chicago: Markham, 1971).

Econometric Gaming: A Kit for Computer Analysis of Macroeconomic Models (with M. K. Evans and M. Hartley) (New York: Macmillan Co., 1969).

Essays in Industrial Econometrics, ed. in 3 vols. (Philadelphia: The Wharton School of Finance and Commerce, 1969).

The Brookings Model: Perspective and Recent Developments, ed. (with Gary Fromm), (Amsterdam: North-Holland, 1975).

Econometric Model Performance, ed. (with E. Burmeister), (Philadelphia: University of Pennsylvania Press, 1976).

An Introduction to Econometric Forecasting and Forecasting Models (with R. M. Young), (Lexington: D.C. Heath, Lexington Books, 1980).

Quantitative Economics and Development, Essays in Memory of Ta-Chung Liu, ed. (with M. Nerlove and S. C. Tsiang), (New York, Academic Press, 1980).

Econometric Models as Guides for Decision Making (New York: The Free Press, 1981), The Charles C. Moskowitz Memorial Lecture Series.

Wykłady z ekonometrii (Warszawa: PWE, 1982).

The Economics of Supply and Demand (Oxford: Basil Blackwell, 1983)

Industrial Policies for Growth and Competitiveness (with F. G. Adams), (Lexington, Mass.: D.C. Heath and Co., 1983).

Lectures in Econometrics (Amsterdam: North-Holland, 1983) (with a chapter by Władysław Welfe).

Capital Flows and Exchange Rate Determination Supplementum 3, Zeitschrift für Nationalökonomie ed. (with W. Krelle), (Wien: Springer-Verlag, 1983).

Economics in Theory and Practice: An Eclectic Approach ed. (with Jaime Marquez, (Dordrecht: Kluwer, 1989).

A History of Macroeconometric Model-Building (with Ronald G. Bodkin and Kanta Marwah), (Aldershot, England: Edward Elgar, 1991).

Comparative Performance of US Econometric Models ed. (New York: Oxford University Press, 1991).

A Quest for a More Stable World Economic System, ed. (Dordrecht: Kluwer Academic Publishers, 1993).

SCHOLARLY PAPERS

(1943) "Pitfalls in the Statistical Determination of the Investment Schedule," Econometrica, Vol. 11, July-October, 246-58.

(1944) "The Statistical Determination of the Investment Schedule: A Reply," Econometrica, Vol. 12, January, 91-92.

(1944) "The Cost of a 'Beveridge Plan' in the United States," The Quarterly Journal of Economics, Vol. LVIII, May, 423-37.

(1946) "Macroeconomics and the Theory of Rational Behavior," Econometrica, Vol. 14, April, 93-108.

(1946) "Dispersal of Cities and Industries," (with J. Marschak and E. Teller), Bulletin of Atomic Scientists, Vol. 1, 13-15, and 20.

(1946) "A Post-Mortem on Transition Predictions of National Product," The Journal of Political Economy, Vol. LIV, August, 289-308.

(1946) "Remarks on the Theory of Aggregation," Econometrica, Vol. 14, October, 303-12.

(1947) "The Use of Econometric Models as a Guide to Economic Policy," Econometrica, Vol. 15, April, 111-51.

(1947) "Theories of Effective Demand and Employment," The Journal of Political Economy, Vol. LV, April, 108-31.

(1948) "Notes on the Theory of Investment," Kyklos, Vol. 2, 97-117.

(1948) "Planned Economy in Norway," The American Economic Review, Vol. XXXVIII, December, 795-814.

(1948) "Economic Planning - Western European Style," Statsokonomisk Tidsskrift, Vols. 3-4, 97-124.

(1947-8) "A Constant-Utility Index of the Cost of Living," (with H. Rubin), The Review of Economic Studies, Vol. XV (2), 84-87.

(1949) "A Scheme of International Compensation," Econometrica, Vol. 17, April, 145-59.

(1950) "Stock and Flow Analysis in Economics," Econometrica, Vol. 18, July, 236-41, and 246.

(1950) "The Dynamics of Price Flexibility: A Comment," The American Economic Review, Vol. XL, September, 605-09.

(1951) "The Life of John Maynard Keynes," The Journal of Political Economy, Vol. LIX, October, 443-51.

(1951) "Estimating Patterns of Savings Behavior from Sample Survey Data," Econometrica, Vol. 19, October, 438-54.

(1951) "Studies in Investment Behavior," Conference on Business Cycles, National Bureau of Economic Research, New York, 233-303.

(1951) "Results of Alternative Statistical Treatments of Sample Survey Data," (with J.N. Morgan) The Journal of the American Statistical Association, Vol. 46, December, 442-60.

(1951) "Assets, Debts, and Economic Behavior," Studies in Incoem and Wealth, Vol. 14, National Bureau of Economic Research, New York, 197-227.

(1952) "Psychological Data in Business Cycle Research," (with G. Katona) The American Journal of Economics and Sociology, Vol. 12, October, 11-22.

(1952-53) "On the Interpretation of Professor Leontief's System," The Review of Economic Studies, Vol. XX, 131-36.

(1953) "National Income and Product, 1929-50," The American Economic Review, Vol. XLIII, March, 117-32.

(1953) "Savings Concepts and Data: The Needs of Economic Analysis and Policy," Savings in the Modern Economy, eds. W.W. Heller, F.M. Boddy & C.L. Nelson (Minneapolis: University of Minnesota Press), 104-107.

(1953) "Negro-White Savings Differentials and the Consumption Function Problem," (with H.W. Mooney) Econometrica, Vol. XXI, 425-56.

(1953) "The Estimation of Disposable Income by Distributive Shares," (with Lenore Frane) The Review of Economics and Statistics, Vol. XXXV, November, 333-37.

(1954) "A 'Mild Down Turn' in American Trade," (with A.S. Goldberger) The Manchester Guardian Weekly, January 7, p. 3.

(1954) "Statistical Studies of Unincorporated Business," (with J. Margolis) The Review of Economics and Statistics, Vol. XXXVI, February, 33-46.

(1954) "Savings and the Propensity to Consume," Determining the Business Outlook, ed. H. Prochnow, New York, Harper and Bros., 109-25.

(1954) "A Quarterly Model of the United States Economy," (with H. Barger) Journal of the American Statistical Association, Vol. 49, September, 413-37.

(1954) "Empirical Foundations of Keynesian Economics," Post Keynesian Economics, ed. K. K. Kurihara, (New Brunswick: Rutgers Univ. Press) 277-319.

(1954) "The Contribution of Mathematics in Economics," The Review of Economics and Statistics, Vol. XXXVI, November, 359-61.

(1955) "The U.S. Economy in 1955," The Manchester Guardian, January 3.

(1955) "British and American Consumers - A Comparison of Their Situations and Finances," The Bankers Magazine, March, 241-46.

(1955) "The Savings Survey 1955 - Response Rates and Reliability of Data," Bulletin of the Oxford University Institute of Statistics, Vol. 17, 91-126 (with T.P. Hill and K.H. Straw).

(1955) "Major Consumer Expenditures and Ownership of Durable Goods," Bulletin of the Oxford University Institute of Statistics, Vol. 17, 387-414.

(1955) "Decisions to Purchase Consumer Durable Goods," (with J.B. Lansing) The Journal of Marketing, Vol. XX, 109-32.

(1955) "Statistical Testing of Business Cycle Theory: The Econometric Method," The Business Cycle in the Post-War World, ed. E. Lundberg, (London; Macmillan).

(1955) "On the Interpretation of Theil's Method of Estimating Economic Relationships," Metroeconomica, Vol. VII, December, 147-53.

(1955) "Patterns of Savings - The Surveys of 1953 and 1954," Bulletin of the Oxford University Institute of Statistics, Vol. 17, 173-214.

(1956) "Insulation of the Modern Economy," The Banker's Magazine, January, 1-5.

(1956) "The Practicability of an Expenditure Tax in the Light of the Oxford Savings Survey," The Banker's Magazine, March, 235-39.

(1956) "Personal Savings and the Budget," The Banker's Magazine, June, 485-89.

(1956) "Econometric Models and the Evidence of Time Series Analysis," The Manchester School, Vol. XXIV, May, 197-201.

(1956)   "Savings and Finances of the Upper Income Classes," Bulletin of the Oxford University Institute of Statistics, Vol. 18, 293-319 (with K.H. Straw and P. Vandome)   .

(1956)   "Quelques Aspects Empiriques du Modele de Cycle Economique de Kaldor," Les Modeles Dynamiques en Econometrie (Paris: Centre National de la Recherche Scientifique), 127-43. [vec la collaboration de A. Buckberg, L. Gyorki et H. Runyon]

(1956-57) "The Interpretation of Leontief's System - A Reply," The Review of Economic Studies, Vol. XXXV (1), 69-70.

(1957)   "The Scope and Limitations of Econometrics," Applied Statistics, Vol. VI, 1-18.

(1957)   "A Note on 'Middle-Range' Formulation," Common Frontiers of the Social Sciences, ed. M. Komarovsky, The Free Press, Glencoe, 383-91.

(1957)   "Sampling Errors in the Savings Surveys," Bulletin of the Oxford University Institute of Statistics, Vol. 19, 85-105 (with Peter Vandome).

(1957)   "The Significance of Income Variability on Saving Behavior," Bulletin of the Oxford University Institute of Statistics, Vol. 19, 151-60, (with N. Liviatan).

(1957)   "Trade of the United Kingdom and the Sterling-Area in Two American Recessions," The Banker's Magazine, Vol. CLXXXV, 426-31 (with R. J. Ball and A. Hazlewood).

(1958)   "The British Propensity to Save," Journal of the Royal Statistical Society, Series A (general), Vol. 121, 60-96.

(1958)   "The Friedman-Becker Illusion," The Journal of Political Economy, Vol. LXVI, 539-45.

(1958)   "The Estimation of Distributed Lags," Econometrica, Vol. 26, October, 553-65.

(1958)   "Measuring Soviet Industrial Growth," Bulletin of the Oxford University Institute of Statistics, Vol. 20, 373-77.

(1958)   "Econometric and Sample Survey Methods of Forecasting," Business Forecasting, No. 3, the Market Research Society, London, 9-18.

(1959)   "Econometric Forecasts for 1959," Bulletin of the Oxford University Institute of Statistics, Vol. 21, 3-16 (with R. J. Ball and A. Hazlewood).

(1959)   "Some Econometrics of the Determination of Absolute Prices and Wages," Economic Journal, Vol. LXIX, Sept, 465-82 (with Ball).

(1959)   "Economic Forecasting," Kyklos, Vol. XII, Fasc. A, 650-57.

(1960)   "The American Balance-of-Payments Problem," The Bankers Magazine, April, 299-305.

(1960)   "Some Theoretical Issues in the Measurement of Capacity," Econometrica, Vol. 28, April, 272-86.

(1960)   "The Efficiency of Estimation in Econometric Models," Essays in Economics and Econometrics, ed. R.W. Pfouts, (Chapel Hill: Univ. of North Carolina Press), 216-32.

(1960)   "Single Equation vs. Equation System Methods of Estimation in Econometrics," Econometrica, Vol. 28, October, 866-71.

(1960)   "Entrepreneurial Saving," Proceedings of the Conference on Income and Saving, Vol. II (eds. I. Friend and R. Jones) (Philadelphia: Univ. of Pennsylvania Press), 297-335.

(1961)   "Re-Estimation of the Econometric Model of the U.K. and Forecasts for 1961," Bulletin of the Oxford University Institute of Statistics Vol. 23, 23-40 (with A. Hazlewood and P. Vandome).

(1961)   "Some Econometrics of Growth: Great Ratios of Economics," The Quarterly Journal of Economics, Vol. LXXV, 173-98 (with R. F. Kosobud).

(1961)   "A Model of Japanese Economic Growth, 1878-1937," Econometrica, Vol. 29, July, 277-92.

(1961)   "An Econometric Analysis of the Postwar Relationship Between Inventory Fluctuations and Changes in Aggregate Economic Activity," Inventory Fluctuations and Economic Stabilization, Part III, Joint Economic Committee, U.S. Congress, Washington, D.C. USGPO, 69-89, (with Joel Popkin).

(1962)   "The Measurement of Industrial Capacity," Hearings Before the Sub-Committee on Economic Statistics, J.E.C. 87th Congress, 2nd Session, May, 53-59.

(1962) "Singularity in the Equation Systems of Econometrics: Some Aspects of the Problem of Multicollinearity," International Economic Review, III, September, 274-99 (with M. Nakamura).

(1963) "An Econometric Model of Japan, 1930-59," International Economic Review, IV, January, 1-29 (with Y. Shinkai).

(1964) "A Postwar Quarterly Model: Description and Applications," Models ofIncome Determination, Studies in Income and Wealth, Vol. 28 (Princeton: Princeton Univ. Press), 11-36.

(1964) "Empirical Aspects of the Trade-Offs Among Three Goals: High Level Employment, Price Stability and Economic Growth," Inflation, Growth and Employment, Commission on Money and Credit (New York: Prentice-Hall), 367-428 (with Ronald G. Bodkin).

(1964) "A Quarterly Econometric Model of Japan, 1952-1959," Osaka Economic Papers, Vol. 22, March, 19-44 (with S. Ichimura, S. Koizumi, K. Sato and Y. Shinkai).

(1964) "Economics as a Behavioral Science," The Behavioral Sciences: Problems and Prospects, Institute of Behavioral Science, Univ. of Colorado, August, 21-26.

(1964) "The Social Science Research Council Econometric Model of the United States," Colston Papers, Vol. XVI, University of Bristol, 129-68.

(1964) "The Keynesian Revolution Revisited," The Economic Studies Quarterly, XV, November, 1-24.

(1964) "The Role of Econometrics in Socialist Economics," Problems of Economic Dynamics and Planning (Warsaw: PWN-Polish Scientific Publishers), 181-91.

(1965) "Stocks and Flows in the Theory of Interest," The Theory of Interest Rates, eds. F. Hahn and F. Brechling, (London: Macmillan), 136-51.

(1965) "The Brookings - SSRC Quarterly Econometric Model of the US: Model Properties," American Economic Review, Papers and Proceedings, LV, May, 348-61 (with Gary Fromm) .

(1965) "What Kind of Macro-Econometric Model for Developing Economies?" The Indian Economic Journal, XIII, 313-24.

(1966) "On Econometric Models and Economic Policy," The Oriental Economist, XXXIV, June, 375-78.

(1967) "Problems in the Estimation of Interdependent Systems," Model Building in the Human Sciences, ed. H. O. A. Wold (Monaco: Union Europeene d'Editions), 51-58.

(1967) "Racial Patterns of Income and Employment in the U.S.A.," Social and Economic Administration, I, January, 32-42.

(1967) "Some New Results in the Measurement of Capacity Utilization," American Economic Review, LVII, March, 34-58 (with R. Preston)

(1967) "Nonlinear Estimation of Aggregate Production Functions," The Review of Economics and Statistics, XLIX, February, 28-44 (with R. G. Bodkin)

(1967) "On the Possibility of Another '29," The Economic Outlook for 1967, (Ann Arbor: University of Michigan, Dept. of Economics), 45-87.

(1967) "Comment on Solving the Wharton Model," Review of Economics and Statistics, XLIX, November, 647-51.

(1967) "Wage and Price Determination in Macroeconometrics," Prices: Issues in Theory, Practice, and Public Policy, eds. A. Phillips and O. Williamson (Philadelphia: University of Pennsylvania Press), 82-100.

(1968) "Simultaneous Equation Estimation," International Encyclopedia of the Social Sciences, (NY: Macmillan and Free Press), vol. 14, 281-94.

(1968) "The Brookings Model Volume: A Review Article, A Comment," Review of Economics and Statistics, L, May, 235-40 (with Gary Fromm)

(1969) "The Role of Mathematics in Economics," The Mathematical Sciences, COSRIMS, (Cambridge: MIT Press), 161-75.

(1969) "Stochastic Nonlinear Models," Econometrica, 37, January, 95-106 (with R. Preston).

(1969) "Estimationof Interdependent Systems in Macroeconometrics," Econometrica, 37, April, 171-92.

(1969) "Econometric Model Building for Growth Projections," Business Economics IV, September, 45-50.

(1969) "On the Possibility of the General Linear Economic Model," Economic Models, Estimation, and Risk Programming, eds. K. Fox, J. K. Sengupta, and G. V. L. Narasimham, (Berlin: Springer-Verlag), (with D. W. Katzner), 343-64.

(1969) "Experience with Econometric Analysis of the U.S., Konjunktur Position," Is the Business Cycle Obsolete? ed. M. Bronfenbrenner, (New York: Wiley-Interscience)   (w i t h Michael K. Evans), 359-88.

(1969) "Specification of Regional Econometric Models," Papers of the Regional Science Association, XXIII, 105-15.

(1969) "Nobel Laureates in Economics," Science, 166, November 7, 715-17.

(1970) "Estimation of Distributed Lags," International Economic Review 11 June, 235-50 (with P. Dhrymes and K. Steiglitz) .

(1970) "Econometric Growth Models for the Developing Economy," Induction, Growth, and Trade, eds. W. Eltis, M. Scott, and J. N. Wolfe, (Oxford: Clarendon Press) (with J. Behrman), 167-87.

(1971) "Forecasting and Policy Evaluation Using Large Scale Econometric Models: The State of the Art," Frontiers of Quantitative Economics, ed. M. D. Intriligator (Amsterdam: North-Holland), 133-64.

(1971) "Estimating Effects within a Complete Econometric Model," Tax Incentives and Capital Spending, ed. Gary Fromm, (Amsterdam: North-Holland)(with Paul Taubman), 197-242.

(1971) "Wither Econometrics?" Journal of the American Statistical Association, 66, June, 415-21.

(1971) "Empirical Evidence on Fiscal and Monetary Models," Issues in Fiscal and Monetary Policy, ed. J. J. Diamond, (Chicago: DePaul University), 35-50.

(1971) "The Role of War in the Maintenance of American Economic Prosperity," Proceedings of the American Philosophical Society, 115, December 30, 507-16.

(1971) "Guidelines in Economic Stabilization: A New Consideration," Wharton Quarterly, VI (Summer), 20-24, (with Vijaya Duggal).

(1971) "The Survey: Lifeblood of the Quantitative Economist," Survey of Current Business, Anniversary Issue, 51, July, Part II, 108-110.

(1972) "The Treatment of Expectations in Econometrics," Uncertainty and Expectations in Economics, eds. C. F. Carter and J. L. Ford (Oxford: Blackwell), 175-90.

(1972) "Short-Run Prediction and Long-Run Simulation of the Wharton Model," Econometric Models of Cyclical Behavior, ed. by Bert. G. Hickman, (New York: Columbia University Press) (with M. Evans and M. Saito), 139-85.

(1972) "Short- and Long-Term Simulations with the Brookings Model," Econometric Models of Cyclical Behavior, ed. Bert G. Hickman, (New York: Columbia University Press) (with G. Fromm and G. Schink), 201-92.

(1972) "Analog Solution of Econometric Models," The Engineering Economist, Vol. 17, No. 2, 115-33, (with Hamid Habibagahi).

(1972) "Computerized Econometric Methods in Business Applications," Journal of Contemporary Business, Vol. 1, Spring, 63-71.

(1972) "Dynamic Properties of Nonlinear Econometric Models," International Economic Review, 13, October, 599-618 (with E. Philip Howrey)

(1972) "Price Determination in the Wharton Model," The Econometrics of Price Determination, ed. Otto Eckstein (Washington: Federal Reserve Board), 221-36.

(1972) "The Brookings Econometric Model: A Rational Perspective," Problems and Issues in Current Econometric Practice, ed. Karl Brunner, (Columbus: Ohio State University), (with Gary Fromm), 52-62.

(1972) "Anticipations Variables in Macro-Econometric Models," Human Behavior in Economic Affairs, ed. B. Strumpel, et al. (Amsterdam: Elsevier) (with F. G. Adams), 289-319.

(1973) "The Precision of Economic Prediction: Standards, Achievement, Potential," The Economic Outlook for 1973, (Ann Arbor: University of Michigan, Deptartment of Economics), 91-111.

(1972-73) "The Wharton Forecast Record: A Self-Examination," The Wharton Quarterly, Winter, 22-28 (with George R. Green).

(1973) "The Treatment of Undersized Samples in Econometrics," Econometric Studies of Macro and Monetary Relations, eds. A. A. Powell and R. A. Williams, (Amsterdam: North-Holland), 3-26.

(1973) "Background, Organization, and Preliminary Results of Project LINK," International Business Systems Perspectives, ed. C.G. Alexandrides, (Atlanta: Georgia State University, School of Business Administration), (with Bert Hickman and Rudolf R. Rhomberg), 19-56.

(1973) "A Comparison of Eleven Econometric Models of the United States," The American Economic Review, Papers and Proceedings LXIII, May, 385-93 (with Gary Fromm).

(1973) "The Impact of Disarmament on Aggregate Economic Activity: An Econometric Analysis," The Economic Consequences of Reduced Military Spending, ed. B. Udis (Lexington, MA: D.C. Heath) (with Kei Mori), 59-77.

(1973) "Introduction," The International Linkage of National Economic Models, ed. R. J. Ball (Amsterdam: North-Holland) (with B. Hickman and R. Rhomberg), 1-5.

(1973) "Forecasting World Trade Within Project Link," The International Linkage of National Economic Models, ed. R. J. Ball (Amsterdam: North-Holland) (with A. Van Peeterssen), 429-63.

(1973) "Dynamic Analysis of Economic Systems," International Journal of Mathematical Education in Science and Technology, 4, July-September, 341-59.

(1973) "Commentary on 'The State of the Monetarist Debate,'" Federal Reserve Bank of St. Louis Review, 55, September, 9-12.

(1973) "Capacity Utilization: Concept, Measurement, and Recent Estimates," Brookings Papers on Economic Activity, 3, 743-56 (with V. Long)

(1974) "Issues in Econometric Studies of Investment Behavior," Journal of Economic Literature, XII, March, 43-49.

(1974) "Notes on Testing the Predictive Performance of Econometric Models," International Economic Review, 15, June, (with E. P. Howrey and M. D. McCarthy), 366-83.

(1974) "LINK Model Simulations of International Trade: An Evaluation of the Effects of Currency Realignment," Journal of Finance, Papers and Proceedings, 29, May, (with K. Johnson), 617-30.

(1974) "Macroeconometric Model Building in Latin America: The Mexican Case," The Role of the Computer in Economic and Social Research in Latin America, National Bureau of Economic Research (New York: Columbia University Press) (with Abel Beltran del Rio), 161-90.

(1974) "Econometrics," Encyclopedia Britannica, 15th Edition, 200-201.

(1974) "Supply Constraints in Demand Oriented Systems: An Interpretation of the Oil Crisis," Zeitschrift fur Nationalokonomie 34, 45-56.

(1974) "An Econometric Analysis of the Revenue and Expenditure Control Act of 1968-69," Public Finance and Stabilization Policy, eds. W. L. Smith and J.M. Culbertson, (Amsterdam: North-Holland), 333-53.

(1974) "Intractability of Inflation," Methodology and Science, 7 (3), 156-73.

(1974)   "The Wharton Mark III Model - A Modern IS-LM Construct," International Economic Review 15, October, (with M. McCarthy and V. Duggal), 572-94.

(1974)   "Estimation and Prediction in Dynamic Econometric Models," Econometrics and Economic Theory, ed. W. Sellekaerts (London: Macmillan),(with H. N. Johnston and K. Shinjo), 27-56.

(1974)   "The Next Generation of Macro Models -- The Present and Steps in Progress," Proceedings of the Inaugural Convention of the Eastern Economic Association, Albany, N.Y., October, 25- 27, 24-33.

(1975)   "Stability in the International Economy: The LINK Experience," International Aspects of Stabilization Policies, eds. A. Ando, R. Herring, and R. Marston (Boston: Federal Reserve Bank of Boston), (with Keith Johnson), 147-88.

(1975)   "Research Contributions of the SSRC -- Brookings Econometric Model Project - A Decade in Review," The Brookings Model: Perspective and Recent Developments, eds. Gary Fromm and L. R. Klein (Amsterdam: North-Holland), 13-29.

(1975)   "The LINK Model of World Trade, with Applications to 1972-73," International Trade and Finance, ed. Peter Kenen (Cambridge: Cambridge University Press),(with C. Moriguchi and A. Van Peeterssen), 453-83.

(1976)   "Long Term Policies and Outlook for World Inflation," The Role of Japan in the Future World, Proceedings of the 2nd Tsukuba International Symposium (The University of Tsukuba, Japan), 99-110.

(1976)   "The NBER/NSF Model Comparison Seminar: An Analysis of Results," Annals of Economic and Social Measurement 5, 1-28 (with Gary Fromm)

(1976)   "Pacific Basin Econometric Research," Central Bank Macroeconomic Modeling in Pacific Basin Countries (San Francisco: Federal Reserve Bank of San Francisco), 29-41.

(1976)   "Five-Year Experience of Linking National Econometric Models and of Forecasting International Trade," Quantitative Studies of International Economic Relations, ed. H. Glejser (Amsterdam: North-Holland, 1-24.

(1976)   "Applications of the LINK System," (with K. Johnson, J. Gana, M. Kurose, and C. Weinberg) in The Models of Project LINK, ed. J. Waelbroeck, (Amsterdam: North-Holland), 1-16.

(1976)   "Statistical Needs for Economic Analysis: A User's Viewpoint," Proceedings of the Business and Economic Statistics Section, American Statistical Association, 110-13.

(1976-7) "Project LINK," The Columbia Journal of World Business, Vol. 11, Winter, 1976, 7-19; reprinted in Economics and Mathematical Methods, Vol. 13, May-June, 1977, 471-88; Academy of Sciences [USSR]; also in Lecture Series, No. 30 (Athens: Center of Planning and Economic Research, 1977)

(1977)   "Early Warning Signals of Inflation," [with Sonia A. Klein] in Economic Progress, Private Values, and Public Policy, ed. B. Balassa and R. Nelson, (Amsterdam: North-Holland), 263-77.

(1977)   "Intermediate Term Outlook for the Housing Market," in The Construction Industry: New Adaptations to a Changing Environment, eds. W. Gomberg and L. Robbins, (Philadelphia: Wharton Entrepreneurial Center, University of Pennsylvania) (with Vincent Su), 5-26.

(1977)   "Econometric Model Building at the Regional Level," in Regional Science and Urban Economics, No. 7, 3-23, (with Norman J. Glickman).

(1977)   "Waiting for the Revival of Capital Formation," The World Economy, No. 1, October, 35-46.

(1977)   "Comments on Sargent and Sims's 'Business Cycle Modeling Without Pretending to Have Too Much A Priori Economic Theory '," New Methods in Business Cycle Research: Proceedings From a Conference, ed. C.A. Sims,(Minneapolis: Federal Reserve Bank of Minneapolis), 203-208.

(1977)   "Econometrics of Inflation, 1965-1974: A Review of the Decade," Analysis of Inflation: 1965-1974, Studies in Income and Wealth, Vol. 42, ed. Joel Popkin. (Cambridge: Ballinger for the National Bureau of Economic Research), 35-64.

(1977)   "The Longevity of Economic Theory," Quantitative Wirtschaftsforschung, eds. Horst Albach, et al. (Tübingen: J.C.B. Mohr), 411-19; reprinted in Cahiers du Seminaire d'Econometrie, No. 20. (Paris: CNRS, 1979).

(1977)   "Economic Policy Formation Through the Medium of Econometric Models," Frontiers of Quantitative Economics, Vol. 3-B, ed. M. Intriligator, (Amsterdam: North-Holland), 765-82.

(1977)   "Comment on a Multiregional Input-Output Model of the World Economy," The International Allocation of Economic Activity, eds. G. Ohlin et al. (London: Macmillan), 531-37.

(1977)   "Some Observations on the World Business Cycle, "International Cooperation and Stabilization Policies; A New Dimension of Keynesian Policy, eds. L. R. Klein, and C. Moriguchi (Tokyo: The Forum for Policy Innovation), 4-16.

(1978)   "Money in a General Equilibrium System: Empirical Aspects of the Quantity Theory," Economie Appliquee, XXXI (1-2), 5-14.

(1978)   "Trade Impact Studies Using the Wharton Annual and Industry Forecasting Model," The Impact of International Trade and Investment on Employment, ed. William G. Dewald (Washington, D.C.: U.S. Department of Labor, Bureau of Labor Affairs), 293-306.

(1978)   "Perspectivas de la economia mundial, 1977-79, "Revista de Economia Latinoamericana, XIII, No. 52, 15-22, Publicada bajo los auspicios del Banco Central de Venezuela.

(1978)   Comment on: "An Overview of the Objectives and Framework of Seasonal Adjustment," by Shirley Kallek, Seasonal Analysis of Economic Time Series, (Washington, D.C., Bureau of the Census, ER-1), 30-32.

(1978)   "Demand Forecasting and Capacity Creation in the Private Sector I," Long Term Economic Planning, ed. by P. K. Mitra (Laxenburg, Austria: IIASA, 1978), 41-59.

(1978)   Ökonometrische Modelle: Empirische Anwendung," Handwörterbuch der Mathematischen Wirtschaftswissenschaften, Ökonometrie ünd Statistik, ed. Gunter Menger (Wiesbaden: Gabler, 1978)   105-18.

(1978)   "The Deterrent Effect of Capital Punishment: An Assessment of the Estimates," [with Brian Forst and Victor Filatov] in Deterrence and Incapacitation: Estimating the Effects of Criminal Sanctionson Crime Rates, eds.   Blumenstein, et al. (Washington, D.C.: National Academy of Sciences), 336-80.

(1978)   "Understanding Inflation," Alternative Directions in Economic Policy, eds. F. Bonello and T. Swartz, (Notre Dame: University of Notre Dame Press), 62-77.

(1978)   "Potentials of Econometrics for Commodity Stabilization Policy Analysis," Stabilizing World Commodity Markets, eds. F. G. Adams and S. A. Klein (Lexington, MA: Lexington Books), 105-16.

(1978)   "The Supply Side," American Economic Review, 68, March, 1-7.

(1978)   "Computer Modeling of Macroeconomic Systems: The State of the Art," Ökonometrische Modelle und Systeme, eds. Schober and Plotzeneder (München: Oldenburg), 25-38 (videotape recording of lecture of December 10, 1975, Ottignies, Belgium).

(1978)   "Oil and the World Economy," Middle East Review, X, Summer, 21-28, and Economic Impact, XXIII, 49-55.

(1979)   "Protectionism: An analysis from Project LINK," Journal of Policy Modeling (January), 5-35, (with V. Su).

(1979)   "Disturbances to the International Economy," After the Phillips Curve: Persistence of High Inflation and High Unemployment (Federal Reserve Bank of Boston) 94-103.

(1979)   "Econometrics," Across the Board, XVI, February, 49-58.

(1979) "The Next Generation of Macro Models: The Present and Steps in Progress," Communication and Control in Society, ed. Klaus Krippendorff (New York: Gordon and Bresch), 293-303.

(1979) "Transportation Demand - "Aggregate and Major Freight Category Demand Estimation," Forecasts of Freight System Demand and Related Research Needs (National Research Council, Assembly of Engineering, Committee on Transportation, Washington, D.C.), 10-25 (with Colin J. Loxley).

(1979) "Political Aspects of Economic Control," Theory for Economic Efficiency: Essays in Honor of Abba P. Lerner, eds. Harry I. Greenfield, et al. (Cambridge: MIT Press),76-91.

(1979) "Managing the Modern Economy: Econometric Specification," Optimal Control for Econometric Models: An Approach to Economic Policy Formulation, eds. Sean Holly, et al. (London: Macmillan), 265-85.

(1979) "Error Analysis of the LINK Model," (with K.N. Johnson),45-71; "Long Run Projections of the LINK World Trade Model," (with Asher Tischler),73-74; also, "Coordination of International Fiscal Policies and Exchange Rate Revaluations," (with Vincent Su and Paul Beaumont), 143-59, in Modelling the International Transmission Mechanism, ed. J. Sawyer, (Amsterdam: North-Holland).

(1979) "Direct Estimates of Unemployment Rate and Capacity Utilization in Macroeconometric Models," International Economic Review, 20, October, 725-40 (with Vincent Su).

(1979) "International Coordination of Economic Policies," Greek Economic Review I, August, 27-47 (with H. Georgiadis and V. Su).

(1979) "International Research Cooperation," Man, Environment, Space and Time, I, Fall, 47-51.

(1980) "Regional Sublinkages of Economic Systems," Proceedings of the Fourth Pacific Basin Central Bank Conference on Econometric Modeling, (Tokyo: Bank of Japan), 3-18.

(1980) "Recent Economic Fluctuations and Stabilization Policies: An Optimal Control Approach," Quantitative Economics and Development, ed. L.R. Klein, M. Nerlove, and S.C. Tsiang (New York: Academic Press), 225-54 (with Vincent Su)

(1980) "Use of Econometric Models in the Policy Process," Economic Modeling, ed. Paul Ormerod (London: Heinemann), 309-29.

(1980) "Money Supply Hard to Control," Controlling Money: A Discussion, intro. by W. R. Allen (Los Angeles: International Institute for Economic Research), 9-14, 39-42.

(1980) "On Econometric Models," Issues and Current Studies, The National Research Council, (Washington, D.C.: National Academy of Sciences,1981), 41-55.

(1980) "Some Economic Scenarios for the 1980s," Nobel Memorial Lecture December 8, Les Prix Nobel (Stockholm: Almqvist & Wiksell, 273-94.

(1981) "Tax Policies and Economic Expansion in the U.S.," Technology in Society, VI. 3, 205-12.

(1981) "Oil Prices and the World Economy," The Middle East Challenge, ed. Thomas Naff (Carbondale: Southern Illinois University Press), 75-85.

(1981) "The LINK Project," International Trade and Multi-Country Models, ed. R. Courbis (Paris: Economica), 197-209.

(1981) "Project LINK: Policy Implications for the World Economy," Knowledge and Power in a Global Society, ed. William M. Evan (Beverly Hills: Sage), 91-106.

(1981) "The Practice of Macro-Econometric Model Building and Its Rationale," in Large Scale Macro-Econometric Models, eds. J. Kmenta and J. Ramsey, (Amsterdam: North-Holland), 19-58, (with E. P. Howrey, M. D. McCarthy, G. Schink).

(1981) "Scale of Macro-Econometric Models and Accuracy of Forecasting," in in Large Scale Macro-Econometric Models, eds. J. Kmenta and J. Ramsey, (Amsterdam: North-Holland), 369-88, (with G. Fromm).

(1981) "Computers in Economics," "Econometrics," "Economic Models," Encyclopedia of Economics, ed. Douglas Greenwald (New York: McGraw-Hill), 303-308.

(1981)   "Equazione per il futuro," Revista IBM, XVII (3), 5-11.

(1981)   "International Aspects of Industrial Policy," Toward a New U.S. Industrial Policy, eds. M.L. and S.M. Wachter. (Philadelphia: University of Pennsylvania Press), 361-77.

(1981)   "Coordinated Monetary Policy and the World Economy," Prevision etAnalyse Economique 2 (July-October), 75-105. (with R. Simes and P. Voisin)

(1981)   "Purchasing Power Parity in Medium Term Simulation of the World Economy," Scandinavian Journal of Economics, 479-96. (with V. Filatov and S. Fardoust) .

(1982)   "The Neoclassical Tradition of Keynesian Economics and the Generalized Model," Samuelson and Neoclassical Economics ed. G.R. Feiwel, (Boston: Kluwer-Nijhoff), 244-62.

(1982)   "The Value of Models in Policy Analysis," Modeling Agriculture for Policy Analysis in the 1980s (Kansas City: Federal Reserve Bank of Kansas City), 1-18.

(1982)   "The World Economy - A Global Model," Perspectives in Computing 2, May, 4-17. (with Peter Pauly and Pascal Voisin).

(1982)   "The Scholarly Foundations of the Econometrics Industry," Economics and the World Around It, ed. S.H. Hymans (Ann Arbor: University of Michigan Press), 111-22.

(1982)   "Industrial Policy in the World Economy: Medium Term Simulations," Journal of Policy Modeling, 4 (2, 1982), 175-89, (with C.A.B. Bollino and S. Fardoust).

(1982)   "Two Decades of U.S Economic Policy and Present Prospects: A View From the Outside," The Political Economy of the United States ed. by Christian Stoffaes (Amsterdam: North-Holland), 111-24.

(1982)   "The Phillips Curve in the US," Prevision et Analyse Economique, 3 (Juillet-Decembre), 17-26.

(1982)   "Economic Theoretic Restrictions in Econometrics," Evaluating the Reliability of Macroeconomic Models, eds. Gregory C. Chow and Paolo Corsi (Chichester: John Wiley), 23-28.

(1982)   "The Supply Side of the Economy: A View From the Perspective of the Wharton Model," Supply Side Economics: A Critical Appraisal, ed. R.H. Fink (Frederick, MD: University Publications of America), 245-52.

(1982)   "Alternative Policies for Stable Non-Inflationary Growth," Supply Side Economics in the 1980s (Westport, Ct.: Quorum Books for the Federal Reserve Bank of Atlanta), 67-75.

(1982)   "The Present Debate about Macro Economics and Econometric Model Specification," Chung-Hua Series of Lectures by Invited Distinguished Economists, The Institute of Economics, Academia Sinica, Taiwan, July, 3-20.

(1983)   "Some Laws of Economics," Bulletin of the American Academy of Arts and Science XXXVI, January, 21-45.

(1983)   "NIPA Statistics: A User's View," The U.S. National Income and Product Accounts: Selected Topics. ed. M.F. Foss (Chicago: University of Chicago Press for NBER), 319-23.

(1983)   "Supply Side Modeling," Large Scale Energy Models: Prospects and Potential, eds. R.M. Thrall, R.G. Thompson, and M.L. Holloway (Washington, D.C.: American Association for the Advancement of Science), 55-75.

(1983)   "Money in the Wharton Quarterly Model," Journal of Money, Credit, and Banking, 15, May, 237-59, (with Edward Friedman and Stephen Able)

(1983)   "Modeling Exchange Rate Fluctuation and International Disturbance," Managing Foreign Exchange Risk, ed. R.J. Herring (Cambridge: Cambridge University Press), 85-109.

(1983)   "International Productivity Comparisons (A Review),Proceedings of the National Academy of Sciences of the U.S.A. 80, July, 4561-8.

(1983)   "Inflation: Its Causes and Possible Cures," Essays in Regional Economic Studies eds. M. Dutta, et. al. (Durham, N.C.: The Acorn Press), 26-32.

(1983) "A Model of Foreign Exchange Markets: Endogenising Capital Flows and Exchange Rates," Capital Flows and Exchange Rate Determination, ed. L.R. Klein and W.E. Krelle, Supplementum 3, Zeitschrift für NationalÖkonomie (Wien: Springer-Verlag), 61-95, (with K. Marwah).

(1983) "Long-term Simulation with the Project LINK System, 1978-85," Global International Economic Models, ed. B.G. Hickman (Amsterdam: North-Holland), 29-51.

(1983) "Identifying the Effects of Structural Change," Industrial Change and Public Policy (Kansas City: Federal Reserve Bank of Kansas City), 1-19.

(1983) "The Deficit and the Fiscal and Monetary Policy Mix," The Economics of Large Government Deficits (Boston: Federal Reserve Bank of Boston),Conference Series No. 27, 174-94.

(1984) "Money in the Wharton Quarterly Model: A Reply," Journal of Money, Credit and Banking, XVI, February, 76-79.

(1984) "The Importance of the Forecast," Journal of Forecasting 3, 1-9.

(1984) "An Interview," in The U.S.A. in the World Economy (San Francisco: Freeman, Cooper & Co.), 30-39.

(1984) "What Makes a Good Forecast?" Economic Forecasts: A Worldwide Survey, August, 19-21.

(1984) "Wage-Price Behavior in the National Models of Project LINK," American Economic Review 74, May, 150-54 (with B. Hickman)

(1985) "Asia and the Pacific Far East: A Region for International Relationships" in Studies in United States-Asia Economic Relations ed. M. Dutta, (NC Durham, The Acorn Press), 6-10.

(1985) "Trade Linkages within the Pacific Far East" (with Vincent Su) in Studies in United States-Asia Economic Relations ed. M. Dutta,(NC Durham, The Acorn Press), 13-34.

(1985) "Perspectives of Future World Trade -- Some Results of Project LINK," Probleme und Perspectiven der Weltwirtschaftlichen Entwicklung (Berlin: Duncker & Humboldt), 469-86.

(1986) "International Aspects of Saving," Savings and Capital Formation, The Policy Options, eds. F.G. Adams and S.M. Wachter (Lexington: D.C. Heath), 195-204.

(1986) "Macroeconometric Modeling As a Background to Development Planning," International Journal of Development Planning Literature, 1, January-March, 39-56, (with Ronald G. Bodkin and Kanta Marwah).

(1986) "Macroeconometric Modeling and Forecasting," Behavioral and Social Sciences. Fifty Years of Rediscovery, eds. Neil J. Smelser and Dean R. Gerstein (Washington, D.C.: National Academy Press), 95-110.

(1986) "Modeling the People's Republic of China and Its International Relationships," (Hong Kong: The Chinese University of Hong Kong, New Asia College), 1-24.

(1986) "Research Strategies in Macroeconometrics," A Tribute to Arvi Leponiemi on His 60th Birthday, eds. J. Kähkönen and J. Ylä-Liedenpohja (Helsinki: Helsinki School of Economics), 9-23.

(1986) "Lawrence R. Klein," Lives of the Laureates: Seven Nobel Economists (Cambridge: MIT Press), 21-41.

(1986) "The Energy Crisis Ten Years Later," The Future of Electrical Energy: A Regional Perspective of an Industry in Transition, eds. Sidney Saltzman and Richard E. Schuler (NY: Praeger), 18-31.

(1986) "Disarmament and Socio-Economic Development," Disarmament IX, No. 1, Spring, 49-63. Reprinted in Spanish in Commercio Externa vol. 36 (Dec. 1986)

(1986) "Keynes and the Origins of Macroeconometric Modeling," Eastern Economic Journal, XII, October-December, 442-50, (with R. G. Bodkin and K. Marwah).

(1987) "The South Asian and Pacific Far East Countries in Project LINK," Trade and Structural Change in Pacific Asia, eds. C. I. Bradford, Jr., and W. H. Branson (Chicago: University of Chicago Press for National Bureau of Economic Research) 157-69.

(1987) "The Restructuring of the American Economy," Pacific Northwest Executive 3, No. 1, January, 16-17.

(1987) "Empirical Aspects of Protectionism: Results from Project LINK," The New Protectionist Threat to World Trade, ed. D. Salvatore (Amsterdam: North-Holland), 69-94. (with Peter Pauly and Christian E. Petersen)

(1987) The Changing World Economic Situation and Its Impact on International Trade," The Export-Import Bank at Fifty, ed. Rita M. Rodriguez (Lexington Heath), 37-53.xxii

(1987) "Growth, Inflation and Employment: An Introduction," Structural Change, Economic Interdependence and World Development, eds. Luigi Pasinetti and Peter Lloyd for the IEA (London: Macmillan), 269-74.

(1987) "Economic Morality and Justice for All," Journal of Commerce and Finance, 6, Spring, 7-11.

(1987) "International Debt of Developing Countries--Any Steps Towards Solution?" Journal of Philippine Development XIV, No. 1, 1-11.

(1987) "Asia-Pacific Economies: Challenges and Prospects," in Asia-Pacific Economies: Promises and Challenges, ed. M. Dutta, vol. 6 Part A (Greenwich, CT: JAI Press), 3-8.

(1987) "The Choice Among Alternative Exchange-Rate Regimes," Journal of Policy Modeling 9, Spring, 7-18.

(1987) "Alcune considerazioni sull'economia umanistica," Fondamenti 8, 29-36.

(1987) "The Arms Race and the Economy," Vorträge des Festkolloquiums aus Anlass des 70 Geburtstages von Wilhelm Krelle (Bonn: Bouvier Verlag Herbert Grundmann), 9-51, (with H. Kosaka).

(1987) "Financial Innovation: Effects on Economic Performance," C.D. Deshmukh Memorial Lecture, Reserve Bank of India, Bombay, 8-23.

(1988) "The Statistical Approach to Economics," Journal of Econometrics 37, 7-26.

(1988) Testimony on "Reform of the Nation's Banking and Financial Systems," Congressional Hearings, The Committee on Banking, Finance, and Urban Affairs, U.S. House of Representatives, February 9, 57-62, and 271-326.

(1988) "The LINK Model and Its Use in International Scenario Analysis," Economic Modelling in the OECD Countries, ed. Homa Motamen (London: Chapman and Hall), 1-10.

(1988) "Global Monetarism," National Income and Economic Progress, eds. D. Ironmonger, J.O.N. Perkins, and Tran van Hoa (London: MacMillan), 168-76.

(1988) "Financial Innovation: Its Implications for Modelling, Policy Formation and Regulation," National Economic Review, No. 9, July, (National Institute of Economic & Industry Research, Melbourne) 10-14.

(1988) "The Medium-Term Outlook for the World Economy and the Implications for East-West Economic Relations," Macroeconomic Management and the Enterprise in East and West, ed. Christopher T. Saunders (London: MacMillan), 33-46 (with Daniel L. Bond).

(1988) "Components of Competitiveness," Science, 241, (15 July), 308-13.

(1988) "Carrying forward the Tinbergen Initiative in Macroeconometrics," Royal Netherlands Economics Association, The Hague. Reprinted in Review of Social Economy, XLVI, December, 231-51.

(1988) "Le Politiche Economiche Keynesiane: Un' Analisi Retrospettiva," Rassegna Economica LII, July-Sept, 501-20. xxiii

(1988) "Econometric Models, Planning and Developing Countries," Development Perspective, Institute of Economic Growth (Silver Jubilee Lectures) Chapter II, 29-40.

(1989) "The Economic Principles of Joan Robinson," Joan Robinson and Modern Economic Theory, ed. G. R. Feiwel (NY: NYU Press), 258-63.

(1989) "Econometric Aspects of Input-Output Analysis," Frontiers of Input-Output Analysis, eds. R. E. Miller, K. R. Polenski, and A. Z. Rose (New York: Oxford University Press), 3-11.

(1989) "Combinations of High and Low Frequency Data in Macroeconometric Models," eds. L. R. Klein and J. Marquez, Economics in Theory and Practice: An Eclectic Approach, (Dordrecht: Kluwer), 3-16, (with E. Sojo)

(1989) "The State of Macroeconomic Theory and Policy in Relation to Econometrics," The State of Economic Science, ed. Werner Sichel (Kalamazoo: W. E. Upjohn Institute), 41-58.

(1989) "The Case for International Coordination of Economic Policy," Business in the Contemporary World, 1, Spring, 11-16.

(1989) "The Restructuring of the American Economy," Inflation and Income Distribution in Capitalist Crisis, ed. J.A. Kregel, (London: MacMillan), 25-45.

(1989) "Désarmement et d'veloppment", Science, Guerre et Paix, ed. J.J. Salomon, (Paris: Economica), 157-67.

(1989) "The Two-Gap Paradigm in the Chinese Case: A Pedagogical Exercise," China Economic Review, 1, Spring, 1-8, (with Youcai Liang)

(1989) "Outward-Looking Policy Scenarios for Africa with LINK" African Development Perspectives, ed. Dominick Salvatore (New York: Taylor and Francis for the United Nations), 234-40.

(1989) "Development and Prospects in Macroeconometric Modeling," Eastern Economic Journal XV, 4, October-December, 287-304.

(1989) "Diagnosis of Contemporary American Economic Problems," Congressional Hearings, The Commitee on Banking, Housing, and Urban Affairs, the U. S. Senate, November 15, 390-95.

(1989) "Global Adjustment and the Future of Asian-Pacific Economies," Global Adjustment and the Future of the Asian-Pacific Economy, eds. Miyohei Shinohara and Fu-chen Lo (Tokyo: Asian and Pacific Development Center and Institute of Developing Economies), 15-27.

(1990) "The Open Economy" Open Economies Review 1, 3-16.

(1990) "Has there Been a Structural Change in the World Economy?" Problems of Building and Estimation of Large Econometric Models (Part A), (Lodz, Poland: Wydawnictwo Uniwersytetu Lodzkiego),eds. W. Welfe and P. Tomczyk, 7-24.

(1990) "Cyclical Indicators in Econometric Models," Analyzing Modern Business Cycles, ed. P.A. Klein (Armonk: M.E. Sharpe), 97-106.

(1990) "Can Export-Led Growth Continue Indefinitely? An Asia-Pacific Perspective," Journal of Asian Economics, Vol. 1, No. 1, Spring, 1-12; reprinted in Asian Economic Regimes: An Adaptive Innovation Paradigm ed. M. Dutta, (Greenwich, CT: JAI Press), 3-15.

(1990) "The Concept of Exogeneity in Econometrics," R.A.L. Carter, J. Dutta, and A. Ullah, Contributions to Econometric Theory and Application (New York: Springer-Verlag), 1-22.

(1990) "Conversion: The Trade-Off Between Military and Civilian Production in Warsaw Pact Countries," Conflict Management and Peace Science, 11, 1, 45-56, (with Miroslaw Gronicki).

(1991) "Econometric Contributions of the Cowles Commission, 1944-47: A Retrospective View," Banca Nazionale del Lavoro Quarterly Review, 177, June, 107-17.

(1991) "Model Building for a Planned System," Economic Modelling, 8, October, 418-23.

(1991) "The Statistics Seminar, MIT 1942-1943," Statistical Science, 6, No. 4, 320-30.

(1991) "Global Markets in International Trade," Toward a North American Common Market, ed. Charles F. Bonser (Boulder: Westview Press), 49-63.

(1991) "Financial Options for Economic Development," The Pakistan Development Review 30, Winter, 369-80.

(1992) "My Professional Life Philosophy," Eminent Economists: Their Life Philosophies, ed. Michael Szenberg (Cambridge: Cambridge University Press) 180-89.

(1992)   "Smith's Use of Data," <u>Adam Smith's Legacy</u>, ed. Michael Fry, (London: Routledge), 15-28.

(1992)   "A Linear Model for Environment and Development," <u>Science and Sustainability</u>, (Laxenburg: IIASA), 213-242.

(1992)   "Economic Restructuring," <u>Journal of Asian Economics</u>, 3, Spring, 87-96.

(1992)   "Some National Income Accounting Issues," <u>The Economic Outlook for 1993</u>, (Ann Arbor: University of Michigan, November 19-20), 85-95.

(1992)   "Economic Restructuring" a Symposium:Restructuring the International Economy, <u>Journal of Asian Economics</u> 3, 1, Spring, 87-96.

(1992)   "Impact of Military Cuts on the Soviet and East European Economies: Models and Simulations," <u>Economics of Arms Reduction and the Peace Process</u> eds. Walter Isard and C. H. Anderton (Amsterdam: North-Holland), 69-87, (with Miroslav Gronicki and Hiroyuki Kosaka).

(1993)   "The Peace Dividend: Will There Be One and How Large Could It Be?" <u>Economic Issues of Disarmament</u> eds. M. Chatterji and J. Brauer (London: Macmillan), 15-25.

(1993)   "Economic Forecasting at High-frequency Intervals," <u>Journal of Forecasting</u>, 12, 301-319, (with J.Y. Park).

(1993)   "Irving B. Kravis: Memoir of a Distinguished Fellow," <u>Journal of Economic Perspectives</u>, 7, Summer, 175-184.

(1993)   "Dévelopment et désarmement: Le signification," <u>Economistes de la paix</u>, ed. Jacques Fontanel, (Grenoble: Presses Universitaire de Grenoble), 81-88.

(1993)   "'Fortress Europe' and Retaliatory Economic Warfare," <u>Protectionism and World Welfare</u> ed. Dominick Salvatore (Cambridge: Cambridge University Press), 99-127, (with Pingfan Hong).

(1993)   "What is Macroeconomics?" <u>Monetary Theory and Thought</u>, eds. Haim Barkai, Stanley Fischer, and Nissan Liviatan, (London: The Macmillan Press), 35-51.

(1994)   "The Mixed Economy: A Practical Target," <u>A World Fit for People</u>, eds. Üner Kirdar and Leonard Silk (N.Y.: NYU Press), 73-82.

(1994)   "Economic Analysis of Inflation", <u>Journal of Asian Economics</u>, 4, No. 2, Fall, 239-247.

Who Innovated the Keynesian Revolution?

Paul A. Samuelson

As a precocious scholar, Lawrence Klein in his MIT Ph.D. dissertation taught the world that there had been a Keynesian Revolution and coined its name. Since then, citation counts list hundreds of demonstrations that there never had been such a thing -- definitive proof that indeed there had been a revolution.

But who perpetrated that Revolution? Few aver that Maynard had naught to do with the event. Michael Postan of the London School of Economics and Cambridge, an erudite economic historian, sometimes floated the innuendo that the ghost inside the machines of John Maynard Keynes and Joan Robinson was Richard Kahn. Joseph Schumpeter (1954, p. 1172), Luigi Pasinetti (1984, 1990), and a larger group of scholars propounded a somewhat less extreme version of this case for Kahn. A still wider circle of economists have attributed a major role in the genesis of the 1936 *General Theory* to the 1930-1931 "Circus", that informal get-together first initiated by Piero Sraffa which included among others, Richard Kahn, Austin and Joan Robinson, Sraffa, Oxonian James Meade, and (later, from a distance) Roy Harrod. This Circus communicated with the Master largely through Kahn, the Messenger Gabriel who carried communications both ways between Keynes and the Circus -- as well as being himself in constant one-on-one communication orally and in writing with Keynes.

To round out the continuum of diverse views, Don Patinkin has devoted articles and books to the documentation of his view that Keynes was himself the principal innovator of the theory of effective demand; that he received some help, stimulus, and criticism from Kahn, the Circus, and from D.H. Robertson and Ralph Hawtrey -- but that their importances should not be exaggerated. *L'Etat, c'est moi*; and in Patinkin's considered summing up, Keynes by 1933 had arrived at the essence of his Thomas Kuhn (1962) scientific revolution, by which time the Circus had ceased to exist.

In agreeable private and public debate with Patinkin, I have over the years been a hold-out to his verdict. It is my contention that a number of outside scholars -- J.M. Clark (1931, 1935), Ragnar Frisch (1934), Michal Kalecki (1933), various heretics -- had as early as Keynes some intuitive but significant perception of that same theory of effective demand. I differ with Patinkin in my view that Kahn's 1931 multiplier article indeed *did contain* that theory of effective demand. (Only more recently did I come to appreciate that James Meade, in communication with Kahn, had at an early date (1931) definite perception that autonomous fluctuations in investment will move income to an equilibrium level where saving is induced to equal that investment.)[1] As early as 1946, in my *Econometrica* obituary article on Keynes, I attributed to Joan Robinson (1933) an early command of the *General Theory*'s paradigm by which the level of (less than full employment) income gets determined by the $S(Y) = I$ cross. On reflection I continue to find unpersuasive Patinkin's doubts that, at times prior to mid-1933, some of Keynes's younger circle did actually run ahead of him in scienific innovations. I do not press this contention because I am more impressed than Patinkin is with Keynes's own early glimpses of effective demand in famous *Treatise* (1930) and Harris Lecture (1931) passages.

Lawrence Klein's MIT dissertation published in 1947, *The Keynesian Revolution*, enunciated and documented a somewhat similar (but not identical) evaluation. I remember with pleasure and gratitude our mutual discussions on these topics in those interrupted war years.

My purpose here is not to rehash my agreements and differences with Don Patinkin. The discussion reaches a stage where the jury knows what the contenders are averring and what their evidences adduced are. The jury must then make up *its* mind. Often the rational result can be only a Scotch verdict: Undecidable. Maybe yes. Maybe no.

Recently, Patinkin (1993) produced what I had thought to be new evidence and argumentation. I had thought that he believed this to weaken the case for Joan, Richard, James, and the Circus, while strengthening the case for Maynard. But I accept what he explained to me in correspondence: that his 1993 paper, and its adapted version in this volume, represent no change in his original position that Keynes deserves the essential credit for arriving at the theory of effective demand around mid-1933.

As I re-read his old and new expositions, I deem them to add slightly to the believability of my eclectic view and subtract slightly from any claim for Keynes's overriding importance in innovating the Keynesian Revolution on effective demand.

In my conclusion, I make an attempt to explain how it can be that Patinkin and Samuelson are so much in agreement on the facts, and yet can sum up somewhat differently. The point is an interesting one for the methodology of the history of science.

My present exposition can be brief. By scholars' agreement, it narrowly focusses on the genesis of the following *heart* of the *General Theory*.

Income = Consumption + Saving

Investment = Saving, as a definitional statistical identify

         and as a stationary-state equilibrium condition

$$\text{Saving} = S(Y), \text{ where } 0 < S'(Y) < 1 \text{ usually} \qquad (1a)$$

$$\text{Investment} = \text{autonomous } \bar{I}\ (\ t\ ) \text{ or} \tag{1b}$$
$$= I\ (Y),\ 0 < I'(Y) < S'(Y)$$
$$\bar{I} = S(Y), \text{ or } I(Y) = S(Y) \tag{1c}$$

$$Y^* = S^{-1}[\bar{I}] = M[\bar{I}],\ M'\ [\bar{I}] = 1 + c' + (c')^2 + \dots \tag{1d}$$
$$= 1/(1-c')$$
$$= 1/S',\ 0 < c' < 1$$

This theory of effective demand, we can provisionally depose, constitutes the most revolutionary break with Say's Law and with Walras-Debreu-Arrow clearing of all micro markets. However, I am prepared to argue on another occasion that what caused the explosion of *enduring* interest in the *General Theory*, and the obliteration of rival contemporary *Weltanschaungs*, was the *General Theory's* 1936 presentation of what has come unfairly to be known as the Hicks-Hansen I-S and L-M behavior equations. Let (M,Y,r) be (money stock, real or nominal income flow, and the interest rate [or surrogate parameter for the vector of interest rates]). Then considerably before John Hicks's 1937 perception, Keynes (1936), David Champernowne (1936), Meade (1937), Brian Reddaway (1936), and most especially Harrod (1937) had published the augmented system:

$$Y = C\ (Y) + I\ (Y, r) \tag{2a}$$
$$\overline{M} = M\ (Y, r) \text{ or } M_1\ (Y) + M_2\ (r) \text{ or } Y/V\ (r);\ M_1' > 0 > M_2',\ V' > 0 \tag{2b}$$

This adds to the "multiplier notion of effective demand," the notion of "Marginal Efficiency of (Investment or) Capital," and the notion of "Liquidity Preference." As Hicks has noted, much of his understanding of his diagramatic 1937 formulation he got from Harrod; Alvin Hansen (1953) was writing up Keynes for students and found the Hicks diagrams useful. Although during 1933-1939 Keynes sometimes got his own system wrong, I believe the weight of the evidence will favor the finding that Keynes's role in completing the (2a)-(2b) augmentation of (1) was more unilaterally autonomous and owed less to his Oxbridge contemporaries than was the case for (1)'s effective demand heart. In particular, from his 1930 *Treatise on Money*, Keynes already in the pre-Kahn and pre-Circus era had perceptions of what was to become his Theory of Liquidity Preference. On this I feel in closer agreement with Patinkin's view.

The importance of this digression to the *complete General Theory* paradigm is neither to change the subject and lessen controversy nor to find something which I can concede to Patinkin in more amiable agreement. Its importance to me is that the basic multiplier doctrine would not in my view have commanded for Keynes the lasting durability and scientific fame that his augmented system did. Frisch (1934) and J.M. Clark, (1931, 1935) garnered insignificant world attention for their respective circulation-planning and public-works models, which resembled Keynesian effective demand. Same goes for the similar trace elements in Sumner Slichter (1934). And for Kahn (1931). The non-lasting impact of Keynes's important 1933 *Means to Prosperity* is another case in point.

Relevant too is what Robert K. Merton (1973) would call "Obliteration by Incorporation", which was the fate of the Stockholm School of Bertil Ohlin and others. The excellent notion that an autonomous new desire to spend will stimulate real output in times of considerable unemployment comes to receive limited attention in more prosperous times.

The fact that Keynesianism is still alive 60 years after 1933, alive as a technique of analysis differing from the Walrasian technique and alive among diverse schools of anti-Keynesians and neo-neo-Keynesians -- this in my view stems from the (2a)-(2b) synthesis and would not have survived from (1) alone. To kill a complete theory, you need not a fact or an important perception; you need an *alternative complete theory*. The (2a)-(2b) corpus, once wealth and other Fisherine variables were added to it, could serve anti-Keynesian as well as Keynesian needs. "We are all Nineteen-Nineties Keynesians now." This even goes for market-clearing New Classicals!

I. The New Evidence

Patinkin (1993) develops the interesting hypothesis that prior to 1933 (or a month earlier) Keynes may not have been in command of the theory of effective demand. He supports this possibility by doubting that the fragments of the early *General Theory* published by editor Donald Moggridge in Volume XIII of Keynes's *Collected Writings* were correctly dated. In particular, on the basis of various students' notes covering Keynes's lectures (Robert Bryce, Lorie Tarshis, Walter Salant,...) and on the basis "of their internal evidence," Patinkin (1975) was led to "suggest that ... the chronological sequence of two of these fragments was actually the reverse of what had been indicated." Also, Patinkin points out:

> It is noteworthy that there is no reference to the equilibrating role of changes in the level of income in the May 1932 `manifesto' which Joan and Austin Robinson and Richard Kahn subsequently submitted to Keynes (Keynes, XXIX, pp. 42-45).

What is the effect on the debate if (as I would doubt) Maynard did not know *his* theory of effective demand before December, or Mid, 1933? It certainly destroys any simple-minded chronology that the Circus unilaterally created that theory and pumped it into Keynes by the autumn of 1931 when James Meade left the Circus for Oxford.

Put Patinkin's hypothesis next to my own 1946 hypothesis (which I modified after Joan wrote to tell me that her 1933 *Economica treatise*-like piece was written eighteen months before its publication). I jokingly put the genesis of the S (Y) = $\bar{I}$ paradigm between June, 1933 and October, 1933 (dates of Joan's *Economica* and *Review of Economic Studies* pieces). I at that time (but did not underline the view) supposed that Keynes had made the breakthrough and Joan newly learned about it. Since December 1933 comes months after Joan wrote the *RES* item, on the new hypothesis Keynes either learned it from Joan (and the inseparable Richard) or by coincidence happened to work it out for himself.

I have seen naught to make me doubt that, leaving Maynard out of it, by mid-1933 at least three scholars had already mastered the theory of effective demand: Richard Kahn, James Meade, and Joan Robinson. How they got it is still to be resolved, but that they had it seems to me -- after I have pondered over Don's evidences and doubts -- not in doubt. My Bayesian probabilities are 99%, which to me represents effective certainty.

I have nothing new to say here about Richard's 1931 perceptions. See Samuelson (1994) for Kahn recollections and commentary. But I should record that the totality of what Kahn (1905-1989) put into the record in the last 15 years of life is, to me, compatible with his believing then that in 1931 he had grasped how the level of income is moved by changes in autonomous investment to a new equilibrium income level at which the induced level of saving comes to match the new level of $\bar{I}$.[2] What he did not get from his $\bar{I}[1 + c' + (c')^2 + ...]$ sums, he surely got from Mr. Meade's Relation. James Meade and I are not young; and Don Patinkin is correct that scholars' memories are less than perfect. I may add: Some scholars give themselves the benefit of the doubt more than Zeus will warrant. However, when an innovator has ever downplayed his own brainchildren as Meade has always done, I see no reason not to take his careful 1993 statements about 1931 at their face value. (What is not relevant, I will nevertheless add. Some of the remembrances of my own about this period of 60 years ago I at first feared would have to be taken with a grain of salt. But when data later turned up -- as in my July, 1993 discovery of Joan Robinson's crucial letter to me about my 1946 Keynes obituary article, it was just as I remembered it. For the record I append that brief letter at the end of this piece.)

Knowing what I know from the published record and from my own personal contacts with Joan and Richard, I have to believe that they shared fundamental insights in those days of her work on imperfect competition and in later decades.

Still, I ought to document my view that Joan's 1933 *RES* article did enunciate the essence of the $S(Y) = \overline{I}$ Keynesian theory of effective demand. Her clearcut statements, indeed, sound just like many of Keynes's clearcut post-publication summaries of the *General Theory*, as in his reply to Ohlin and in separate prefaces to foreign translations of the *General Theory*. Robinson never says that she, in 1933, is expounding new stuff due to John Maynard Keynes. She repeatedly professes that Keynes had (in the 1930 *Treatise*) worked out a revolutionary saving-investment theory of output that he had not realized, confusing it with his explicit saving-investment theory of the price level. Mrs. Robinson did not always know when she was seeming to be condescending, but a Freudian stranger could be forgiven for reading into her 1933 text trace elements of put-down for Keynes. (The letters they wrote tell an opposite story: she is the friendly and admiring disciple throughout, content to devote hours to painstaking proof criticisms.) Here are significant quotations from Robinson (*RES*, 1933):

> ...Mr. Keynes himself overlooked the fact that he was writing the analysis of output, as these examples show... (p. 26)
>
> A second example of Mr. Keynes's failure to realise the nature of the revolution that he was carrying through is to be found in the emphasis which he lays upon relationship of the quantity of investment to the quantity of saving. He points out that if savings [*Treatise* definition] exceed investment consumption goods... output will consequently decline until the real income of the population is reduced to such a low level that savings are perforce reduced to equality with investment. But he completely overlooks the significance of this discovery, and throws it out in the most casual way without pausing to remark that he has proved that output may be in equilibrium at any number of different [less than full employment] levels, and that while there is a natural tendency towards equilibrium between savings and investment (in a very long run) there is no natural tendency towards full employment of the factors of production. (pp. 24-25)
>
> ... it was only with disequilibrium positions that Mr. Keynes was consciously concerned with when he wrote the *Treatise*. He failed to notice that he had incidentally evolved a new theory of the long-period analysis of output. (p. 25)
>
> But suppose that over a certain range the supply of goods is perfectly elastic? Then whatever happens prices cannot rise or fall. Since Mr. Keynes's truisms [I = S] must be true, a rise or fall in demand for goods, which will be met by an increase or decrease of output without any change in prices, must necessarily be accompanied by changes in savings and

> investment which keep the two in equality. When an increase
> in output is brought about by an increase in investment, if
> prices do not alter, the increase in output must bring about an
> increase in savings (as defined by Mr. Keynes) equal to the
> initial increase in investment, for Mr. Keynes's truisms must be
> true. (p. 25)

Don Patinkin has taught me how delicious it is to appreciate the
intricacies of human discovery, innovation, perception, understanding and
misunderstanding. When and how Isaac Newton discovered the universal
theory of gravitation is not a question like, "When did Goethe go to Weimar?"
In Patinkin's own spirit, I want to register some impressions I recently formed
when re-reading his analyses of the path from the *Treatise* to the *General
Theory*. The bottom line of my reflections is to attribute to Keynes himself
imperfect and incomplete, but psychologically important, perceptions in the
1930-1931 period that would ripen and evolve into the 1932-1936 theory of
effective demand.

In *Treatise* (1930, pp. 158-60), Keynes writes his parable of the banana
plantation. There he says (my paraphrase): A thrift campaign raises desire to
save while investment opportunity is unchanged. Glossing over Keynes's 1930
discussion of how entrepreneurial losses are the mechanism which causes
employment, output and incomes to contract, we can see that this lowers Y
either (a) to zero, where all actual saving and activity ceases; or (c) to a time
when $\bar{I}$ fortuitously and compensatingly happens to rise enough to match the
new thrift; or (b) the thrift propensity is self-destroyed by the community's
becoming impoverished enough so as not to end up with any achieved new
saving but only with the level of saving that matches the pre-thrift achieved
level (which is the unchanged $\bar{I}$ level).

Purposely, I have altered Keynes's words to put them into post-1936
language. This is not in a foolish attempt to give 1930 Keynes post-1936
wisdom. It is to test the cogency of Don Patinkin's logic of (a), (b), (c). My
wording of (b) is perfectly compatible with a diagram of a rising SS line
intersecting a horizontal $\bar{I}\bar{I}$ line at $(\bar{Y}, \bar{I})$ equilibrium. (The slope of $S = a + bY$
would be $0 < b < 1$.) Patinkin (1993, pp. 649-58) says he does not read
Keynes's words for case (b) as having "the connotation of a systematic
equilibrating process." But he does say that some others have regarded them
"as describing a systematic influence of the decline in income on savings," I
being cited as one such. So we can together investigate how Keynes's case of
(a) and (c) throw light on case (b) as being a systematic case.

Patinkin says, in my paraphrase: if Keynes understood (b) in the
Samuelson sense, he would not have listed (a) with its completely contradictory
assumption? I must ask, "How can one agree that (a) contradicts (b)?" On the

contrary, when we make the upward shift in thriftiness large -- when we raise *a* enough in a + bY -- the $\overline{Y}$ equilibrium moves to zero and because of the physical impossibility of Y ever going negative, any post-1936 mathematical economist would have agreed precisely with Keynes's case (a) being a necessary entailment of his (b) case analysis. (In my early MIT classes, I often gave tests involving boundary equilibria: that alerted students to the fact that S = a + bY relations were short run, and would shift endogenously when people used up wealth resources.) I vote against Patinkin's *logic* of rebuttal on this point.

Now turn to any alleged incompatibility of (c) with (b). It shows no lack of belief in (b), or in understanding of it, if Keynes or I add specifically to the categories of what can happen chronologically subsequent to a thrift campaign an alternative scenario in which upward shift in *SS* is matched (or overmatched) by an upward shift in $\overline{II}$. Indeed, back in the Great Depression many economists advised President Roosevelt to offset an excess of saving over feasible investment by government spendings and deficits; and, as I write in 1993, many of us economists are advocating that a deliberate contrived increase in U.S. thriftiness be matched by incrementally stimulating monetary policy that will effectively increase I. Once again my vote has to go against this part of Patinkin's 1993 rebuttal.

This puts a more favorable light on my reading into Keynes's June 1931 Harris Foundation Lecture in Chicago of a positive-sloping S(Y) schedule. Patinkin counters this, in part, by pointing out that Keynes in Chicago is defending the 1930 *Treatise*. Yes, but as we have just seen, the trace element that appeared in 1930 does reappear in 1931, and that should make us date much earlier than 1933 Keynes's glimpse of the theory of effective demand. If a rise in thrift can push real Y down until other forces cause it at some time to recover, why does that betray some failure to understand plateaus of Y?

In this brief paper I fall far short of documenting the evidence on which I base my summarized judgments, and fall short of Don Patinkin's magisterial analyzing of all the available pieces of evidence. I require readers' indulgence and beg to cite other writings where I have commented in somewhat greater depth on Kahn, Joan Robinson, and Keynesianism. See Samuelson (1946; 1964; 1977, pp.72-89; 1988; 1989; 1991). For Don Patinkin's many publications on this subject see the References in his recent 1993 *Economic Journal* article, and its revised version in the present volume.

II. Conclusions

To sum up. My own suspicion is that by 1932, Keynes, Kahn, Joan Robinson, James Meade, and perhaps others in their circle had each at one time or another thought through the fundamentals of the theory of output as determined by propensity to save (and consume) and variability of investment (and of other autonomous spending elements, such as government items).[3] If Maynard never tried these notions out on undergraduates before December 1933, that is a surprise; but it is not the kind of surprise, which taken with all the other evidences and plausible surmises, compels one to judge that Maynard knew not of it. Don is right to insist that a task of the historian is to document as best can be done just when certain things can be dated to have happened. As a non-antiquarian, I am right to apply high Bayesian probabilities to scenarios and *gestalts* that the wide weight of the evidence dominates. A fallible procedure that could go wrong? Yes, of course. And, in the light of still newer data, a new set of Bayesian probabilities will be formed -- which go ahead of the documentable facts and which are not valueless because they too are inconclusive and vulnerable to still later factual corrections.

Before my final summing up, I wish to illuminate how Don Patinkin and I can agree so much on the chronological facts and still differ so profoundly on our interpretations. Patinkin, I hypothesized in my first draft of this paper, had been unduly attracted to the methodological view point of Yehuda Elkana, a history-of-science scholar at Tel Aviv University. In a justly famous classic article, (Thomas Kuhn, 1959), *Energy Conservation as an Example of Simultaneous Discovery,* Kuhn argues, persuasively, that along with the well-known triad of Robert Mayer, Hermann Helmholtz, and James J. Joule as independent discoverers of the Law of Conservation of Energy, must be included another nine scientists with legitimate claims for independent analysis. None of the dozen began from the same viewpoint, all differed in their emphases, but with the legitimate hindsight of scientific understanding, each can be granted a legitimate role as independent discoverer. What does Dr. Elkana make of Dr. Kuhn's agreed upon facts? With skill and insight, Elkana wants to argue that it was not so much a case of twelve who arrive at the same one North Pole. Rather, each has found a somewhat different thing: each sets out on an idiosyncratic journey and (what is not the same thing) arrives at a distinguishable destination.

My reply to Dr. Elkana in correspondence could, I originally opined, throw light on the dissonance between Patinkin and me on the primacy of Keynes as the formulator of the theory of effective demand. For, Patinkin seemed to be stressing the Elkana wrinkle while I was stressing the peculiar prospectives of optimal history of science (as against antiquarian history of kings, politics, customs, and events; and as against antiquarian rambles over the meandering hits and misses of science in the making).

I replied to the question in the 1970 title: "The Conservation of Energy: A Case of Simultaneous Discovery?" as follows:

> Yes..... . That is not because I know facts you don't know. Or disbelieve in facts you have reported. If we are left with a difference in judgment, it has to do with what verbal useage we

would each prefer to countenance.

I agree that different scientists start out with (possibly) different notions of a problem to be `solved.' It is the duty of the historian of science to report on such differences when they can be observed. People who start out with different perceptions may end up with [somewhat] similar [or even identical] results. And it may well be only later that they, or someone else, explicitly perceive the similarity (or identity) of their "results." Again, the meticulous historian will want to report on these phenomena. Just as I never take at face value what a scholar records as *his* conscious *ex ante* problem, neither do I take at face value what he reports *ex post* on what he did accomplish. I do not have in mind [conscious or unconscious] false claims so much as continued fuzziness concerning interrelated concepts prior to a late stage of a paradigm's development.

After I sent my draft to Patinkin, he informed me and I inform readers: In 1978 in open lectures at the University of Chicago, *before* he knew of Elkana's 1970 paper, he had fully enunciated the view I am criticizing.

Let me connect up with economics. The world believes, rightly believes, that Mayer, Helmholtz, and Joule deserve full marks each as discoverers of the First Law of Thermodynamics. I believe that. And in the same sense I affirm that Keynes, Kahn, and Robinson deserve credit for the theory of effective demand. I understand the position of Patinkin and Elkana; but it, they, do not tempt me out of my view. (To digress, I like Helmholtz's analysis best of the three, just as Patinkin likes 1933 Keynes better than 1931 Kahn. But still, Mayer and Joule deserve their laurels: Clausius's superiority to Kelvin, Rankine, and Sadi Carnot as an ultimate synthesizer of classical thermodynamics does not dull their medals. I do not disregard Kahn's many words on what he meant to do and did do; but knowing all his career words, I do not give them the weight and interpretation that Patinkin does and I deny any charge of Humpty-Dumptyism.)

Here is my bottom line.

1. Keynes is first among equals. Kahn and Joan Robinson rank second among equals. Austin Robinson, James Meade, and Roy Harrod (especially in the 1934-1936 wind-up) are important equals. Piero Sraffa, never much involved with macroeconomics (save in his 1932 polemic with Friedrich Hayek), Austin Robinson reports as taking a less active part.

2. Keynes influenced the others and they influenced him. He put questions to them; and all provided answers, conjectures, and questions to one another.

3. Keynes is *first above equals* when we go beyond the effective-demand heart of the *General Theory* system to its complete structure.

4. It is the genius of Keynes that gives the *General Theory* its panache, brilliance, width, and depth -- not so much his *analytical* genius as his *joie de vivre*, practical insight, command of the history of ideas, and Shavian power of pen. Austin Robinson is surely correct in saying that those who knew the two men well could not be in doubt as to which of Keynes and Kahn was capable of writing a classic.

5. It is childish to harp unduly on how credit for a scientific breakthrough is to be divided up among contemporaries. When these remarkable half dozen minds worked separately and together, the whole became more than the sum of its parts. Without the whole team, John Maynard Keynes could not have written the book he did write; and with the whole team, only he could have written the book that got written. Compare the *General Theory* with Keynes's other comparable books: *The Tract on Monetary Reform* and *The Treatise on Money*.

6. Richard and Joan provided for Maynard the important ingredient of dogmatism, a factor not to be underestimated and which they both possessed in abundant degree. Not that Keynes was lacking in arrogance and self-confidence. But when you are part of a claque, convinced of the stupidity of conventional fogies and of the truth and novelty of your own novel ideas, your energies are enhanced resonantly. And of course Cambridge University was in those days a bully pulpit from which to persuade a generation of economists.

If my point here had only to do with the issue of *rhetoric* in shaping the history of scientific belief, it would still be important. But I have in mind something basic to the content of economic science itself. One of the obstacles to embracing a new paradigm like the theory of effective demand was its manifest lack of microfoundations in the market-clearing pre-existing orthodoxy of Walrasian neoclassicism. (I can testify that my own delays and agonizings in the months immediately after the February 1936 *General Theory* publication had precious little to do with intricacies of the new paradigm and difficulties in learning to understand it. For my teacher Joseph Schumpeter that was more crucial; only years later did he begin to understand its differences from simple-minded hoarding theories; and by his death in 1950, during his sixty-seventh year of life, Schumpeter, in my observation and on the testimony of Harvard colleagues like Alvin Hansen, had still not attained to sophisticated mastery of the apparatus. That, and not simple jealousy of Keynes, helps to explain his indignation that the brilliant young went whoring after what he thought to be only Keynesian leftisms and policy do-goodism.)

I was right to be concerned about the new Gospel's microfoundations. And what was crucial for my Road to Damascus was the moment when I resolved that, *A good handle on real-world macrohappenings, whatever its microfoundations (such as price rigidities and imperfect competitions), was better than being stuck with a fine logical system that failed to apply to the contemporaneous world.* No such soul-searchings and torment as I have described seems ever to have afflicted Richard and Joan. And I must believe that this would play a role in egging Maynard on to boldness in innovation.[4]

7. When I was a 16-year-old beginner in economics, I was told by my University of Chicago tutor (Eugene Staley) that John Maynard Keynes was then the most famous economist in the world. Probably he was. But after 1940 I could tell my students that Keynes was the leading economist of the Century, ranking with eighteenth-century Adam Smith, and nineteenth-century Léon Walras. Alone and working out of Chernowitz, Keynes could not have achieved that pinnacle.

Paul A. Samuelson, Institute Professor Emeritus, Massachusetts Institute of Technology, Department of Economics, E52-383C, Cambridge, Massachusetts 02139.

NOTES

1. A referee has suggested that I have overestimated Meade's importance and underestimated that of Hawtrey. She or he may be right that "Mr. Meade's relation" is an obvious corollary of Kahn's dynamic multiplier analysis. However, economists back then were obsessed by saving-investment identities and discrepancies; therefore, Meade put to rest a potential qualm; and also, incidentally helped to clarify the essential identity between 1931 Kahn and 1933 Robinson, thereby casting doubt on Patinkin's repeated attempts to deny that the theory of effective demand is already there in 1931.

On Hawtrey's influence, I now feel I should have noted explicitly that in 1930 when the *Treatise* was just published, Hawtrey pointed out to Keynes its relative neglect of systematic induced *output* effects, an anticipation of what the Circus would later do.

Score one for the referee. However, I am less impressed by the referee's citation of Hawtrey's command of the multiplier analysis in unpublished late-1920s writing (see Skidelsky, 1993, p. 445). This puts Hawtrey in the list of Frisch, J.M. Clark, Kalecki, and innumerable heretics who worked out spending and respending paradigms that never influenced the Keynes circle. However, in the published correspondence between Hawtrey and Keynes, *Collected Writings of John Maynard Keynes*, Vol. XIII (pp. 565-633), Hawtrey's influence on Keynes's formulation of effective demand was unimportant and probably negative. I seem to recall, too, that in the workings of the Macmillan Committee, Hawtrey argued against Keynes, often in Say's Law terms concerning fiscal policy's impotency apart from money and credit changes. Keynes owed thanks to Hawtrey for much time spent in criticizing proofs and he expressed those formal thanks; he should have done the same for Dennis Robertson but his irritation with him exceeded his irritation with Hawtrey. Keynes well illustrated the truth: One learns better from friends than from opponents.

It appears, from valuable on-going work by Robert W. Dimand (1988) that Hawtrey's pre-1930 multiplier sequence had for its only leakage induced imports: for a closed economy this implied a Say's Law world and no anticipation of a Keynesian theory of effective demand. Dr. Dimand has also documented that, in 1931, Mr. Meade's relation involved a mélange of leakage components; until Meade's 1933 pamphlet, the only systematic income-saving relation he used was the propensity of corporations to save part of the profits induced as a fraction of enlarged income. Partially supported is Patinkin's suspicion that an account by James Meade (1907-   ) as octogenarian of his 24-year-old's understanding might be a bit tainted by some hindsight. As qualified, I still judge Meade's 1931 insight to have been important.

2. I have pondered over Patinkin's account of his correspondence with Kahn and his reading of Kahn's 1984 memoirs, and remain of this view. My reading of Kahn's quoted remarks

on his role are those of a proud man, who will not put forward what he believes to be his just claims because that would be to sully the memory of his cherished mentor, John Maynard Keynes; it is for others to press his just claims. I do not believe for a moment that, because of the passage of time, he had lost interest in such petty problems of priorities. In his place, where I was convinced of the paramount importance of what *I had* done, I could still have written in the exact vein of Kahn. Also, it could well have been the case that Kahn had considered the main purpose of his 1931 article to be something different from the $Y - C(Y) = \bar{I}$ paradigm, but *that* does not persuade me that in 1931 he had fallen short of this paradigm. Often in my own work, I started out with one goal as consciously paramount only to arrive at important results that others concentrated on almost totally. I have never considered Richard Kahn an ideal formulator of "Keynesian economics." He was too emotionally impatient. Often he wrote and talked as if

$S(Y) = \bar{I}$ is a cogent rebuttal to the Quantity Theory of Money. In my view *it* is not. But when Kahn says he had much trouble weaning Keynes from the Quantity Theory, Samuelson's perception is irrelevant and Kahn's is relevant. I do not believe that Keynes got all his understanding of effective demand from Kahn -- in 1931 or later. But if I did believe that, Kahn's remarks about his influence in weaning John Maynard Keynes from the Quantity Theory would be cogent ammunition for such a view. (Digression: $MV = PQ$, when rewritten to be $MV(i) = PQ$ with $V'(i) > 0$, is a version of Keynes's Liquidity Preference. Kahn's baleful effect on the 1959 Radcliffe Committee could have been avoided had he realized that; and English Keynesianism could have been spared the contempt with which it was regarded around the world after 1959. $MV(i) = PQ$, P a slowly-rising Marshallian function of Q, and $Q = C(Q) + \bar{I}$ is actually one of many valid 1936 formulations of the Model T *General Theory*. Thus, the Quantity Theory, when properly formulated to take into account realistic short- and intermediate-run relations, is compatible with the *General Theory*. Only in liquidity trap contexts, when i is so low that almost any V can obtain (depending upon $\bar{I}$), does the Kahn view I have caricatured make any sense. Alas, 1959 was two decades past 1931-1939!)

3. So as not to be misunderstood, let me emphasize that I attribute no perfection of understanding concerning the theory of effective demand by any of these early writers or by Keynes. Indeed, too long *after* the 1936 *General Theory*, Joan, Richard, Maynard, Abba, Oskar Lange, and other "Keynesians" remained woefully confused about the nature of tautological $S \equiv I$ and "equilibrium $S(...,...,...) = I(...,...,...)$ relations." Thus, we are sometimes told that S must always equal I, only to be also told that Y is driven to bring $S \rightarrow I$. Sometimes Joan would say, if Y equilibrates I and S then the interest rate cannot -- a less than whole truth. As late as 1938, when Lange had demonstrated his mastery of the augmented (2a)-(2b) system, he said to me: "I understand that $S'(Y) > I'(Y)$ is a necessary and sufficient condition for stability of equilibrium. But for the life of me, since $Y = I + C$, I can't grasp why $I'(Y)$ is not identical to $1-C'(Y)$ and to $S'(Y)$." The primitive theory of effective demand, described in equations (1) here and exposited in Joan's post-1936 publications, is not an intricate and difficult paradigm for a 1932-1933 writer to grasp -- a consideration Patinkin and I must keep in mind when we study the delicious intricacies of the human mind as it grapples with the new.

4. Earlier I spoke of certain shakiness in their post-1936 understanding of $I = S$ as a tautology and as an equilibrating condition determinative of aggregate output. My words were deliberately made to be condescending toward them, as if superior mathematical chaps like me were free of the errors and perplexities of Lerners, Kahns, Robinsons, and Langes. Let me point out here, that purists in market-clearing methodologies will still in 1993 have legitimate worries on how to interpret a 1937-like relation: $dY(t)/dt = |\alpha| \{\bar{I} - S(Y)\}$ and $\lim t \rightarrow \infty$, $Y(t) \rightarrow M[\bar{I}] = S^{-1}[\bar{I}]$; how does measured saving or investment compare at each t with $\bar{I}$ and $S(t)$? So much for condescension and mathematical superciliousness!

REFERENCES

Champernowne, David. (1936). Unemployment, basic and monetary: The classical analysis and the Keynesian. *Review of Economic Studies*. *3*(June), 201-16.

Clark, John M. (1931). *The Costs of the World War to the American People*. New Haven: Yale University Press; London: H. Milford, Oxford University Press, for the Carnegie Endowment for International Peace, Division of Economics and History.

Clark, John M. (1935). *Economics of Planning Public Works*. A study made for the National Planning Board, Federal Emergency Administration of Public Works, Washington, D.C.: Government Printing Office.

Dimand, R. W. (1988). *The Origins of the Keynesian Revolution*. Aldershot: Edward Elgar, and Stanford, CA: Stanford University Press.

Elkana, Yehuda. (1970). The conservation of energy: a case of simultaneous discovery?. *Archives D'histoire des Internationales Sciences*. *90-91*(Janvier-Juin), 31-60.

Frisch, Ragnar. (1934). Circulation planning: proposal for a national organization of a commodity and service exchange. Parts I-II. *Econometrica*. *2*(July, October), 258-336; 422-35.

Hansen, Alvin H. (1953). *A Guide to Keynes*. New York: McGraw-Hill Book, Inc.

Harrod, Roy. (1937). Mr. Keynes and traditional theory. *Econometrica*. *5*(January), 74-86.

Hawtrey, Ralph. (1973). See comments in Moggridge, D., (Ed). *The General Theory and After*, Part I, *op. cit.*.

Hicks, John R. (1937). Mr. Keynes and the 'Classics', a suggested interpretation. *Econometrica*. *5* (April), 147-49.

Kahn, Richard. (1931). The relation of home investment to unemployment. *Economic Journal*. *41*(June), 173-98. Reprinted in Richard Kahn *Selected Essays on Employment and Growth*. Cambridge: The Cambridge University Press, 1972.

Kahn, Richard. (1984). *The Making of Keynes's General Theory*. Raffaele Mattioli Lectures, Cambridge: Cambridge University Press.

Kalecki, Michal. (1933). Outline of a theory of the business cycle. Translated into English as Chapter 1 in *Studies in the Theory of Business Cycles 1933-1939*. New York: Augustus M. Kelley, 1966.

Keynes, John Maynard. (1923). *A Tract on Monetary Reform*. London: Macmillan. Reprinted in Keynes, *The Collected Writings*. Vol. IV. London: Macmillan, for the Royal Economic Society, 1971.

Keynes, John Maynard. (1930). *A Treatise on Money*. Vols. I and II, London: Macmillan & Co., Ltd.. Reprinted in *The Collected Writings*. Vols. V and VI. London: Macmillan, for the Royal Economic Society, 1971.

Keynes, John Maynard. (1933). *The Means to Prosperity*, London: Macmillan & Co., Ltd.. Reprinted in *The Collected Writings*. Vol. IX, London: Macmillan, for the Royal Economic Society, 1972.

Keynes, John Maynard. (1936). *The General Theory of Employment, Interest and Money*, London: Macmillan & Co., Ltd.. Reprinted in *The Collected Writings*. Vol. VII, London: Macmillan, for the Royal Economic Society, 1973.

Klein, Lawrence R. (1947). *The Keynesian Revolution*. New York: The Macmillan Co..

Kuhn, Thomas. (1959). Energy conservation as an example of simultaneous discovery. In M. Clagett, (Ed.) *Critical Problems in the History of Science*. Madison: University of Wisconsin Press. Reproduced in *The Essential Tension Selected Studies in Scientific Tradition and Change*, Chicago: the University of Chicago Press, 1977.

Kuhn, Thomas. (1962). *Structure of Scientific Revolution*. Chicago and London: The University of Chicago Press.

Meade, James. (1933). *Public Works in Their International Aspect*. London: New Fabian

Research Bureau.

Meade, James. (1937). A simplified model of Mr. Keynes's system. *Review of Economic Studies.* 4(February), 98-107.

Meade, James. (1993). The relation of Mr. Meade's relation to Kahn's multiplier. *The Economic Journal. 103*(May), 664-65.

Merton, Robert K. (1973). *The Sociology of Science Theoretical and Empirical Investigations,* Chicago: The University of Chicago Press.

Moggridge, Donald, (Ed.). (1973). The general theory and after. Part I, Preparation, in Keynes, *The Collected Writings.* Vol. XIII. London: Macmillan, for the Royal Economic Society.

Moggridge, Donald, (Ed.). (1979). The general theory and after: A supplement. In Keynes, *The Collected Writings.* Vol. XXIX, London: Macmillan for the Royal Economic Society.

Pasinetti, Luigi. (1984). Comments In Richard F. Kahn, *The Making of Keynes's General Theory, op. cit..* (pp. 222-25, 240-41, 249).

Pasinetti, Luigi. (1990). Richard Ferdinand Kahn 1905-1989. *Proceedings of the British Academy, Lectures and Memoirs. 76,* 423-44.

Patinkin, Don. (1975). The collected writings of John Maynard Keynes: From the 'Tract' to the 'General Theory'. *Economic Journal. 85*(June) 249-71.

Patinkin, Don. (1990). On different interpretations of the 'General Theory'. *Journal of Monetary Economics. 26*(October), 205-43.

Patinkin, Don. (1993). On the chronology of the 'General Theory'. *Econometrica. 103*(May), 647-63.

Patinkin, Don. and Clark, Leith J., (Eds.). (1977). *Keynes, Cambridge and the General Theory: the Process of Criticism and Discussion Connnected with the Development of the General Theory.* London: Macmillan.

Reddaway, Brian. (1936). The general theory of employment, interest and money. *Economic Review. 12*(June), 28-36.

Robertson, D. H. (1973). Comments, in D. Moggridge (Ed), *The General Theory and After., Part I. op. cit..*

Robinson, Joan. (1993). A parable on savings and investment. *Economica. 13*(February), 75-84.

Robinson, Joan. (1993). The theory of money and the analysis of output. *Review of Economic Studies. 1*(October), 22-26.

Samuelson, Paul A. (1946). Lord Keynes and the general theory. *Econometrica. 14*(July), 187-200. Reproduced as Chapter 114 in J. Stiglitz (Ed.) *The Collected Scientific Papers of Paul A. Samuelson,* Vol. 2, under the title "The General Theory" and with its original Footnote 15 omitted, Cambridge, Mass.: The MIT Press, 1966.

Samuelson, Paul A. (1964). A Brief Survey of Post-Keynesian Developments. In R. Lekachman (Ed.) *Keynes's General Theory: Reports of Three Decades.* New York: St. Martin's Press. Reproduced as Chapter 114 in *the Collected Scientific Papers of Paul A. Samuelson,* Vol. 2, Cambridge, Mass.: The MIT Press, 1966.

Samuelson, Paul A. (1977). See Patinkin and Leith, pp. 72-89.

Samuelson, Paul A. (1988). The Keynes-Hansen-Samuelson Multiplier-Accelerator Model of Secular Stagnation. *Japan and the World Economy, 1*(October), 3-19.

Samuelson, Paul A. (1989). Remembering Joan. In George Feiwel (Ed.) *Joan Robinson and Modern Economic Theory.* New York: New York University Press.

Samuelson, Paul A. (1991). Thoughts on the Stockholm School and on Scandinavian economics. In Lars Jonung (Ed.) *The Stockholm School of Economics Revisited.* Cambridge: Cambridge University Press.

Samuelson, Paul A. (1994). Richard Kahn: His welfare economics and lifetime achievement. *Cambridge Journal of Economics.* Forthcoming.

Samuelson, Paul A. Patinkin, Don. and Blaug, Mark. (1991). On the historiography of economics: A correspondence. *Journal of the History of Economic Thought. 14*(Fall), 144-58.

Schumpeter, Joseph. (1954). *History of Economic Analysis.* New York: Oxford University Press.
Skidelsky, Robert. (1994). *John Maynard Keynes, Volume Two The Economist as Saviour 1920-1937.* New York: Viking. Forthcoming.
Slichter, Sumner. (1934). *Towards Stability the Problem of Economic Balance.* New York: Henry Holt & Co..

ENCLOSURE
The 1946 Joan Robinson Letter

Below is the missing footnote from my 1946 *Econometrica* obituary article on Keynes. Now that I have found the enclosed October 22, 1946 letter of Joan Robinson commending this piece, my belief is confirmed that I omitted it from reprintings of the article because the news that her *Economica* piece was delayed 18 months before publication put a damper on the notion that the time-slot June-to-October in 1933 contained the Cambridge breakthrough on the theory of income determination. Joan did not mention how long her second article was delayed between writing and publication: if we assume only a few months, that makes December 1931 to July 1933 as possibly an interesting time slot. Patinkin's contention that the Circus was mainly engaged in shedding skins of the *Treatise* is, if anything, confirmed by the *Economica* piece's date of conception.

Here is the footnote, Joan's letter, (see p. 19) and a comment on it.

Samuelson's Footnote, *Econometrica* (1946), p. 200.

[15] I should like at this point to pass a clue on to the future historian of economic thought. What was happening in Cambridge in the months between Mrs. Robinson's patient elucidation of an aspect of the *Treatise* entitled, "A Parable on Savings and Investment," *Economica*, Vol. 13, February, 1933, pp. 75-84, and her publication of "The Theory of Money and the Value of Output," *Review of Economic Studies*, Vol. 1, October, 1933, pp. 22-26? Could it be that Mrs. Robinson was let in on a little secret in between?

A Comment

When the reader chases down Mrs. Robinson's reference to Keynes's 1939 *Preface to the French Edition (The Collected Writings,* Vol. VII, pp. xxxi-xxxv), no real light is cast on our present problem: Who helped to write the *General Theory*? In some respects the French Preface overlaps with the book's original 1935 Preface and with the 1936 German and Japanese Prefaces. Keynes sounds his own horn on his brainchild's great importance, and how he had to grow out of the orthodox notions of mainstream pre-1936 Anglo-Saxon economics. A trace element of what my Footnote 1 describes as Kahn's idiosyncratic view on (against!) the Quantity Theory surfaces here.

Telephone: 2248.

62a GRANGE ROAD,
CAMBRIDGE

Oct 22

Dear Professor Samuelson

I much enjoyed your amusing and penetrating article on Keynes in Econometrica. You hit a lot of nails on the head.

I can very simply clear up for you the mystery you point out in your last footnote. My Economica article arose from the immediate post-Treatise discussions, but Economica allowed about 18 months to elapse between accepting + publishing it. So that the change you point out took place over two years (two years of furious & continuous discussion) not a few months.

Did you ever see the Preface which Keynes wrote for the French translation of the General Theory? He there made some rather interesting comments on his book + the way it was written

Yours sincerely
Joan Robinson

On the Chronology of the *General Theory*[1]

Don Patinkin

The workings of the human mind are a thing of wonder, and all the more so in the case of a great mind. It is therefore not surprising that in his pioneering 1947 study of *The Keynesian Revolution*,[2] Lawrence Klein, and students of the development of Keynes's thought after him, have devoted much attention to tracing the steps by which Keynes moved from the *Treatise on Money* to the *General Theory*. Indeed, in recent years that chronology has been one of the most heatedly debated issues of Keynesiology.

In order to discuss this chronology in a meaningful way, one must of course first define the meaning of "the *General Theory*" — that is, one must first identify the theoretical novelty of the book and its central message. Needless to say, on this question there is a wide difference of opinion (see Patinkin 1990). In this note I shall restrict myself to the definition that I argued for in my *Keynes's Monetary Thought* (1976, Ch. 8): namely, that the theoretical novelty of the book is its theory of effective demand *cum* equilibrating role of changes in output; that is, a theory of aggregate demand and supply for output in which changes in the level of output themselves act as the equilibrating force that brings the economy to an equilibrium level of output, which need not be one of full employment. In Keynes's words in his 1937 reply to Ohlin, "the novelty in my treatment of saving and investment consists, not in my maintaining their necessary aggregate equality, but in the proposition that it is, not the rate of interest, but the level of incomes which (in conjunction with certain other factors) ensures this equality" (*JMK* XIV, pp. 211-12; see also *GT*, pp. 31 [lines 16-23] and 179 [lines 2-6]).

I find it convenient to divide the question of the chronology into four separate ones:

(1)   The "analytical distance" between the *Treatise* and the *General Theory*.

(2)   The relation between Kahn's 1931 multiplier article and the *General Theory*.

(3)   The stage reached in the discussions of the "Cambridge Circus" (in the sense of "Circle," or what might today be called the "Cambridge Colloquium").

(4)   The subsequent stages.

The simplest way of answering the first question is to note that the analytical framework of the *Treatise* is constructed about its so-called fundamental equations, the full name of which is "The Fundamental Equations *for the Value of Money*" (title of chapter 10 of the *Treatise*, with italics added). Thus the central message of the *Treatise* has to do with the price level. More specifically, it has to do with the relation between per-unit price and cost of production as presented in the fundamental equations; consequently with the per-unit profit (loss) that occurs when the former exceeds (falls short of) the latter; and with the consequent increase (decrease) in output that is generated by this profit (loss). Thus changes in output in the *Treatise* are derivative from changes in prices; furthermore, the *Treatise* specified only the direction of change in output, and not its extent. In Keynes's words in his preface to the *General Theory*,

> My so-called "fundamental equations" were an instantaneous picture taken on the assumption of a given output. They attempted to show how, assuming the given output, forces could develop which involved a profit-disequilibrium, and thus required a change in the level of output. But the dynamic development, as distinct from the instantaneous picture, was left incomplete and extremely confused. This book, on the other hand, has evolved into what is primarily a study of the forces which determine changes in the scale of output and employment as a whole.

At the same time, I feel that when Keynes precedes these words with the statement that

> The relation between this book and my *Treatise on Money*, which I published five years ago, is probably clearer to myself than it will be to others; and what in my own mind is a natural evolution in a line of thought which I have been pursuing for several years, may sometimes strike the reader as a confusing change of view.

— when he does that, I feel that Keynes is succumbing to the natural temptation of an author to present himself as having been consistent over time, and that he accordingly attempts to minimize the difference between the *Treatise* and the *General Theory*. And an even clearer attempt to minimize this difference occurs at a later point in the *General Theory* (pp. 77-78) where Keynes states that "change in the excess of investment over saving [in the fundamental equations of the *Treatise*] was the motive force governing changes in the volume of output" — and does not even mention that in the first instance this "excess" affects prices.

In any event, the basic analytical difference between the *General Theory* and the *Treatise* is nicely illustrated by Keynes's famous "parable" in the *Treatise* of a simple "banana plantation" economy in an initial position of full-employment equilibrium which is disturbed because (in Keynes's words) "into this Eden there enters a thrift campaign". Making implicit use of his "first fundamental equation" (see Patinkin 1976, p. 67), Keynes then explains that the resulting increased savings, unmatched by increased investment, will cause entrepreneurs to suffer losses and that as a result they will

> seek to protect themselves by throwing their employees out of work or reducing their wages. But even this will not improve their position, since the spending power of the public will be reduced by just as much as the aggregate costs of production. By however much entrepreneurs reduce wages and however many of their employees they throw out of work, they will continue to make losses so long as the community continues to save in excess of new investment. Thus there will be no position of equilibrium until either (a) all production ceases and the entire population starves to death; or (b) the thrift campaign is called off or peters out as a result of the growing poverty; or (c) investment is stimulated by some means or another so that its cost no longer lags behind the rate of saving (*Treatise* I, pp. 158-60).

Though we should not take this first alternative seriously (a clear example of Keynes's propensity to shock the reader and to strive for the paradoxical), for our present purposes it is nevertheless most significant: for it is implicitly based on the assumption that the decline in output does not exert any endogenous equilibrating force. Nor, clearly, is such a force in operation in the third alternative, where the decline in output is exogenously brought to an end by an increase in investment "by some means or another." And I would say that the same is true of the second alternative, in which the "thrift campaign is called off"; nor do I think that "peters out as a result of the growing poverty" has the connotation of a systematic equilibrating process. In sum, none of these alternatives indicates that Keynes of the *Treatise* understood that the decline in output itself acts directly as a systematic endogenous force to bring the economy into equilibrium at a specific level of employment (and unemployment).

I must, however, admit that some commentators have contended that "peters out" might be interpreted as describing a systematic influence of the decline in income on savings (see Samuelson as cited by Patinkin [1977, p. 23, n. 10]). But if this were Keynes's intention in alternative (b), it seems to me that no matter how much he enjoyed shocking the reader, he would not have listed alternative (a) with its completely contradictory assumption. Similarly, if he had regarded (b) as a systematic influence that always equilibrates saving and investment, there would have been no need for him to have added alternative (c), in which equilibrium is achieved as a result of a fortuitous exogenous increase in investment. And it is significant that when, in what has been identified as a draft-fragment of the *General Theory*, Keynes essentially considers again the case of an initial equilibrium disturbed by an increase in savings, and this time emphasizes that the decline in income ultimately exerts a negative effect on savings that will restore equilibrium, he implicitly refers to — but then puts aside — the other two alternatives that he had described in his aforementioned "parable." In particular, he puts aside alternative (a) on the grounds that it is based on "somewhat stringent assumptions"; and he rules out "for the moment" what can be regarded as an example of alternative (c) — namely, the possibility of an exogenous increase in investment as a result of "the deliberate effort either of the banking system or of the government" (*JMK* XIII, p. 385; see n. 11 below).

I must also admit that in her 1933 article on "The Theory of Money and the Analysis of Output," Joan Robinson also interprets alternative (b) of the "parable" as a description of an endogenous equilibrating process. Most significantly, however, she also cites it as supporting evidence for her contention that "when he published the *Treatise*, [Keynes] had no very clear perception of the fact that the subject with which he was dealing was the Analysis of Output". As she went on to explain,

> A second example[3] of Mr. Keynes's failure to realize the nature of the revolution that he was carrying through is to be found in the emphasis which he lays upon relationship of the quantity of investment to the quantity of saving. He points out that if savings exceed investment, consumption goods can only be sold at loss. Their output will consequently decline until the real income of the population is reduced to such a low level that savings are perforce reduced to equality with investment. [At this point there is a footnote reference to p. 178 of the original edition of the *Treatise*, which is the page on which the three alternatives of Keynes's parable appears]. But he completely overlooks the significance of this discovery, and throws it out in the most casual way without pausing to remark that he has proved that output may be in equilibrium at any number of different levels, and that while there is a natural tendency towards equilibrium between savings and investment (in a very long run), there is no natural tendency towards full employment of the factors of production. (Robinson, 1933b, pp. 55-6)

Further evidence on the *"distance"* between the *Treatise* and the *General Theory* is provided by Keynes's 1936 letters to Roy Harrod and Abba Lerner on their respective reviews of the *General Theory*. In these letters, Keynes described the three basic analytical components of the book as being the theory of effective demand, the theory of liquidity preference, and the theory of the marginal efficiency of capital (*JMK* XIV, pp. 83-6; *JMK* XXIX, pp. 214-16). And what is significant for our purpose is that of these three components, the only one that Keynes of the *General Theory* associates with his earlier discussion in the *Treatise* is the theory of liquidity preference. Specifically, in chapter 13 on *"The General Theory* of the Rate of Interest" the *"bull-bear"* discussion of the *Treatise* is related to the speculative demand for money (*GT*, pp. 169 [n. 1] and 173). Similarly, chapter 15 on "The Psychological and Business Incentives to Liquidity" explains the relationship between the transactions, precautionary, and speculative balances of the *General Theory*, on the one hand, and the income-deposits, business-deposits, and savings-deposits of the *Treatise*, on the other (*GT*, pp. 194-5).

In sharp contrast, there is no reference to the *Treatise* in Keynes's presentation of the theory of effective demand in chapter 3 of the *General Theory*. Nor (with one irrelevant exception [*GT*, p. 124]) is there any such reference in the chapters of the *General Theory* (8-10) in which Keynes presents the consumption-function component of aggregate demand. Nor, finally, is there any reference to the *Treatise* in his chapter 11 on the marginal efficiency of capital, which underlies the investment-function component. Indeed, in another basic difference from the *General Theory*, the *Treatise* is devoid of marginal analysis (Patinkin, 1976, pp. 47, 94).

Let me turn now to the question of the relationship between Richard Kahn's 1931 multiplier article and the *General Theory* — and with this to a friendly ongoing disagreement of mine with Paul Samuelson that began almost twenty years ago (see Patinkin and Leith, eds. 1977, pp. 18-20, 81-7). At that time Samuelson demonstrated that Kahn's multiplier formula is logically equivalent to the simplest case of Keynes's theory of effective demand as presented in the familiar equation $C(Y) + I_0 = Y$, where Y is national income, $C(Y)$ the consumption function, and I_0 the exogenous level of investment — which equation is then solved for the equilibrium level of Y. I, however, contended that logical equivalence does not imply chronological equivalence: that at the time Kahn was not aware of this implication and that his article (whose analytical framework is explicitly that of the *Treatise*) actually has no concern whatsoever with the question of the determination of the equilibrium level of income. Indeed, the term "equilibrium" appears only once in his article (p. 183, n. 1), and even then in a different context.

Instead, the purpose of Kahn's article was to provide a rigorous basis for Keynes and Henderson's contention in their 1929 *Can Lloyd George Do It?* that an increase in public-works expenditures would generate an increase in income and hence in the savings needed to finance it, and that accordingly there was no basis for the "Treasury View" that such an increase in government investment would leave unchanged the total amount of savings in the economy, hence would simply be offset by a corresponding decrease in private investment, and therefore would not generate any net increase in employment. Kahn accomplished this purpose not by means of any macroeconomic-equilibrium equation, but by summing up an infinite geometric series of additional savings over time as generated by the multiplier, and showing that this sum equaled the initial increase in investment. He also described this as "a particular case of a general relation, due to Mr. Meade," which relation also took account of the negative effect of a possible increase in prices on savings (Kahn 1931, p. 188). The sharp distinction between the purpose of Kahn's multiplier article and the *General Theory* can be brought out most clearly by noting that Kahn refers to a possible concurrent upward shift of the consumption function (as we would term it today), and consequent reduction in savings, as an "aggravation" of the initial problem created by the public-works expenditures, and the additional savings generated by the multiplier process as the "alleviation" (ibid., pp. 179-80) — hardly the terminology that reflects the respective roles of these magnitudes in the *General Theory*.

To this evidence I would like to add that of Kahn's own view of his multiplier article in a letter that he wrote me in March 1974 in reply to a query of mine as to what he saw its "primary importance" to have been. His answer was:

> I regard the main importance of my 1931 article as: (1) finally disposing of the `Treasury view' that at a time of unemployment, an increase in one kind of investment will be at the expense of another kind. I demonstrated how the whole of the necessary finance is provided in the form of an increase in saving. The point is that in *Can Lloyd George Do It?*, Keynes and Hubert Henderson seem to imply that only part of the additional finance would be provided as a result of the investment.... (2) Finally disposing of the idea that the price level is determined by the quantity of money.... (3) I had not altogether escaped from Keynes's *Treatise*, and I thought it necessary to translate my ideas into terms of the *Treatise*. Although I was not conscious of that at the time, the effect was to demonstrate how unsuitable the terminology and assumptions of the *Treatise* were.[4]

I might finally note that in September 1931, Keynes wrote Kahn a letter with some indications of the equilibrating mechanism, but Kahn replied that he "[had] not been able to follow the steps" (*JMK* XIII, pp. 373-5). Nor (according to the materials reproduced in *JMK* XIII and XXIX) did Keynes at the time follow up the approach of this letter.[5]

So all of the above confirms Klein's (1947, pp. 36 and 38) conclusion from his discussion of the multiplier article that Kahn "obviously did not see the great theoretical implications of his work."

Another passage in Keynes's writings that Samuelson (1946, p. 200) has cited as an adumbration of the *General Theory*[6] is the one in his June 1931 Harris Lecture at the University of Chicago in which, after explaining how an excess of savings over investment generates a decline in output, Keynes went on to say:

> Now there is a reason for expecting an equilibrium point of decline to be reached. A given deficiency of investment causes a given decline in profit. A given decline in profit causes a given decline in output. Unless there is a constantly increasing deficiency of investment, there is eventually reached, therefore, a sufficiently low level of output which represents a kind of spurious equilibrium. (*JMK* XIII, pp. 355-6)

In order to place this passage in its proper context, let me first of all point out that this lecture was first and foremost a Song of Praise to the *Treatise on Money*, which had appeared less than a year before. In Larry Klein's words (1947, pp. 35-6), "the theoretical views [of this lecture] were basically unchanged" from those of the *Treatise*. Thus Keynes began his analysis of the slump in it with a verbal rendition of the fundamental equations and with the accompanying proclamation, "That is my secret, the clue to the scientific explanation of booms and slumps (and of much else as I should claim) which I offer you" (*JMK* XIII, p. 354). And what this passage (which appears in the discussion which follows) describes is not a continuing state of unemployment equilibrium, but the transitory stationary point at the trough of the business cycle. That is, what Keynes is analyzing here is the cause of the eventual elimination of the original excess of saving over investment which (according to the analysis of the *Treatise*) generated the slump, and its replacement at the turning point of the cycle by an opposite excess which then begins to generate the recovery.

I come now to the discussions that took place in the famous Cambridge Circus during the period of its activity (viz., January-May 1931). In the early 1970s, as part of his editorial task of editing volume XIII of Keynes's *Collected Writings*, Donald Moggridge prepared a summary of these discussions that was based on a meeting with its erstwhile participants. Moggridge prefaced this summary with the proviso that "it must be read with a full awareness of the fallibility of memories, particularly with regard to exact dates, over so long a period" (*JMK* XIII, p. 338). In reply to my query, Moggridge also provided the following description of how this summary was prepared:

> Austin Robinson and I took the initiative in organising the
> discussion. Joan Robinson, Austin Robinson, Richard Kahn,
> James Meade and Piero Sraffa attended. I then compiled a
> note of what transpired, circulated it to those involved for
> correction, and, when I got an agreed version, put it in the
> volume (letter of 1 December 1990, cited with Moggridge's
> kind permission; see also Austin Robinson's [1985, p. 52]
> account of the preparation of this summary).

In my discussion in *Keynes' Monetary Thought* (1976, pp. 55-60) of
the role of the Cambridge Circus, I concluded (on the basis of this summary,
as well as of other considerations to which I shall return below) that its main
contribution lay in its criticisms of the notions of the *Treatise*, and not in the
development of the theory of effective demand of the *General Theory*. And it
may not be irrelevant to point out that though they referred to this conclusion
of mine, and indicated their disagreement with it, neither Richard Kahn (1984,
p. 105; 1985, p. 44)[7] nor Austin Robinson (1985, p. 53; see also his comment
in Patinkin and Leith, eds., 1977, p. 81) attempted to refute it. It is also
noteworthy that there is no reference to the equilibrating role of changes in the
level of income in the May 1932 "manifesto" which Joan and Austin Robinson
and Richard Kahn subsequently submitted to Keynes (*JMK* XXIX, pp. 42-5).

Moggridge's summary also reports that "James Meade, an active
participant in the discussions [of the Cambridge Circus], returned to Oxford in
the autumn of 1931...[and] is cautiously confident that he took with him back
to Oxford most of the essential ingredients of the subsequent system of the
General Theory" (*JMK* XIII, p. 342). Since these "essential ingredients" were
not defined, it is difficult to know what "ingredients" Meade had in mind, and
whether in particular they included that "ingredient" that I consider to be the
central message of the *General Theory*. Indeed, for reasons that will be
indicated below, I doubt if they did.

Let me at this point digress for a moment to say that though the
contribution of the Cambridge Circus as such to the development of the *General
Theory* has been exaggerated, two of its members — Richard Kahn and Joan
Robinson — subsequently played an important role (together with Roy Harrod
at Oxford) in providing Keynes with critical comments on the successive drafts
of the book.[8] There is little if any evidence that the *Treatise on Money* was
effectively subjected to such pre-publication criticism. Thus Austin Robinson
has told us that the "fundamental equations of the *Treatise* appeared for the first
time, I understand, in Keynes's Lectures of 1928-29" and that "they were still
relatively new and relatively undigested even by the people in Cambridge in
closest touch with his work when the *Treatise* appeared in 1930" (A. Robinson

1947, p. 53). Furthermore, because of the time pressures generated by the other commitments that he had undertaken (notably, as a member of the Macmillan Committee as well as of the Economic Advisory Committee), Keynes was not able to take account of the detailed criticisms that he did receive a few months before the publication of the book from Ralph Hawtrey, who in them described Keynes's fundamental equations as tautologies and emphasized that the book did not contain a theory of output — points that were to be stressed in the subsequent critical reviews of the book. Similarly, the discussions of the Cambridge Circus shortly after its publication quickly pointed out the erroneous nature of various conclusions in the book to which Keynes had attached importance (e.g., the paradox of "the widow's cruse" [*Treatise* I, p. 125]).[9]

I would conjecture that it was this experience that was in the back of Keynes's mind when in the Preface to the *General Theory* he wrote that "it is astonishing what foolish things one can temporarily believe if one thinks too long alone, particularly in economics (along with the other moral sciences), where it is often impossible to bring one's ideas to a conclusive test either formal or experimental" (*GT*, p. xxiii). So if, as Austin Robinson (1947, p. 55) tells us, it was a fortunate coincidence that there was in the early 1930s "a remarkable younger generation in Cambridge" who could supply criticisms to Keynes as he developed his *General Theory*, I suspect that an equally important aspect of this fruitful coincidence was the fact that — as a result of his unhappy experience with the *Treatise* — Keynes then had a demand for such criticism!

I return now to the concern of this paper. When Paul Samuelson wrote his 1946 obituary article on Keynes, the only documentary evidence on the chronology available to him was material that had been published. Accordingly, as a "clue...to the future historian of economic thought," he pointed out the contrast between Joan Robinson's February 1933 *Economica* article on "A Parable on Savings and Investment," which was concerned with "an aspect of the *Treatise*," and her October 1933 *Review of Economic Studies* article on "The Theory of Money and the Analysis of Output," and playfully asked, "What was happening in Cambridge in the [intervening] months?...Could it be that Mrs. Robinson was let in on a little secret in between?" Larry, who explained in the Foreword to *The Keynesian Revolution* (p. viii) that it originated in a 1944 doctoral dissertation at M.I.T. and who expressed his indebtedness to Samuelson, took up on the latter's suggestion and concluded that "the difference in theoretical structure between these two articles...should lead us to suspect the occurrence of the [Keynesian] revolution in Cambridge during 1933" (Klein 1947, pp. 38-40). However, in her introduction to her 1951 *Collected Economic Papers* (p. ix), Joan Robinson — as "an awful warning to historians" — wryly commented that Samuelson and Klein should not have based their chronology on the publication date of the first of these articles, for this was actually written in the summer of 1931 and that for some reason its publication was delayed.[10]

That is where the matter rested until the publication in 1973 of Volume XIII of Keynes's *Collected Writings* with its various draft-fragments of the *General Theory*, dated as having been written at various times from 1932-1935. However, in the process of writing my 1975 review article on this as well as other volumes of the *Collected Writings*, I wrote to Donald Moggridge, the editor of Volume XIII, and asked him what the basis was for the dating of the aforementioned fragments. In reply, he wrote:

> The dating of the early *General Theory* fragments: In the [Keynes] papers, all the surviving fragments are in a single bundle, which was not in any particular order. The ordering chosen reflects a mixture of guesses, plus conversations with Austin and Joan Robinson and Richard Kahn. (from Moggridge's letter to me of 18 October 1974, cited with his kind permission).

In the light of this reply, I felt free in my review article (1975, p. 252, n. 1) — as well as in subsequent publications (e.g., 1976, p. 71, n. 7) — to question the dating of two of these fragments[11] and to suggest that on the basis of their internal evidence, their chronological sequence was actually the reverse of what had been indicated.

Because of the problematic dating of the draft-fragments, I also decided to base my conclusion about the chronology of the *General Theory* on a typescript of Robert Bryce's precisely dated student notes of Keynes's lectures for the years 1932, 1933, and 1934.[12] It is also most significant that in his own detailed study of the chronology, Moggridge (1973, p. 82, n. 34) depended not only on the materials of Volume XIII that he himself had edited, but also made "great use" of Bryce's notes. At a 1975 conference on "Keynes, Cambridge, and the *General Theory*" in which, in addition to Bryce, Lorie Tarshis also presented a paper, it was discovered that Tarshis also had notes of the lectures, in his case for the years 1932-35[13]. And the fact that these notes provide two independent and similar observations on these lectures can only increase their credibility (Patinkin 1977, pp. 14-16; see also Moggridge 1992, p. 556).

It was, then, from the fact that I did not find the theory of effective demand *cum* equilibrating role of changes in the level of output in Bryce's notes of Keynes's lectures in the fall of 1932, but did find it in those for the fall of 1933, that I concluded that "Keynes formulated his theory of effective demand during 1933, and in all probability during the first half of that year" (1976, p. 79; see also Patinkin 1977, pp. 14-16, which also takes account of similar evidence from Tarshis' notes). And this conclusion was — and remains — my main reason for rejecting the traditional description of the role of the 1931 Cambridge Circus in the development of the *General Theory* (Patinkin 1976, pp. 59-60; 1977, p. 17).

On the basis of an examination of the lecture notes, Robert Dimand (1988, pp. 147-67, especially pp. 166-67) also concluded that the theory of effective demand had not yet appeared in Keynes's 1932 lectures, and accordingly concluded that Keynes formulated his theory "between his 1932 and 1933 lectures." A similar conclusion was reached by Moggridge (1992), who agrees that the theory of effective demand is not to be found in the lecture notes of November 1932, and who concludes that "we can date Keynes's adoption of his output equilibrating model to late 1932 or early 1933 *at the latest*" (ibid., p. 562, italics in original) — where in this context "late 1932" means what remained of that year after the lectures that Keynes gave in November. But though this constitutes a high degree of agreement about the chronology of the *General Theory*, I would not want to create the impression that resort to the lecture notes has eliminated all differences of opinion about it. Thus on the basis of his examination of these notes, Peter Clarke (1988, pp. 262-4) has concluded that the theory of effective demand manifests itself already in the fall 1932 notes, and even contends that this theory was developed in the summer before that.[14]

And so we have come full circle. For though the basis for Larry Klein's original conclusion that "The Keynesian Revolution" occurred in 1933 turned out to be unacceptable, the weight of opinion today, based on the evidence of the lecture notes, accords with his conclusion.

Don Patinkin, The Israel Academy of Sciences and Humanities, Albert Einstein Square, 43 Jabotinsky Road, P.O. Box 4040, Jerusalem 91040, Israel.

ACKNOWLEDGEMENTS

This is a modified version of an article with the same title which appeared in the May 1993 issue of the *Economic Journal*. I am indebted to that journal for its permission to publish this version here.

This research was supported by the Israel Science Foundation administered by the Israel Academy of Sciences and Humanities, and by the Latsis Foundation, to which I express my appreciation.

NOTES

1. All references to Keynes's writings are to the form in which they appear in the volumes of the Royal Economic Society's edition of his *Collected Writings*, referred to henceforth (unless otherwise indicated) as, e.g. *JMK* XIII, *JMK* XXIX, and so forth. The *Treatise on Money* is sometimes referred to as the *Treatise*, and the *General Theory* as *GT*.

2. My acqaintance with this book began in 1946-47, when as a graduate student at the University of Chicago, I was a research assistant at the Cowles Commission. That was the year that Larry joined Cowles in order to begin his seminal work on an econometric model of the United States. Part of his time, however, was devoted to putting the finishing touches on the manuscript of the book, which I then read with great interest and even excitement. In my possession, I have

a much-worn and appreciated presentation copy of the book. As acknowledged in my *Money, Interest, and Prices* (1956, p. 234, n. 4; 1965, p. 336, n. 4) —- and as I am happy to acknowledge again — it has had an important influence on my own interpretation of the *General Theory*.

3. The first was Keynes's discussion of the "widow's cruse," which Robinson pointed out was based on the implicit assumption of an unchanged level of output — a point that she had already made in her 1933 *Economica* article on "A Parable on Savings and Investment" (p. 82).

4. This is part of the excerpts from this correspondence that were reproduced — with Kahn's permission — in Patinkin and Leith (1977, pp. 146-7).

5. For the most recent round in the disagreement between Samuelson and myself about the role of Kahn's multiplier article, see Samuelson's 1991 article and my comment thereon. See also Patinkin (1976, pp. 69-71; 1982, pp. 26-9, 189-99).

6. Samuelson's discussion appears on p. 330 of his article as reprinted in Lekachman (1964). The following reproduces part of my discussion of this passage in Patinkin (1976, pp. 68-9 and 1982, pp. 23-6), which should be consulted for further details.

7. The relevant passage from p. 105 of Kahn's 1984 book appears in a section entitled "The Cambridge `Circus'" and reads as follows:

> Don Patinkin disputes the importance commonly attributed to us in assisting Keynes to write the *General Theory*. Insofar as he relies on documents, he is fully entitled to make his case. Others can judge. For my own part I feel myself unable to arouse any feeling of passion over events which took place so long ago.

A similar statement appears in his 1985 paper. It is also noteworthy that in a later section of his 1984 book entitled "My part in the `*General Theory*,'" Kahn does not mention the discussions of the Cambridge Circus.

8. The first extant correspondence dealing with such criticism refers to that which Keynes received from Kahn in March-April 1934, on which Keynes reported to Joan Robinson that Kahn "is a marvelous critic and suggester [sic] and improver — there never was anyone in the history of the world to whom it was so helpful to submit one's stuff" (*JMK* XIII, p. 422). There is a brief postcard from Keynes to Joan Robinson dated May 1933 which may be a reference to her criticism of a draft of the book (ibid., p. 419). Harrod's criticisms were of a very advanced draft of 1935 (ibid., p. 526; see also *JMK* XIV, p. 351).

9. For details, see Patinkin (1976, pp. 29-32, 54-57).

10. Samuelson's footnote just referred to does not appear in his article as reproduced in the 1947 Harris volume (pp. 145-60) and subsequently in the 1964 Lekachman volume (pp. 315-31); and since it was from the latter volume that this article was reproduced in volume 2 of his *Collected Scientific Papers* (1966, pp. 1517-33), it does not appear there either. In reply to a query of mine on this point, Samuelson replied that Joan Robinson had written him a letter on his 1946 obituary article in which she explained that the *Economica* article had been published with a long delay, and that this led him to delete the footnote when the article was reprinted.

11. One of which was the fragment referred to on p. 24 above.

12. An unedited typescript of Bryce's notes is in the Keynes Papers at Cambridge. I learned of their existence from one of Moggridge's editorial notes in *JMK* XIII (p. 343, n. 1). In response to my request, Moggridge then kindly provided me with a copy. Interestingly enough, Klein (1947, p. 42, n. 10) noted that a copy of these notes had "been made available in mimeographed form," but he does not seem to have made use of them in his attempt to trace the chronology of the *General Theory*.

13. A synthesis of these two sets of notes, as well as those of several other students who attended Keynes's lectures during various years in the period 1932-35, has recently been published by Thomas Rymes (who also attended the 1975 conference) in his *Keynes's Lectures, 1932-35:*

Notes of a Representative Student (1989).

14. Note that even according to Clarke's chronology, Keynes developed the theory of effective demand a year after the discussions of the Cambridge Circus had come to an end.

REFERENCES

Reprinted works are cited in the text by year of original publication; the page references to such works in the text are, however, to the pages of the reprint in question.

Clarke, Peter. (1988). *The Keynesian Revolution in the Making, 1924-1936*. Oxford: Clarendon Press.

Dimand, Robert W. (1988). *The Origins of the Keynesian Revolution: the Development of Keynes's Theory of Employment and Output*. Aldershot: Edward Elgar.

Harris, Seymour E. (Ed.). (1947). *The New Economics: Keynes' Influence on Theory and Public Policy*. New York: Alfred A. Knopf.

Kahn, R. F. (1931). The Relation of Home Investment to Unemployment. *Economic Journal. 41*(June), 173-98.

_____ (1984). *The Making of Keynes' General Theory*. Cambridge: Cambridge University Press.

_____ (1985). The Cambridge 'Circus'. In G.C. Harcourt (Ed.), *Keynes and His Contemporaries*. (Pp. 42-51). London: Macmillan.

Keynes, John Maynard. (1930). *A Treatise on Money*. Vol. I: *the Pure Theory of Money*, Vol. II: *the Applied Theory of Money*. As Reprinted in the *Collected Writings of John Maynard Keynes*. Vols V and VI.

_____ (1931a). An Economic Analysis of Unemployment. *Unemployment as a World Problem*. edited by Quincy Wright, Vol. XI. Chicago: University of Chicago Press, for the Harris Memorial Foundation., Pp. 3-42. As Reprinted in the *Collected Writings of John Maynard Keynes*. Vol. XIII, pp. 343-67.

_____ (1931b). *Essays in Persuasion*. As reprinted with additions in *the Collected Writings of John Maynard Keynes*, Vol. IX.

_____ (1936). *The General Theory of Employment, Interest and Money*. As reprinted with prefaces to foreign editions and three appendices in the *Collected Writings of John Maynard Keynes*, Vol. VII.

_____ (1937). Alternative theories of the rate of interest. *Economic Journal, 47*(June), 241-52. As reprinted in the *Collected Writings of John Maynard Keynes*, Vol. XIV, Pp. 201-15.

_____ (1973). *The General Theory and After: Part I, Preparation*. Vol. XIII of the *Collected Writings of John Maynard Keynes*, edited by Donald Moggridge.

_____ (1973). *The General Theory and After: Part II, Defence and Development*. Vol. XIV of the *Collected Writings of John Maynard Keynes*, edited by Donald Moggridge.

_____ (1979). *The General Theory and After: a Supplement*. Vol. XXIX of the *Collected Writings of John Maynard Keynes*, edited by Donald Moggridge.

_____ *The Collected Writings of John Maynard Keynes*, Vols. I-VI (1971), Vols. VII-VIII (1973), Vols. IX-X (1972), Vols. XI-XII (1983), Vols. XIII-XIV (1973), Vols. XV-XVI (1971), Vols. XVII-XVIII (1978), Vols. XIX-XX (1981), Vol. XXI (1982), Vol. XXII (1978), Vols. XXIII-XXIV (1979), Vols. XXV-XXVII (1980), Vol. XXVIII (1982), Vol. XXIX (1979), Vol. XXX (Bibliography and Index, 1989), London: Macmillan, for the Royal Economic Society.

_____ (1929). and Hubert Henderson, *Can Lloyd George Do It?: An Examination of the Liberal Pledge*. As reprinted in the *Collected Writings of John Maynard Keynes*, Vol. IX, Pp. 86-125.

Klein, Lawrence R. (1947). *The Keynesian Revolution*, New York: Macmillan.

Pp. 86-125.

Klein, Lawrence R. (1947). *The Keynesian Revolution*, New York: Macmillan, 1947.

Lekachman, Robert. (Ed.). (1964). *Keynes's General Theory: Reports of Three Decades*, New York: St. Martin's Press.

Moggridge, Donald. E. (1973). From the *Treatise* to *The General Theory*: An exercise in chronology. *History of Political Economy.* 5(spring), 72-88.

_____ (1992). *Maynard Keynes: an Economist's Biography*, London: Routledge.

Patinkin, Don. (1965). *Money, Interest, and Prices: an Integration of Monetary and Value Theory.* 1st Edition. Evanston, Ill.: Row, Peterson, 1956; 2nd edition. New York: Harper and Row.

_____ (1975). The collected writings of John Maynard Keynes: From the *Tract* to the *General Theory. Economic Journal.* 85(June), 249-71.

_____ (1976). *Keynes's Monetary Thought: a Study of its Development*, Durham, N.C.: Duke University Press.

_____ 1977). The process of writing the *General Theory*: A critical survey. In D. Patinkin and J. C. Leith (Eds.), *Keynes, Cambridge and the General Theory: the Process of Criticism and Discussion Connected with the Development of the General Theory.* (Pp. 3-24). London: Macmillan.

_____ (1982). *Anticipations of the General Theory? And Other Essays on Keynes*, Chicago: University of Chicago Press, 1982.

_____ (1990). On different interpretations of the *General Theory. Journal of Monetary Economics.* 26(October), 205-43.

_____ (1991). Comment [On Samuelson]. In L. Jonung (Ed.), *The Stockholm School of Economics Revisited.* (pp. 407-410). New York: Cambridge University Press. (Proceedings of a 1987 conference.)

Robinson, Austin. (1947). John Maynard Keynes, 1883-1946. *Economic Journal.* 57(March), 1-68. As reprinted in R. Lekachman (Ed.), *Keynes's General Theory: Reports of Three Decades.* (Pp. 13-86). New York: St. Martin's Press, 1964.

_____ (1985). The Cambridge `Circus'. In G. C. Harcourt (Ed.), *Keynes and His Contemporaries.* (Pp. 52-57). London: Macmillan.

Robinson, Joan. (1933a). A parable on savings and investment. *Economica.* 13(February), 75-84.

_____ (1933b). The theory of money and the analysis of output. *Review of Economic Studies. 1*(October), 22-26. As reprinted in Joan Robinson, *Collected Economic Papers*, Vol. I, Oxford: Basil Blackwell, 1951, pp. 52-58.

_____ (1951). *Collected Economic Papers.* Vol. I, Oxford: Basil Blackwell.

Rymes, Thomas K. (Ed.). (1989). *Keynes's Lectures, 1932-35: Notes of a Representative Student.* London: Macmillan.

Samuelson, Paul A. (1939). A synthesis of the principle of acceleration and the multiplier. *Journal of Political Economy.* 47(December), 786-97. As reprinted in J. E. Stiglitz (Ed.), *The Collected Scientific Papers of Paul A. Samuelson.* (Pp. 1111-22) Vol. II.

_____ (1946). Lord Keynes and the General Theory. *Econometrica.* 14(July), 187-200.

_____ (1966). *The Collected Scientific Papers*, Vol. II, Edited by Joseph E. Stiglitz, Cambridge, Mass.: MIT Press.

_____ (1991). Thoughts on the Stockholm School and on Scandinavian economics. In L. Jonung (Ed.), *The Stockholm School of Economics Revisited* (pp. 391-407). New York: Cambridge University Press.

Rereading the 'Keynesian Revolution'

Wilhelm Krelle

It is *nobile officium* and a pleasure to contribute to a "Festschrift" for Lawrence Klein. We have known each other since 1954 when we met in Ann Arbor. Larry represents a type of economist which I think is in scarce supply: he knows economic theory and econometrics and works actively at the front line of research in this area, but understands practical economic and social policy as well as institutional and political constraints. The "Klein-Goldberger Model" was the prototype for almost all models of national economies which are now in use. All this began with his book *The Keynesain Revolution* (Klein, 1950). The econometric models of today still preserve many Keynesian features. So it might be interesting and a fitting present for the jubilee to look back at the beginning of Larry's work. This inevitably leads to the problem: what is left of Keynesianism? Some thoughts on this might be interesting for Larry as well as for a broader audience.

I. Lawrence Klein's Understanding of Keynes's *General Theory*

Keynes's *General Theory* (Keynes, 1936) is a difficult book. It is not well organized, it comprises different approaches and it is often not clear which assumptions are made. It is thus not very surprising that the first reviews of the *General Theory* by eminent scholars such as Pigou, Knight, Cassel, Viner and others were more or less negative. Keynes's reply (Keynes, 1937) consists mostly of a reformulation and clarification of the basic ideas of the *General Theory*. Lawrence Klein's interpretation of Keynes apparently rests on this self-interpretation by Keynes and on Hicks' (1937) article "Keynes and the 'Classics'" where he introduced the famous LM-SI diagram which Klein (p. 88) adopts.

The charateristic feature of Lawrence Klein's presentation of Keynes is the sovereignty and assurance with which he judges Keynes. Klein throws out everything that does not fit into the logical scheme of the hard core basic propositions of the *General Theory*, and blames the author for not having understood his own work correctly (Klein, 1950 p. 83). Thus, Lawrence Klein was the personification of the real Keynes, and I think that most economists of the time accepted this and learned and taught the *General Theory* along the lines of Lawrence Klein. He made the *General Theory* comprehensible. The influence of this book is still visible in many textbooks, e.g. in one of the leading textbooks in Germany (Felderer and Homburg, 1985). The neoclassical model of Felderer and Homburg (p. 86) is practically identical to the classical model of Lawrence Klein (p. 200). The Keynesian system as presented by

Lawrence Klein (p. 202) is also practically identical to the first Keynesian
model of Felderer and Homburg (p. 134). But Felderer and Homburg also
present two variants of this Keynesian model (p. 141 and p. 151). This is not
astonishing since Clower and Leijonhufvud presented other versions of the
Keynesian system.

Lawrence Klein understands the classical model as one where labor
supply and labor demand depend on real wages. Equality of labor supply and
demand determines the real wage. This yields full employment and determines
real production simultaneously, since demand for goods equals supply. The
interest rate is determined by the equality of savings and investment, and the
price level by the quantity equation (p. 200). This, of course, is a highly
aggregated model of the classical (or neoclassical) system. I would prefer to
take the Walrasian system as a representative of the classical or neoclassical
approach. But Lawrence Klein's version conforms better to the Keynesian
system. In the Keynesian system labor supply depends on the nominal wage
rate, labor demand on the real wage rate. Consumption (in nominal terms)
depends on the interest rate and nominal income, investment and money
demand depend on the same variables. Money supply is exogenous. Real
production is a function of the employed labor force. This system does not
necessarily yield full employment. The same applies for similar systems which
use consumption and investment functions which depend on real production and
a liquidity function which determines the real money demand. It is difficult to
tell what Keynes really had in mind. According to his formulation in Ch. 3, pp.
28 of the *General Theory* in connection with his ideas on price determination
in Ch. 21 where he states that the price level depends on the wage level and on
employment, one would rather say that Keynes was thinking of consumption
and investment functions in real terms. At least in his refomulation of the
General Theory in Keynes (1937) this seems to be the right interpretation (see
p. 221). I think that Lawrence Klein uses this interpretation in order to
demonstrate that it is not the inflexibility of wages and prices that leads to unemployment but the lack of demand. He criticizes Keynes for not being clear
on that point: "Again, as in the 'Treatise', Keynes did not really understand what
he had written" (p. 83). I think that the solution of the problem lies in our
assumption concerning the relation between price and wage movements. If
prices are proportional to wages, then workers cannot reduce their real wage by
reducing the nominal wage rate. By cutting nominal wage rates, we would thus
get a downward spiral and there would be no way out of unemployment. This
is what Lawrence Klein had in mind. If there is no such link the result is more
of neoclassical type.

Keynes's really revolutionary idea was the principle of effective
demand, according to Lawrence Klein. He did not consider the Keynesian
liquidity preference theory of the interest rate to be a major contribution. He
writes (p. 117):"Interest is not a very important variable in the modern

economic world". I think that Lawrence Klein does not hold this opinion any more.

Summing up, I would like to say that Lawrence Klein delivered an invaluable service to the scientific community by interpreting the Keynesian system in a logically consistent way and thus made it usable both for theory and for practical applications in econometric forecasting models.

II. Is there a Keynesian Revolution?

Lawrence Klein says so. His book holds this title. He states in the foreword (p. VII): "The Keynesian theory is viewed in the following pages as a revolutionary doctrine...The 'revolution' discussed here is a revolution in thought, not in the economic policies of governments". But what exactly is the revolution? Lawrence Klein states two contributions. On p. 56 he says: "The revolution was solely the development of a theory of effective demand; i.e., theory of the determination of the level of output as a whole". Then on p. 76 Lawrence Klein explains this in theoretical terms: "We know that Keynes's procedure was to do away with a savings-investment theory of interest and replace it with a savings-investment theory of output. When this had been done the revolution was a *fait accompli*". There is still another statement as to what the revolution is: Keynes "gave up Say's law that makes savings flow automatically into investment regardless of the level of income. He also developed a theory which did not make flexible wages such a powerful tool, always ensuring full employment. The incompatibility of the simplest classical system with the simplest Keynesian model shows clearly what is meant by a revolution" (p. 79/80). I think that most young economists at that time felt the same way, at least I did so. In view of millions of unemployed people at that time the attitude of *laissez faire* seemed unacceptable. If one wishes to preserve the whole system one has to find out what to do about it. But this means first to understand the functioning of the system and the reasons for unemployment. The classical system offered little help. One could not imagine that only some frictions or rigidities could be responsible for such a malfunctioning of the system. Keynes seemed to offer this explanation. The classical and neoclassical view was that the economy is comparable to a big ship with an automatic steering system so that it does not need a captain. The Keynesian view was that the ship will run off course if there is no captain and no helmsman who corrects the course by using the appropriate levers and other steering devices. Keynes showed that in his system government and money supply would be the main instruments to keep employment at an acceptable level. Since the beginning of the sixties the general philosophy of the German economic policy was very much influenced by these ideas. A special law (the *'Wachstumsgesetz'*) was enacted in 1967 which gave the government the right to change the general tax level and the general level of expenditure according to the actual state of the economy without going through the time-consuming procedure of enacting a new budget. In fact this law has never been applied by the government but it shows the spirit and the general philosophy behind the economic policy of that time. This policy worked quite well in the sixties and in the early seventies but it failed completely later. This shows that there are hidden assumptions on the economic behaviour of the economic agents behind the Keynesian systems which were taken for granted and not considered as crucial and which did not remain valid after the middle of the seventies, at least in Germany.

These assumptions refer to the behavior of the trade unions, of the employers and of the central bank. If the central bank pursues a monetary policy which keeps the rate of inflation below a certain level, if the trade unions press for wages above the increase of labor productivity and the tolerated rate of inflation and if the employers accept that, then there will be unemployment. The same result will come out of a system where the central bank does not consider inflation and the government finances budget deficits by money creation and the trade unions know this and ask for wages which compensate for the expected rate of inflation in the future. Prices will follow wages, or in the situation of a runaway inflation they will even precede them. Today one would say: if all economic agents have rational expectations and if the trade unions are representatives of the interests of the employed workers, the Keynesian instruments for steering the economy are inefficient. Government deficits and money creation only increase the rate of inflation but leave unemployment untouched. When the rate of inflation exceeds a certain threshold the employment effect even becomes negative because the basis for investment calculation disappears eventually. Keynesianism thus became discredited. It was substituted by the so-called supply policy which tries to increase investment by improving the conditions for it, by increasing technical progress, improving the infrastructure, reshaping the tax system in favor of investment and the like. Thus there was some sort of counter-revolution led by experiences from practical economic policy.

But also on the theoretical side there was no real revolution in the sense that the old classical or neoclassical system was crowded out by the new Keynesian one. It very soon became clear that Keynes only dealt with one type of economic problem, namely employment, and that other problems could not be dealt with within the system. There was no real wage and price theory in the Keynesian system, the theory of allocation was absent as well as theories of money supply, foreign trade, exchange rates and so on. The Keynesian theory, thus, could not replace the whole body of classical or neoclassical economics which is far larger and comprises many more fields than just employment theory.

These theories could fit into the general neoclassical approach which explains the economic activities by the interaction of independent decisions of economic agents where the decisions are coordinated by the price system. But eventually the Keynesian system became incorporated into a new extended classical approach by the introduction of time lags into the neoclassical system. If wages and prices are sticky and do not change much in the short run the Keynesian results remain valid in the short run. This is perhaps what Keynes had in mind because, as he put it, "in the long run we all are dead". But this is not the standpoint of an economist who feels responsible for the well-being of a society in the long run.

Though we cannot know the future in detail, we can at least try to discover principles which will improve the economic situation of the society irrespective of the specific political or social situation in the future. Thus, we may say: Keynes opened up a new and fruitful area of research and developed economic concepts which are of lasting value for economics, e.g. the concept of the consumption or savings function, the liquidity preference function and so on. He also made it clear why the stability conditions for the classical or neoclassical economic system may not be fulfilled in the short run. It remains open whether they will be fulfilled in the long run. Looking back one may say: There was no Keynesian revolution, but an important step forward in understanding economics. The largest part of the neoclassical theory remained untouched. Keynesianism is thus largely incorporated into the general macroeconomic theory of classical or neoclassical descent.

In econometric forecasting systems - the special field of Lawrence Klein in latter years - the neoclassical feature becomes visible by the introduction of price and wage functions, of functions determining the capital accumulation and the development of capacity, and by introducing substitutions between different products, especially between homemade and imported products and in other fields. There are still some few dissenters around King's College in Cambridge, England, where Joan Robinson pursued the Keynesian line as she understood it. This school adheres to fixed input coefficients and minimizes the price substitutability. I think that this is an inferior approach to economics, at least in general. Only a minority of economists follow this line now and Lawrence Klein is not among them.

III. The Problem of an Optimal Economic System

Keynes saw the shortcomings of capitalist systems. In his opinion the main shortcomings are unemployment and unjust distribution of income and wealth. He was in favor of this system and tried to improve it by conceiving instruments to cure unemployment. With respect to the distribution of wealth and income he thought that this problem would be solved more or less automatically. The rate of interest would fall, the rentier would disappear eventually but the true entrepreneur would stay (p. 375/376). Keynes liked freedom and self-determination and was opposed to authoritarian systems even if they could ensure full employment more successfully than democracies. But to preserve the system it must be changed in such a way that a high level of employment can be guaranteed. This also would lead to peace among nations because one could avoid cut-throat competition between national economies on world markets, competition which can lead to wars.

Lawrence Klein did not follow Keynes in this respect. Unemployment is such a social evil for him that he considered a change of the economic system a real alternative. A socialist system where all investment decisions are made by the central planning board and are not based on principles like profit maximization seemed to be an attractive alternative for him (p. 77). He considered a socialist planning system as a superior economic organization which Keynes failed to see in the years preceding the *General Theory* (p. 79). He was not alone in this opinion. There were some illusions prevalent in intellectual circles at that time as to what a central planning system could achieve. Surely it could guarantee full employment and that seemed to be the most important goal. The dark sides of a socialist system were less visible at the time. "Economists have to ... reconsider ... their social philosophy" (p. 166). Keynesianism or socialism seemed to be the only alternatives to save a society from fascism which is the worst form of capitalism (p. 167).

Certainly unemployment is the most negative feature of a free market system. We cannot be content with how the market economy is working now. The Keynesian instruments are blunt, new instruments must be developed and social rules changed. Economists must think along these lines. Socialism or Keynesianism are no longer alternatives. The cost of getting rid of unemployment in one of these ways is too high. Collective wage bargaining for large areas and industries does not seem to be an appropriate method for finding the full employment wage level, given price stability. The instruments of supply policy are not fully developed. Budget deficits of government have their limits and cannot be extended indefinitely. Thus, supply policy seems to be the way out, but this field is not sufficiently instrumentalized so that it could be easily handled by government decisions. Lawrence Klein recommended a high-level consumption economy in the long run as a hope for capitalism, connected with the redistribution of income and development of a social security system. He did not accept the counter-arguments against the Keynesian policy (inflation, indebtedness of the government, loss of freedom). All this is still better than unemployment (p. 179) but intelligent planning should yield price stability and full employment simultaneously. Price controls could be used to curb inflation. Government debts are no evil and no burden because the society owns the debts to itself (p. 181/182).

I think that many young economists who really felt the suffering of people under mass unemployment were rather near to the political opinion of Lawrence Klein at that time. Keynes was much more cautious in this respect. He did not like the Soviet experiment and did not like Marx. He referred to the "Capital" as "an obsolete economic textbook ... not only scientifically erroneous but without interest or application for the modern world" (quoted by Lawrence Klein, p. 130). I agree with that. Larry Klein disagreed at the time that he wrote his "Keynesian Revolution".

There have been theories on the functioning of a socialist economic system, see e.g. Barone, von Mises and others. In Germany there was (and still is) an economic school which is concerned with the economic order as a whole (*Wirtschaftsordnung* in German). Eucken was one of the most representative scholars in this field. The German economic policy was heavily influenced by these ideas. The new institutional economics tries to develop this field by applying the normal economic concepts. The large scale experiment of communism in the Soviet Union and Eastern Europe provided results which were forecast by many economists. Unfortunately, this experiment was done at the expense of the well-being of at least two generations of people in these countries, unfortunately. Now I think people know that we have to stay in the realm of market economies. But there are enough variations possible, and we have to find one which minimizes the grim problem of unemployment.

IV. Keynes's Position in the Development of Economic Thought

In Chapter 23 of his *General Theory*, Keynes deals with his predecessors and related ideas of other authors. This chapter is perhaps the weakest in his book. He finds that in mercantilism economic principles and economic policies have been pursued which he thinks are also appropriate now for fighting unemployment. Of course Mandeville and Malthus are mentioned (perhaps Keynes saw his fight against "orthodox theory" anticipated in the controversy between Malthus and Ricardo). But he also deals with Silvio Gesell. Lawrence Klein in his Chapter 5 where he reviews the anticipations of the *General Theory*, would classify him as one of the cranks, heretics, demagogues and amateur economists "who usually sensed what is wrong, ... but have usually been far wide off the mark in the formulation of their fantastic theoretical systems" (p. 31/32). Of course there were also respectable economists among those who tried to explain unemployment and partly held similar views to Keynes. Lawrence Klein deals with them, e.g. Hobson, Foster and Catchings, N. Johannsen, and Preiser in Germany, to name only a few. Keynes and his secluded circle of economists in King's College did not much consider economic ideas outside Great Britain. Not even American writers on the subject of unemployment were admitted into the halls of King's College; see the review "Keynesianer vor Keynes" by Garvy, in Bombach, Ramser, Timmermann (1976), p. 22. Lawrence Klein surveys the more important forerunners of Keynes - forerunners in the sense that they had similar ideas in one or another respect. Nobody anticipated the Keynesian system as a whole. But also Lawrence Klein leaves out a real duplication of Keynes's ideas by Carl Föhl, see Föhl (1935).

This book written by a then unknown German scholar in German had no effect on the economic science whatsoever and had to be rediscovered by Erich Schneider and others in the 40's. Föhl used the concepts of Keynes's *Treatise* and thus always has profits as a decisive variable in his system. His recommendations for fighting unemployment are the same as those to be found in Keynes (1936). Demand determines employment. The rate of interest is explained as a price to induce the owner of money to lend it and thus is the consequence of the limitation of the amount of money in the economy (Föhl, 1935 p. 204-205). This is very near, if not identical, to Keynes's liquidity preference theory. But of course the book by Föhl had no chance against the book of the world-famous and outstanding economist Keynes. Lawrence Klein did not know the book either. But I think Föhl's contribution which is so much clearer than Keynes should not be forgotten.

V. Conclusions

Keynes's contributions to economics will stay and form an integral part of the body of general economic theory. And Keynes will be understood much better in the light and the interpretation of Lawrence Klein. This interpretation is now used more or less explicitly in many textbooks. Its weakness from the theoretical point of view is the lack of connection with microeconomics. But this is true for the whole macroeconomic approach. Lawrence Klein is well aware of this point (pp. 57/58) but underestimates the difficulties connected with this problem. I think that we have to take macroeconomics as an approximation and give up the idea that we can derive the macroeconomic laws and regularities in a direct way from theories of individual behavior. We should rather consider the distribution of preferences and production coefficients. These distributions might give us the macroeconomic regularities which we need if we want to understand the economic development as a whole and give some guidelines for economic policy.

The "Keynesian Revolution" by Lawrence Klein will remain a classical piece of economic literature. His interpretation of Keynes has shaped the understanding of Keynes. The scientific community is indebted to him for this service.

Wilhelm Krelle, Institut für Gesellschafts-und Wirtschaftswissenschaften der Universität Bonn, Adenauerallee 24-42, 53113 Bonn, Germany.

REFERENCES

Bombach, Gottfried, Ramser, Hans-Jürgen, Timmermann, M., and Wittmann, W. (1976). *Der Keynesianismus II.* Berlin-Heidelberg-New York: Springer.
Felderer, Bernhard, and Homburg, Stefan. (1985). *Makroökonomik and Neue Makroökonomik,* Berlin-Heidelberg-New York: Springer, 5th. ed. 1991.

1955.

Hicks, John R. (1937). Mr. Keynes and the "Classics": A suggested interpretation. *Econometrica.* 5, 147-159.

Keynes, John Maynard. (1936). *The General Theory of Employment, Interest and Money.* London: Macmillan., repr. 1955.

Keynes, John Maynard. (1937). The general theory of employment. *The Quarterly Journal of Economics.* February, 209-223.

Klein, Lawrence R. (1950). *The Keynesian Revolution.* London: Macmillan.

The Discussion among Future Nobel Laureates Becker, Friedman and Klein on Macro Models and Consumption Functions, in 1957 and 1958

Ronald G. Bodkin

More than a third of a century ago, when the bulk of their careers still lay ahead of them, three very promising young men, Professors Gary Becker, Milton Friedman and Lawrence R. Klein, engaged in an exchange of views on the appropriate approach to macroeconomic modelling, joined by Jack Johnston and the late Edwin Kuh. Milton Friedman had won the John Bates Clark Award of the American Economic Association in 1951, while Lawrence R. Klein was to win the same medal in 1959; thus it seems fair to argue that the talents of these two scholars were already recognized at the time, even though their most prestigious awards were to arrive only two decades in the future. (Friedman's Nobel Memorial Prize in Economic Science was awarded in 1976, while Klein's occurred in 1980.) As for Gary Becker, he was to win the John Bates Clark Medal a decade and a half (in 1974) after the discussions surveyed here, while his Nobel Memorial Prize in Economic Science was awarded just after the first draft of this paper was written, in the fall of 1992. Thus a point of disagreement (or at least of discussion) among these leading economists would appear to be of some interest, even in retrospect. Indeed, in some sense, as a historian of thought would emphasize, a retrospective view can often be advantageous, as new insights may be available as a result of developments in the interim. Moreover, sometimes (but not always) the passage of time can provide an objectivity (or at least a sort of distance) on the hot issues under discussion.

We may begin our review of this episode by noting that this discussion was a bit unusual (over and above the quality of the participants involved). Indeed, the nature of this exchange was one of the points under contention. Friedman and Becker contended that there was no essential disagreement between them and Klein (and also with J. Johnston and Edwin Kuh), and on this point of view the participants to the exchanges all accepted a common IS-LM framework and were merely discussing the detailed specification of the system. On the other hand, Klein, Kuh, and Johnston certainly thought that they were taking strong exception to what Friedman and Becker had to say. Thus, in evaluating this discussion, one of the points that a commentator should reasonably take into account is whether there was indeed a genuine disagreement, or whether the issues among the participants were principally semantic (or originated in some other superficial form of non-understanding).

The plan, then, of this paper is as follows. In the next section the original note of Friedman and Becker is reviewed, while Klein's vigorous defence of (then) Keynesian macroeconometric methodology is summarized in Section 3. The interventions of Johnston and Kuh are summarized in Section 4, and Friedman's and Becker's remarks (protesting, as already indicated, that their critics had misunderstood them and that there was no essential disagreement among them) are reviewed in Section 5. In the final section, the present writer attempts, with a certain measure of trepidation, a disinterested evaluation of this fascinating incident.[1]

I. Friedman's and Becker's Critical Sally

The opening salvo in this exchange was fired by Friedman and Becker (1957) in an eleven page article published in the February issue of the *Journal of Political Economy*, with the somewhat emotive title, "A Statistical Illusion in Judging Keynesian Models." Friedman and Becker adopted a very simple Keynesian system with a consumption function, autonomous investment, and a national income accounting identity; thus their model might be written:

$$C_t = \alpha + \beta Y_t + \dots + \epsilon_t \quad \text{(consumption function)} \qquad (1)$$
$$Y_t = C_t + I_t \quad \text{(National income accounting identity).} \qquad (2)$$

Here, C_t represents consumption at time t, Y_t represents income at time t, and I_t (strictly all non-consumption expenditures that enter national product) was interpreted as investment demand. (The symbol ϵ_t is the stochastic perturbation in the consumption function, while α and β represent parameters.) It should be noted that, then as now, this system represents a pedagogical model, rather than one that a working econometrician would use for practical forecasts. Friedman and Becker then addressed the issue of whether a researcher should focus on relative error in predicting consumption level (on the basis of the single equation, equation (1) above or its analogue) or in predicting the level of national product (on the basis of the simultaneous equation system). In particular, the forecast of national product, on the basis of this simple system, would be:

$$Y_t = (\alpha + I_t + \epsilon_t + \dots)/(1 - \beta). \qquad (3)$$

It was claimed that the former (examining the predictions of consumption level) was standard practice, but that this was a dangerously misleading approach, at least if our ultimate objective was to predict real national income. Indeed, Friedman and Becker considered this practice to be so misleading that they termed this notion a *"statistical illusion."*

This contention was then illustrated with an example involving two consumption functions based on a version of the absolute income hypothesis, three "naive" models of consumption, and a consumption function with a distributed lag in income that was based on Friedman's own "Permanent Income Hypothesis" (Friedman, 1957).[2] (The first naive model postulated that current consumption was equal to the previous year's consumption, while the second naive model was an exponential, in which consumption was predicted to grow by a constant rate of growth. The third naive model was a modified exponential which allowed for a constant level of "autonomous" consumption.) While the distributed lag consumption function had a slightly lower mean square error in predicting consumption (in comparison to both the naive models and the two functions based on the absolute income hypothesis), its mean square error was much, much lower in predicting national income. By contrast, while the "standard" consumption functions based on the absolute income hypothesis had lower errors in predicting consumption levels than the naive models, the naive models were superior in their predictions of real national product. Friedman and Becker's conclusion, based on this illustrative example, was:

"The relatively small errors made by functions (12) and (13) [the consumption functions based on the two versions of the absolute income hypothesis] are in part a statistical illusion." (Friedman and Becker, 1957, p. 70.)

The remainder of the article is devoted to examining the implications of their principal point. In particular, they concluded that high measured marginal propensities to consume were particularly dangerous, as they both force the fit and give rise to a high measured multiplier, which "... multiplies the error as well as investment proper." (Friedman and Becker, 1957, p. 72.) Indeed, there is the intriguing suggestion, to which I shall return below, that the researcher should check the accuracy of the predictions of his/her consumption function, but not principally against income (whatever concept); rather, it is suggested that other explanatory variables be used to check how well the model predicts the level of consumption. Of course, in retrospect, this article was a precursor of the great debate in macroeconomics of the 1960s, in which Friedman and his associates (e.g., Friedman and Meiselman, 1963) argued that the stock of money as a key exogenous explanatory variable was a more reliable explanation of macroeconomic movements than some "Keynesian" notion of autonomous investment. (Another way of putting their assertions would be that the demand for money was somewhat more central or "stable" than the Keynesian analogue, the consumption function.) Interpreted in this light, some of the more puzzling assertions of the article fall into place.

However, for the immediate forecasting problem, another issue arises immediately; couldn't errors in forecasting autonomous investment also be a source of prediction error? Indeed, in the Main Stream Keynesian view, one might expect these to be the *major* source of forecasting errors. But no, at least according to Friedman and Becker, this would not be the case, as the error in predicting investment expenditures is likely to be only a minor portion of the

.errors in predicting GNP, at least as judged by the portion of GNP that autonomous investment typically represents (implicitly, for the US economy). Accordingly, Friedman and Becker argued that a good research strategy would be to improve the statistical consumption function. Indeed, in this connection, they argued that even autonomous consumption expenditures would impart a certain measure of stability to the macroeconomy, and in the process help improve the forecastability of real national product.

Following up on these considerations, Friedman and Becker argued that the Main Stream Keynesian research strategy was inappropriate: "Regarding investment as having no multiplier effects is a much better first approximation than regarding investment as the prime mover, at least with respect to the changes in real per capita income from 1905 to 1951 [for the USA]...". (Friedman and Becker, 1957, p. 74.) The issue was apparently joined. Although the last page of the article lists some minor qualifications, the message that the Main Stream Keynesian approach is inappropriate (because it is based, in part, on "statistical illusion") persists. Apparently, we are in the midst of a sharp and essential disagreement about the *substance* of knowledge of the macroeconomy. Notice, however, that there is an interesting issue of *methodology* that has certainly been raised by Friedman and Becker: how can applied macroeconometricians validate (or, if this word is too strong, corroborate) their tentative hypotheses? Friedman and Becker appear to be suggesting an alternative method of testing a structural relationship embedded in the context of a system of simultaneous equations (though this may not be apparent on a first reading of their article, especially given the controversial nature of its contents).

II. Klein's Spirited Defence of then Current Main Stream Research

Although Kuh's and Johnston's comments appeared a few months earlier, we shall turn next to Klein's comments on this article because they are at the centre of this note. (Also, Friedman and Becker in their reply focused mainly on Klein's comments.) Before launching into the substance of Klein's remarks, we may remark that the tone of this communication is one of righteous indignation, one that reflects a feeling that the individual in question (or his close associates) have been the victim of a less than fair attack.[3] This certainly coincides with the recollections of the present author as well.

Klein begins his critique by arguing that there has been *no* statistical illusion in judging Keynesian macroeconomic (or macroeconometric) models. Indeed, he asserts (as in the title of his comment) that the illusion has been all on the part of Friedman and Becker. Moreover, he argues that the two principal points of Friedman and Becker, which he takes to be the existence of simultaneous equations biases in econometric estimation and the use of reduced forms (in place of structural equations) in a forecasting context, have been known for some years (more than a decade), say by the early work of T. Haavelmo (another distinguished researcher whose life's work received the accolade of the Nobel Prize in economics in 1989.)

Warming to his critique, Klein argues that the so-called "Permanent Income" consumption function that Friedman and Becker fitted was really a version of the late T. M. Brown's (1952) "Habit Persistence" hypothesis. Another supplementary point is that there are, according to Klein, a number of alternative methods of evaluating macroeconometric models (or a key structural relationship, like a consumption function), and that the mean square error criterion is only one of them. Klein points to the ability of a model to forecast turning points as another useful criterion, and argues that the fact that the Keynesian models predicted the 1953-54 downturn is certainly a point in their favour.

The bulk of the article, however, is devoted to Klein's fitting a consumption function based on a modified absolute income hypothesis, in which wealth is an important supplemental explanatory variable. Klein then puts this consumption function through its paces, including a study of its relative root mean square error in the GNP forecasting (reduced form) context (as in equation (3) above). For him, it was obviously a matter of some satisfaction that, even under the Friedman-Becker criterion, the relative root mean square error of this modified absolute income hypothesis consumption function was 4.95 per cent, as contrasted to the 5.1 per cent score obtained by the Friedman-Becker function (based on a distributed lag version of the Permanent Income Hypothesis) for the same sample period.

In conclusion, then, Klein argues that (then) current research is (was) on the right track and that future work will (would) probably include new predetermined variables in the consumption function. Furthermore, according to Klein, such progress will (would) *not* be illusory.

III. The Interventions of Kuh and Johnston

As indicated earlier, the late Edwin Kuh and J. Johnston also participated in this debate. As this brief review of their contributions to the discussion will indicate, both of them tended to divide their remarks between the methodological issue of how to evaluate a macroeconometric model and the substantive issue of whether the Keynesian approach to macroeconomic research is the best one possible (or, indeed, a legitimate approach).

Kuh seems a bit confused by whether to focus on the methodological issues or on the substantive issues, as though he were uncertain as to what is the principal point of Friedman and Becker. He spends a lot of time discussing what is "autonomous" consumption, a point that seems (to this author) secondary to the principal discussion. He begins by pointing out that the version of the absolute income hypothesis chosen by Friedman and Becker is a very simple one, one that would be used (even at the time of this discussion) only for pedagogical or classroom purposes. He continues by asserting that, if one employs the consumption function favoured by the late Alvin Hansen (current consumption a linear function of income lagged one period), then investment (strictly speaking, the whole path of investment) does indeed determine the path of national income, so that the Keynesian view would appear to be validated. He then continues by arguing that, if the point of Friedman and Becker is that the forecast variance (the variance of current real national product in the reduced form of equation (3) above) is larger than the structural variance (that of consumption from the consumption function), then this is trivially true and easily follows from the simple theory of the multiplier. [Note a definite uncertainty, on the part of Kuh, about the objectives of the Friedman-Becker critique.] Finally, and most surprisingly, he begins to attack the criterion suggested by Friedman and Becker, namely a low root mean square error for key jointly determined variables from the macroeconometric model. Kuh goes on to state that, if all one wishes is accurate prediction, *perhaps the money-income relationship will do quite nicely.* (Emphasis added.) To the present writer, this is amazing, especially in retrospect with the 1960s debate (e.g., Friedman-Meiselman, 1963) kept in mind. If this exchange represents a debate, then Kuh has inadvertently conceded Friedman's and Becker's principal substantive point. No wonder that they felt that, in their reply, they could be quite brief in their comments on Kuh's contribution.

J. Johnston, on the other hand, takes issue directly with Friedman and Becker on *both* the substantive and the methodological issues. On the methodological issue, Johnston challenges the criterion, arguing that a number of alternative criteria may well be used reasonably to evaluate macroeconometric models. Johnston points to apparent randomness of forecast residuals from a macroeconometric model as another quite reasonable criterion.

On the substantive issue, Johnston argues the general point that Friedman's and Becker's results, obtained for a period of a pronounced trend in U.S. economic growth, may well not generalize to other historical periods and other societies. But he also raises a specific objection as well. He claims that Friedman's and Becker's results may well not hold with an alternative consumption function as the point of comparison; as his alternative consumption function, he chooses two versions of Franco Modigliani's previous peak hypothesis as an explanation of aggregative consumption. (There is a fine irony here, because already by the end of the 1950s, Franco Modigliani had abandoned this explanation of macro consumption behaviour in favour of his "life cycle" hypothesis, a theory that is quite similar to Friedman's Permanent Income Hypothesis. But this irony need not concern us here; conceivably, Modigliani was ill-advised to abandon his previous formulation.) Johnston carries out the calculations and contends that his suspicion is confirmed, namely that, even with the Friedman and Becker criterion, the alternative consumption function has a tighter fit of forecast residuals in the analogue of equation (3) above (as judged by the mean square error criterion) than the distributed lag aggregative consumption function employed by Friedman and Becker, which was based in principle on the Permanent Income Hypothesis. Finally, Johnston contends that, even granting Friedman and Becker their consumption function for modelling purposes, there are indeed multiplier effects with their consumption function; these are just not so pronounced, especially upon impact.

IV. The Reply of Friedman (and Becker)

Friedman's and Becker's replies to their critics may be found (briefly) at the end of the notes by Kuh and Johnston in the August, 1958, issue of the *Review of Economics and Statistics*, and at greater length in two separate comments (one joint, one signed only by Friedman) in the December 1958 issue of the *Journal of Political Economy* (right after Klein's spirited communication). In general, they save their detailed arguments for the December reply, and so this review of the discussion will focus on their December reply, except where explicitly indicated otherwise.

As indicated in the introduction, Friedman and Becker really refused to admit that they were in the middle of a full scale debate on how to conduct macroeconomic research. In the August reply, they argue that their critics have misunderstood them or have missed the point. While they were more deferential toward Lawrence R. Klein, the point that he is attacking them inappropriately (in their view) remains; two quotes will illustrate this amply, in my view:

"To us, Lawrence Klein's comment seems less a criticism of our article than a valuable supplement." (P. 545.)

and [at the very end of Friedman's individual reply]

"From this point of view, Klein's empirical results, far from casting doubt on the general theory [Friedman's Permanent Income Hypothesis, asserted to be more general than Friedman's and Becker's distributed lag consumption function] add further evidence in support of it. His functions, in particular the function that includes 'wealth' as a variable, can all be regarded as special applications of the permanent income hypothesis. There is every reason to hope that a better specific formulation of it than the expected income equation can be found. I regard Klein's work as a useful step in that direction."

Moreover, in the middle of their reply, Friedman and Becker assert, somewhat surprisingly, that they are not advocating any particular consumption function, only "improvement in the consumption function." The present writer is reminded of the debating technique in which one attempts to incorporate the opposition's solid arguments (i.e., those not easily refutable) into one's own case!

As for the issues under discussion, they reiterate their criterion, which they consider the "correct" criterion for judging a macroeconomic or macroeconometric model. Moreover, they feel entitled to some originality on this point, because (in their view) Haavelmo's reduced forms were primarily concerned with structural estimation (designed to obviate simultaneous equations biases), rather than being concerned with the problem of forecasting (whether within the sample period or outside of it). Friedman and Becker seem relatively resistant to suggestions of other criteria that might be used to evaluate macroeconometric models. They do concede, however, that ordinary goodness-of-fit may be a sort of test of predictive accuracy, especially if the theory under evaluation were propounded before the full sample became available.

Turning to the substantive issues, they do largely maintain that "investment" has little impact on national income, at least if "investment" is interpreted to mean "current" investment. (In their August reply, they conceded that the history or entire time path of investment could well influence current national product.) In this sense, they assert that the multiplier notion of Main Stream Keynesian theory is somewhat illusory. In particular, as they spell out in their August reply, the economy is still rather stable because variations in average investment over an entire decade are much less important than year-to-year variations.[4] ("In the latter case, the year-to-year fluctuations in current investment lose much of their importance for short-run as well as long run analysis." [P. 298, Friedman and Becker, "Reply to Kuh and Johnston," August 1958.])

On the issue of the choice of a consumption function, Friedman refuses to concede much to Klein (or to Johnston, for that matter). Friedman and Becker assert that Klein's criticism of Friedman's consumption function is simply beside the point. In his detailed reply, Friedman explains why. In his view, the permanent income approach is a very general one, as it unifies a number of approaches to aggregative consumption theory. Moreover, it makes sense of a host of isolated explanatory variables. The time series approach (pp. 142-152 of Friedman, 1957) leads to "expected income" as a key explanatory variable for aggregative consumption, but this "expected income" variable is certainly less general than "permanent income," the theoretical construct. For this reason, Friedman feels much more confident about the theoretical concept than the operational counterpart (as indeed was spelled out in the 1957 book), and this is also why he feels that Klein's research on this substantive issue is compatible with his overall framework. (As noted above, it must be stated that Friedman's time series version of the consumption function based on the permanent income hypothesis would not be regarded by many as his major contribution in this field.)

Finally, we note that Friedman did make a substantive concession to

Klein on the issue of T. M. Brown's consumption function as an antecedent to Friedman's own work. (Whether Friedman should have conceded on this point is examined in the next section.) Friedman asserts, that, at the time of his writing of the 1957 book, "I [Friedman] did not recognize that this procedure was equivalent to Brown's use of the consumption of the previous year, and Klein is quite right in criticizing me for this error of omission." Indeed, Friedman goes on to state that Klein's linking of Friedman's formulation of an aggregative consumption function to Koyck's work on distributed lags, with its implications for efficient estimation, is highly valuable. Thus, the reply of Friedman ends on a conciliatory note.

V. Evaluation of the Exchange

There would appear to be several issues to evaluate, particularly as the present writer has organized the presentation. The most obvious issue is the semantical one of whether the exchange between Friedman and Becker, on the one side, and Klein, Kuh, and Johnston, on the other. is better characterized as a "debate" or a "discussion." In fact, the evaluation of the present writer is mixed. As for the issues involving the evaluation of macroeconometric models, this part of the discussion might reasonably characterized as "discussion," and Friedman's and Becker's remark that Klein's comments were a fruitful addition seems reasonable. On the other hand, the issue of how best to study the macroeconomy, in particular the question of whether or not to use a Main Stream Keynesian approach to macroeconometric modelling or, rather, to use what would come to be called a monetarist approach, was surely a debate and not a discussion. (Friedman was to pursue this debate vigorously in the decade to come, in such contributions as Friedman and Meiselman [1963], with the goal of reversing some of the principal tenets of the Main Stream Keynesian approach.) In addition, the use of a term such as "statistical illusion" would appear to be tendentious;[5] it is certainly oriented toward arguing that an entire approach is in error, rather than toward discussing a methodological point dispassionately. In the present writer's view, it is hardly surprising that Klein and his associates in this exchange thought that they were engaging in a full-scale debate, rather than a tame discussion of the methodology of the evaluation of macroeconometric models!

Nevertheless, the methodological issues were very important throughout this discussion, and here the basic point of Friedman and Becker appears only to gain force, from a current perspective. Basically, Friedman's and Becker's argument that one should evaluate the error predictions of aggregate income (rather than aggregate consumption) in a simple Keynesian model of income determination can be rephrased to argue that full model simulations (rather than single equation predictions) are the critical element in the evaluation of a structural equation embedded in a simultaneous equations econometric model of the macroeconomy. (Even if one should like to examine single equation properties in a balanced evaluation of a medium-sized to large macroeconometric model, this point would appear to retain considerable force.) In other words, this point could be interpreted to assert that model-builders *should give major weight to the simultaneous equations characteristics of a structural equation (particularly its ability to generate satisfactory predictions in a full system context)* and only secondary emphasis to its single equation characteristics (including isolated predictions on the basis of this equation alone). Today, this point seems unexceptional and widely accepted, but of course it was not so in 1957. At present, almost all macroeconometric model builders give considerably more weight to the capacity of a structural equation to perform well in full model simulations (or, in some cases, in block simulations of a large, complex model) rather than to the single equation characteristics of the structural equation. However, the use of full model simulations was to become prevalent only a decade or so later, when the computational hardware and software would permit such "capital-intensive" study. (Documentation of these assertions may be found in our recent history of macroeconometric model-building, Bodkin, Klein, Marwah, 1991.) Moreover, the connection between one structural equation and the reduced form for an endogenous variable other than its dependent variable will in general be far more complicated than in the simple example that Friedman and Becker discussed. But this takes nothing away from the justice of their methodological point. In this regard, they deserve full marks for anticipating an important development in the art of evaluating macroeconometric models, especially later macroeconometric models of some complexity. It also seems fair to add that Klein, Kuh, and Johnston did not fully appreciate their methodological point, because their attention was strongly focused on the debatable portion of the Friedman and Becker paper (i.e., was there really statistical illusion [a serious flaw] in the Main Stream Keynesian approach to macroeconometric model-building?).

This leads one immediately into the issue of how one might judge the substantive debate between the Main Stream approach to macroeconometric modelling and the alternative advocated by Friedman and Becker, which was later to be called Monetarism. In this reviewer's judgement, Klein and Johnston did an adequate job of showing that, even using the Friedman-Becker criterion, a modified consumption function along Main Stream Keynesian lines could perform just as well (if not slightly better) than Friedman's candidate, the distributed lag consumption function developed on the basis of the Permanent Income Hypothesis. (Kuh was surely right in asserting that the version of the Main Stream Keynesian consumption function employed by Friedman was a straw man, one that would have been used by a Main Stream Keynesian [even in 1957] only for classroom purposes.) So in a narrow sense Klein, Kuh, and Johnston did not lose the debate. However, in a broader sense they did not win it either, as Monetarism (as it was later to be called) survived and flourished. We have already referred to the much broader debate of the 1960s (e.g., Friedman and Meiselman, 1963), where these substantive issues were discussed at much greater length (and probably greater depth). It is the feeling of this reviewer that that debate was equally indecisive, as both approaches to macroeconometric modelling (and more generally, to macroeconomics) survived and flourished. Indeed, the two approaches (with a number of variants and nuances) have survived through the 1970s and the 1980s right to the present. While the Main Stream Keynesian approach has survived, it is certainly on the defensive today, and governments in North America seem extremely reluctant to employ the suggested policies for attempting to pull the economy out of the current recession (1990-1993), which has been extraordinarily stubborn and long-lasting. It is considerations such as these which lead the present reviewer to argue that this debate has been, from the point of view of the discipline (if not necessarily from the view-point of the reputation of the participants!), a stalemate.

From some very general points, we turn to a set of specifics, which will close this article. It will be recalled that in his personal reply to Klein, Friedman conceded the point that T. M. Brown's formulation of the aggregate consumption function was an important anticipation of Friedman's own work on the subject. So Klein would have appeared to have won this portion of the debate. However, roughly two decades later, the late Balvir Singh and Aman Ullah reexamined this question. Basically, they argued that the two underlying theoretical frameworks (habit persistence and the permanent income hypothesis) are quite different in their descriptions of consumption behaviour. The Permanent Income Hypothesis is based on the highly rational consumer who optimizes over the entire horizon period (presumably the lifetime, under full certainty), while Brown's typical consumer is sluggish and lazy, unwilling to carry out changes that are certainly in his/her best interest from any longer term perspective. It would thus be surprising if the statistical implications of the two theories were the same. Fortunately [from this point of view], they are not the same; Singh and Ullah (1976) show that, although the two theories give rise to the same reduction from the point of view of the deterministic variables of the regression, the error or disturbance terms have a quite different structure. This is in turn implies that different estimators are appropriate depending upon the origin of the Koyck-like reduced form; if one uses the same estimator (say an instrumental variable estimator based on an approach developed by Liviatan), the biases will be quite different in the two cases, as Singh and Ullah show. Thus, Singh and Ullah argue that "Brown's model is by no means 'truly a complete anticipation of Friedman'," and it would appear that, even in the spirit of a dispassionate portion of the exchange, Friedman conceded this point too readily.[6] Perhaps it is encouraging for the rest of us that even giants can make mistakes, particularly in the throes of a passionate discussion of controversial issues of our discipline.[7]

Ronald G. Bodkin, Faculty of Social Sciences, Department of Economics, 550 Cumberland, University of Ottawa, Ottawa, Ontario, KIN 6N5, Canada.

ACKNOWLEDGEMENTS

For very helpful comments on earlier drafts of this manuscript, I am indebted to Gary Becker, Jan Dutta, Milton Friedman, Lawrence R. Klein, David Laidler, Kanta Marwah, and Michael C. McCracken. The standard caveat applies with particular force in this case, as Milton Friedman, Lawrence R. Klein, and also Kanta Marwah had serious reservations about my interpretation of the articles and notes under discussion. Indeed, both Professors Friedman and Klein indicated, on separate occasions, that their recent rereading of the relevant materials a third of a century earlier would lead to a negligible modification in their respective earlier positions. Nevertheless, their frank comments on these earlier drafts were much appreciated.

NOTES

1.   The reader will note that I have not claimed objectivity; in light of my own history, that would be impossible, as Professor Friedman reminded me in his comment on an earlier draft. Professor Klein was my professor in several courses at the time that this exchange was occurring; he was later to be the supervisor of my doctoral dissertation. Accordingly, I find it amusingly ironic that, toward the end of both of our careers, I find myself in a position to evaluate his work. As for Professor Friedman, I began my career as a researcher in criticizing one aspect of his permanent income hypothesis, a matter under discussion (at least tangentially) in the present exchange. This could give rise to at least two sources of bias, which fortunately would appear to be at least partially offsetting. On the one hand, given his magnificent professional success subsequently, an early critic might attempt to make amends. On the other hand, in that the criticism got the critic a certain amount of attention and professional recognition, the critic might be tempted to continue in the same vein. Of course, there could be other sources of bias as well. The reader will have to decide which (if either) of these two (or more) tendencies is the dominant one, in the present case.

2.   David Laidler has argued, in his comment on an earlier draft of this paper, that Friedman's formulation of a time series version of the aggregate consumption function was not his principal contribution to this literature, and that this should be kept firmly in mind in evaluating this exchange.

3.   And, more than a third of a century later, it seems fair to report that this sense of being wronged had not dissipated, at least as judged by Klein's comments at a session of the Canadian Economics Association, at which an earlier version of this paper was presented.

4.   In this context, it could be argued, along with David Laidler, that Friedman and Becker are hinting at the issue of whether monetary or fiscal policy is the more useful instrument for macroeconomic stabilization, which, as indicated above, was to be the subject of a much larger literature at a later date. In particular, if it were decided, on the basis of solid research results, that monetary policy were the more potent instrument, this would reverse the results of Main Stream Keynesianism (particularly of that earlier period).

5.   When Klein was commenting on an earlier draft of this paper, at the aforementioned Canadian Economics Association meetings in June 1993, he seized on this point. Looking the present writer straight in the eye, with a hard glint of stern debate partially relieved by a twinkle of humour, he cried, "You call this term 'tendentious'! I call it 'provocative'!".

6. It is interesting to note that Gary Becker, in a recent article (1992), has conceded as much. In his discussion of addictions and traditions, he draws a sharp distinction between the forces of habit persistence, which are essentially backward- looking and depend on the whole personal history, and the utility-maximizing influences from looking forward over the range of all possibilities, which influences are stressed in the usual formulation of the permanent income hypothesis.

7. Interestingly, even in 1993, Klein would not concede much on this point. He still considers that the similarity of the deterministic form of the reduced forms of the two theories is, in a sense, more interesting than the difference in the disturbance terms of these reduced forms.

REFERENCES

Becker, Gary S. (1992). Habits, Addictions, and Traditions. *Kyklos. 45*(Fasc. *3*), 327-345.

Bodkin, Ronald G., Klein, Lawrence R. and Marwah, Kanta. (1991). *A History of Macroeconometric Model-building*. Aldershot, England: Edward Elgar Limited.

Brown, T. M. (1952). Habit persistence and lags in consumer behaviour. *Econometrica. 20*(3, August), 207-233. Reprinted in J. W. Hooper and M. Nerlove, eds., *Selected Readings in Econometrics from Econometrica*. Cambridge, Mass.: M.I.T., 1970).

Friedman, Milton. (1957). A theory of the consumption function. Princeton, New Jersey: Princeton University Press [for the National Bureau of Economic Research].

_____(1958). Supplementary Comment. *Journal of Political Economy 66*(6, December), 547-549.

Friedman, Milton, and Becker, Gary S. (1957). A Statistical Illusion in Judging Keynesian Model. *Journal of Political Economy. 65*(1, February), 64-75.

_____(1958). Reply to Kuh and Johnston. *Review of Economics and Statistics. 40*(3, August), p. 298.

_____ (1958). Reply. *Journal of Political Economy. 66*(6, December), 545-547.

Friedman, Milton, and Meiselman, David. (1963). The relative stability of monetary velocity and the investment multiplier in the United States, 1897-1958. pp. 165-268 of Commission on Money and Credit. *Stabilization policies*. Englewood Cliffs, New Jersey: Prentice-Hall, Inc..

Johnston, J. (1958). A Statistical Illusion in Judging Keynesian Models: Comment. *Review of Economics and Statistics. 40*(3, August), 296-298.

Klein, Lawrence R. (1958). The Friedman-Becker Illusion. *Journal of Political Economy. 66*(6, December), 539-545.

Kuh, Edwin. (1958). A Note on Prediction from Keynesian Models. *Review of Economics and Statistics. 40*(3, August), 294-295.

Singh, Balvir, and Ullah, Aman. (1976). The consumption function: The permanent income versus the habit persistence hypothesis. *Review of Economics and Statistics. 58*(1, February), 96-103.

Macroeconomics - Supra-national Macroeconomic Core

M. Dutta

To successive generations of students of economics, *The Keynesian Revolution* (Klein, 1947) has been the treatise on Keynes's *General Theory*, which presents an erudite and comprehensible education in macroeconomics. In its appendix the author offers a forthright exposition of a macroeconomic structure in terms of a set of simultaneously interdependent behavior equations, given the definitional national income identity and the price-income accounting identity. An exciting initiation in the education of macroeconomics begins.

Samuelson (1983) refers to Klein's "Macroeconomics and the Theory of Rational Behavior" (1946) and cites him as "Columbus" in the voyage of exploration of macroeconomics. Research shows, Samuelson adds, that de Wolff (1941) reported his microeconomic and macroeconomic interpretations of income elasticity of demand five years earlier.

The Keynesian Revolution (Klein 1947) indeed revolutionized studies in macroeconomics with a global readership including members of the United States Congress. A study of macroeconomics immediately warrants an appreciation of the issue of aggregation, - aggregation of behavior of individual micro-units in an economy - households, private firms and corporations. An alternative scheme would be to formulate the study of macroeconomics in terms of the Marshallian "representative" firm and household. If any such shortcut or oversimplification of the crucial issue was ignored, aggregation became a dominant topic in macroeconomic studies.

Many expressed serious reservations and questioned the scientific purity of aggregation of microeconomic actors - households and business firms. The rationale of macroeconomic theory continues to be debated though it has become much more grounded in microeconomic foundations *à la* theories of expectation. The basic issue in aggregation is one of distribution and can be approximated in terms of the theory of distribution. Let students of economics continue their debate why and how markets do and don't clear and the logic of equilibrium and anti-equilibrium.

Samuelson (1983) closes the debate: "..it is gratifying to see that this most elegant paradigm of nature, thermostatics and thermodynamics, can be antiseptically formulated solely in terms of the *observable measurements* (italics ours). It is now clear that, by an independent path, economic theory has arrived at an isomorphic logical structure." (p.35)

Thus, with the problem of aggregation understood in the context of approximation, this paper will briefly discuss specific issues - micro foundations of macroeconomics in section 1, macroeconomic structure and its parameters in section 2, macroeconomic core and microeconomic optimization in section

3, the concept of a supra-national macroeconomic core in section 4, and then finally present in section 5 some conclusions.

I. Micro Foundation of Macroeconomics

Following Tinbergen (1939), Klein undertook the challenging task of explaining economic fluctuations in the United States 1921-1941 by first specifying a macroeconomic model and then estimating its parameters by using econometrics. The macroeconomic model Klein (1950) specified drew upon microeconomic theory, "classical in its methodology", based on the two groups of actors - households and business firms, each group following "specific types of behavior patterns.....the profit or utility maximization equations of the business firms, the utility maximizing equations of the households, and the interactions of these two groups in the market to determine prices, wages, rents, etc." (p. 13), with a forthright acknowledgement: "It is a very difficult problem to pass from the theories of microeconomics to the theories of macroeconomics." In a later study, it is stated, that the model specification most certainly used "some cruder interpretations of macroeconomic structure derived from the Keynesian theory of employment." (Klein and Goldberger, 1955 p. 2).

More recently, Fair (1974) has revisited the problem as he outlines the four characteristics of a macroeconomic model, one of the four being: "The model should be based on solid microeconomic foundations in the sense that the decisions of the main behavioral units in the model should be derived from the assumption of maximizing behavior." (p .3)

Literature in mathematical economics presents much discussion on micro foundations of macroeconomics. It is a difficult and complex relationship. In a conceptual sense, the two-fold reality of all economic agents - individual(s) and the group are very real and they must belong to the system as a whole. Micro maximization behavior, if totally and absolutely unconstrained, will result in chaos and hence a state of non-maximization of utility /profit/ welfare for all individuals belonging to the group. Adam Smith's teaching of the "invisible hand" and the market was completed only by his forceful warning that what was good for the East India Company was not necessarily good for England. Pareto optimality can be meaningful only in the context of a given set of institutional conditions. On the other hand, total regimentation by the group will result in an authoritarian, command economy, which may be inefficient in allocation of resources and consequently unable to maximize gains for all its members - individual units of households and business firms belonging to the group. We continue to assume that there exist micro foundations of macroeconomics in an idealized sense. Individual households and firms belonging to a given macroeconomy share a common group-experience and their individual behavior for maximization of gains cannot be achieved without a macroeconomic togetherness - a macro core. The optimization behavior of individuals and expectations of the future are very much conditioned by the group membership.

II. Structure of a Macroeconomy and its Parameters

Klein's teachings centered on a basic theme. Macroeconomic experience is real, and a macroeconomy has a structure with its parameters constant. The maximizing behavior of member-individuals - households and firms - defines the parameters which are "true" and as such constant, albeit unknown and unknowable. Numerical values of these behavior parameters can, however, be estimated by using probability theory and statistical methods of estimation. It follows that the behavioral relations of an economic structure will be stochastic (Haavelmo, 1944).

The task begins with the specification of the model, an approximation of a real economy, "a schematic simplification," indeed a description of the "skeleton" of the body-politic of an economic system, as it were. The system consists of a body of interdependent behavioral relations, stated by a set of mathematical equations, parameters of which are jointly determined. "Everything depends on everything else" and the system is closed by a set of relationships that are definitional, accounting, institutional, and technological.

Klein-Goldberger (1955) econometric model of the United States was a pioneering exercise in macroeconomic model-building. Since then, Klein has authored/co-authored several other econometric models of the United States and also of many other countries. They are of various dimensions - very small, small, medium, large, very large - with the number of behavior equations in a model ranging from a few to a few hundred. One of the largest in size of Klein's national models, including various generations of Wharton Quarterly Econometric Models of the United States, *The Brookings Model* was constructed in early 1960s (Duesenberry et al 1965). The recent version of the Wharton Model is, of course, larger by a considerable margin. Many economists in many countries have followed Klein and joined the ranks. Indeed, econometric model-building has emerged as a full-fledged multi-million dollar industry round the world. The LINK model, led by Klein and Hickman, is an extended exercise to model the interdependence of various national economies of the world - a macro-economic model of the world economy, as it were.

The Klein-type macroeconomic models draw strictly upon economic theory, "teachings of great masters", encompassing micro foundations of behavior maximization in the specification of behavior equations of the system, with a straightforward assertion that they are well defined *on average.*

A very serious challenge to the process of econometric model construction has been referred to as the problem of identification (Haavelmo, 1944; Koopmans, 1953; Fisher, 1966; Hatanaka , 1975). The task becomes even more baffling when one is taught that economic theory is relatively limited by the omnibus qualification, "other things being equal" - a catch-all phrase. So many other things that impact on our economic behavior in the real world seldom remain *equal*. True, research in economic theory has progressively pushed out the frontier of our knowledge. Even so, limitations remain pressing. Thus, the identification of each member equation of a jointly determined simultaneous system of equations remains a challenge. The commonly used practice of "zero restriction", deciding which variables enter a linear behavior relationship, has been subject to much criticism.

Data limitations further add to the identification problem if the model specified be undersized in the sense that the sample size is not large enough to obtain statistically consistent and unbiased estimates of the parameters. One wonders if application of the "zero restriction" rule results in the inclusion of so many of exogenous variables in many different equations of the model that the sample is undersized. If so, a singularity occurs, and the estimation of parameters of the model by simultaneous equation methods becomes impossible.

Research efforts proposing alternative estimation methods have presented limited results (Dutta and Lyttkens, 1974). We may then revisit Herman Wold's recursive model, where each equation in the simultaneous system is to be specified in a strict recursive order. Or, we return to Franklin Fisher's reformulation of Wold's recursive model, not by each single equation, but by forming fairly small blocks of equations in a recursive order, the equations in each single block remaining jointly determined. This does help solve the estimation problem when the sample is undersized.

Klein invites economists to accept challenges and work harder to develop solutions to problems as they are confronted. Notwithstanding the familiar problems of model specification, identification of its behavior equations, estimation of parameters of an interdependent simultaneous system, especially when the sample is undersized, Klein has held to his firm belief that there exists a well-defined macroeconomic structure, rooted in the teachings of economic theory with its micro foundations, and that its parameters are estimable.

Klein has recently proposed "a new approach" for macroeconomic studies more rigorously "through the accounting structure," and "*neutral* (italics ours) as far as doctrine is concerned" - doctrines of supplyside/ demandside, or doctrinaire tilts of ideologues. He adds: "The theory of employment and output determination developed in the 1930s formed the basis of abstract and, later, statistical model building for the economy *as a whole*. The *real meaning of the Keynesian Revolution became clear* (italics ours) when model comparisons of alternative systems - the classical, neoclassical, Keynesian, Marxian - were formulated side by side in mathematical equation systems." (Klein, 1983, p. 2)

If earlier attempts to construct macroeconomic models were very much based on national income accounts (NIAs), recently attempts have been made to integrate modeling work with another set of national accounts, namely the

Input-Output (I-O) Tables, *"à tableau économique*, a matrix lay-out showing how outputs are shipped from producing to using sectors of the economy." .. the end product resulting in a much larger system. "The first version of the Input-Output/National Income Account Wharton Model consisted of 346 equations, while the newest version has about 2000." (*op cit.* pp. 33-34, see also Preston, 1976). Władysław Welfe (1983) reported that modeling of socialist economies began with Input-Output Tables, and subsequently became integrated with macroeconomic models based on national income accounts (NIAs).

Yet another set of accounting information relates to the flow-of-funds (F/F) models, with focus on changes in financial assets and liabilities and there is "a natural linkage" between these accounts and the national income accounts. Construction of macroeconomic models establishing "a formal relationship" between the two accounts is yet another extended exercise (Klein, 1983, 36-46; see also Klein et al., 1983).

Modeling macroeconomies based on sovereign nation-states leaves some questions unanswered in an interdependent global economy. In 1968, the Project LINK, with L.R. Klein and Bert Hickman as chief investigators, was launched to find some answers in this regard, and the LINK Model, initially based primarily on a trade matrix of 28 countries or country-groups that seeks "to endogenize export volumes and import prices", was constructed (Klein 1983 pp.175-196). The system now has 79 interlinked national economic models and consists of more than 30,000 equations. The LINK Model is indeed a bold step toward viewing the world economy as one macroeconomic unit. We shall return to this later.

Sims (1980) questions if it is at all possible to specify a structure of a macroeconomy as the problem of identification of behavioral equations in a given system remains a debated issue. In standard macroeconomic modeling, he argues, equation specification is very much dependent on extra-economic information, not on "economic theory".

Sims proposes econometric work in "a non-standard style" by emulating "the frequency-domain time series theory.. in which what is being estimated (e.g., the spectral density) is implicitly part of an infinite-dimensional parameter space. After the arbitrary "smoothness" or "rate-of-damping" restrictions have been used to formulate a model which serves to summarize the data, hypotheses with *economic content* (italics ours) are formulated and *tested* at *a second stage* ....." (p.15).

His guideline to select a class of multivariate time series models which will serve as the *unstructured* (italics ours) first-stage modeling remains to be answered. Selection of a vector of economic variables, compilation of time series data on them, and defining their appropriate lag structure, must involve an act of arbitrary decision-making for the researcher. The gain in statistical analyses of a body of contemporaneous time-series data must be balanced against the compromise on the theoretical specification of the vector of variables. Are we then back to the problem of Franklin Fisher's reformulation of Herman Wold's recursive modeling by selecting blocks of sub-systems of an interdependent simultaneous macroeconomic system, which is by its nature otherwise too large?

To recall Tinbergen's word of caution, a suggestion for arbitrary selection of "unstructured" first-stage models, postponing analyses of "economic content" to the second stage, may involve a detour into the familiar "jungle land" of fact. True, much of Sims's critique has been discussed in the literature and authors of "standard" structural macro-models are *fully* aware of them. The jury has not yet been presented with sufficient, robust evidence of Sims's "non-standard type" macroeconomic modeling, *à la*, vector autoregressive (VAR) models.

III. Macroeconomic Core

We return to the "standard" macroeconomic models. A model, specified by a jointly interdependent simultaneous system of behavioral equations plus necessary identities to close the system, has its "structure" with its constant parameters which are "true". Behavioral relationships of the model are derived from optimal decision rules of micro economic actors - households / business firms, constituent members of the macro-system. The data-based numerical estimates of the true parameters, with their estimated errors, are only statistical approximations to the true values.

The first generation of these models, anchored in the *Keynesian Revolution*, were given empirical contents in terms of national income accounts (NIAs). Later, the art of macromodeling has been further expanded to make use of alternative sets of accounts - input-out (I-O) tables and flow-of-fund (F/F) accounts. Thus, recent macro-modeling, more targeted to the "accounting structure", has been viewed as doctrine-neutral and welcomed as a "more rewarding approach".

Samuelson (1975) presents a survey of the art and science of macroeconomic models over 50 years. We have argued in section 1 that the micro foundation of macroeconomics constitutes a core argument and remains *apriori* for macroeconomic studies. It is equally important to state clearly that micro behavior optimization can be meaningfully defined only in the context of a macro core. Indeed, the macro-core is *apriori* too.

Social consumption in the national income accounts (NIAs) is a crucial entry. It is so too in input-output (I-O) tables and in flow-of-fund (F/F) accounts. The government of the sovereign nation-state becomes the operational agent for managing social consumption in a macroeconomic system. A lawfully constituted monetary authority - a central bank - is a hallmark of sovereignty of each such system (unusual cases of exceptions apart). A sovereign macroeconomic system must have the authority to supply the quantity of money that the level of its economic activity warrants. Money demand is a behavioral function. Is money supply truly exogenous? Money supply has now been considered endogenous in macromodels, especially when using the flow-of-funds accounts.

In a set of typical national income accounts, social consumption includes "pure" public goods and semi-public goods, such as infrastructure, human capital in the form of health care, education and basic scientific research, environmental protection programs, civil rights and affirmative action programs to ensure barrier-free access to the job market to all micro agents who have been hitherto denied access by institutional handicaps, and so on. All of these have an impact on the macro model specified. Many of these entries, once conveniently left out as exogenous, have progressively been defined as endogenous variables in the macro-economic models.

Environmentalists would prefer the term "ecosystem", which encompasses people - households and business, nature and the green, wild lives and the endangered species. They argue that optimization behavior of micro agents is conditioned by the totality of the environment. Indeed, there exists a set of social, cultural, and religious values defining a given macro-system, which must govern micro-behavior optimization, often so forcefully argued by social anthropologists and sociologists. Of course, many major economies are multi-dimensional with increasing references to a modern macro-system as a "gorgeous mosaic" or "a salad bowl". Historians argue that sharing a common history provides the core sense of belonging together in a macro system. Politicians prefer to refer to the historically defined political borders. Lawyers and jurists draw their inspiration from the constitution of the land and the legal codes that have come into force.

Let us return to the discipline of economics and limit our analyses to its lessons. The point to underscore here is that micro behavior optimization can be real only in the context of a well-defined macroeconomic core. Two principal elements defining this core are monetary and fiscal authorities and their policies, imbedded in textbooks as the IS-LM model, the geometric presentation of *The Keynesian Revolution*, which John Hicks offered much earlier. The IS-LM analytic tool of macroeconomic policy management continues to be an integral part of macroeconomic education.

True, the intensity of debate over the oversimplification of the Hicksian geometry has been overwhelming and at times baffling. The fact remains that monetary and fiscal policies in a given macroeconomic system are crucial conditions for micro behavior optimization - for utility maximization by households and also for profit maximization by business firms. Optimal balancing of monetary and fiscal policies alone can provide the macroeconomic core for micro optimization.

In modern macro systems, social consumption is managed by the government, and it is the prerogative of the government to define fiscal policy.

The monetary authority is exercised by a complex social institution, socially owned and/or managed - the central bank. The basic authority in both cases is derived from the sovereignty of a nation-state. Thus, the macro core of an economy is viewed as a political arrangement - government-cum-central banking authority. No wonder, macroeconomic studies have in reality been limited to political debates over fiscal and monetary issues, and the political agenda of participants in the debate has historically been much too colored by partisan considerations.

IV. Supra-national Macroeconomic Core

Macroeconomic cores have historically been defined by the maps of sovereign nation-states. Thus, it has been a common practice to include a foreign sector in all national income accounts (NIAs), input-out (I-O) tables and flow-of fund (F/F) accounts. Given the heroic assumption under the familiar rubric that other things are "equal," relative prices and economic activity levels are expected to determine economic transactions - trade and capital flows - among the sovereign nation-state-based macroeconomies.

The LINK Model is one of the first systematic approaches to interlink sovereign macro economies and thus to endogenize export volumes and import prices (Klein, 1983). The trade matrix in the LINK Model is based on goods - merchandise trade - defined consistently with national income accounts of the member countries as constituent national macro models are built on their respective national income accounts (NIAs). To minimize the impact of structural divergences of member economies, the LINK Model was originally limited to developed industrialized economies, with the rest of the world grouped into three major categories - developing countries, socialist economies and all others. LINK progressively broadened its scope, and many individual countries hitherto in the three residual groups have been individually incorporated as constituent LINK members.

Two basic points are crucial for the present study. First, there should be a critical re-evaluation of the concept of sovereignty in the present world where member states are very much inter-related. Macroeconomies, based on sovereign nation-states, are increasingly interdependent, given the state of technological development following what has been characterized by some as the Fifth Industrial Revolution and the consequent high-speed mode of international communication. If they subscribe to the view that free trade among them is the only way to maximize global welfare - maximize gains for all micro agents in all member states, as they do - they must accept limitations on their sovereign authority.

The concept of absolute sovereignty is at best anachronistic.
Nation states are at best "accidents of history" as has been forcefully stated by some. This is currently being illustrated in the break-up of the Soviet Union and

formerly associated countries in Eastern Europe. Unrestricted sovereignty of the nation states, however legal, cannot optimize economic welfare of their members - households and business firms.

In the context of the LINK Model's scheme to endogenize export volumes and import prices, a crucial issue for examination is the exchange rate, - value of a national currency in terms of the currencies of other nations. Is it determined competitively in the free market? Its management by the sovereign member economies will have an impact on export/import prices (Dornbusch, 1980, 1987).

Management of exchange rates has been known to be a problem even among a select group of developed, industrialized sovereign national economies all located in the common geographical locale of Western Europe. Italy and United Kingdom, two of the eight major nations of the twelve member-economies of the European Community (EC), opted in 1992 to take leave of the Exchange Rate Mechanism (ERM) to which many of the EC countries belonged. Currency wars have been known to be a part of international economics. Exercising their respective sovereign authority, nation-state-based macroeconomies have managed their currencies and exchange rates, thus distorting export/import prices in the world market. The sovereign authority of a nation-state in this regard can hardly remain unlimited. The argument for *absolute* sovereignty in a world economy where nation-state-based economies are so interdependent must be rejected.

Second, the export/import prices among trading economies can, and often are, exposed to fiscal policies exercised by nation-states, by virtue of their legal sovereignty. The trading partners of the United States have recently raised their voice in a chorus to the effect that prolonged deficit financing by the U.S. government and growth of its national debt in the 1980s have reached a limit, increasingly contributing to a global misallocation of resources. The cumulative effect of fiscal policy over the past twelve years have also hurt optimization of gains for all micro agents - households and businesses - in the U.S. macroeconomy.

The case for a supra-national macroeconomic core is argued here. Otherwise, a sovereign nation-state will be left with its "fortress" economy, to exercise its absolute sovereign power in isolation, in its own Robinson Crusoe world. The impact of exchange rate management is not limited to export/import prices. Exchange rates also affect international capital flows and interest rates. Thus, appreciation and/or depreciation of the value of a currency, albeit sovereign, can hardly be a unilateral process.

I have argued elsewhere (Dutta, 1992a and 1992b) that a return to the idealized paradigm of international trade, wherein n sovereign economies engage in trade in m final goods and services with their respective sovereign currencies irreversibly linked to a 100-percent gold standard and settle their respective national accounts with the rest of the world by transfers of pure gold, is hardly an option. To facilitate trade and investment flows in the absence of a common macroeconomic core of n nations, ad hoc binational/multinational institutional arrangements, in the form of International Monetary Fund (IMF), General Agreement on Trade and Tariff (GATT), Free Trade Areas (FTAs), and the Group of Seven nations (G-7) have been part of the post-WWII experience.

That optimization of microeconomic gains is facilitated by international trade and investment has long been a received theorem *à la* the Ricardian theory of comparative advantage. Its pre-WWII expression was too crude. Individual macroeconomies set their goals for shares of the world market. The result was the founding of several economic regimes, each anchored to a specific currency - the British pound sterling, the Dutch guilder, the U.S. dollar, the Japanese yen. The respective "Home" governments established monetary-fiscal policies and thus defined the macroeconomic core for its satellite economies.

Indeed, the result was the establishment of several super-national economic hegemonies, *not* a supra-national macroeconomic core for the member states of a given regime. The system collapsed under its own weight.

The post-WWII experience was unique, an approximation to a supra-national macroeconomic core, defined by U.S. dollar at a fixed gold value of U.S. $ 35 an ounce. On August 15, 1971 that arrangement was formally liquidated. The Kindleberger critique argued, and rightly so, that the U.S. could not indefinitely continue to provide free "international public goods" in the form of both monetary and financial stability and defense and security of the free world.

The pious wish of some that this presented an opportunity to return to an idealized pure 100-percent gold standard under the free float of all currencies proved vain. The Plaza Accord came in September 1985 and the era of the managed exchange rate regime was ushered in. First, the group of five nations, the G-5, then the group of seven, the G-7, joined forces together to manage exchange rates for themselves and also for the rest of the world. Some economists question if the G-7-led managed exchange rate regime has been successful. Others argue that the G-3 with U.S. dollar, Japanese yen and European Community's Ecu (European Currency Unit anchored to German mark), rather than the G-7, would have been a more effective forum for the stated purpose. Or, will the "invisible hands" in the "market" work to set an optimum free float of exchange rates *in the long run*?

The two polar options, one of absolute sovereign authority of a nation-state based macroeconomy to define the value of its currency, and another, a universal global currency, fixed and stable in terms of its pure gold value, proved to be unsuccessful. The supernational hegemony was never a true economic option because micro gains in this system were found to be tilted favorably to households and business firms belonging to the respective "Home" macroeconomies. The micro optimization process was thus seen to work in a one-sided way. Yet another form of ad hoc regionalism became manifest during the Cold War regime and they could at best be described as forms of "strategic," not economic, regionalism.

A more workable option is a supra-national regional macroeconomic core. The Triffin treatise on "deep" integration as a core for economic regionalism initiated the dialogue. Experiments in Western Europe present a model. The European Community (EC), founded in 1958, following the signing of the Treaty of Rome a year earlier, and progressively advancing through successive regional agreements - the Single European Act of 1985 and the Maastricht Treaty of 1992, has attracted much attention. The EC offers a paradigm of economic regionalization and its distinguishing features merit attention (Hesse, 1993; Forrestal, 1993).

The EC seeks to provide a supra-national macroeconomic core for its twelve members - each a sovereign nation-state - by way of developing a regional institutional mechanism to forge common monetary and fiscal policies. To this end, the Treaty of Maastricht explicitly states four major guidelines (Hesse, 1993):

(i) coordination of the average inflation rate by relating increases in consumer price indices in member countries to an average of the best performance by any three member states;

(ii) coordination of current national budget deficit and national debt as percentages of gross domestic product (GDP), i.e. current deficit/ GDP and national debt/GDP ratios by a regionally accepted code of fiscal discipline;

(iii) control of unilateral changes in values of currencies of respective nation-states and the management of exchange rates of 12 member economies in terms of a historical bench-mark; and

(iv) limiting fluctuations of national interest rates on long- term bonds within the band set by an average rate of any three member-states with the best inflation performance.

The above four-point agenda would work toward the setting up of a regional supra-national macroeconomic core as we have defined above. However, the EC leadership is aware of the practical constraints and has set fairly elaborate escape clauses to enable each individual member - sovereign nation-state - to mend its "home" macroeconomic system, as may be necessary.

The agenda seeks to impose a common monetary-fiscal code for regional discipline with sufficient flexibility for adjustments for one or more individual members of the 12-member regional compact in Western Europe. The time table is well defined. Indeed, several member nations have taken advantages of escape clauses. I would argue that EC presents a successful model for a regional macroeconomic core, wherein sovereign nation-state-based macroeconomies have elected to accept limitations of the concept of absolute sovereignty.

Forrestal (1993) draws upon the historical experience of the United

States in this regard and forthrightly reminds students of recent difficult EC developments that although "... we (the U.S.) had achieved a single currency shortly after the Civil War in the mid-1800s, we did not have full monetary union until the establishment of a central bank, the Federal Reserve System, in 1913." He proceeds to add: "Even after the Fed was founded with its 12 Reserve Banks, the United States did not have a common discount rate for some time." (pp. 132-133) He argues that "economic union" facilitates "increased factor mobility" and thus "typically precedes monetary union".

Indeed, the founding of the European Monetary Institute (EMI) as of January 1994 in Frankfurt is a progressive step towards EC monetary union with one central bank and one currency. The landmark decisions by the highest judicial authorities in the United Kingdom and in Germany in 1993, upholding regional integration of sovereign-nation states in the EC add further momentum to the Western European experiment in economic regionalization. No wonder, in 1993 the Netherlands reversed its one-year old negative referendum on its EC membership.

There is a *third* important point in the Maastricht Treaty which must be emphasized. The EC paradigm of regional macroeconomic core is firmly rooted in the economics of location theory (Lorenz 1993). Belonging to a common economic space of a natural geographic continuity is its base and the gain in transportation costs further aids economic mobility of factors/ products.

The Maastricht Treaty stipulates a very explicit provision for comparative economic development of the region's member-states. If the twelve member states in the EC are arranged on a scale of economic development, one observes, eight are richer and four are poorer. The provision requires the setting up of a joint regional fund to help accelerate economic development of the poorer four. The shared goal is to bring all twelve member-economies to the same approximate level. This is an achievable goal with increased intra-regional factor mobility and trade/ capital flows. This unique provision will help cement the regional economic togetherness.

Whether the EC paradigm of a regional macroeconomic core will be replicated in other continental regions - North America or the Americas, Asia/Pacific, Africa - remains to be seen. It is, however, important to note that some follow-up activities in other regions, especially in North America and Asia/Pacific, are in progress. One notes the progressive institutionalization of the Asia-Pacific Economic Cooperation (APEC) in 1993 providing a forum to seventeen sovereign-nation state-based macro economies in the region. A further crucial step followed as the North American Free Trade Agreement (NAFTA) bringing Canada, the United States and Mexico into a regional free trade compact became a reality in 1993.

I have argued that if gains in intra-regional economic activities in the EC do not result in the loss of inter-regional global economic activity, there is no evidence of substitution and the case for Pareto optimality remains true. The

EC is no "fortress" Europe and the argument for "regionalism" and "globalism" involves no contradiction. Indeed, with the regional macroeconomic frameworks in place, the post WWII international institutions for global economic cooperation - IMF, World Bank and GATT - may become more effective in working out their respective agenda.

The LINK Model is a bold attempt to bring the economies of the world to a global macroeconomic specification. Its efforts to endogenize export volumes and import prices in the absence of an idealized, perfectly free competitive world market with total free float of currency exchange rates will, however, be a challenging task.

Extra-economic diversities of the EC members and their shared history of hatred and hostility, wars and destructions, may be a lesson to draw upon. Optimization of micro economic gains - economic gains for the individual households and business firms in all twelve member sovereign nation-states - is the base for the supra-national regional macroeconomic core.

Whether the EC will fulfil its agenda on schedule or if the EC will reach out to all nation-states in Europe, remains to be seen. For the present, the EC member-states are seen to be experiencing a rocky road ahead. Even so, the EC presents the unique paradigm for an intra-regional, supra-national macroeconomic core. An optimal international economic order of "regionalism" and "globalism" may follow.

V . Conclusions

Aggregation is a knotty problem but one that can be handled approximately. Micro foundation of macroeconomics is a given. There exists a structure of a given macroeconomy, defined by a system of jointly interdependent simultaneous equations - behavioral, technological and institutional, accounting identities/ definitions and model-closing relations. The vote is for the "standard" macroeconomic modeling, as opposed to the "non-standard" modeling approach, recently proposed. The LINK Model, endogenizing international trade volumes and prices, is seen as an extension of the modeling exercise to specify the macro core of the world economy.

The case for absolute sovereignty is rejected. In an increasingly interdependent world economy, the "fortress" sovereign nation-state, based on an exclusive macroeconomic core, does not ensure microeconomic optimization and warrants regional/global internationalization. The global macro core of one world economy remains far out to be achieved. A supra-national regional macroeconomic core, following the EC paradigm, is an option to examine. Open and competitive cooperation among the regional macroeconomic cores in *a free market* will contribute to maximization of global welfare.

M. Dutta, Department of Economics, Faculty of Arts and Sciences, New Jersey Hall, Rutgers, The State University, 75 Hamilton Street, New Brunswick, N.J. 08903-5055.

ACKNOWLEDEMENTS

My sincerest appreciation to Robert J. Alexander, H. W. Arndt, Ronald G. Bodkin, Norman S. Fieleke, Detlef Lorenz, Kazuo Sato, Anthony M. Tang and Paul Taubman for their generous and helpful comments.

REFERENCES

de Wolff, Pieter. (1941). Income elasticity of demand, a microeconomic and a macro-economic interpretation. *Economic Journal. 51*, 140-145.

Dornbusch, Rudiger. (1987). Exchange rates and prices. *American Economic Review. 77*(1, March), 93-106.

_____ (1980). *Open Economy Macroeconomics.* New York: Basic Books.

Duesenberry, James. et al. (1965). *The Brookings Quarterly econometric model of the United States.* Amsterdam: North-Holland.

Dutta, M. (1992.b). Economic regionalism: macroeconomic core and microeconomic optimization. *mimeo.* A paper presented at the First APEC: ASEAN/SAARC Conference, Bangkok, Thailand, December 16-18.

_____ (1992.a). Economic regionalization in western europe: Asia-pacific economies - Macroeconomic core: Microeconomic optimization. *American Economic Review, Papers and Proceedings*, 82(2, May), 67-73.

Dutta, M. and Lyttkens, Ejnar. (1974). Iterative instrumental variables (IIV) Method and estimation of a large simultaneous system. *Journal of American Statistical Association. 69*(348), 977-986.

Fair, Ray C. (1974). *A Model of Macroeconomic Activity.* Vol. I. *The Theoretical Model.* Cambridge, MA: Lippincot, Ballinger.

Fisher, Franklin M. (1966). *The Identification Problem in Econometrics.* New York: McGraw-Hill.

Forrestal, Robert P. (1993). The role of central banks in a global marketplace, exchange rates and policy coordination. *Journal of Asian Economics*, 4(1, Spring), 131-137.

Haavelmo, Trygve. (1944). The probability approach in econometrics, *Econometrica. 12*(Supp.).

Hatanaka, M. (1975). On the global identification of the dynamic simultaneous equation model with stationary disturbances. *International Economic Review. 16*, 545-554.

Hesse, Helmut. (1993). Ecu - Now and later, Some considerations after maastricht. *Journal of Asian Economics*, 4(1, Spring), 139-142.

Klein, Lawrence R. (1983). *Lectures in Econometrics.* Amsterdam: North-Holland.

_____ (1950). *Economic Fluctuation in the United States 1921-1941, Cowles Commission for Research in Economics.* New York: John Wiley & Sons.

_____ (1947). *The Keynesian Revolution*, Macmillan, New York.

_____ (1946). Macroeconomics and the theory of rational behavior, *Econometrica, 14*, 93-108.

Klein, Lawrence R. and Goldberger, A. S. (1955). *An Econometric Model of the United States, 1929-1952*, Amsterdam: North-Holland.

Klein, Lawrence R., Friedman, B. and Abbe, S. (1983). Money in the wharton quarterly model. *Journal of Money, Credit and Banking, XVL*, May, 237-259.

Koopmans, Tjalling C. (1953). Identification problems in economic model construction. In W.C. Hood and T.C. Koopmans *Studies in Econometric Method.* (pp. 27-48). New York: Wiley & Sons, .

Lorenz, Detlef. (1993). Europe and East Asia in the context of regionalization, theory and economic policy. *Journal of Asian Economics.* 4(2, Fall), 255-270.

Preston, R. S. (1976). The Wharton Long-Term Model: Input-output within the context of a macro forecasting model. In L. R. Klein and E. Burmeister. *Econometric Model Performance* (pp. 271-287). Philadelphia: University of Pennsylvania Press.

Samuelson, Paul A. (1983). Rigorous observational positivism: Klein's envelope aggregation: Thermodynamics and economic isomorphism, In Adams and Hickman (Eds.) *Global Econometrics* (pp. 1-38). Cambridge, MA: The MIT Press.

_____ (1975). The art and science of macromodels over 50 years. In Gary Fromm and Lawrence R. Klein (Eds.). *The Brookings Model: Perspective and Recent Developments* (pp. 3-10). Amsterdam: North-Holland.

Sims, Christopher A. (1980). Macroeconomics and reality, *Econometrica, 48*(1, January), 1-48.

Tinbergen, Jan. (1939). *Statistical Testing of Business-cycle Theories.* Vols. I and II, Geneva: League of Nations Economic Intelligence Service.

Welfe, Władysław. (1983). Models of the socialist economy, An appendix in Lawrence R. Klein (Ed.) *Lectures in Econometrics* (pp. 197-227).

An Asymptotic Approximation to the Probability Density Function of the Durbin Watson Test Statistic

A. L. Nagar and P. D. Sharma

Durbin and Watson (1950 and 1951) considered the problem of testing the independence of disturbances in a linear regression model with fixed regressors and an intercept term. They proposed a test statistic **d** and observed that the exact null distribution of **d** is difficult to obtain analytically and, therefore, it is not possible to derive exact critical values of **d** to carry out any test of significance. The bounds test proposed by them is well known and it has the advantage that it does not depend on the values taken by the regressors, and the bounds are attained when the columns of the regressor matrix are certain eigenvectors. In that case the test is also uniformly most powerful against suitable alternative hypotheses.

Several procedures for approximating the true critical value of **d**, when the bounds test is inconclusive, were reviewed by Durbin and Watson (1971), Maddala (1977) and King (1987). It has been observed that the DW beta and $a + bd_u$ approximations are adequately accurate for large samples, were d_u is the upper bound of the DW bounds test.

Fast methods of computing the distribution of the Durbin-Watson statistic have been proposed in the literature; for example, see Shiveley et al. (1990). The method of Shiveley et al. involves numerical inversions of characteristic functions for which computer algorithms have been developed; for example, see Koerts and Abrahamse (1969), Imhof (1961), Farebrother (1980) etc. This method provides numerically good approximations to the exact critical values, but the exact evaluation of the distribution function is prohibitively costly for most applications when the sample size is moderate or large.

The method adopted by us for deriving asymptotic approximation to the probability density function of DW statistic is theoretically the same as that of Shiveley et al. However, we obtain an analytical expression for the asymptotic approximation of the probability density function in a fairly easy and straightforward manner.

Since the upper and lower bounds of the Bounds Test, i.e., d_u and d_L, are, in fact, the exact critical values if the columns of the matrix of regressors are certain eigenvectors (to be specified in the text), we compare the asymptotic approximations to the critical values and power of the test obtained by our method with d_u and d_L. It turns out that our approximations tend to be the same as the exact values as sample size increases, but they agree up to only one decimal place when the sample size is smaller than 15.

I. The Linear Regression Model and Some Properties of Durbin-Watson Statistic

We may write the linear regression model as

$$y = X \beta + u \tag{1.1}$$

where y is a T x 1 vector of observations on the dependent variable, X is a T x K matrix of observations on K explanatory variables, β is a K x 1 vector of coefficients and u is a T x 1 vector of disturbances. We assume that the elements of X are fixed in repeated samples, rank of X is K < T and the first column of X consists of unit elements only. The disturbances are assumed to have a normal distribution with

$$Eu = 0 \text{ and } Euu' = \Sigma \tag{1.2}$$

where Σ is a positive definite matrix.

The Durbin-Watson statistic is

$$d = \frac{\hat{u}' A \hat{u}}{\hat{u}' \hat{u}} \tag{1.3}$$

where

$$\hat{u} = y - Xb \text{ and } b = (X'X)^{-1} X'y \tag{1.4}$$

are the Ordinary Least Squares estimators of u and β, respectively, and

$$A = \begin{bmatrix}
1 & -1 & & & & & & \\
-1 & 2 & -1 & & & & & \\
. & -1 & 2 & -1 & & & & \\
. & . & . & . & . & & & \\
. & . & . & . & . & & & \\
. & . & . & . & -1 & 2 & -1 \\
. & . & . & . & . & -1 & 1
\end{bmatrix} \tag{1.5}$$

is a T x T matrix with zeroes in vacant places.

The exact mean and variance of d were obtained by Durbin and Watson (1950). It can be shown that

$$Ed \to 2 \qquad \text{and} \qquad Var\ d \to 0 \tag{1.6}$$

as $T \to \infty$; therefore, the limiting distribution of d is degenerate, under the null hypothesis of independence of regression disturbances. Therefore, we consider the limiting distribution of the standardized statistic

$$\delta = \frac{d - 2}{\sqrt{4 / T}} = \sqrt{T} \left(\frac{d - 2}{2} \right) \tag{1.7}$$

where 4/T is the asymptotic variance of d.

In order to obtain the distribution function of δ, we require

$$P(\delta < \delta_O) = P [\sqrt{T}(\frac{d - 2}{2}) < \delta_o]$$
$$= P [d < 2 (1 + \frac{\delta_o}{\sqrt{T}})] \tag{1.8}$$

for any preassigned value δ_o of δ. Since û = Mu, and $M = I - X(X'X)^{-1} X'$ is an indempotent symmetric matrix, we can express (1.8) as

$$P(\delta < \delta_o) = P(u'_* Q u_* < 0) \tag{1.9}$$

where $u_* = R'u$, $M = RR'$ and $R'R = I$, so that R is a T x (T - K) transformation matrix and the columns of R are orthonormal; and

$$Q = R'AR - 2 (1 + \frac{\delta_o}{\sqrt{T}}) I \tag{1.10}$$

I being an unit matrix of size (T - K) x (T - K).

The characteristic function of $u'_* Q u_*$ can be expressed as the determinant

$$| I - 2it \theta |^{-\frac{1}{2}}, \quad where \quad \theta = R' \Sigma MAR - 2 (1 + \frac{\delta_o}{\sqrt{T}}) R' \Sigma R. \tag{1.11}$$

We may derive the first four moments, about the origin, of $u'_* Q u_*$ by straightforward differentiation of the characteristic function; and hence obtain the first four cumulants of $u'_* Q u_*$ as

$$\kappa_1 = tr \ \theta, \ \kappa_2 = 2 \ tr \ \theta^2, \ \kappa_3 = 8 \ tr \ \theta^3 \ \text{and}$$

$$\kappa_4 = 48 \ tr \ \theta^4. \tag{1.12}$$

It should be noted that the values of these cumulants do not depend on the choice of the transformation matrix R, and that they are all of the same order of magnitude T. Therefore, we consider the standardized statistic

$$z = \frac{u'_* Q u_* - E(u'_* Q u_*)}{\sqrt{var(u'_* Q u_*)}} \tag{1.13}$$

so that $\kappa_1(z) = 0$, $\kappa_2(z) = 1$ \hfill (1.14)

$$\kappa_3(z) = 2\sqrt{2}\, \frac{tr\,\theta^3}{(tr\,\theta^2)^{3/2}} \tag{1.15}$$

and $\qquad \kappa_4(z) = 12\, \dfrac{tr\,\theta^4}{(tr\,\theta^2)^2}$ \hfill (1.16)

We note that $\kappa_3(z)$ is $O(1/\sqrt{T})$ and $\kappa_4(z)$ is $O(1/T)$ and rest of the cumulants of z are of decreasing order of magnitude.

It follows that the characteristic function of z can be expressed as

$$\phi_z(t) = \exp[\,\frac{it}{1!}\kappa_1(z) + \frac{(it)^2}{2!}\kappa_2(z) + \dots\,]$$

$$= e^{-\frac{1}{2}t^2}[\,1 + \frac{(it)^3}{3!}\kappa_3(z)$$

$$+ \{\frac{(it)^6}{2!(3!)^2}\kappa_3^2(z) + \frac{(it)^4}{4!}\kappa_4(z)\} + o(\frac{1}{T})\,] \tag{1.17}$$

where $o(1/T)$ represents terms of lower order than $1/T$.

Using the Inversion Theorem (for instance, see Cramer 1962) we obtain, up to order $1/T$, the asymptotic approximation to the probability density function of z as

$$f(z) = \frac{1}{\sqrt{2\pi}}e^{-\frac{1}{2}z^2}[1 + \frac{1}{6}(z^3 - 3z)\kappa_3(z)$$

$$+ \{\frac{1}{24}(3 - 6z^2 + z^4)\kappa_4(z)$$

$$- \frac{1}{72}(15 - 45z^2 + 15z^4 - z^6)\kappa_3^2(z)\} + o(\frac{1}{T})]. \tag{1.18}$$

In order to test the null hypothesis of independence of regression disturbances we determine the critical value δ_o or $d_o = 2 + 2\,\delta_o/\sqrt{T}$ by solving the equation

$$P \ (\delta < \delta_o) = P(d < 2 + \frac{2\delta_o}{\sqrt{T}}) = P(z < - \frac{tr \ \theta}{\sqrt{2 \ tr \ \theta^2}}) = \alpha \qquad (1.19)$$

where α is any preassigned level of significance and the probability density function of z is given by (1.18), where the range of z is - ∞ to ∞. One-sided alternatives have been considered. Durbin & Watson also restricted themselves to one-sided alternative as it is known that an uniformly most powerful test does not exist when the alternative is two-sided (see for instance, T.W. Anderson (1962), Koerts et al. (1969).

The test procedure requires that we specify the numerical values of X and Σ as the values of $tr \theta$ and $tr \theta^2$ depend on them. However, for determining the critical values, we must assume that the null hypothesis of independence of regression disturbances is true and, therefore, specify $\Sigma = \sigma^2$ I, where σ^2 is the disturbance variance in (1.1) and I is a T x T unit matrix. It is not necessary to specify the value of σ^2 because that will cancel out in the ratio

$tr \ \theta / \sqrt{2 \ tr \ \theta^2}$

II. Numerical Comparison of Exact and Approximate Critical Values

In order to numerically evaluate the critical values, we choose the following specifications of X.

Case (a). The columns of X are the same as the K eigenvectors corresponding to the K smallest eigen values of A, defined in (1.5).

It is well known that in this case the critical value is exactly equal to d_u , i.e. the upper bound of the Bounds Test of Durbin and Watson (1950), p.416.

Case (b). The last K-1 columns of X correspond to K-1 largest eigen values of A, but the first column consists of unit elements only.

In this case the exact critical value is equal to d_L i.e. the lower bound of the Durbin-Watson's Bounds Test.

We have chosen X as above in order to be able to compare our approximations to the critical values with the exact values available in the literature.

The exact critical values, d_u and d_L, have been computed by Koerts and Abrahamse (1969), p.178; and Durbin and Watson (1951) for some values of T and K.

The cases (a) and (b) are analyzed for

K = 2, 3, 4, 5 and 6 $\qquad (2.1)$

and

T = 15, 20, 30, 40, 50 and 60. $\qquad (2.2)$

The critical values of Durbin and Watson statistic

$$d(\delta_a) = 2 + 2\frac{\delta_a}{\sqrt{T}} \tag{2.3}$$

and

$$d(\delta_b) = 2 + 2\frac{\delta_b}{\sqrt{T}} \tag{2.4}$$

have been tabulated in Table 1, along with the 5 per cent values of d_u and d_L as reported by Koerts and Abrahamse (1969), p.178. For details regarding computation of critical values, δ_a and δ_b, please refer to Module I of the Appendix. The critical values computed by our method tend to be the same as the exact values for large sample sizes. For samples of size less than 15 they agree up to one decimal point only.

Critical Region

An alternative way of comparing our critical values with the exact ones is the following. As noted above the exact critical value of d, in case (a), is d_u. Then $d_u = 2 + \dfrac{2\delta_0}{\sqrt{T}}$ gives $\delta_0 = \sqrt{T}\dfrac{d_u - 2}{2}$. Using this value of δ_0 and X as in (a), we compute

$$P(\delta < \delta_0) = P(z < -\frac{tr\,\theta}{\sqrt{2tr\,\theta^2}}). \tag{2.5}$$

These probabilities are reported in Table 2.

We also consider case (b) when d_L is the exact critical value and derive probabilities in the same way. Again our approximations are fairly close to the true ones for samples of size 20 and more. The approximations are poor when sample size is 15 or less.

Power of the Test

The power of the test is calculated by evaluating the probability $P(\delta < \delta_0) = P(z < -\dfrac{tr\,\theta}{\sqrt{2tr\,\theta^2}})$ for given δ_0, in cases (a) and (b), when the alternative hypothesis is that disturbances follow a stationary first order autoregression with parameter values ρ = 0.2, 0.4, 0.6, 0.8, respectively. The results are tabulated in Tables 3 and 4.

Table 1 Comparison of 5 per cent Critical Values of $d(\delta_a) = 2+2\ \delta_a\ /\ \sqrt{T}$ with d_u and $d(\delta_b) = 2+2\ \delta_b\ /\ \sqrt{T}$ with d_L

Sample Size		Explanatory Variables (including intercept)				
T		2	3	4	5	6
15	$d(\delta_a)$	1.349	1.529	1.733	1.956	2.193
	d_u	1.361	1.543	1.750	1.977	2.220
	$d(\delta_b)$	1.063	0.926	0.789	0.653	0.522
	d_L	1.077	0.946	0.814	0.685	0.562
20	$d(\delta_a)$	1.410	1.535	1.674	1.825	1.987
	d_u	1.411	1.537	1.676	1.828	1.991
	$d(\delta_b)$	1.200	1.100	0.992	0.887	0.782
	d_L	1.201	1.100	0.988	0.894	0.792
30	$d(\delta_a)$	1.491	1.569	1.652	1.741	1.834
	d_u	1.489	1.567	1.650	1.739	1.833
	$d(\delta_b)$	1.354	1.286	1.216	1.144	1.072
	d_L	1.352	1.284	1.214	1.143	1.071
40	$d(\delta_a)$	1.546	1.602	1.661	1.723	1.788
	d_u	1.544	1.600	1.659	1.721	1.786
	$d(\delta_b)$	1.444	1.393	1.340	1.287	1.232
	d_L	1.442	1.391	1.338	1.285	1.230
50	$d(\delta_a)$	1.586	1.629	1.675	1.722	1.772
	d_u	1.585	1.628	1.674	1.721	1.771
	$d(\delta_b)$	1.505	1.464	1.422	1.379	1.336
	d_L	1.503	1.462	1.421	1.378	1.335
60	$d(\delta_a)$	1.617	1.653	1.690	1.728	1.768
	d_u	1.616	1.652	1.689	1.727	1.767
	$d(\delta_b)$	1.549	1.515	1.480	1.445	1.409
	d_L	1.549	1.514	1.480	1.444	1.408

Note: The 5 percent d_u and d_L values are as reported by Koerts and Abrahamese (1969, p.178).

Table 2: Size of the Critical Region[*]

Sample Size		Explanatory Variables (including intercept)				
		2	3	4	5	6
15	$\alpha(d_u)$	0.053	0.054	0.055	0.056	0.058
	$\alpha(d_L)$	0.054	0.056	0.058	0.061	0.066
20	$\alpha(d_u)$	0.050	0.051	0.051	0.051	0.051
	$\alpha(d_L)$	0.050	0.051	0.049	0.052	0.054
30	$\alpha(d_u)$	0.049	0.050	0.049	0.050	0.050
	$\alpha(d_L)$	0.049	0.050	0.050	0.050	0.050
40	$\alpha(d_u)$	0.049	0.049	0.049	0.049	0.049
	$\alpha(d_L)$	0.049	0.049	0.049	0.049	0.049
50	$\alpha(d_u)$	0.050	0.049	0.050	0.049	0.050
	$\alpha(d_L)$	0.049	0.049	0.050	0.049	0.050
60	$\alpha(d_u)$	0.050	0.050	0.050	0.050	0.050
	$\alpha(d_L)$	0.050	0.049	0.050	0.049	0.049

[*] If d_u is the exact critical value $P[d < d_u] = 0.05$. Similarly if d_L is the exact critical value $P[d < d_L] = 0.05$.

Table 3: Power of the Test for Case δ_a (corresponding to d_u) for Varying ρ when Alternative Hypothesis is $u_t = \rho\ u_{t-1} + \epsilon_t$, $|\rho| < 1$, where ϵ_t is N (0, 1)*

T	ρ	Explanatory Variables (including intercept)				
		2	3	4	5	6
15	0.0	0.05	0.05	0.05	0.05	0.05
	0.2	0.14	0.13	0.11	0.10	0.09
	0.4	0.30	0.24	0.19	0.16	0.15
	0.6	0.49	0.37	0.28	0.23	0.24
	0.8	0.64	0.47	0.35	0.33	0.47
20	0.0	0.05	0.05	0.05	0.05	0.05
	0.2	0.18	0.17	0.15	0.14	0.12
	0.4	0.42	0.36	0.30	0.25	0.21
	0.6	0.67	0.57	0.46	0.37	0.29
	0.8	0.81	0.70	0.56	0.44	0.36
30	0.0	0.05	0.05	0.05	0.05	0.05
	0.2	0.25	0.24	0.22	0.20	0.19
	0.4	0.61	0.56	0.51	0.46	0.41
	0.6	0.87	0.82	0.75	0.68	0.60
	0.8	0.96	0.92	0.87	0.79	0.70
40	0.0	0.05	0.05	0.05	0.05	0.05
	0.2	0.31	0.30	0.29	0.27	0.25
	0.4	0.74	0.71	0.67	0.63	0.59
	0.6	0.96	0.93	0.90	0.86	0.82
	0.8	1.00	0.99	0.98	0.95	0.91
50	0.0	0.05	0.05	0.05	0.05	0.05
	0.2	0.37	0.36	0.35	0.33	0.32
	0.4	0.83	0.81	0.78	0.75	0.72
	0.6	0.99	0.98	0.97	0.95	0.93
	0.8	1.00	1.00	1.00	1.00	0.99
60	0.0	0.05	0.05	0.05	0.05	0.05
	0.2	0.43	0.42	0.40	0.39	0.38
	0.4	0.90	0.88	0.86	0.84	0.82
	0.6	1.00	1.00	0.99	0.99	0.98
	0.8	1.00	1.00	1.00	1.00	1.00

* Five percent level of significance.

Table 4: Power of the Test for Case δ_b (corresponding to d_L) for Varying ρ when Alternative Hypothesis is $u_t = \rho \, u_{t-1} + \epsilon_t$, $\mid \rho \mid$ < 1, where ϵ_t is N (O, 1)*

T	ρ	Explanatory Variables (including intercept)				
		2	3	4	5	6
15	0.0	0.05	0.05	0.05	0.05	0.05
	0.2	0.14	0.13	0.12	0.10	0.09
	0.4	0.32	0.28	0.24	0.20	0.16
	0.6	0.57	0.56	0.46	0.39	0.31
	0.8	0.74	0.70	0.66	0.60	0.53
20	0.0	0.05	0.05	0.05	0.05	0.05
	0.2	0.18	0.17	0.16	0.14	0.13
	0.4	0.45	0.41	0.38	0.34	0.30
	0.6	0.73	0.70	0.66	0.62	0.57
	0.8	0.87	0.86	0.84	0.81	0.78
30	0.0	0.05	0.05	0.05	0.05	0.05
	0.2	0.25	0.24	0.23	0.22	0.20
	0.4	0.63	0.61	0.59	0.56	0.53
	0.6	0.90	0.89	0.87	0.86	0.84
	0.8	0.98	0.98	0.97	0.97	0.96
40	0.0	0.05	0.05	0.05	0.05	0.05
	0.2	0.32	0.31	0.29	0.28	0.27
	0.4	0.76	0.74	0.73	0.71	0.69
	0.6	0.97	0.96	0.96	0.95	0.94
	0.8	1.00	1.00	1.00	1.00	1.00
50	0.0	0.05	0.05	0.05	0.05	0.05
	0.2	0.38	0.37	0.36	0.35	0.33
	0.4	0.85	0.84	0.83	0.81	0.80
	0.6	0.99	0.99	0.99	0.99	0.99
	0.8	1.00	1.00	1.00	1.00	1.00
60	0.0	0.05	0.05	0.05	0.05	0.05
	0.2	0.43	0.42	0.41	0.40	0.39
	0.4	0.90	0.90	0.89	0.88	0.87
	0.6	1.00	1.00	1.00	1.00	1.00
	0.8	1.00	1.00	1.00	1.00	1.00

* Five percent level of significance.

APPENDIX

A. Computational Algorithm

The algorithm underlying the results reported in Section II was developed in a DOS environment using the PC RATS compiler (version 3.0, large memory) designed by VAR Econometrics (1988). It comprises two modules which are described below.

Module I.

First, we note that the integral of f(z) (cf. 1.18) over the range (- ∞, z_o)

$$z_o = \frac{-tr\theta}{\sqrt{2tr\theta^2}}$$ (A.1)

is

$$F(z_o) = \frac{1}{\sqrt{2\pi}} \int_{-\infty}^{z_o} e^{-\frac{1}{2}z^2} dz + \kappa_3(z) \left[\frac{1}{6} - \frac{1}{12}\frac{tr\theta}{tr\theta^2} \right] \frac{1}{\sqrt{2\pi}} e^{-\frac{(tr\theta)^2}{4tr\theta^2}} + \left[\kappa_4(z) \left(-\frac{1}{8}\frac{tr\theta}{\sqrt{2tr\theta^2}} + \frac{1}{24}\frac{(tr\theta)^3}{(2tr\theta^2)^{3/2}} \right) \right.$$

$$\left. + \kappa_3^2(z) \left(\frac{5}{24}\frac{tr\theta}{\sqrt{tr\theta^2}} - \frac{5}{36}\frac{(tr\theta)^3}{(2tr\theta^2)^{3/2}} + \frac{1}{72}\frac{(tr\theta)^5}{(2tr\theta^2)^{5/2}} \right) \right] \frac{1}{\sqrt{2\pi}} e^{-\frac{(tr\theta)^2}{4tr\theta^2}}$$ (A.2)

where terms up to $O(1/T)$ have been retained.

Module I determines critical values δ_o and d_o,

$$d_o = 2 + \frac{2\delta_o}{\sqrt{T}}$$ (A.3)

to test the null hypothesis of independence of regression disturbances (i.e. $\Sigma = I$), for given regressor matrix X.

It is required to solve the equation

$$P(\delta < \delta_o) = P(d < d_o) = P(z < z_o) = F(z_o) = \alpha$$ (A.4)

where α is a preassigned level of significance. It should be noted that z_0 is a function of δ, X and Σ. Having specified X and Σ the algorithm iteratively determines the values of δ such that $F(z_0)$ = 0.05. When X is as specified in case (a) or (b) the algorithm determines δ_a, δ_b or $d(\delta_a)$, $d(\delta_b)$. The iterative procedure is as follows.

Let superscript j denote successive iterative passes. Starting from an initial critical value

$$d^{(o)} = d_u$$ (A.5)

and convergence criterion

$$c^{(j)} = \left| \frac{F^{(j)} - \alpha_o}{\alpha_o} \right| < 0.0002$$ (A.6)

where α_o = 0.05 is the preassigned level of significance. Table 5 provides the correction

$$\Delta d^{(j)} = d^{(j+1)} - d^{(j)}$$ (A.7)

for different ranges of $c^{(j)}$ and also the order of discrepancy between actual d_o and iterate $d^{(j)}$. Variable correction factors $\Delta d^{(j)}$, for different ranges of $c^{(j)}$, substantially reduce the number of iterations, starting with a large step and providing flexibility to reduce the increment step as necessary. The concomitant risk of infinite oscillation within a range of $c^{(j)}$ is prevented by assigning unequal absolute corrections for positive and negative values of $c^{(j)}$, respectively.

In all critical value calculations reported in Table 1 convergence was achieved in less than eight iterations on a PCAT-386 with a processing speed of 30Mhz. The time taken varied between 45 seconds and 2.5 minutes for sample sizes between 20 and 60. It may be noted that the first pass typically took 95 per cent of the time as it performed all necessary matrix operations which were not repeated in subsequent iterations. This time could not be reduced as matrix operations are compiler dependent and outside program control. Also, experimentation with an initial value of $d^{(o)} = 2$ did not perceptibly affect accuracy of results or the time. However, the maximum number of iterations increased to about 10.

Module II

Power computations are done in this module which is a subroutine of Module I. Starting with the critical value computed in Module I, specification of Σ is reset according to the alternative hypothesis and integral (A.2) evaluated.

Table 5: Convergence Control

Condition $\mid c^{(j)} \mid$	Correction $= \Delta d^{(j)}$		
	Positive $c^{(j)}$	Negative $c^{(j)}$	Order $\mid d^{(j)} - d_o \mid$
A. $\mid c^{(j)} \mid \geq 2 \times 10^{-1}$	-4×10^{-2}	5×10^{-2}	$O(10^{-1})$
B. $[\mid c^{(j)} \mid < 2 \times 10^{-1}] \cap [\mid c^{(j)} \mid \geq 2 \times 10^{-2}]$	-4×10^{-3}	5×10^{-3}	$O(10^{-2})$
C. $[\mid c^{(j)} \mid < 2 \times 10^{-2}] \cap [\mid c^{(j)} \mid \geq 2 \times 10^{-3}]$	-4×10^{-4}	5×10^{-4}	$O(10^{-3})$
D. $[\mid c^{(j)} \mid < 2 \times 10^{-3}] \cap [\mid c^{(j)} \mid \geq 2 \times 10^{-4}]$	-4×10^{-5}	5×10^{-5}	$O(10^{-4})$
E. $\mid c^{(j)} \mid < 2 \times 10^{-4}$	convergence		

A. L. Nagar, Pro-Vice Chancellor, Delhi University, Delhi-110007, India, and P. D. Sharma, Institute of Economic Growth, University Enclave, Delhi-110007, India.

ACKNOWLEDGEMENTS

We wish to record our thanks to Professor Angelo Melino of University of Toronto for useful discussions and Mr. K. Lal of Institute of Economic Growth, Delhi, for computational help.

Work on this paper was initiated when the first author was Visiting Professor at the Department of Economics of University of Windsor, Canada. He wants to record his thanks to the department for providing all facilities for research during the tenure of his assignment at Windsor.

REFERENCES

Anderson, T. W. (1962). Least squares and best unbiased estimates. *Annals of Mathematical Statistics. 33*, 266-272

Cramer, H. (1962). *Mathematical methods of statistics.* Bombay: Asia Publishing House.

Durbin, J. and Watson, G. S. (1950). Testing for serial correlation in Least Squares Regression I. *Biometrika. 37*, p.409.

Durbin, J. and Watson, G. S. (1951). Testing for serial correlation in Least Squares Regression II. *Biometrika. 38*, p.259.

Durbin, J. and Watson, G. S. (1971). Testing for serial correlation in Least Squares Regression III, *Biometrika. 58*, p.1.

Farebrother, R. W. (1980). Alg. AS 153: Pan's procedure for the tail probabilities of the Durbin-Watson statistic. *Applied Statistics. 29*, pp.224-227. Correction (1981). *Applied Statistics. 30*, p.189.

Imhof, J. P. (1961). Computing the distribution of quadratic forms in normal variables. *Biometrika. 48*, 419-426.

King, Maxwell, L. and Giles, David. (Eds.). (1987). *Specification analysis in the linear model: In honour of Donald Cochrane*. London: Routledge and Kegan Paul.

Koerts, J. and Abrahamese, A.P.J. (1969). *On the theory and application of the general linear model*. Rotterdam University Press.

Maddala, G.S. (1977). *Econometrics*, McGraw-Hill.

Shiveley, T.S., Ansley, C.F., and Kohn, R. (1990). Fast evaluation of the distribution of the Durbin Watson and other invariant test statistics in time series regression. *Journal of the American Statistical Association*, 676-685.

Stochastic Prediction in Dynamic Nonlinear Systems: Econometric Forecasting Applications in Models for Developing Economies

Roberto S. Mariano and Wesley W. Basel

The use of stochastic simulators for prediction in nonlinear systems has been suggested frequently to avoid the nonlinearity bias of the standard deterministic predictor. The asymptotic behavior of these predictors has been extensively studied in a series of papers by Mariano & Brown.[1] Several new variations and techniques have also been developed recently, on the theoretical level, for validation and inference in such models. One group of techniques utilizes stochastic simulations to address complexities caused by nonlinearities in the model: see Mariano and Brown (1991a,b) and Lin and Granger (1992) for example. Fair (1986) also applies stochastic simulations of estimated models to evaluate their predictive accuracy. A second group deals with nonparametric and semiparametric estimation of conditional means and derivatives to reduce sensitivity of inference to misspecification in the nonlinear model. See Pagan and Ullah (1991) for a comprehensive reference.

While this theoretical progress has been made, little work has yet been done on the practical applications of such techniques to the analysis and use of econometric models under small sample conditions. Preliminary finite sample results indicate that stochastic predictors retain their bias advantage in small-sample, but increasing effect of the simulation and parameter estimation components of variation may lead to their mean-squared error approaching that of the deterministic predictor.

This paper focuses further on small sample applications, with particular emphasis on data conditions present when modeling developing economies. First a summary of definitions and asymptotic results is presented, along with discussion of likely finite sample behavior and implementation difficulties. Next, a simple two-equation model is examined in a Monte-Carlo experiment to examine our conjectures in a simple context. Finally, the procedures are used to generate predictions from a moderately large scale annual model of the Philippine macroeconomy. Practical problems that arise when implementing these stochastic procedures are examined and their respective advantages or disadvantages are discussed.

II. Definitions and Summary of Asymptotic Results

Consider a dynamic nonlinear system with structural equations:

$$f(y_t, y_{t-1}, x_t; \alpha) = u_t, \quad t = 1, 2, \ldots, T \quad (2.1)$$

where $f(\cdot)$ is an $n \times 1$ vector of functions of the endogenous variables y_t ($n \times 1$), the lagged endogenous variables y_{t-1}, the vector of exogenous variables x_t ($m \times 1$) and the vector of parameters α^0 ($p \times 1$). The disturbances u_t are assumed to be intertemporally independently distributed with mean zero and identity covariance matrix.[2]

As this model represents the structural form of the y_t generating process given the exogenous and predetermined variables and the disturbances, it is assumed that a locally unique solution is implicitly defined:

$$y_t = g(u_t, y_{t-1}, x_t; \alpha) \qquad (2.2)$$

This solution of the model is generally unavailable in closed form. The object of ex ante one-period-ahead prediction is the prediction of:

$$y_{T+1} = g(u_{T+1}, y_T, x_{T+1}; \alpha) \qquad (2.3)$$

and for two-period-ahead prediction:

$$y_{T+2} = g\,[u_{T+2}, g\,(u_{T+2}, y_T, x_{T+1}; \alpha), x_{T+2}; \alpha] \qquad (2.4)$$

where the prediction in either case is conditional on y_T, x_{T+1}, and x_{T+2}.

Taking the case of one-period-ahead prediction, given the assumption that the first two conditional moments of y_t exist, the estimate of the minimum mean-squared error predictor should be the conditional first moment, $E(g_{T+1})$ evaluated at a consistent estimator of α. However, a closed form expression for this moment is seldom obtainable. This leads to the standard definitions for the deterministic, Monte Carlo stochastic, and residual-based stochastic predictors:

$$\hat{y}^{(d)}_{T+1} = g(0, y_T, x_{T+1}; \hat{\alpha}) \qquad (2.5)$$

$$\hat{y}^{(m)}_{T+1} = \sum_{s=1}^{s} g(\tilde{u}_s, y_T, x_{T+1}; \hat{\alpha}/S \qquad (2.6)$$

$$\hat{y}^{(r)}_{T+1} = \sum_{t=1}^{T} g(\hat{u}_t, y_T, x_{T+1}; \hat{\alpha}/T \qquad (2.7)$$

where

$\tilde{u}_s$ are independent draws from $N(0, I_n)$, $s = 1, 2, \ldots, S_t$ $\qquad (2.8)$

$$\hat{u}_t = f(y_t, y_{t-1}, x_t; \hat{\alpha}), \qquad t = 1, 2, \ldots, T. \qquad (2.9)$$

Given conditions on the solution form, (2.2), the conditional moments, and the asymptotic behavior of the parameter estimator, Brown & Mariano (1989) derive expressions for the asymptotic bias and mean-squared prediction error for each of the above predictors. The basic bias results are that the bias of the Monte Carlo and residual-based stochastic predictors are of order $O(1/T)$, while the asymptotic bias of the deterministic predictor contains a $O(1)$ term, i.e. a non-vanishing component:

$$B^{(d)}_1 = g(0, y_T, x_{T+1}; \alpha^0] - E[g(u_{T+1}, y_T, x_{T+1}; \alpha^0)|y_T] + 0(1/T) \quad (2.10)$$

The asymptotic mean-squared prediction error (AMSPE) expressions give the usual result that the conditional covariance matrix of y_{T+1} is a lower bound for AMSPE. Both the residual-based and Monte Carlo stochastic predictors attain this lower bound asymptotically (assuming the number of draws S is the same order as T). However, the deterministic predictor has a further $O(1)$ term -- namely the outer product of the bias. Thus, both the Monte Carlo and residual-based predictors are asymptotically efficient relative to the deterministic predictor. The residual-based predictor dominates the Monte Carlo predictor if $S \le T$, but otherwise the ranking is indeterminate.

Preliminary results have indicated that similar relationships between the alternative predictors exist in small samples as well. The stochastic predictors are biased, but the bias of the deterministic predictor generally increases more proportionally relative to the stochastic predictors. However, in the expansion for AMSPE the terms due to estimator variance and simulation variance may come to dominate in small samples. Thus, the MSPE of stochastic and deterministic predictors in small samples may be much closer.

Besides the distributional differences that occur in finite samples, several practical difficulties can arise in implementing stochastic predictions. Most of these problems involve the estimation of the various covariance matrices used in generating the predictors and drawing inferences from them.

Two difficult factors to compute arise in the AMSPE expressions due to estimator variation. First is $\Gamma_1' \Psi \Gamma_1$, where Γ_1 is the conditional expected value of the partial first derivatives of $g(\cdot)$ and Ψ is the asymptotic covariance matrix of $\hat{\alpha}$. This term is present in both the bias and AMSPE expansions for all three predictors. Also present is a factor derived from the second order term in the estimator's asymptotic expansion. Any analytical approximation of the predictor's MSPE or derivation of a prediction region requires estimation of these factors. Analytical derivation of either factor can be difficult in a large or complicated model. Expansions yielding the second-order factor exist but are difficult to implement.[3] It is unlikely that Γ_1 will be available in closed form for most nonlinear models and thus must be estimated numerically.

Several numerical alternatives to such a difficult derivation exist. First, Γ_1 could be generated through simulation utilizing the estimated covariance of

u_t (See below for difficulties arising from this estimation). However, initial attempts have not been useful. A second alternative, and very simple to implement, is to recognize that the simulated disturbances drawn for the Monte Carlo predictor, or the estimated disturbances utilized in residual based prediction, represent pseudo-samples that generate a simulated distribution for y_{T+1}. Quantiles and thus prediction regions can be estimated directly from this pseudo-sample. Brown and Mariano (1984) show that these quantile estimates are consistent and asymptotically normal. In practical terms, the quantile estimate is chosen directly from the sorted sample, as the value attained by the first observation for which the empirical distribution function is greater than or equal to the desired probability value. Results of this procedure are shown for the Philippine model below.

A more fundamental problem in implementing the Monte Carlo stochastic predictor is estimating the covariance matrix of the disturbance term. Further research will consider utilizing a heteroskedasticity and autocorrelation consistent covariance estimator.[4]

III. A Simulation Study of a Simple Model

The following simple model is defined to study finite sample behavior more directly:

$$y_{1t} = \alpha + \gamma x_t + \lambda y_{1,t-1} + u_{1t} \tag{3.1}$$

$$\ln(y_{2t}) = \delta + \beta \ln(y_{1t}) + u_{2t} \tag{3.2}$$

with (u_1, u_2) independently and identically normally distributed with mean zero and covariance matrix. This model was chosen because the reduced form can be analytically determined and closed form expressions for the mean and variance of y_2 can be found for positive integer values of β_2. The reduced form for equation 2 is given below as equation (3.3), and the mean and variance expressions for y_2 are shown in equations (3.4)-(3.9):

$$y_2 = \exp(\delta) \, y_{1t}^{\beta} \exp(u_{2t}) \tag{3.3}$$

Expressions for Mean and Variance of y_{2t} for $\beta = 1$:

$$E(y_{2t}) = \exp(\delta)\exp(\sigma^2_2/2)(m_{1t} + \sigma_{12}) \tag{3.4}$$

$$\mathrm{Var}\,(y_{2t}) = \exp(2\delta)[\exp(2\sigma^2_2)(\sigma^2_1 + (m_{1t} + 2\sigma_{12})^2) - \exp(\sigma^2_2)(m_{1t} + \sigma_{12})^2] \tag{3.5}$$

Expressions for Mean and Variance of y_{2t} for $\beta = 2$:

$$E(y_{2t}) = \exp(\delta)\exp(\sigma^2_2/2)[\sigma^2_1 + (m_{1t} + \sigma_{12})^2] \tag{3.6}$$

$$\text{Var}(y_{2t}) = \exp(2\delta)\{\exp(2\sigma^2_2)[(3\sigma^2_1 + 6\sigma^2_1(m_{1t} + 2\sigma_{12})^2 \\ + (m_{1t} + 2\sigma_{12})^4] - \exp(\sigma^2_2)[\sigma^2_1 + (m_{1t} + \sigma_{12})^2]^2\} \quad (3.7)$$

Expression for Mean and Variance of y_{2t} for $\beta = 3$:

$$E(y_{2t}) = \exp(\delta)\exp(\sigma^2_2/2)[3\sigma^2_1(m_{1t} + \sigma_{12})^2 + (m_{1t} + \sigma_{12})^3] \quad (3.8)$$

$$\text{Var}(y_{2t}) = \exp(2\delta)\{\exp(2\sigma^2_2)[(15\sigma^6_1 + 45\sigma^4_1(m_{1t} + 2\sigma_{12})^2 + 15\sigma^2_1(m_{1t} + 2\sigma_{12})^4 \\ + (m_{1t} + 2\sigma_{12})^6] - \exp(\sigma^2_2)[3\sigma^2_1(m_{1t} + \sigma_{12})^3]^2\} \quad (3.9)$$

where σ^2_1, σ^2_2 and σ_{12} are the variances of u_1 and u_2 and their covariance respectively and m_{1t} is the conditional mean of y_{1t}:

$$m_{1t} = \alpha + \gamma x_t + \lambda y_{1,t-1} \quad (3.10)$$

The zero-order term in the asymptotic expansion of the bias of the deterministic predictor shown in equation (2.10) is $[g_1(0) - E(y_{T+1}|y_T)]$ and represents the component of nonlinearity bias that does not tend to zero as T increases. For y_2 in this mode, $g_1(0)$ is simply equation (3.3) with the disturbance factor set equal to 1 and the conditional expectation is shown in equations (3.4), (3.6) and (3.8) for three values of β. This component of the bias, and its partial derivatives with respect to underlying parameters, can then be analyzed. The first conclusion that can be made is that the zero-order bias will always be negative in this model. Second, by comparing the conditional means, it is seen that the bias increases (in absolute terms) with increasing values of β. Derivation of higher-order moments show that the skewness and kurtosis of y_2 also increase with β. Finally, it is also readily apparent that the bias is increasing with respect to σ_{21}, σ^2_2 or σ_{12} for fixed values of the other parameters, which represents a decrease in the relative fit[5] of each equation and increasing correlation between $\ln(y_1)$ and u_2 in the second equation.

Two further points are evident from the bias and mean squared error expansions. First, the second-order terms also differ between the stochastic and deterministic predictors, with the difference again arising from the difference between functions of the reduced form evaluated at zero disturbance and conditional expectation of these functions. This factor will affect the bias in small samples as well. Second, from the asymptotic mean-squared prediction error, it appears likely that the components caused by parameter estimation and simulation variance may come to dominate as T becomes small.

To further examine the properties of the three predictors in a finite sample context, a Monte Carlo study was performed. Besides the sample size, the effect of three other attributes of the model were considered in successive experiments. These factors were the value of β, the correlation between $\ln(y_1)$ and u_2, and the relative fit of the second equation (the relative fit of the first was kept fixed). From the preceding analysis, these factors are expected to have, respectively, a positive, positive, and negative effect on the deterministic bias; while the effect on asymptotic mean-squared prediction error is undetermined.

Tables 1-9 give the results of these experiments. The values of X and the parameters of equation 1 remain fixed throughout the study. To reduce simulation variation, the $(x_t, y_{1t}, y_{2t}; t = 1, 2, . . ., T)$ sample was first generated for the maximum T considered and then successive subsamples taken for smaller sample sizes. A T + 100 i.i.d. sample for X_t was drawn from a normal distribution of mean 100 and variance 625, and these values then remained fixed throughout the study. u_1 and u_2 were T + 100 observations drawn jointly from a normal distribution with mean zero and $\sigma^2_1 = 375$ and σ^2_2, σ_{12} varying.[6] Y_{1t} observations were then generated from the first equation with $\alpha = 0$, $\gamma = 1$, and $\lambda = 0.5$. The first 100 observations were then discarded to remove the initial condition effect. This generates a sample of Y_1 with unconditional mean 200, and variance 1000.[7] Finally, Y_{2t} was generated from x_t, y_{1t} and u_2 with β varying between experiments, and δ adjusted to maintain a similar scale.

For each subsample of the generated data, T = 50, 100, 200 and 400, the parameters of the model were estimated. Equation 1 was estimated by ordinary least squares. Equation 2 estimates were two-stage least squares with instruments x_t, $\ln(x_t)$, $y_{1,t-1}$ and $\ln(y_{1,t-1})$. Given these estimates (and error variance estimates), the deterministic, residual-based and Monte Carlo-based predictors for were computed for $y_{2,T+1}$. This process of generating the endogenous variables, estimation and prediction was repeated 2500 times to obtain the simulated distribution of the predictors.[8] The mean prediction error, and root mean-squared prediction error are then reported in the succeeding tables. It is evident from these results that the general ranking of the predictors in terms of bias and mean-squared error remains the same in small samples. However, for certain values of the parameters, namely β small, high relative fit, and low endogenous correlation, the bias of the deterministic estimator becomes negligible (1-2% of the mean). Thus for these parameter values, there is little difference between predictors in terms Table 10 of their root mean percentage errors (RMSPE). Note also that for increasing sample size (looking down the table, in a given column), we can discern no clear trend in either bias or RMSPE. The number of repetitions will be increased in future research to obtain better pseudo-sample estimates.

IV. Application of Stochastic Prediction to a Large Model of the Philippine Economy

To study the applicability of stochastic prediction techniques in a real data situation, the residual-based and Monte Carlo-based predictors were implemented for a moderately large scale annual macroeconomic model of the

	Table 1: Sampling Distribution of Alternative Predictors $\beta = 1, \rho = 0.26$								
	Eq. 2 Fit = 0.50			Eq. 2 Fit = 0.70			Eq. 2 Fit = 0.90		
	Mean	ME	RMSE	Mean	ME	RMSE	Mean	ME	RMSE
PREDICTOR				T = 50					
Actual	11268	0	2732	11032	0	1913	9896	0	1237
Determ.	10889	-379	2806	10887	-145	1929	9845	-51	1246
Res-Based	11125	-143	2786	10989	-44	1925	9873	-24	1245
MC-Based	11118	-136	2795	10996	-37	1928	9872	-25	1247
				T = 100					
Actual	11190	0	2627	11059	0	1859	9929	0	1254
Determ.	10926	-263	2623	10928	-132	1853	9888	-40	1249
Res-Based	11170	-19	2611	11032	-27	1848	9917	-11	1248
MC-Based	11170	-19	2621	11034	-25	1851	9918	-11	1253
				T = 200					
Actual	11228	0	2665	10975	0	1880	9938	0	1229
Determ.	10930	-298	2677	10927	-47	1876	9887	-51	1229
Res-Based	11177	-51	2661	11034	59	1877	9917	-22	1228
MC-Based	11180	-48	2664	11033	59	1884	9918	-20	1229
				T = 400					
Actual	11114	0	2665	11058	0	1905	9955	0	1243
Determ.	10920	-194	2666	10916	-142	1911	9881	-74	1244
Res-Based	11168	54	2658	11023	-35	1906	9911	-45	1243
MC-Based	11172	58	2657	11021	-37	1907	9911	-45	1243

	Table 2: Sampling Distribution of Alternative Predictors $\beta = 1, \rho = 0.36$								
	Eq. 2 Fit = 0.50			Eq. 2 Fit = 0.70			Eq. 2 Fit = 0.90		
	Mean	ME	RMSE	Mean	ME	RMSE	Mean	ME	RMSE
PREDICTOR				T = 50					
Actual	10196	0	2484	10019	0	1774	9985	0	1769
Determ.	9854	-342	2532	9856	-163	1797	9845	-139	1797
Res-Based	10086	-109	2512	9957	-62	1791	9947	-38	1794
MC-Based	10088	-108	2515	9962	-57	1793	9951	-34	1802
				T = 100					
Actual	10173	0	2443	10000	0	1757	9980	0	1779
Determ.	9892	-281	2442	9890	-110	1752	9890	-90	1765
Res-Based	10133	-40	2425	9995	-5	1749	9995	15	1762
MC-Based	10138	-35	2434	9997	-3	1753	9998	19	1767
				T = 200					
Actual	10122	0	2540	10014	0	1747	9906	0	1672
Determ.	9888	-235	2544	9888	-126	1746	9884	-22	1669
Res-Based	10132	9	2532	9995	-20	1741	9990	84	1671
MC-Based	10130	7	2541	10001	-14	1743	9990	84	1671
				T = 400					
Actual	10214	0	2503	10000	0	1738	9994	0	1751
Determ.	9879	-335	2521	9880	-120	1741	9885	-110	1749
Res-Based	10125	-90	2499	9987	-13	1737	9992	-3	1745
MC-Based	10123	-91	2499	9989	-11	1738	9991	-4	1750

Table 3: Sampling Distribution of Alternative Predictors $\beta = 1$, $\rho = 0.46$								
	Eq. 2 Fit = 0.50			Eq. 2 Fit = 0.70			Eq. 2 Fit = 0.90		
	Mean	ME	RMSE	Mean	ME	RMSE	Mean	ME	RMSE
PREDICTOR					T = 50				
Actual	10109	0	2385	9945	0	1835	9917	0	1321
Determ.	9854	-255	2427	9843	-102	1856	9833	-84	1336
Res-Based	10095	-14	2416	9953	8	1854	9866	-51	1335
MC-Based	10101	-8	2420	9961	15	1858	9864	-53	1339
					T = 100				
Actual	10087	0	2500	10034	0	1855	9891	0	1285
Determ.	9888	-200	2474	9884	-150	1849	9877	-14	1277
Res-Based	10138	51	2465	9998	-36	1844	9912	20	1277
MC-Based	10136	49	2475	9998	-36	1851	9912	20	1278
					T = 200				
Actual	10026	0	2439	9979	0	1802	9929	0	1292
Determ.	9883	-143	2442	9886	-93	1797	9884	-45	1286
Res-Based	10137	111	2440	10002	23	1795	9920	-9	1286
MC-Based	10137	110	2443	10001	22	1797	9920	-9	1289
					T = 400				
Actual	10078	0	2511	10068	0	1837	9862	0	1262
Determ.	9876	-202	2516	9879	-189	1845	9877	16	1259
Res-Based	10133	55	2509	9996	-72	1837	9913	52	1260
MC-Based	10132	54	2509	9997	-71	1837	9914	53	1261

Table 4: Sampling Distribution of Alternative Predictors $\beta = 2$, $\rho = 0.47$								
	Eq. 2 Fit = 0.50			Eq. 2 Fit = 0.70			Eq. 2 Fit = 0.90		
	Mean	ME	RMSE	Mean	ME	RMSE	Mean	ME	RMSE
PREDICTOR					T = 50				
Actual	10017	0	5162	10468	0	3892	10122	0	2554
Determ.	8881	-1136	5328	9809	-659	3999	9793	-329	2590
Res-Based	9903	-114	5224	10343	-126	3950	10013	-109	2573
MC-Based	9947	-70	5243	10355	-113	3951	10022	-100	2578
					T = 100				
Actual	10102	0	5233	10481	0	3862	10139	0	2566
Determ.	8938	-1164	5342	9881	-600	3889	9869	-270	2567
Res-Based	9992	-110	5212	10434	-48	3843	10094	-45	2553
MC-Based	10012	-90	5209	10439	-42	3841	10095	-44	2560
					T = 200				
Actual	9982	0	5125	10434	0	3821	10033	0	2599
Determ.	8938	-1044	5220	9874	-560	3847	9868	-165	2602
Res-Based	10012	31	5110	10437	3	3804	10098	65	2598
MC-Based	10023	41	5112	10440	7	3812	10103	70	2605
					T = 400				
Actual	10005	0	5275	10373	0	3811	10115	0	2556
Determ.	8928	-1077	5380	9860	-514	3846	9856	-259	2568
Res-Based	10008	3	5269	10426	53	3811	10087	-28	2556
MC-Based	10009	4	5278	10426	52	3811	10089	-26	2561

	Table 5: Sampling Distribution of Alternative Predictors $\beta = 2, \rho = 0.365$								
	Eq. 2 Fit = 0.50			Eq. 2 Fit = 0.70			Eq. 2 Fit = 0.90		
	Mean	ME	RMSE	Mean	ME	RMSE	Mean	ME	RMSE
PREDICTOR	T = 50								
Actual	9880	0	5084	10538	0	3786	10055	0	2577
Determ.	8877	-1003	5214	9789	-749	3897	9790	-265	2613
Res-Based	9828	-52	5129	10300	-238	3838	10005	-50	2600
MC-Based	9852	-29	5135	10304	-234	3848	10009	-46	2605
	T = 100								
Actual	9952	0	4991	10371	0	3751	10137	0	2585
Determ.	8942	-1010	5072	9872	-499	3771	9870	-267	2584
Res-Based	9924	-28	4972	10398	27	3738	10092	-46	2571
MC-Based	9948	-3	4973	10401	30	3743	10099	-39	2578
	T = 200								
Actual	9902	0	4919	10452	0	3845	10043	0	2547
Determ.	8936	-966	4999	9877	-575	3877	9870	-173	2553
Res-Based	9932	30	4900	10412	-40	3833	10096	53	2548
MC-Based	9930	28	4910	10419	-32	3840	10105	61	2555
	T = 400								
Actual	9899	0	4985	10501	0	3652	10168	0	2586
Determ.	8923	-976	5078	9865	-636	3702	9861	-307	2597
Res-Based	9928	28	4982	10405	-96	3649	10089	-80	2580
MC-Based	9930	31	4982	10409	-93	3652	10087	-82	2584

	Table 6: Sampling Distribution of Alternative Predictors $\beta = 2, \rho = 0.465$								
	Eq. 2 Fit = 0.50			Eq. 2 Fit = 0.70			Eq. 2 Fit = 0.90		
	Mean	ME	RMSE	Mean	ME	RMSE	Mean	ME	RMSE
PREDICTOR	T = 50								
Actual	9963	0	4990	10571	0	4003	10042	0	2650
Determ.	8886	-1078	5151	9804	-767	4115	9793	-249	2687
Res-Based	9819	-144	5051	10340	-231	4058	10016	-26	2676
MC-Based	9854	-110	5069	10367	-205	4071	10032	-10	2680
	T = 100								
Actual	9862	0	4859	10561	0	3735	10153	0	2608
Determ.	8948	-915	4904	9876	-686	3771	9869	-284	2601
Res-Based	9909	46	4818	10430	-131	3710	10099	-54	2585
MC-Based	9915	52	4824	10441	-120	3725	10099	-54	2598
	T = 200								
Actual	10014	0	5133	10433	0	3847	10137	0	2646
Determ.	8938	-1076	5237	9879	-551	3882	9868	-269	2653
Res-Based	9914	-100	5126	10442	9	3844	10103	-34	2638
MC-Based	9922	-92	5138	10446	13	3854	10098	-39	2639
	T = 400								
Actual	9782	0	4901	10381	0	3813	10101	0	2636
Determ.	8917	-865	4971	9862	-520	3845	9855	-245	2645
Res-Based	9901	119	4895	10429	47	3809	10091	-9	2633
MC-Based	9902	121	4902	10435	54	3811	10092	-8	2634

Table 7: Sampling Distribution of Alternative Predictors $\beta = 3$, $\rho = 0.27$								
	Eq. 2 Fit = 0.50			Eq. 2 Fit = 0.70			Eq. 2 Fit = 0.90		
	Mean	ME	RMSE	Mean	ME	RMSE	Mean	ME	RMSE
PREDICTOR				T = 50					
Actual	9269	0	7482	9163	0	4788	9520	0	3545
Determ.	7252	-2008	7761	8004	-1158	4964	8804	-716	3661
Res-Based	9013	-247	7534	8968	-195	4849	9281	-239	3603
MC-Based	9087	-173	7564	9010	-153	4857	9293	-227	3617
				T = 100					
Actual	9268	0	7467	9203	0	5159	9466	0	3585
Determ.	7312	-1957	7691	8102	-1102	5247	8907	-560	3610
Res-Based	9133	-136	7439	9097	-107	5129	9393	-74	3568
MC-Based	9175	-93	7450	9109	-95	5145	9410	-56	3579
				T = 200					
Actual	9273	0	6915	9243	0	4884	9370	0	3523
Determ.	7307	-1966	7191	8090	-1153	5007	8908	-462	3544
Res-Based	9159	-114	6923	9104	-139	4877	9402	32	3513
MC-Based	9169	-104	6931	9100	-143	4876	9409	38	3520
				T = 400					
Actual	9192	0	7033	9147	0	4982	9463	0	3587
Determ.	7296	-1897	7285	8067	-1079	5091	8898	-564	3626
Res-Based	9160	-32	7035	9086	-61	4973	9395	-67	3582
MC-Based	9169	-24	7030	9086	-61	4967	9395	-67	3579

Table 8: Sampling Distribution of Alternative Predictors $\beta = 3$, $\rho = 0.37$								
	Eq. 2 Fit = 0.50			Eq. 2 Fit = 0.70			Eq. 2 Fit = 0.90		
	Mean	ME	RMSE	Mean	ME	RMSE	Mean	ME	RMSE
PREDICTOR				T = 50					
Actual	9292	0	7445	9069	0	4933	9485	0	3591
Determ.	7263	-2029	7730	7998	-1071	5087	8817	-668	3691
Res-Based	9229	-62	7482	8983	-87	4988	9322	-163	3641
MC-Based	9308	16	7510	9010	-59	4999	9339	-146	3659
				T = 100					
Actual	9333	0	7530	9250	0	5037	9527	0	3598
Determ.	7323	-2010	7742	8093	-1157	5139	8928	-599	3726
Res-Based	9346	14	7473	9108	-142	5009	9449	-78	3680
MC-Based	9403	70	7497	9126	-125	5009	9456	-71	3693
				T = 200					
Actual	9294	0	7213	9124	0	4862	9479	0	3672
Determ.	7316	-1978	7474	8081	-1043	4966	8923	-556	3706
Res-Based	9362	69	7205	9114	-10	4855	9454	-25	3665
MC-Based	9378	85	7205	9118	-6	4867	9461	-18	3674
				T = 400					
Actual	9367	0	7648	8949	0	4818	9566	0	9566
Determ.	7289	-2078	7930	8060	-883	4895	8906	-661	3774
Res-Based	9346	-21	7651	9100	156	4818	9440	-127	3718
MC-Based	9354	-12	7664	9112	168	4822	9444	-123	3723

	Eq. 2 Fit = 0.50			Eq. 2 Fit = 0.70			Eq. 2 Fit = 0.90		
	Mean	ME	RMSE	Mean	ME	RMSE	Mean	ME	RMSE
PREDICTOR					$T = 50$				
Actual	9163	0	7282	8435	0	4566	9422	0	3516
Determ.	7285	-1878	7563	7234	-1201	4759	8821	-602	3596
Res-Based	9172	8	7375	8182	-253	4622	9324	-98	3554
MC-Based	9222	58	7398	8216	-218	4638	9339	-83	3559
					$T = 100$				
Actual	9171	0	7048	8275	0	4744	9399	0	3641
Determ.	7238	-1833	7243	7318	-957	4811	8918	-480	3643
Res-Based	9280	108	7018	8294	19	4710	9434	36	3610
MC-Based	9313	141	7031	8309	34	4719	9449	50	3622
					$T = 200$				
Actual	9429	0	7529	8246	0	4650	9405	0	3453
Determ.	7323	-2105	7808	7309	-937	4734	8918	-487	3476
Res-Based	9298	-130	7515	8304	58	4641	9446	41	3442
MC-Based	9315	-113	7536	8317	71	4653	9445	41	3452
					$T = 400$				
Actual	9423	0	7259	8457	0	4816	9512	0	3714
Determ.	7300	-2123	7567	7291	-1167	4949	8900	-612	3765
Res-Based	9288	-135	7256	8295	-163	4811	9430	-82	3717
MC-Based	9289	-134	7272	8294	-163	4817	9430	-82	3716

Table 9: Sampling Distribution of Alternative Predictors
$\beta = 3$, $\rho = 0.47$

Philippines. This model consists of three basic modules: production, expenditure and monetary. The production sector has stochastic equations determining three components of gross domestic product: agriculture, industry and services; as well as their associated implicit price deflators. The expenditure sector models private and government consumption, components of domestic capital formation, and components of exports and imports. There are numerous links, both ways, between these two sectors. Finally, the monetary sector determines the base interest rate and money supply, again with 2-way links to the other sectors. The major exogenous variables are total population, external price indices for export and import components, US and Japanese GNP and price indices, and the exchange rate. In total there are 30 stochastic equations and 19 identities.

Annual data were available for 1967-1991 for most variables. However, many variables were missing data at the beginning of the sample. For this reason, the system was estimated equation by equation. Ordinary least squares estimates were obtained for each equation, with first-order autocorrelation corrections made when necessary. Given these estimates, the residual terms were computed and then corrected for autocorrelation using the estimated coefficient. These corrected residuals were then used as the basis for stochastic prediction.[9]

Due to the missing data problem, only 17 residuals were available for each equation. These 17 corrected residuals were used directly in computing the residual-based predictor. However, likely due to the missing data and misspecification of the system covariance structure, problems were encountered in obtaining a positive-definite estimate of the error covariance for use in Monte Carlo-based prediction. For future research, calculation of a heteroskedasticity and autocorrelation consistent estimator[10] corrected for the missing data is planned. However, for the results reported below, only the variances of each equation's disturbance was estimated and the covariances set to zero.

Table 10: Deterministic & Stochastic Predictors' For Philippine Macroeconomic Model, T+1 = 1992								
	Actual Values		Determin.		Residual Based		Monte Carlo Based	
VARIABLE	t=1991	t=1992	Pred.	% Err	Pred.	% Err	Pred.	% Err
GDP	710530	712300	717149	0.7	717185	0.7	717796	0.8
GNP	718497	727164	732011	0.7	732046	0.7	732658	0.8
PGDP	174.4	188.4	204.5	8.5	204.0	8.3	204.7	8.7
GNPN	1253172	1370000	1495199	9.1	1490909	8.4	1497553	9.3
GNPGR	-0.06	1.21	1.88	0.59	1.89	0.68	1.97	0.76
TBILL	21.48	16.02	17.29	1.29	17.77	1.75	16.86	0.84
DEFG	27200	.	13777	.	14468	.	13436	.
INFL	16.80	8.03	17.29	9.26	16.96	8.93	17.40	8.37
TL	445164	.	566607	.	564046	.	569271	.
WAGE	210.2	.	236.8	.	236.0	.	237.0	.
CET	22181	.	22409	.	22385	.	22371	.
UERA	12.02	.	13.53	.	13.63	.	13.68	.
CP	543690	539798	567550	5.1	567532	5.1	567282	5.1
CG	53615	55648	54316	-2.4	54466	-2.1	54405	-2.2
GDCF	140519	148935	155902	4.7	155318	4.3	156972	5.4
EXPORT	229822	242176	223873	-7.6	223277	-7.8	223580	-7.7
IMPORT	260853	302096	295312	-2.2	293328	-2.9	294734	-2.4
XAGRRL	19027	.	17096	.	17140	.	17359	.
XNAG	129587	.	128983	.	128951	.	128240	.
XSV	77591	.	74177	.	73566	.	74364	.
MFUEL	47597	.	53174	.	53048	.	53023	.
MNF	192416	.	221177	.	219232	.	220813	.
MSV	20840	.	20961	.	20984	.	20898	.
CONSGO	28002	.	28832	.	29095	.	28806	.
CONSPR	29573	.	30697	.	30585	.	31426	.
IDER	70497	.	83926	.	83192	.	84293	.
VAR	161859	.	160686	.	161001	.	161328	.
PVAR	162.1	.	199.8	.	199.1	.	200.2	.
VIR	250190	.	243579	.	243558	.	243901	.
PVIR	170.3	.	197.8	.	197.7	.	197.9	.
VSER	298481	.	312822	.	312625	.	312567	.
PSER	184.5	.	212.2	.	211.4	.	212.4	.

Given the estimated parameters and corrected residuals for the sample period 1967-1991, 1992 one-step ahead and 1993 two-step ahead predictors were obtained. The 1992 predictions for selected variables are reported in Table 10 along with 1991 and 1992 actual values (where available). Variable definitions are summarized in Table 11. Both 1992 and 1993 predictions are available from the author upon request. For most variables, there is little difference between the alternative predictors. The stochastic predictors do appear to yield a better forecast of the T-bill rate.

Table 11: Variable Definitions		
VARIABLE	Definition	Units
GDP	Gross Domestic Product, Real	Million pesos
GNP	Gross Natioanl Product, Real	Million pesos
PGDP	Implicit Price Deflator for Gross Domestic Product	1985 = 100
GNPN	Gross National Product, Nominal	Million pesos
GNPGR	GNP Growth Rate	percent
TBILL	90 day T-bill interest rate	percent
DEFG	Government budget deficit, Nominal	Million pesos
INFL	Inflation rate (based on GDP imp. price deflator)	percent
TL	Money Supply (M3)	Million pesos
WAGE	Nominal Wage Rate Index (Industrial unskilled)	1985 = 100
CET	Total Employment	Thousands
UERA	Unemployment Rate	percent
CP	Private Consumption, Real	Million pesos
CG	Government Consumption, Real	Million pesos
GDCF	Gross Domestic Capital Formation, Real	Million pesos
EXPORT	Total Exports, Real	Million pesos
IMPORT	Total Imports, Real	Million pesos
XAGRRL	Agricultural Exports, Real	Million pesos
XNAG	Non-Agricultural Goods Exports, Real	Million pesos
XSV	Services Exports, Real	Million pesos
MFUEL	Fuel Imports, Real	Million pesos
MNF	Non-Fuel Goods Imports, Real	Million pesos
MSV	Services Imports, Real	Million pesos
CONSGO	Investment: Government Construction, Real	Million pesos
CONSPR	Investment: Private Construction, Real	Million pesos
IDER	Investment: Durable Equipment, Real	Million pesos
VAR	Value-Added: Agriculture, Real	Million pesos
PVAR	Implicit Price Deflator for Agricutural Value-Added	1985 = 100
VIR	Value-Added: Industrial, Real	Million pesos
PVIR	Implicit Price Deflator for Industrial Value-Added	1985 = 100
VSER	Value-Added: Services, Real	Million pesos
PSER	Implicit Price Deflator for Services Value-Added	1985 = 100

Given the difficulties in obtaining a reliable covariance matrix estimate and the impracticality of estimating functions of the reduced form, estimating the asymptotic mean-squared error from the expansions in Section 1 was not attempted. However, recognizing that the residuals form a pseudo-sample of the disturbance terms, it is possible to obtain consistent quantile, and thus prediction region, estimates as mentioned in Section I. These estimates are for 1992 and 1993 are available from the author upon request.

V. Conclusions

In conclusion, although we can demonstrate the relative efficiency of stochastic predictors in a simple nonlinear context, we have yet to show this for the more general model utilized in the econometric study of developing economies. Still, stochastic techniques do show an advantage in drawing inferences, such as prediction regions. Further research is focusing on utilizing these techniques in the validation of nonlinear simultaneous systems.

Roberto S. Mariano and Wesley W. Basel, Department of Economics, University of Pennsylvania, 9718 Locust Walk, Philadelphia, PA 19104-6297.

ACKNOWLEDGEMENTS

Partial support from the University of Pennsylvania Research Foundation is gratefully acknowledged.

NOTES

1. Mariano & Brown (1983), Brown & Mariano (1984), Mariano & Brown (1989), Brown & Mariano (1989).
2. As stated in Mariano & Brown (1983), no loss of generality results from assuming the identity covariance matrix, since the general model with general covariance structure can be converted to this model through a nonsingular linear transformation, thus embedding the covariance matrix parameters in θ.
3. See for example Rothenberg (1984).
4. See Newey and West (1987) for example.
5. By relative fit we mean the variance due to right-hand side variables relative to the total variation of the left-hand variable. In a classical regression model, this is one less the disturbance variance relative to total variance of y. However, in the presence of endogenous right-hand variables the covariance effect must be included with the disturbance variance.
6. These parameter values for equation 1, generates a fixed relative fit equal to 0.70, and varying degrees of relative fit and correlation between $\ln y_{1t}$ and u_{2t} in the second equation -- denoted in Tables 1-9 by FIT and ρ respectively.
7. The data in rows labeled "Actual" are the first two moments of the generated y_{T+1}, sample, $\Gamma = 1, 2, \ldots, 2500$. Those are the values for which the errors were calculated for each repetition. The mean value is an estimate of the mean of y_{T+1}, which is included for comparison with the bias estimates, and the standard deviation (in the RMSE) column represents an estimate of the lower bound for the predictors' RMSE.

8. To be able to interpret the simulated distribution obtained as conditional on Y_{1T}, Y_{1T} (as well as X_T and X_{T+1}) was fixed at unconditional mean for all samples generated.

9. Solution of the system for any given error term was by the standard Gauss-Seidel algorithm with the appropriate autocorrelation correction applied.

10. The simple estimator proposed by Newey and West (1987) appears to be the most useful in this context.

11. The data in rows labeled "Actual" are the first two moments of the generated y_{T+1}, sample, $r = 1, 2, \ldots, 2500$. These are the values for which the errors were calculated for each repetition. The mean value is an estimate of the mean of y_{T+1}, which is included for for comparison with the bias estimates, and the standard deviation (in the RMSE) column represents an estimate of the lower bound for the predictors' RMSE.

12. The data in rows labeled "Actual" are the first two moments of the generated y_{T+1}, sample, $r = 1, 2, \ldots, 2500$. These are the values for which the errors were calculated for each repetition. The mean value is an estimate of the mean of y_{T+1}, which is included for for comparison with the bias estimates, and the standard deviation (in the RMSE) column represents an estimate of the lower bound for the predictors' RMSE.

13. Missing values are for unavailable t = 1992 data. For variables in percentage units, INFL, GNPGR, UERA, & TBILL, the error itself is reported rather than percent error.

REFERENCES

Brown, B. W. and Mariano, R. S. (1984). Residual-based procedures for prediction and estimation in a nonlinear simultaneous system. *Econometrica. 52*, 321-343.

Brown, B. W. and Mariano, R. S. (1989). Predictors in dynamic nonlinear models: Large-sample behavior. *Econometric Theory. 5*, 420-452.

Fair, R. C. (1986). Evaluating the predictive accuracy of models. In Z. Griliches and M.D. Intriligator, (Eds.), *Handbook of Econometrics*, Vol. III. Elsevier Science Publishers BV, 1979-1995.

Lin, J. L. and Granger, C. W. J. (1992). Forecasting from non-linear models in practice. Unpublished Manuscript.

Mariano, R. S. and Brown, B. W. (1983). Asymptotic behavior of predictors in a nonlinear simultaneous system. *International Economic Review. 24*, 523-536.

Mariano, R. S. and Brown, B. W. (1991a). Stochastic-simulation tests of nonlinear econometric models. In L.R. Klein (Ed.), *Comparative performance of U.S. econometric models.* Oxford University Press. pp. 250-259.

Mariano, R. S. and Brown, B. W. (1991b). Interval and quantile prediction in nonlinear simultaneous systems. Unpublished discussion paper.

Newey, W. K. and West, K. D. (1987). A simple, positive semi-definite, heteroskedasticity and autocorrelation consistent covariance matrix. *Econometrica. 55*, 703-708.

Pagan, A. and Ullah, A. (1991). *Nonparametric and semiparametric methods in econometrics.* Cambridge University Press.

Rothenberg, T. J. (1984). Approximating the distributions of econometric estimators and test statistics. In Z. Griliches and M.D. Intriligator. (Eds.), *Handbook of econometrics.* Vol. II. Elsevier Science Publishers BV; 881-935.

On the Distribution of Horse Qualities at Racetracks: An Analysis of Cournot-Nash Equilibria

Richard E. Quandt

It has been noted in numerous studies that in parimutuel betting at racetracks the fraction of all win-bet moneys wagered on the various horses bears a close relationship to their objective winning probabilities (Ziemba and Hausch, 1987; Asch, Malkiel, Quandt, 1982, 1984; Fabricand, 1979; Ali, 1977). If we denote the fraction of all moneys bet to win on horse i by s_i and its objective winning probability by p_i, the s_is are, in general, excellent predictors of the p_is and can be used effectively as estimates of the p_is.[1].

In this paper we concern ourselves with a somewhat different, albeit related, question. If one examines the fractions of the total win betting pool that are bet on the various horses in a given race, one is struck by the fact that certain patterns of the s_i in a given race are much more frequent than others. Thus, one would expect to encounter races only infrequently in which all horses have the same percentage of the win pool bet on them; by the same token, we do not often see races in which 99.9 percent of the win pool is bet on a particular horse, with the remaining 0.1 percent bet on the remaining horses.

What are the typical patters of the s_is? We examine two datatsets[2] representing 712 thoroughbred races at the Atlantic City racetrack in 1978 and 706 harness races at Meadowlands in 1984. In the Atlantic City sample there are races containing from 5 to 12 horses, whereas at Meadowlands the number of horses per race ranged from 6 to 12. Each sample is partitioned into subsets with exactly n ($n = 5, \ldots, 12$ for Atlantic City and $n = 6, \ldots, 12$ for Meadowlands) horses. In each race with n horses we rank the s_is from highest to lowest. We then average within each subset the highest s_is, separately average the second highest s_is, third highest s_is, etc., and also compute the standard deviations of these s_is. The mean s_is and the standard deviations of the s_is are displayed from least favored to most favored in Tables 1 and 2 for

Table 1: Mean s_i for Atlantic City

Races	Mean s_i s $n =$											
	1	2	3	4	5	6	7	8	9	10	11	12
44	.066	.127	.180	.244	.383							
141	.046	.076	.115	.161	.238	.363						
145	.033	.060	.086	.117	.162	.214	.328					
114	.025	.040	.060	.086	.119	.154	.205	.311				
94	.020	.031	.045	.066	.088	.117	.146	.191	.297			
89	.016	.024	.035	.048	.067	.088	.111	.140	.186	.285		
34	.011	.017	.027	.036	.048	.068	.085	.114	.148	.182	.263	
50	.010	.014	.019	.026	.035	.049	.068	.094	.116	.142	.179	.248

Atlantic City and in Tables 3 and 4 for Meadowlands. Thus, for example, 0.383 in row 1 of Table 1 is the average over the sample of the fractions of moneys bet on the favorite in five-horse races. The means exhibit certain fairly striking patterns. Thus for Atlantic City, the average percentage of the win-pool bet on the favorite declines monotonically with the number of horses in the race. For Meadowlands the decline is also monotonic, except for races with 11 or 12 horses (categories in which there were few observations). The ratio of the average percentage on the favorite to that on the second-favorite ranges from 0.64 to 0.72 at Atlantic City and from 0.74 to 0.81 at Meadowlands; the

Table 2: Standard Deviations of s_i for Atlantic City

Races	Standard Deviations $n =$											
	1	2	3	4	5	6	7	8	9	10	11	12
44	.024	.033	.039	.048	.090							
141	.021	.027	.030	.035	.045	.079						
145	.016	.024	.028	.025	.032	.041	.074					
114	.012	.019	.022	.025	.027	.028	.041	.078				
94	.011	.013	.016	.020	.021	.023	.026	.036	.074			
89	.007	.010	.013	.016	.020	.018	.021	.024	.034	.061		
34	.005	.007	.011	.014	.013	.018	.021	.022	.024	.030	.057	
50	.003	.005	.008	.010	.011	.014	.016	.022	.019	.020	.027	.043

Table 3: Mean s_i for Meadowlands

Races	Mean s_is $n =$											
	1	2	3	4	5	6	7	8	9	10	11	12
10	.046	.061	.094	.166	.230	.404						
16	.031	.051	.066	.100	.150	.227	.376					
60	.024	.040	.056	.079	.109	.149	.206	.337				
157	.016	.026	.039	.056	.080	.107	.143	.199	.333			
429	.013	.020	.030	.042	.060	.081	.107	.140	.193	.315		
20	.007	.009	.013	.019	.028	.045	.061	.097	.141	.199	.381	
12	.007	.010	.012	.017	.026	.039	.055	.071	.102	.124	.191	.346

corresponding figures for the ratio of the second-favorite to the third-favorite are 0.52--0.61 and 0.65--0.72, respectively. This suggests a certain order in the pattern of the s_is which is incompatible with such extreme outcomes as those consisting of all horses being bet on nearly equally or of having one horse bet on to the (almost) complete exclusion of the others.[3]

The question to which we address outselves is this: are there simple assumptions and models concerning how horses are entered in races that are at least broadly compatible with the observed facts?

Table 4: Standard Deviations of s_i for Meadowlands

Races	Standard Deviations $n =$											
	1	2	3	4	5	6	7	8	9	10	11	12
10	.018	.018	.034	.061	.061	.117						
16	.018	.028	.026	.038	.051	.058	.095					
60	.015	.021	.021	.026	.032	.031	.048	.111				
157	.009	.012	.017	.021	.024	.025	.030	.041	.093			
429	.007	.011	.014	.017	.020	.021	.026	.031	.043	.089		
20	.003	.004	.008	.010	.014	.020	.022	.033	.036	.056	.099	
12	.003	.006	.008	.012	.016	.018	.027	.025	.033	.035	.045	.126

In Section 2 we outline a simple model of how horses are entered in races. In Section 3 we make the model more specific so as to allow computer simulations of it. Section 4 presents some results from the simulations. Section 5 contains brief conclusions.

II. A Model of Optimal Entry

We begin by assuming that each racehorse owner possesses a large stable of horses, which have intrinsic abilities denoted by α_i. The measure of ability is a nonnegative continuous variable for each stable. We may think of α_i as the decision variable for the stable that owns the ith horse, and indicates whether the stable wishes to enter in the race a relatively more or less "able" horse.[4]

The problem of entering horses in a given race will have the following stylized form. The racetrack is assumed to have decided that a particular race will be an n-horse race ($n = 5, \ldots, 12$ typically) and to have invited n stables to enter one horse each. Each stable is prepared to enter a horse and has to decide on the optimal horse to enter in terms of ability in the light of the purse and claiming price set for that race.[5]

We now introduce some notation and make some assumptions.

1. The objective probability that horse i wins the race is denoted by p_i. In general, p_i depends on the abilities of all the horses in the race, denoted by $p_i = p_i(\alpha_1, \ldots, \alpha_n)$, with (a)$\partial p_i / \partial \alpha_i > 0$, $\partial^2 p_i / \partial \alpha_i^2 < 0$ and $\partial p_i / \partial \alpha_j < 0$ for $i \neq j$. It is further reasonable to require that (b) if $\alpha_i = \alpha_j$, then $p_i = p_j$, (c) if $\alpha_i = 0$, then $p_i = 0$, and (d) for any set of finite α_js, ($j \neq i$), as $\alpha_i \to \infty$, $p_i \to 1$.

2. The subjective probability s_i that horse i will win is measured by the fraction of the win betting pool that is bet on horse i. If D_i denotes the odds on horse i and if t is the track "take" (17--23 percent at most U.S. racetracks), then

$$D_i = \frac{1-t}{s_i} - 1$$

3. We denote by P the purse and by L the claiming price (relevant in claiming races only). The purse is the amount of money distributed among the top finishers. In the analysis that follows in the present Section, we assume that the purse accrues entirely to the winner. In the empirical analysis we also consider the case when δP goes to the winner and $(1-\delta)P$ to the runner-up. The analysis of this case is in the Appendix.[6]

4. The cost of entering a horse with ability α_i in a race is denoted by $c_i(\alpha_i)$. This includes entry fees, jockey's fees, transport cost to and from the racetrack, insurance, the expected cost of injury, training costs, amortization of the purchase price, etc. At least some of these, such as amortization of purchase price and insurance cost, are likely to depend on the horse's ability.[7] We assume that $c'_i > 0$ and $c''_i > 0$.

5. We denote by b_i the amount bet by the owner on his or her own horse.[8] We assume that b_i also depends on the abilities of all horses in the race, with $b_i = b_i(\alpha_1, \ldots, \alpha_n)$, where $\partial b_i/\partial \alpha_i > 0$, $\partial b_i/\partial \alpha_j < 0$ for $j \neq i$, and $b_i \geq 0$.

6. We finally denote by $q(\alpha_i)$ the probability that horse i will be claimed, with $q' > 0$, and $q'' < 0$.

The stable owner is assumed to be an expected profit maximizer. Let π_{it} denote the profit of the ith stable in th t^{th} race (with one race occurring per time period), and let V_{it} be the expected profit from so doing. Then, letting β be the discount factor,

$$\pi_{it} = \begin{cases} P + D_i b_i - c_i(\alpha_i) + L, & \text{if } i \text{ wins and is claimed;} \\ P + D_i b_i - c_i(\alpha_i) + \beta V_{it+1}, & \text{if } i \text{ wins and is not claimed;} \\ -b_i - c_i(\alpha_i) + L, & \text{if } i \text{ does not win and is claimed;} \\ -b_i - c_i(\alpha_i) + \beta V_{it+1}, & \text{if } i \text{ does not win and is not claimed.} \end{cases}$$

$$(2.1)$$

The probabilities of the four events are pq, $p(1-q)$, $(1-p)q$, $(1-p)(1-q)$. Hence, abbreviating $p_i(\alpha_1, \ldots, \alpha_n)$ by $p_i(\alpha)$ and $b_i(\alpha_1, \ldots, \alpha_n)$ by $b_i(\alpha)$, we have

$$V_{it} = p_i(\alpha)[P + D_i b_i(\alpha)] - c_i(\alpha_i)$$
$$- (1 - p_i(\alpha))b_i(\alpha) + q(\alpha_i)L + (1 - q(\alpha_i))\beta V_{it+1}. \quad (2.2)$$

If the stable is in long-run equilibrium, $V_{it} = V_{it+1}$, and

$$V_{it} = [1 - \beta(1 - q(\alpha_i))]^{-1} [p(\alpha_i)[P + D_i b_i(\alpha)]$$
$$- c_i(\alpha_i) - (1 - p_i(\alpha))b_i(\alpha) + q(\alpha_i)L] \quad (2.3)$$

For simplicity we now assume that $b_i = 0$; i.e., the ith stable does not bet at the track. Assuming that all α_j, $j \neq i$, are given, the ith stable optimizes by setting $\partial V_{it} / \partial \alpha_i = 0$. We have

$$\frac{\partial V_{it}}{\partial \alpha_i} = [1 - \beta(1-q(\alpha_i))]^{-2} \left\{ \left[\frac{\partial p_i}{\partial \alpha_i} P - c_i'(\alpha_i) + q_i'(\alpha_i)L \right] [1 - \beta(1 - q(\alpha_i))] \right.$$

$$\left. -\beta q'(\alpha_i)[p_i(\alpha)P - c_i(\alpha_i) + q(\alpha_i)L] \right\} = 0. \qquad (2.4)$$

We also have

$$\frac{\partial^2 V_{it}}{\partial \alpha_i^2} = [1 - \beta(1-q(\alpha_i))]^{-2} \left\{ \left[\frac{\partial^2 p_i}{\partial \alpha_i^2} P - c_i''(\alpha_i) + q_i''(\alpha_i)L \right] [1 - \beta(1 - q(\alpha_i))] \right.$$

$$\left. -\beta q''(\alpha_i)[p_i(\alpha)P - c_i(\alpha_i) + q(\alpha_i)L] \right\}. \qquad (2.5)$$

which must be negative; this is automatically satisfied if q does not depend on α (or if q'' is sufficiently small).[9]

From (2.3) and the second order condition we immediately obtain certain comparative statics results which are summarized in the following.

Proposition 1. *If the second order condition is satisfied,*
(a) $\partial \alpha_i / \partial L > 0$; (b) *If* $\partial \log p_i / \partial \alpha_i > \partial \log q / \partial \alpha_i$, *then* $\partial \alpha_i / \partial P > 0$;
(c) *If* $\partial^2 p_i / \partial \alpha_i \partial \alpha_j > 0$, *then* $\partial \alpha_i / \partial \alpha_j > 0$ *for* $j \neq i$.

Proof. See Appendix.

Thus an increase in the claiming price will increase the optimal ability of the ith entry (holding the abilities of the other horses constant). An increase in the purse will have the same effect if the proportionate effect on the winning probability from a unit increase in ability is greater than the proportionate effect on the probability that the horse will be claimed. It follows that in non-claiming races an increase in the purse unambiguously increases the optimal ability. Finally, the last result says that the ability-reaction-functions are positively sloped. These results are what one would expect and justifies the common view that higher purses and claiming prices make for better quality races.

Proposition 2. If the cost function of the ith stable is a function of a parameter θ, say $c_i (\alpha_i, \theta)$, such that $\partial^2 c_i / \partial \alpha_i \partial \theta > 0$, *then in non-claiming races* $\partial \alpha_i / \partial \theta < 0$.

Proof.

Under the assumptions, (2.4) is

$$(1 - \beta)^{-1} \left[\frac{\partial p_i(\alpha)}{\partial \alpha_i} P - \frac{\partial c_i(\alpha_i, \theta)}{\partial \alpha_i} \right] = 0.$$

Differentiating totally,

$$\left[\frac{\partial^2 p_i(\alpha)}{\partial \alpha_i^2} P - \frac{\partial^2 c_i(\alpha_i, \theta)}{\partial \alpha_i^2} \right] d\alpha_i - \frac{\partial^2 c_i(\alpha_i, \theta)}{\partial \alpha_i \partial \theta} d\theta = 0. \quad (2.6)$$

The bracket in (2.6) is negative by previous assumptions. It follows that $\partial \alpha_i / \partial \theta < 0$.

While an increase in marginal costs diminishes horse quality in non-claiming races, a corresponding proposition does not hold for claiming races unless $q(\alpha_i) = 0$, since a (potential) decrease in quality also affects the probability that the horse will be claimed.

We conclude this Section by considering some relevant n-player solution concepts. The first of these is the Cournot-Nash equilibrium in horse quality. We write the value to the ith stable of entering a horse in the race as V_i $(\alpha_1, \ldots, \alpha_i, \ldots, \alpha_n)$. A Cournot-Nash equilibrium is then an n-tuple of qualities $(\alpha_1^*, \ldots, \alpha_n^*)$ with the property that

$$V_i(\alpha_1, \ldots, \alpha_i, \ldots, \alpha_n) \leq V_i (\alpha_1^*, \ldots, \alpha_i^*, \ldots, \alpha_n^*), \quad (2.7)$$

for $i = 1, \ldots, n$; i.e., α_i^* is the best response by the ith stable when the other stables are employing $\alpha_1^*, \ldots, \alpha_{i-1}^*, \alpha_{i+1}^*, \ldots, \alpha_n^*$ and this is true for all stables. We consider the properties of the Cournot-Nash equilibrium with a simplified model.

An alternative solution concept is the maximin solution. According to this, the ith player chooses α_i^* so that

$$\alpha_i^* = \arg \max_{\alpha_i} \{ \min_{\substack{\alpha_j \\ j \neq i}} V_{it}(\alpha_i, \ldots, \alpha_n) \}. \quad (2.8)$$

We briefly compare this solution with the Cournot-Nash solution.

III. A Particular Model

In order to investigate the model further, we now make some specific assumptions concerning the functions $p_i(\alpha)$, $c_i(\alpha_i)$ and $q(\alpha_i)$.

The Function $p_i(\alpha)$. A function that satisfies the assumptions of 1 in Section 2 is

$$p_i(\alpha) = \frac{\alpha_i}{\sum_{j=1}^{n} \alpha_j} \tag{3.1}$$

The Function $c_i(\alpha_i)$. We assume that the cost function is quadratic and is given by

$$c_i(\alpha_i) = k_i \alpha_i^2, \tag{3.2}$$

where the k_i are positive constants.

The Function $q(\alpha_i)$. We assume that the probability that a horse with ability α_i is claimed is

$$q(\alpha_i) = 1 - e^{-\gamma \alpha_i}, \tag{3.3}$$

where $\gamma > 0$, which has the properties required of q.

The Cournot-Nash Solution. Under the assumption of (3.1)--(3.3), the first order condition for the ith stable, $i = 1, \ldots, n$, (Eqn. (2.4)) is

$$\frac{\partial V_{it}}{\partial \alpha_i} = [1 - \beta e^{-\gamma \alpha_i}]^{-2} \left\{ \left[P \sum_{j \neq i} \alpha_j \Big/ \left(\sum_{j=1}^{n} \alpha_j \right)^2 - 2 k_i \alpha_i + L \gamma e^{-\gamma \alpha_i} \right] (1 - \beta e^{-\gamma \alpha_i}) \right.$$

$$\left. - \beta \gamma e^{-\gamma \alpha_i} \left[P \alpha_i \Big/ \sum_{j=1}^{n} \alpha_j - k_i \alpha_i^2 + L (1 - e^{-\gamma \alpha_i}) \right] \right\} = 0, \tag{3.4}$$

If no claiming is allowed and the purse is not split among several runners, we have

$$V_{it} = (1 - \beta)^{-1} \left[\frac{\alpha_i P}{\sum_{j=1}^n \alpha_j} - k_i \alpha_i^2 \right],$$

(3.5)

and (3.4) simplifies to

$$P \sum_{j \neq i} \alpha_j = 2 k_i \alpha_i \left(\sum_{j=1}^n \alpha_j \right)^2, \qquad i = 1, \ldots, n.$$

(3.6)

The Cournot-Nash solution is given by the simultaneous solution of Eqn. (3.6) for α_j, $j = 1, \ldots, n$. Obviously, $\alpha_j = 0$ is one such solution; however, we rule out such "null races," since the first order and second order conditions are not defined at that point.

Proposition 3. (a) *If* $(\alpha_1^*, \ldots, \alpha_n^*)$ *is a Cournot-Nash solution of (3.6), and P changes to* ρP *($\rho > 0$), then the Cournot-Nash solution becomes* $(\sqrt{\rho} \, \alpha_1^*, \ldots, \sqrt{\rho} \, \alpha_n^*)$. *(b) If all k_i change to* ρk_i, *the solution becomes* $(\alpha_1^* / \sqrt{\rho}, \ldots, \alpha_n^* / \sqrt{\rho})$. *(c) In both cases the optimal probabilities remain unchanged.*

The proposition is obvious by substitution. It follows that a proportionate increase in all costs causes a reduction in the abilities of all horses and an increase in the purse causes an increase in all abilities. Exactly the same result is obtained if the purse is split between the winner and the runner-up (see Appendix for the derivation).

The behavior of the solution is more complicated if only one cost changes. This is easily shown for a two-horse race. Eqn. (3.6) becomes

$$P\alpha_2 = 2 k_1 \alpha_1 (\alpha_1 + \alpha_2)^2, \quad \text{and} \quad P\alpha_1 = 2 k_2 \alpha_2 (\alpha_1 + \alpha_2)^2.$$

Hence the solution for α_1 is

$$\alpha_1 = \left[\frac{P \sqrt{k_1 / k_2}}{2 k_1 \left(1 + \sqrt{k_1 / k_2} \right)} \right]^{1/2}$$

(3.7)

The effect of the changes in P or k_i considered in Proposition 3 are easily verified. Now consider a change in k_1 alone. It is clear that as k_1 increases, α_1 falls. However, when k_2 increases, it is easily shown that α_1 increases or decreases, depending on whether $k_1 > k_2$ or $k_1 < k_2$. The intuition for this is as follows: when $k_2 > k_1$, α_1 is already greater than α_2; hence a further increase in

k_2 allows stable 1 to reduce the ability of the horse it enters in the race. Corresponding analyses in races with more than two horses, races with claiming and races where the purse is split are substantially more difficult and we shall attempt to shed some light on these by numerical simulations.

The Maximin Solution. In the simple case in which the purse is not split and no claiming is allowed, it is clear that $\min_{\alpha_j, \, j \neq i} V_{it}$ is achieved if the α_j, $j \neq i$, are as large as possible. We assume that the players observe a zero-profit constraint; hence the α_js that minimize V_{it}, consistent with (3.5) yielding zero-profit, are given, for given α_i^*, by the simultaneous solution of

$$ k_j \, \alpha_j \, (\alpha_1 + \ldots + \alpha_i^* + \ldots + \alpha_n) = P, \quad j \neq i \tag{3.8} $$

for the variables $(\alpha_1, \ldots, \alpha_{i-1}, \alpha_{i+1}, \ldots, \alpha_n)$. For this equation system we have for α_l, $l \neq$ i,

$$ \alpha_l = \frac{k_l \, \alpha_i^* + \sqrt{(k_l \alpha_i^*)^2 + 4 \, P k_1 \, s_{il}}}{2 \, k_1 \, s_{il}}, \tag{3.9} $$

where $s_{il} = \sum_{j \neq 1} (k_l / k_j)$. Hence $\alpha_j = k_l \alpha_l / k_j$, $j \neq$ i, and definining

$A_i = \sum_{j \neq 1} \alpha_j$, the ith stable maximizes min V_{it} by solving for α_i the equation[10]

$$ A_i \, P - 2 \, k_i \, \alpha_i^* \, (A_i + \alpha_i^*)^2 = 0. \tag{3.10} $$

IV. Simulation Results

We need to assume particular values for the various parameters in the model in order to perform numerical simulations. We first consider non-claiming races.

Non-claiming Races, Cournot-Nash Solutions. In the base-case scenario we assume that the purse $P = 8.0$. (We need not make any assumptions here concerning L, β or γ). The costs associated with the n horses in a race are assumed to be independently distributed as lognormal. Each cost is drawn as me^u, where m is a parameter and u is distributed as $N(0, \sigma^2)$, with $\sigma^2 = 1.0$ in the base case. The parameter m is set at $(r_1/2 + r_2)e^{-\sigma^2/2}$, where $r_1 = 1.0$ and $r_2 = 0.1$ in the base case. The expected value of costs then is $(r_1 / 2 + r_2)$ and their variance is $(r_1 / 2 + r_2) (e^{\sigma^2/2} - 1)$.[11] We generate 10 sets of races with different n and δ values (where δ is the fraction of the purse paid to the winner and $1 - \delta$ the fraction paid to the runner-up), and average over the comparable set of races the ordered probabilities for the various horses. The means and standard deviations are displayed in Tables 5 and 6.

Both the means and the standard deviations are remarkably similar to the corresponding figures in Tables 1 and 2 (Atlantic City). If we let the figures in Table 1 be denoted by p_i and the corresponding figures in Table 5 be e_i, we may test the null hypothesis that the two sets of figures are the same by testing in the regression $p_i = a + be_i + u_i$ the null hypothesis $H_0:$, $a = 0$, $b = 1$. Using $n = 8$ and $\delta = 0.7$ we obtain the estimates $\hat{\alpha} = -0.0056$, $\hat{b} = 1.043$ and $R^2 = 0.994$. For the F-statistic of the test of H_0 we obtain $F_{2,6} = 1.043$, not permitting rejection of H_0 at any reasonable level of significance. Comparable results hold for the other comparisons. It thus seems that Cournot-Nash behavior, together with the assumption of lognormal costs, is sufficient for generating data that look very much like those observed in reality.

The abilities of the horses in the Cournot-Nash equilibrium with $\delta = 1$ are uniformly larger than in the case when the purse is split. For example, when n = 8, the sums of the abilities over the replications for $\delta = 0.7$ are 6.11, 9.49, 9.82, 9.96, 10.52, 10.88, 10.93, 11.30, 12.28, 12.35, while for $\delta = 1$ they are 6.36, 10.02, 10.19, 10.32, 10.93, 11.30, 11.32, 12.81, 12.95. This does not mean, of course, that all the abilities in a given race increase when δ goes from 0.7 to 1.0. In fact, in a given race, only the largest or the largest 2–3 abilities increase and do so markedly, while the remainder exhibit no particular pattern. The implication is clear that with $\delta = 1$ low-cost stables have a much enhanced incentive to enter able runners while the incentives of higher cost stables are not much affected.

Table 5: Base Case Average Probabilities

n	δ				Probabilities						
6	0.7	.055	.084	.113	.154	.232	.361				
6	1.0	.048	.075	.104	.145	.235	.394				
8	0.7	.028	.052	.061	.077	.117	.160	.208	.298		
8	1.0	.024	.047	.055	.071	.111	.157	.213	.322		
10	0.7	.024	.035	.044	.056	.074	.085	.106	.133	.173	.269
10	1.0	.022	.032	.041	.053	.070	.082	.102	.132	.176	.291

We also consider some alternative scenarios.

1. When the purse increases from 8 to 12, the probabilities, and hence the relative odds, remain unchanged as shown by Proposition 3; however, the abilities of all horses increase by about 22.5 percent.

Table 6: Base Case Standard Deviations

n	δ	Standard Deviations								
6	0.7	.025	.024	.034	.033	.044	.084			
6	1.0	.024	.026	.037	.039	.051	.101			
8	0.7	.016	.015	.017	.022	.023	.025	.042	.058	
8	1.0	.015	.015	.017	.023	.025	.030	.048	.070	
10	0.7	.016	.014	.017	.014	.013	.019	.018	.016.	.036	.051
10	1.0	.015	.014	.017	.015	.014	.020	.020	.018	.040	.060

2. We consider variations in the mean costs and in the variance of the costs. The variance alone can be increased by raising σ^2. The mean alone can be increased by raising r_1 and r_2 and simultaneously diminishing σ^2. Alternatively, we raise r_1 to 2.0 and r_2 to 0.2 and lower σ^2 to 0.62011. Raising σ^2 has the effect of substantially increasing average abilities and substantially increasing the winning probability of the favorite: for $n = 8$ and $\delta = 0.7$ the probabilities are 0.0140, 0.0325, 0.0409, 0.0576, 0.1032, 0.1606, 0.2281, 0.3632, making this case resemble slightly more the figures for 8-horse races at the Meadowlands (Table 3). Raising mean costs only reduces, as expected, the average abilities of all horses and also reduces the winning probability of the favorite.

3. We generate costs from the uniform distribution from r_1 to $r_1 + r_2$. For values of $r_1 = 0$ and $r_2 = 0.5$ the average probabilities that emerge in ten replications are 0.0570, 0.0654, 0.0713, 0.0826, 0.0944, 0.1775, 0.3174, which are similar to the corresponding figures at Meadowlands. The null hypothesis that $a = 0$ and $b = 1$ in the regression $p_i = a + be_i + u_i$ yields an F-statistic of 2.8329, not permitting rejection of H_0. An increase in mean costs (with variance constant) diminishes average abilities and tends to make horses more alike; this has qualitatively exactly the same effect as in the lognormal case. In order to have favorites that stand out, there must be sufficient variability in costs.

4. We also consider cost variations that affect a single horse only. We accomplish this by generating a set of costs for n horses and then alternately consider lower and higher costs for a particular, designated horse, keeping all others constant. The general conclusion is that as the costs associated with a particular horse increase, the abilities and winning probabilities of all other horses increase. If the designated horse is an originally low-cost horse, the effect on the others is somewhat more pronounced than when the designated horse is already a high-cost horse.

5. Finally, we note that an endogenous determination of n is possible in the framework of the model. It is necessary to assume that there is a fixed cost of entering a horse in a race in addition to the variable cost. It is then the case that as the number of horses increases, eventually one or more V_i-values associated with the Cournot-Nash solution become negative. This limits the size of the race, since horses with negative optimal V_i's would not be entered.

Claiming Races, Cournot-Nash Solution. For simplicity, we consider in this category only lognormally distributed costs with $\sigma^2 = 1.0$, $r_1 = 1.0$, $r_2 = 0.1$, $p = 8.0$ and $n = 8$, making these cases comparable with the base-case scenario considered earlier. We start by setting $L = 8.0$, $\gamma = 0.25$ and $\beta = 0.95$. The value of β is not directly interpretable because the time units in the model refer to the time that elapses between successive races; if this time were taken to be, say, one month, the chosen value of β would represent a high discount rate. The significance of γ cannot be judged in the abstract either. In the previous, nonclaiming races the range of abilities is approximately from 5 to 13; for horses in this ability range, a γ-value of 0.25 implies a probability of being claimed ranging from 0.71 to 0.96 which are rather high values. We also consider $\gamma = 0.1$ and 0.01; the claiming probabilities then range over 0.39--0.72 for the former and over 0.05--0.12 for the latter.

Comparing the base case with no claiming with the base case with claiming ($\beta = 0.25$) shows that claiming substantially reduces the average abilities of all horses. This is to be expected when the probability is high that a horse will be lost by the owner because of a claim; other things being equal, it is preferable to lose a less able horse to claiming than a more able one. As one might expect, the effect is much attenuated when β is small. A reduction in the claiming price from 8.0 to 4.0 reduces the abilities of all horses, as it should, since now the "compensation" for losing the horse due to a claim is diminished. In all cases with claims, the probability of the favorite is reduced; however, as before, an increase in σ^2 increases the winning probability of the favorite. Thus, for example, with $L = 8.0$ and $\sigma^2 = 2.0$, the average probabilities over ten replications become 0.026, 0.051, 0.060, 0.077, 0.115, 0.156, 0.206, 0.309, which is nearly indistinguishable from the 8-horse races in the Atlantic City sample.

Maximin Solutions. We consider only nonclaiming races with the entire purse going to the winner ($\delta = 1.0$) and with $n = 8$. For the standard case the equilibrium probabilities are 0.097, 0.121, 0.122, 0.122, 0.122, 0.122, 0.122, 0.173, which differs markedly from the facts as well as from the Cournot-Nash solution. Regressing the p_is for Atlantic City on the e_is as before yields $\hat{a} = -0.3557$, $\hat{b} = 3.8454$, $R^2 = 0.7196$ and testing H_0: $a = 0$, $b = 1$ yields an F-statistic $F(2,6) = 5.647$ which is significant at the 0.05 level. We have to conclude that the maximin solution does not yield predictions that agree with the observed facts.[12] Increasing σ^2 has an effect that is in the same direction as before, i.e., the probability of the favorite winning increases, but the effect is small and even for $\sigma^2 = 3.0$ we obtain an array of probabilities that is not compatible with the data.[13]

V. Conclusions

We have attempted to shed light on some empirical regularities in the array of odds, or what is equivalent, winning probabilities in horseraces. Assuming that costs are lognormally distributed and that we observe Cournot-Nash equilibria yields predictions of these probabilities that are quite close to the observed facts. By contrast, maximin behavior, which does not represent a "best reply" to the other players, does not produce good agreement with the facts. This does not, of course, mean that our results *prove* that Cournot-Nash behavior is actually producing the results. Even simpler assumptions, such as the assumption that abilities are lognormally distributed can produce broadly similar results (although extensive experimentation suggests that no *single* assumption about the mean and variance of the generating lognormal is capable of producing well-fitting results for all values of n). What our results do show is that a relatively simple model with some economic content can produce good agreement with the facts.

Of course, the model is extremely simple and abstracts from a number of important features. In reality, horse qualities are multidimensional and are not continuously variable within a stable, and the assumption that a long-run Cournot-Nash equilibrium is achieved is not an innocuous one. In the simulation experiments we abstract from the owner betting on his own horse and give only limited role to the splitting of the purse among several horses. The parameter values assumed for the simulations are purely illustrative and do not purport to represent empirical reality. Furthermore, the empirical data that we compare with our simulation results represent a mixture of claiming and nonclaiming races. Nevertheless, it is noteworthy that even a minimalist model of the type employed can produce the observed results.

Appendix

Proof of Proposition 1.

(a) Denote $[1 - \beta(1 - q(\alpha_i))]^{-2}$ by D, which is clearly positive. Differentiating (2.4) totally with respect to α_i and L and solving for $d\alpha_i/dL$ we obtain

$$\frac{\partial \alpha_i}{\partial L} = -\frac{q'(a_i)(1-\beta)D}{\partial^2 V_{it}/\partial \alpha_i^2} \tag{A.1}$$

which is positive if the second order condition holds.

(b) Differentiation with respect to α_i and P yields

$$\frac{\partial \alpha_i}{\partial P} = \frac{D\left[(1-\beta)\frac{\partial p_i}{\partial \alpha_i} + \beta\left(\frac{\partial p_i}{\partial \alpha_i} q(\alpha_i) - p(\alpha)q'(\alpha_i)\right)\right]}{\partial^2 V_{it}/\partial \alpha_i^2} \tag{A.2}$$

The first term in the bracket is positive by assumption. The second term will be positive if $\partial \log p_i / \partial \alpha_i > \partial \log q(\alpha_i)/\partial \alpha_i$. This is a sufficient condition for the result.

(c) Differentiating with respect to α_i and α_j, we obtain

$$\frac{\partial \alpha_i}{\partial \alpha_j} = - \frac{D\left[\frac{\partial^2 p_i}{\partial \alpha_i \partial \alpha_j} [1 - \beta (1 - q(\alpha_i))] - \beta q (\alpha_i) \frac{\partial p_i}{\partial \alpha_i} P \right]}{\partial^2 V_{it} / \partial \alpha_i^2}. \tag{A.3}$$

Since $\delta \, p_i / \, \delta \, \alpha_j < 0$ by assumption, the result follows if $\partial^2 p_i / \, \partial \alpha_i \partial \alpha_j > 0$.

The Case of Split Purses. We assume that the winner get $\delta \, P$ and the runner-up $(1 - \delta)P$, $0 \le \delta \le 1$; let p_i^1 be the probability that horse i wins and p_i^2 that it is second. Then

$$\pi_{it} = \begin{cases} \delta P + D_i b_i - c_i(\alpha_i) + L, & \text{if } i \text{ wins and is claimed;} \\ \delta P + D_i \, b_i - c_i(\alpha_i) + \beta V_{it+1}, & \text{if } i \text{ wins and is not claimed;} \\ (1 - \delta) P - c_i(\alpha_i) + L, & \text{if } i \text{ is second and is claimed;} \\ (1 - \delta) P - c_i(\alpha_i) + \beta V_{it+1}, & \text{if } i \text{ is second and is not claimed;} \\ -b_i - c_i(\alpha_i) + L, & \text{if } i \text{ is not first or second and is claimed;} \\ -b_i - c_i(\alpha_i) + \beta V_{it+1}, & \text{if } i \text{ is not first or second and is not claimed;} \end{cases}$$

Proceeding as before,

$$V_{it} = [1 - \beta(1 - q(\alpha_i))]^{-1} \left[[p_i^1 \delta + p_i^2 (1-\delta)] P + b_i [D_i p_i^1 - (1 - p_i^1 - p_i^2)] - c_i + Lq(\alpha_i)\right] \tag{A.4}$$

Finally, we simplify by assuming that $b_i = 0$ and that there is no claiming, i.e., that $q(\alpha_i) = 0$. Using the Harville (1973) approximation, we approximate p_i^2 as $\sum_{j \ne 1} (p_i \, p_j) / (1 - p_j)$ and Eqn. (3.7) becomes

$$P\left[\delta \sum_{k=1} \alpha_k + (1 - \delta) \sum_{j \ne i} \frac{\alpha_j \sum_k \alpha_k \sum_{k \ne j} \alpha_k - \alpha_j \alpha_i \left(\sum_{k \ne j} \alpha_k + \sum_k \alpha_k \right)}{\left(\sum_{k \ne j} \alpha_k \right)^2} \right] = 2 \, k_i \alpha_i \left(\sum_k \alpha_k \right)^2.$$

Richard E. Quandt, Department of Economics, Princeton University, Princeton, New Jersey 08544-1021.

ACKNOWLEDGEMENTS

I am grateful to P. Asch, M. La Manna, B. G. Malkiel, G. Norman and J. Swinkels for helpful comments.

NOTES

1. Note, however, the well-known underbetting/overbetting bias according to which small but systematic deviations exist for very large and very small values of p_i, with $s_i - p_i$ being predominantly negative when p_i is large and predominantly small in the reverse case. See Asch and Quandt (1987).

2. These were previously employed in Asch, Malkiel, & Quandt (1982, 1984).

3. This latter event does occur, rarely. See Hausch and Ziemba (1989).

4. We disregard the possibility that ability need not be measurable as a scalar variable.

5. Entering a horse in a claiming race is equivalent to giving a call option at the claiming price with expiration equal to the start of the race. Even with this refinement, the stylized scenario

clearly represents a vast oversimplification. In reality, the number n is not predetermined, nor are only particular stables eligible to enter horses. Furthermore, certain races are restricted (so-called "maiden races" are for horses that have never won before), while others are handicap races in that if $\alpha_i > \alpha_j$, horse i has to carry some extra weight.

     6. The percentage going to the winner depends on the track and the particular race. On the basis of casual examination, the fraction of the purse going to the winner in recent years has been in the 0.55-0.77 range at Churchill Downs and Santa Anita and the fraction of the purse going to the winner as a fraction of the sum going to the winner and the runner-up has been in the 0.75-0.85 range. See Ziemba and Hausch (1987), Mitchell (1989).

     7. See, for example, MacCormack (1978). On balance, owners in Ireland lost substantial amounts of money on racing operations, with training costs accounting for the bulk of the outlays. The amount of such costs attributable to a particular race depends, of course, on how often the horse is entered. We implicitly assume that some long-run average percentage can be so attributed and bypass the problem that these costs may themselves depend on the frequency of being entered.

     8. It is implicitly assumed that an owner will never bet on horses other than his own, thus bypassing a whole range of difficult moral hazard problems.

     9. The probability q would not depend on α if potential claimers could not observe the ability of a horse directly. In that case, q may depend on L; i.e., claimers take the "class" of the race as expressed in L as the signal on which to base their claiming behavior.

     10. It is obvious that in (3.8) we have a positive solution if and only if the positive squareroot is taken. Moreover, it is obvious that (3.10) has a unique positive solution. It should be noted that solving for α_i^* is not the same as solving a cubic in α_i, because c_i is itself a function of α_i^*. However, since there is a unique positive root, the solution is easily obtained by the method of *regula falsi* (see the nonlinear optimization package GQOPT).

     11. As an alternative, we considered generating costs uniformly on the $(r_2, r_1 + r_2)$ interval. The particular way of generating costs described in the text makes it easy to ensure, if desired, that the uniform and lognormal cases have identical means.

     12. For an interesting empirical comparison of Cournot-Nash and maximin behavior in a different context, see Holler and Høst (1990).

     13. Namely 0.080, 0.116, 0.116, 0.117, 0.117 0.117, 0.117, 0.220.

REFERENCES

Ali, M. (1977). Probabilities and Utility Estimates for Racetrack Betting. *Journal of Political Economy*. *85*, 803-815.

Asch, P., Malkiel, B. G., Quandt, R. E. (1982). Racetrack Betting and Informed Behavior. *Journal of Financial Economics*. *10*, 187-194.

_____(1984) Market Efficiency in Racetrack Betting. *Journal of Business*. *57*, 165-175.

Asch, P. and Quandt, R. E. (1987). *Racetrack Betting*. Dover, MA: Auburn Publishing.

Fabricand, B. P. (1979). *The Science of Winning*. New York: Whitlock Press.

Harville, D. A. (1973) Assigning probabilities to Outcomes in Multi-Entry Competitions. *Journal of the American Statistical Association*. *68*, 312-316.

Hausch, D. B. and Ziemba, W. T. (1989). *Locks' in Racetrack Minus Pools*. University of British Columbia, Working paper.

Holler, M. J. and Høst, V. (1990). Maximin vs. Nash Equilibrium: Theoretical Results and Empirical Evidence. In R. E. Quandt and Dušan Tříska (Ed.), *Micromodels for Planning and Markets*. Boulder, CO: Westview Press.

Ziemba, W. T. and Hausch, D. B. (1987). *Dr. Z's Beat the Racetrack*. New York: William Morrow and Co.

On Asymmetry in Economic Time Series

Francis X. Diebold

There is an important nonlinear tradition in theoretical macroeconomics, ranging from the early work of Hicks, Haberler, Goodwin, and Chenery, to the recent work of Brock, Scheinkman, Grandmont and others. A parallel nonlinear tradition in empirical macroeconomics, ranging from the early work of Burns and Mitchell to the recent work of Neftci, Hamilton, and others, highlights the potential importance of regime-switching behavior in macroeconomic systems.

In this regard, Markov processes have played a key role in empirical business cycle analysis. Typically, two-state processes are used, with switching between expansion and contraction states. For example, Hamilton (1989) proposes a Markov-switching model for the dynamics of U.S. real GNP, and Diebold and Rudebusch (1994) argue that a latent factor model with a Markov-switching factor provides a succinct reduced-form empirical characterization of business cycle dynamics. Within the Markov-switching context, long-standing issues such as business cycle asymmetry (Neftci, 1984; Sichel, 1993) and business cycle duration dependence (McCulloch, 1975; Diebold and Rudebusch, 1990; Diebold, Rudebusch and Sichel, 1993) are readily addressed.

The first of these issues, business cycle asymmetry, is the topic of this paper. Asymmetry refers to the idea that economic time series may behave differently on the upswing than on the downswing. It is related to observations like that of Klein (1983), who remarks upon the "relaxation oscillations" which are commonly observed: upswings tend to be long and gradual, while downswings are shorter and steeper.

In this paper, I extend Neftci's (1984) analysis of business cycle asymmetry and put forward some caveats related to the use of Markov models in the study of business cycle asymmetry. The caveats are linked to (1) the method used to map observed time series into Markov processes, and (2) the presence of trend and the necessity of trend removal. In section I, I set up the model, as well as estimation and testing procedures that make use of an approximate likelihood function that yields a drastic simplification of Neftci's procedures. In section II, I discuss the effects of the method used to map observed time series into 0-1 realizations of a Markov process, as well as the effects of trend on symmetry analyses. The results are illustrated in section III using Neftci's unemployment data, and section IV concludes.

I. Assessing Asymmetry: Estimation and Testing

A. A Two-State Second-Order Markov Model

Let $\{y_t\}_{t=1}^T$ be the sample path of a stationary time series, and let $\{\Delta y_t\}_{t=2}^T$ be the corresponding sample path of first differences. Consider defining the indicator series

$$\{I_t\}_{t=2}^T = \begin{cases} 1 & \text{if } \Delta y_t > 0 \\ -1 & \text{if } \Delta y_t \leq 0 , \end{cases}$$

and modeling $\{I_t\}$ as a second-order two-state Markov process. In this regard, it is helpful to define the transition probabilities

$$\lambda_{11} = P\ (I_t = 1 \mid I_{t-1} = I_{t-2} \quad = 1)\ ,\ t = 3, 4, \ldots, T$$

$$\lambda_{00} = P\ (I_t = -1 \mid I_{t-1} = I_{t-2} \quad = -1)\ ,\ t = 3, 4, \ldots, T$$

$$\lambda_{01} = P\ (I_t = -1 \mid I_{t-1} = -1, I_{t-2} = 1)\ ,\ t = 3, 4, \ldots, T$$

$$\lambda_{10} = P\ (I_t = 1 \mid I_{t-1} = 1, I_{t-2} = -1)\ ,\ t = 3, 4, \ldots, T$$

$$\pi_0 = P\ (I_1, I_2\ )$$

The likelihood for the indicator series is easily formed using the transition probabilities and the steady-state probability of I_1 and I_2, as

$$L(I_1, \ldots, I_T) = P(I_T = i_T \mid I_{T-1} = i_{T-1}, I_{T-2} = i_{T-2})$$

$$\bullet\ P(I_{T-1} = i_{T-1} \mid I_{T-2} = i_{T-2}, I_{T-3} = i_{T-3})$$

$$\ldots$$

$$\bullet\ P(I_3 = i_3 \mid I_2 = i_2, I_1 = i_1) \bullet P(I_2, I_1)$$

$$= \pi_0\ \lambda_{11}^{n_{11}}\ (1 - \lambda_{11})^{T_{11}}\ \lambda_{00}^{n_{00}}\ (1 - \lambda_{00})^{T_{00}}\ \lambda_{01}^{n_{01}}\ (1 - \lambda_{01})^{T_{01}}\ \lambda_{10}^{n_{10}}\ (1 - \lambda_{10})^{T_{10}}\ ,$$

where

				t	$t-1$	$t-2$
n_{11}	$=$	number of occurrences of		1	1	1
T_{11}	$=$	"		-1	1	1
n_{00}	$=$	"		-1	-1	-1
T_{00}	$=$	"		1	-1	-1
n_{01}	$=$	"		-1	-1	1
T_{01}	$=$	"		1	-1	1
n_{10}	$=$	"		1	1	-1
T_{10}	$=$	"		-1	1	-1 .

B. Estimation

Note that there are four possible values of π_0, corresponding to $P(1, 1)$, $P(-1, -1)$, $P(1, -1)$, and $P(-1, 1)$, which I denote by π_{11}, π_{00}, π_{10}, and π_{01}. Neftci (1984) obtains the steady-state values of these, which are

$$\pi_{11} = \frac{(1-\lambda_{00})\,\lambda_{10}}{(1-\lambda_{11})(1-\lambda_{00} + \lambda_{01}) + (1-\lambda_{00})(1-\lambda_{11} + \lambda_{10})}$$

$$\pi_{00} = \frac{(1-\lambda_{11})\,\lambda_{01}}{(1-\lambda_{00})(1-\lambda_{11} + \lambda_{10}) + (1-\lambda_{11})(1-\lambda_{00} + \lambda_{01})}$$

$$\pi_{01} = \frac{(\lambda_{10} - \lambda_{00})}{(2-\lambda_{10})}\,\pi_{00} + \frac{(1-\lambda_{10})}{(2-\lambda_{10})}(1-\pi_{11})$$

$$\pi_{10} = 1 - \pi_{11} - \pi_{00} - \pi_{01}.$$

Given these, the partial derivatives of the likelihood with respect to the transition probabilities are easily calculated.   They are complicated nonlinear functions, however, and an analytic expression for the maximum-likelihood estimator appears to be nonexistent.

In contrast, consider the approximate likelihood function formed by ignoring the information contained in the first two observations,

$$L^*(I_1, \ldots, I_T) = \lambda_{11}^{n_{11}}\,(1-\lambda_{11})^{T_{11}}\,\lambda_{00}^{n_{00}}\,(1-\lambda_{00})^{T_{00}}\,\lambda_{01}^{n_{01}}\,(1-\lambda_{01})^{T_{01}}\,\lambda_{10}^{n_{10}}\,(1-\lambda_{10})^{T_{10}}.$$

Under stationarity and ergodicity, neglect of π_0 is of no asymptotic importance because the impact of the first two observations becomes negligible as sample size increases. Immediately,

$$\ln L^*(I_1, \ldots, I_T) = n_{11} \ln \lambda_{11} + T_{11} \ln (1-\lambda_{11})$$
$$+ n_{00} \ln \lambda_{00} + T_{00} \ln (1-\lambda_{00})$$
$$+ n_{01} \ln \lambda_{01} + T_{01} \ln (1-\lambda_{01})$$
$$+ n_{10} \ln \lambda_{10} + T_{10} \ln (1-\lambda_{10}) ,$$

so that

$$\frac{\partial \ln L^*}{\partial \lambda_{11}} = \frac{n_{11}}{\lambda_{11}} - \frac{T_{11}}{1-\lambda_{11}}$$

$$\frac{\partial \ln L^*}{\partial \lambda_{00}} = \frac{n_{00}}{\lambda_{00}} - \frac{T_{00}}{1-\lambda_{00}}$$

$$\frac{\partial \ln L^*}{\partial \lambda_{01}} = \frac{n_{01}}{\lambda_{01}} - \frac{T_{01}}{1-\lambda_{01}}$$

$$\frac{\partial \ln L^*}{\partial \lambda_{10}} = \frac{n_{10}}{\lambda_{10}} - \frac{T_{10}}{1-\lambda_{10}}$$

Equating these to zero yields the solution:

$$\hat{\lambda}_{11} = \frac{n_{11}}{n_{11} + T_{11}}$$

$$\hat{\lambda}_{00} = \frac{n_{00}}{n_{00} + T_{00}}$$

$$\hat{\lambda}_{01} = \frac{n_{01}}{n_{01} + T_{01}}$$

$$\hat{\lambda}_{10} = \frac{n_{10}}{n_{10} + T_{10}} .$$

Thus, the approximate maximum-likelihood estimator exists in closed form, is computationally trivial, and is asymptotically equivalent to the exact maximum-likelihood estimator.[1]

C. Testing

Call a second-order two-state Markov process *first-order symmetric* if $\lambda_{11} = \lambda_{00}$, and call it *second-order symmetric* if it is first-order symmetric and $\lambda_{01} = \lambda_{10}$.[2] Thus, the null hypothesis of first-order symmetry is simply H_0^1: $\lambda_{11} = \lambda_{00}$. and the null hypothesis of second-order symmetry is

H_0^2: $\lambda_{11} = \lambda_{00}$ and $\lambda_{01} = \lambda_{10}$.

The usual testing procedures apply, and the likelihood ratio test is particularly convenient. Let ω refer to the likelihood function maximized over the parameter space as constrained by the null, and let Ω refer to the unconstrained maximization. Then under H_0^1

$$-2 \ln \left(\frac{L_\omega}{L_\Omega}\right) \xrightarrow{d} \chi_1^2,$$

and because L^* is asymptotically equivalent to L,

$$-2 \ln \left(\frac{L_\omega^*}{L_\Omega^*}\right) \xrightarrow{d} \chi_1^2.$$

Similarly, under H_0^2,

$$-2 \ln \left(\frac{L_\omega^*}{L_\Omega^*}\right) \xrightarrow{d} \chi_2^2.$$

The tests are easily made operational, by virtue of the closed-form expressions for the approximate unconstrained estimators derived above, and similarly simple expressions for the approximate constrained estimators, given by

$$\hat{\lambda}_{11}^\omega = \hat{\lambda}_{00}^\omega = \frac{n_{11} + n_{00}}{n_{11} + n_{00} + T_{11} + T_{00}} \quad \text{and} \quad \hat{\lambda}_{01}^\omega = \hat{\lambda}_{10}^\omega = \frac{n_{01} + n_{10}}{n_{01} + n_{10} + T_{01} + T_{10}}.$$

II. On Generation of the Indicator Series and Trend Removal

A. Generation of the Indicator Series

Define an indicator series to be generated by *Method 1* if

$$I_t = \begin{cases} 1 & \text{if } \Delta y_t > 0 \\ -1 & \text{if } \Delta y_t \leq 0, \end{cases}$$

and generated by *Method 2* if

$$I_t = \begin{cases} 1 & \text{if } \Delta y_t \geq 0 \\ -1 & \text{if } \Delta y_t < 0. \end{cases}$$

Method 1 is Neftci's method. It is apparent, however, that the treatment of observations for which $\Delta y_t = 0$ is arbitrary. It is important to be aware of the conditions under which the choice of method has an effect on the generated indicator series, and hence has an effect on the estimated transition probabilities. Let $\{I_t^1\}_{t=2}^T$ be the indicator series generated by Method 1, and $\{I_t^2\}_{t=2}^T$ be generated by Method 2. It is apparent that $\{I_t^1\}_{t=2}^T = \{I_t^2\}_{t=2}^T$ if and only if $\Delta y_t \neq 0, \forall t = 2, \ldots, T.^3$

It is common in the analysis of economic time series to encounter situations in which $\Delta y_t = 0$, as for example when data are available only to one decimal place. In fact, two of the three quarterly series which Neftci studies suffer from this problem (average weekly insured unemployment and unemployment 15 weeks and over). The change in the first of these has 41 zeros (out of 140 observations), and the change in the second has 17 zeros (out of 136 observations).[4]

As an illustration, consider the following quarterly total unemployment series for the period 1980-1982, reported in the March 1983 Business Conditions Digest:

t	y_t	Δy_t	I_t^1	I_t^2
1980.1	6.3	---	---	---
1980.2	7.3	1.0	1	1
1980.3	7.6	.3	1	1
1980.4	7.5	-.1	-1	-1
1981.1	7.4	-.1	-1	-1
1981.2	7.4	0.0	-1	1
1981.3	7.4	0.0	-1	1
1981.4	8.3	.9	1	1
1982.1	8.8	.5	1	1
1982.2	9.5	.7	1	1
1982.3	9.9	.4	1	1
1982.4	10.7	.8	1	1

Assume that the series is trend free and stationary for illustrative purposes. Approximate maximum likelihood produces the sharply contrasting estimates $\hat{\lambda}_{11}^1 = .75$, $\hat{\lambda}_{00}^1 = .66$ (Method 1) and $\hat{\lambda}_{11}^2 = .83$, $\hat{\lambda}_{00}^2 = .00$ (Method 2), which illustrate that the method used to map $\{y_t\}$ into $\{I_t\}$ can matter very much. Method 1 tends to introduce -1's into the indicator series, and method 2 introduces $+1$'s. Thus, one expects that $\hat{\lambda}_{11}^1 < \hat{\lambda}_{11}^2$ and $\hat{\lambda}_{00}^1 > \hat{\lambda}_{00}^2$, which is in fact observed.

To circumvent this problem, the following randomized procedure can be adopted. Let e_t be a discrete random variable such that

$$\varepsilon_t = \begin{cases} 1 & \text{with probability} = 1/2 \\ -1 & \text{with probability} = 1/2. \end{cases}$$

Define a *randomized method* of generating the indicator series by

$$I_t^R = \begin{cases} 1 & \text{if } \Delta y_t > 0 \\ \varepsilon_t & \text{if } \Delta y_t = 0 \\ -1 & \text{if } \Delta y_t < 0, \, t = 2, \, \ldots, \, T. \end{cases}$$

Many $\{\varepsilon_t\}_{t=2}^T$ series may be generated, thus leading to many $\{I_t^R\}_{t=2}^T$ series, and the resulting estimated transition probabilities may be averaged; call this the *average randomized method*. Let $\hat{\lambda}_{ij}^{RK}$ denote the average randomized estimator of λ_{ij} after K randomizations. At each randomization, the randomized estimator lies between $\hat{\lambda}_{00}$ and $\hat{\lambda}_{11}$, so that the average randomized estimator also has that property. The average randomized estimates after 1001 randomizations are given by $\hat{\lambda}_{11}^{R1001} = .78$ and $\hat{\lambda}_{00}^{R1001} = .29$.

B. Trend Removal

The intuition behind the problem of trend is clear. The concept of asymmetry discussed above is relevant to *covariance stationary* time series, which, among other things, requires the absence of trend. A monotone increasing trend tends to pull the indicator series upward, producing more $+1$'s than would be the case if the trend were properly removed. Similarly, a monotone decreasing trend tends to produce more -1's than would otherwise be the case. A nonmonotone, nonlinear trend will also affect the indicator series and hence the transition probability estimates, but the direction of the change cannot be characterized a priori due to competing influences.[5] The upshot is simply that it appears desirable to remove trend from economic time series before examining their symmetry.

III. Illustration of the Results Using the Neftci Unemployment Data

The results are nicely illustrated using one of Neftci's series, the quarterly unemployment rate, 15 weeks and over, 1948-1981. Following Neftci, I take the data from the February 1981 issue of the Business Conditions Digest, except for the 1981 values which I take from the February 1982 issue. The data are seasonally adjusted, so no further analysis of seasonality is undertaken. To maintain further conformity with Neftci, I also adopt the assumption that a second-order Markov process is adequate to model the indicator series.[6]

The results of symmetry analyses for the original series and the detrended series are given in Table 1. First, consider the original non-detrended series. The Method 1 approximate maximum likelihood estimates of λ_{11} and λ_{00} are extremely close to Neftci's, indicating that use of an exact likelihood function makes little difference. The likelihood ratio test indicates that the hypothesis H_0^1: $\lambda_{11} = \lambda_{00}$ can be rejected at approximately the 30% level, which again conforms with Neftci's results. The estimates of λ_{01} and λ_{10} (not reported by Neftci) are also interesting. Note that λ_{01} is much larger than λ_{10}, and the likelihood ratio test of H_0^2: $\lambda_{11} = \lambda_{00}$ and $\lambda_{01} = \lambda_{10}$ produces rejection at the 5% level, and particularly highlights the value of the information contained in λ_{01} and λ_{10}.

The estimates of the same parameters obtained by Method 2 are much different. In fact, Method 2 causes the direction of the estimated asymmetry to be reversed. The likelihood-ratio tests, however, indicate that neither H_0^1 nor H_0^2 can be rejected at any plausible level.

The average randomized estimates lie between their Method 1 and Method 2 counterparts and indicate that the series (in non-detrended form) does not display much asymmetry, both because $\hat{\lambda}_{00}^{R1001} = \hat{\lambda}_{11}^{R1001}$, and because $\hat{\lambda}_{00}^{R1001} = \hat{\lambda}_{11}^{R1001}$ are quite close.

Table 1: Estimation and Testing Results

	$\hat{\lambda}_{11}$	$\hat{\lambda}_{00}$	$\hat{\lambda}_{01}$	$\hat{\lambda}_{10}$	H_0^1	H_0^2
Method 1	0.73	0.82	0.93	0.50	0.99	7.88
Method 2	0.70	0.62	0.46	0.71	0.49	3.61
Average Randomized	0.71	0.71	0.64	0.54	X	X
Neftci	0.74	0.81	X	X	X	X
Detrended	0.65	0.76	0.83	0.61	1.30	3.56

Notes: The λ are estimated transition probabilities, where $\lambda_{11} = P (I_t = 1 \mid I_{t-1} = I_{t-2} = 1)$, $\lambda_{00} = P (I_t = -1 \mid I_{t-1} = I_{t-2} = -1)$, $\lambda_{01} = P (I_t = -1 \mid I_{t-1} = -1, I_{t-2} = 1)$, and $\lambda_{10} = P (I_t = 1 \mid I_{t-1} = 1, I_{t-2} = -1)$. The last two columns are likelihood ratio statistics for the null hypotheses H_0^1: $\lambda_{11} = \lambda_{00}$ and H_0^2: $\lambda_{11} = \lambda_{00}$; $\lambda_{01} = \lambda_{10}$. The first statistic is distributed χ_1^2 under the null, and the second statistic is distributed χ_2^2 under the null.

As discussed earlier, however, the series should be trend-free if the results are to have a meaningful interpretation. Examination reveals a significant third-order polynomial trend, while no higher-order terms are significant. The estimated trend is[7]

$$P_t = .425 + .041 \, t - .0037 \, t^2 + .000003 \, t^3.$$
$$(2.13) \quad (3.27) \quad (-3.15) \quad \quad (3.35)$$

Although not monotone, it is nearly so.

After removing the trend, as expected, both Method 1 and Method 2 yield the same results. $\hat{\lambda}_{11}$ and $\hat{\lambda}_{00}$ are somewhat farther apart, which increases the first-order contribution to asymmetry, as indicated by a likelihood-ratio test statistic for H_0^1 that is significant at the 25% level. However, the likelihood-ratio test statistic for H_0^2 is insignificant at any plausible level.

IV. Conclusions

I have put forth two caveats related to the study of asymmetry in economic time series, and illustrated their relevance with two datasets. First, the method used to map covariance stationary, observed time series into indicator series can severely affect subsequent inferences. Second, when trend is present, failure to detrend can produce similarly misleading inferences.

Francis X. Diebold, Department of Economics, University of Pennsylvania, 3718 Locust Walk, Philadelphia, PA 19104-6297.

ACKNOWLEDGEMENTS

This paper is dedicated to L. R. Klein, my teacher and friend. The first draft was written in June 1984, when I was a Penn graduate student. I thank Patrick DeGraba, Salih Neftci, Peter Pauly, Robin Sickles, Asad Zaman and an anonymous referee for thoughtful comments. Remaining errors are mine.

NOTES

1. Moreover, the approximate procedure generalizes to the pth-order process in a straightforward and computationally tractable way. For example,

$$\hat{\lambda}_{1,1,\ldots,1} = \frac{n_{1,1,\ldots,1}}{n_{1,1,\ldots,1} + T_{1,1,\ldots,1}}$$

is the maximum-likelihood estimator of $\lambda_{1,1,\ldots,1}$ for a process of arbitrary order. This stands in sharp contrast to the exact procedure, which rapidly becomes intractable as the order of the process grows.

2. The definitions can be extended in the obvious fashion up to p^{th}-order symmetry for a pth-order process.

3. While it is *possible* that two different $\{I_t\}$ series might lead to identical parameter estimates, it is highly unlikely. Hence, I focus on the issue of when method affects $\{I_t\}$ rather than when method affects $\hat{\lambda}_{ij}$ i, j = 0, 1.

4. Neftci was able to circumvent the problem with the *total* unemployment series by constructing it as unemployed/total rather than using the published total unemployment rate.

5. A set of theorems characterizing the conditions under which trend removal affects the indicator series is given in an earlier version of this paper, Diebold (1984).

6. The results of Anderson and Goodman (1959) could be used to test for the proper order.

7. t-statistics appear in parentheses.

REFERENCES

Anderson, T. W. and Goodman, L. A. (1957). Statistical inference about Markov chains. *Annals of Mathematical Statistics. 28*, 89-110.

Diebold, Francis X. (1984). On asymmetry in economic time series. Manuscript, Department of Economics, University of Pennsylvania.

Diebold, Francis X. and Rudebusch, G. D. (1990). A nonparametric investigation of duration dependence in the American business cycle. *Journal of Political Economy, 98*, 596-616.

_____ (1994). Measuring business cycles: A modern perspective. NBER Working Paper Series # 4643. Cambridge, MA: National Bureau of Economic Research.

Diebold, Francis X., Rudebusch, G. D. and Sichel, D. E. (1993). Further evidence on business cycle duration dependence. In J. H. Stock and M. W. Watson (Eds.), *Business Cycles, Indicators, and Forecasting.* (pp. 255-284). Chicago: University of Chicago Press for NBER.

Hamilton, J. D. (1989). A new approach to the economic analysis of nonstationary time series and the business cycle. *Econometrica. 57*, 357-384.

Klein, Lawrence R. (1983), *Lectures in Econometrics.* New York: North-Holland.

McCulloch, J. H. (1975). The Monte-Carlo cycle in business activity, *Economic Inquiry, 13*, 303-321.

Neftci, S. N. (1984). Are economic time series asymmetric over the business cycle?, *Journal of Political Economy. 92*, 307-328.

Sichel, D. E. (1993). Business cycle asymmetry: A deeper look, *Economic Inquiry. 31*, 224-236.

An Open Economy Analysis of the Dynamic Properties of the Michigan Quarterly Econometric Model of the U.S. Economy

E. Philip Howrey and Saul H. Hymans

The Michigan Quarterly Econometric Model of the U.S. Economy (MQEM) is maintained and operated by the Research Seminar in Quantitative Economics at the University of Michigan. Michigan's first model was the now-classic Klein-Goldberger model [Klein and Goldberger (1955)] -- an annual model of 20 equations constructed more than four decades ago, and one of the first macroeconometric forecasting models to be developed. Today MQEM contains 213 equations, 108 of which are stochastic.

Michigan Models have been operational, real-time forecasting models since the first serious use of the Klein-Goldberger model in 1952. MQEM, *today's* Michigan Model, is used to produce quarterly forecasts with a horizon of 2 to 3 years at least eight times annually, a schedule which has been in effect for about a decade. A summary view of the degree of success of Michigan's econometric forecasting program is shown in Figure 1. Our "forecast year" begins each November at the University of Michigan's Annual Economic Outlook Conference. The data plotted in Figure 1 show actual Real GNP growth rates for each calendar year for the period 1971-1991, compared with the corresponding Michigan Model forecast released at the immediately preceding November Conference. The forecasting operation based on MQEM has twice (1984 and 1987) received the national Theodore H. Silbert Award for "accuracy, timeliness, and professionalism in economic forecasting."

I. The Economic Structure of MQEM

MQEM is best thought of as an eclectic model. The model is structured around the aggregate demand national income equation, GDP = C + I + G + X - IM, which is a core element of Keynesian macroeconomics. And, quarter-by-quarter, the model forecasts real GDP precisely by adding up the predicted components of real aggregate demand. In its very short term behavior, then, the model is clearly Keynesian -- as that term is most commonly interpreted these days to signify that output is primarily demand-determined in the short run.

In addition, however, the model contains all the elements of the classical quantity theory of money equation, MV=PQ, but not in the naive deterministic and mechanistic form in which monetarism is often caricatured in the textbooks. In particular, velocity (V) in MQEM is implicitly and endogenously determined by a complex interaction involving the level and term

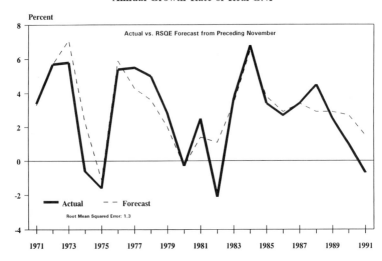

Figure 1: RSQE Forecast Accuracy: 1971-1991
Annual Growth Rate of Real GNP

structure of interest rates, as well as the rate of inflation. The fundamental long-run connection between money growth and inflation, however, which is perhaps the hallmark contribution of monetarist theory, is present in the Michigan Model.

Controversial neutrality restrictions such as the neutrality of money, the superneutrality of money, a vertical long-run Phillips Curve, or a unitary long-run relationship between the rate of inflation and the nominal interest rate are not imposed *a priori*. As in King and Watson (1992), the neutrality of money or the extent to which an increase in government spending is crowded out in the long run are regarded as hypotheses to be tested, not restrictions that are automatically imposed on the model.

II. The Block Structure of MQEM

The behavioral or substantive segmentation of the model is represented in its block structure which is summarized in Table 1. Block #1 (Wages & Prices) and Block #2 (Productivity and Employment) together comprise the Aggregate Supply sector of the model. The core wage rate (hourly compensation in the private nonfarm sector of the economy) is modeled as an expectations-augmented Phillips Curve. Price expectations are adaptively determined, and the wage/price elasticity varies from 4/10 in the short run to 3/4 in the long run. The corresponding core price level is determined by a variable mark-up on the level of wage costs in excess of trend productivity, but depends additionally on crude materials prices and interest rates. Productivity growth depends on output growth in the short run, and on capital accumulation

Table 1: Block Structure of MQEM

Block		Principal Endogenous Variables
(1)	Wages & Prices	Compensation Per Hour
		Private Nonfarm GDP Deflator
		24 GDP Component Deflators
		3 Energy Price Deflators
		Index of the Exchange Value of the Dollar
		Automobile & Truck Price Indexes
(2)	Productivity & Employment	Output per hour
		Employment Rate, Males 20 and over
		Aggregate Unemployment Rate
(3)	Expenditures, Purchases	Unit Vehicle Sales
		U.S. Cars & Trucks
		Imported Cars & Trucks
		Consumption
		Autos, News
		Autos, Net Used & Parts
		Trucks
		Furniture & Household Equipment
		Other Durables
		5 Nondurables Components
		Services
		Business Fixed Investment
		Structures
		Equipment
		Information Processing
		Industrial Equipment
		Autos
		Trucks
		Other
		Residential Building
		Housing Starts
		Inventory Investment
		New Autos
		Trucks
		Nonfarm, Nonvehicle
		Imports
		Non-petroleum
		Autos, Trucks
		Exports
(4)	Income Flows	Private Wages & Salaries
		Profits
		Interest Income
		Dividends
		Other Labor Income
		Nonfarm Proprietor Income
		Farm Proprietor Income
		Govt. Unemployment Benefits
		Taxes
		Capital Consumption Allowances
		Interest on Government Debt
(5)	Monetary Sector	M1
		M2
		3-Month Certificate of Deposit Rate
		3-Month Treasury Bill Rate
		Government Budget Identity
		6 Term Structure Equations
		Monetary Base
		Govt. Demand Deposits
(6)	Output Composition	Services Component of Real GDP
		Manufacturing Index of Industrial Production
		Index of Available Capacity in Manufacturing
		Gross Auto Product
		Gross Truck Product
(7)	Rest-of-World Sector	Index of Industrial Production
		2 Price Indexes

in the long run.

The Aggregate Demand sector of MQEM is represented in Blocks #3-#5. Block #3 determines purchases of goods and services given incomes, prices, and interest rates; while Block #4 determines the flow of private and public sector incomes consistent with fiscal policy and the employment level required to produce the goods and services demanded. Blocks #3 and #4 can therefore be thought of as generating the IS function implicit in the econometric model; i.e., the production/purchase, or supply/demand, equilibrium given fiscal policy, prices and interest rates. Block #5 is the Monetary Sector of the model, and can be thought of as generating the LM function implicit in MQEM; i.e., the level of interest rates at which the demand for and supply of liquidity are in balance given output, prices, and monetary policy.

The principal behavioral characteristics of the IS and LM functions in the Michigan Model will become clear in the context of the policy simulations to be discussed below. For an explicit derivation and discussion of the dynamic properties of the IS and LM curves implicit in the Michigan Model, the reader is referred to Green, Hickman, Howrey, Hymans, and Donihue (1991).

In a previous version of this paper [Hymans (1990)], simulations were performed on a version of the model that allowed, through changes in the exchange rate, for only minimal international reaction to domestic monetary and fiscal policy changes. The sensitivity of the results to exchange rate responses suggested that a more complete characterization of international feedbacks is important, especially for simulations longer than four or five years. A more complete rest-of-world sector was therefore added to the model to capture what were thought to be the most important feedbacks that should be considered.

The principal international linkages are captured by equations for U.S. imports, U.S. exports, the exchange value of the dollar, rest-of-world (ROW) output, traded-goods prices, and consumer-goods prices. Seven important equations are involved, and they appear in several blocks of the model as shown in Table 1, including Blocks #1, #3, and #7. Because of their special significance in understanding some of the open economy aspects of subsequent simulations, the equations directly linking the U.S. and rest-of-world economies are presented here.

U.S. Imports. The important relevant aspects of the import equation are given by

$$\ln \text{MNOIL87} = [2\{1 - .6936\} + .1383 \; \Delta \; \ln \text{SINV87}] \ln \text{GDP87} \qquad (1)$$
$$\qquad\qquad (.0558) \qquad (.0468)$$

$$- .4178 \ln (PMNOIL/PPNF)_{-1} + .6936 \ln MNOIL87_{-1}$$
$$\quad (.0798) \qquad\qquad\qquad (.0558)$$

$$+ \quad . \quad . \quad . \quad . \quad .$$

$R^2 = .996$ $SE = .0206$ $DW = 1.76$ $FP = 1976{:}1 - 1989{:}4$

where

MNOIL87 =	Non-petroleum imports of goods and services (1987 \$),
SINV87 =	Stock of business inventories (end of quarter, 1987 \$),
GDP87 =	Gross domestic product (1987 \$),
PMNOIL =	Implicit deflator for non-petroleum imports of goods and services (1987=100),
PPNF =	Implicit deflator for private nonfarm GDP (1987=100).

Expenditure on imports is modeled as a function of gross domestic product, with a coefficient that depends on the change in the stock of inventories, and the relative price of imported goods. The short-run income elasticity of imports is approximately 0.6 and the long-run income elasticity is 2. The short-run price elasticity of demand for imports is -0.4 and the long-run elasticity is -1.4.

U.S. Exports. The export equation is

$$) \quad \Delta \ln X87 = .4612 \, \Delta \ln JIPROW - .1409 \, \Delta \ln (PX{\cdot}JEXR/PCROW)_{-1} \quad (2)$$
$$\quad\quad\quad\quad (.2565) \qquad\qquad\quad (.0810)$$

$$+ .2390 \, \Delta \ln X87_{-1}$$
$$\quad (.1275)$$

$R^2 = .188$ $SE = .0237$ $DW = 2.09$ $FP = 1976{:}1 - 1989{:}4$

where

X87 =	Exports of goods and services (1987 \$),
JIPROW =	Index of rest-of-world industrial production (weighted average of West Germany [.125], United Kingdom [.125], Japan [.25], and Canada [.5]),
PX =	Implicit deflator for exports (1987 = 100),
JEXR =	Index of trade-weighted exchange value of the U.S. dollar against currencies of G-10 countries plus Switzerland (1973=100),

PCROW   =   Rest-of-world consumer prices (weighted average of West Germany [.125], United Kingdom [.125], Japan [.25], and Canada [.5]).

The equation for the demand for U.S. exports is the mirror image of the U.S. import demand equation: export demand depends on output in the rest of the world and the relative price of exports. It is interesting to note that the estimated rest-of-world price and income elasticities are smaller than the corresponding U.S. price and income elasticities; thus, the U.S. import and export equations are qualitatively but not quantitatively symmetric. Note also that the export equation has been estimated in terms of first differences of logarithms rather than the logarithms themselves; the intent was to develop an export equation that would be capable of capturing induced changes in exports at the margin.

<u>Exchange Value of the U.S. Dollar</u>. The exchange rate equation is

$$\Delta \ln JEXR = .8458 \; \Delta \ln (PMROW/PX) + .0626 \ln (X'_{-1}/IM') \qquad (3)$$
$$\qquad\qquad (.1260) \qquad\qquad\qquad\qquad (.0180)$$

$$+ .0277 \ln JUS.ROW$$
$$(.0105)$$

$R^2 = .870$ SE = .042   DW = 1.89   FP = 1973:2 - 1989:4

where JUS.ROW = RTB/RROW3 and

PMROW   =   Implicit deflator (denominated in foreign currencies) for non-petroleum goods and services imported by the U.S.,
X'   =   Exports of goods and services plus capital grants received by the U.S. (net) in current dollars,
IM'   =   Imports of goods and services plus personal and federal government transfers to foreigners plus government interest payments to foreigners,
RTB   =   U.S. 3-month treasury bill rate,
RROW3   =   trade-weighted 3-month foreign interest rate.

Note that with a balanced U.S. current account $(X'_{-1}/IM' = 1)$ and no differential between U.S. and foreign interest rates (JUS.ROW = 1), the value of the dollar moves only in response to discrepancies between foreign and domestic inflation; i.e., toward restoration of purchasing power parity (PPP) among currencies. However, the presence of a differential between domestic and foreign interest rates or a current account imbalance can lead currency values persistently away from PPP requirements. (For reasons explained in the Appendix, this equation was estimated using the method of instrumental variables.)

ROW Output. The equation for output produced in the rest of the world is

$$\Delta \ln JIPROW = .0815 \ \Delta \ln (MNOIL87/X87) \qquad\qquad (4)$$
$$ (.0351)$$

$$- .0042 \ (\Delta \ RROW3_{-1} + \Delta \ RROW3_{-2})/2$$
$$(.0019)$$

$$- .3006 \ (\Delta \ln PCROW_{-1} + \Delta \ln PCROW_{-2})/2$$
$$(.1684)$$

$$+ .6232 \ \Delta \ln JIPROW_{-1}$$
$$(.0956)$$

$R^2 = .995$ $SE = .0092$ $DW = 1.92$ $FP = 1974:1 - 1989:4.$

The first term in this equation captures the positive effect of U.S. imports (ROW exports) and the negative effect of U.S. exports (ROW imports) on foreign output. The second term captures the negative effect of an increase in foreign interest rates on foreign output. The third term is a proxy for real balance effects: an increase in the rate of inflation reduces the rate of growth of output. Notice that a one-time change in the level of the interest rate results in a temporary rather than a permanent change in the *rate of growth* of output. This equation should be viewed as a type of reduced form equation which is intended to provide a link at the margin between U.S. imports and U.S. exports: an increase in U.S. imports stimulates foreign production which in turn increases the demand for U.S. exports. The price and interest rate terms are included to estimate the secondary effects of U.S. policy changes on ROW output and hence U.S. exports.

ROW Consumer Prices. The equation for rest-of-world consumer prices is

$$\Delta \ln PCROW = .0326 \ \Delta \ln JIPROW_{-1} - .0004 \ t \qquad\qquad (5)$$
$$ (.0169) \qquad\qquad\quad (.0001)$$

$$+ .4027 \ \Delta \ln PCROW_{-1}$$
$$(.1221)$$

$R^2 = .999$ $SE = .0047$ $DW = 2.14$ $FP = 1976:1 - 1989:4.$

The idea behind this equation is that the rest-of-world price level adjusts with a lag to the gap between actual and full employment or trend output. This implies that at the margin, with other things equal, an increase in the rate of growth of output will result in an increase in the inflation rate.

U.S. Import Price Denominated in ROW Currency. The equation for the rest-of-world price of U.S. imports is

$$\Delta \ln \text{PMROW} = 2.5440 \ \Delta \ln \text{PCROW} - .6908 \ \Delta \ln \text{PCROW}_{-1} \qquad (6)$$
$$ (.7494) (.7291)$$

$$+ .1286 \ \Delta \ln \text{JEXR}_{-2}$$
$$(.1036)$$

$$+ .0499 \ \Delta \ln \text{JUS.ROW}_{-1} + .1457 \ \Delta \ln \text{PMROW}_{-1}$$
$$(.0333) (.1310)$$

$R^2 = .990$ $SE = .0314$ $DW = 2.06$ $FP = 1973{:}2 - 1989{:}4.$

This equation translates the rest-of-world rate of inflation of consumer goods into a rate of inflation for traded goods imported by the U.S. The rather surprising implication of the parameter estimates of this equation is that rest-of-world traded goods prices are much more volatile than rest-of-world domestic prices.

U.S. Import Prices. The price of U.S. imports in dollar terms is related to PMROW by the identity

$$\text{PMNOIL} = 100 \cdot \text{PMROW/JEXR}. \qquad (7)$$

III. Policy Simulations

One especially interesting way to study the dynamic properties of an econometric model is through the use of carefully designed policy simulations. Over the years, simulation experiments have been used to study the responses of MQEM and other well-known U.S. models to perturbations of monetary and fiscal policies. Many of these studies were carried out under the auspices of the Model Comparison Seminar of the Conference on Econometrics and Mathematical Economics sponsored by the National Science Foundation with the cooperation of the National Bureau of Economic Research and chaired by Lawrence Klein during the 1970s and '80s and the results are summarized in Klein and Burmeister (1976) and Klein (1991).

Except for studies carried out to determine the impact of supply shocks deriving from sharp movements in the price of imported oil [Hickman,

Huntington, and Sweeney (1987)], most of these studies of model properties have been closed-economy analyses, quite properly so since most U.S. econometric models have had little or no international sector -- with the common exception of an import equation -- until relatively recent years. Model-based studies of fiscal policy, for example, generally focussed on what difference it might make whether or not domestic monetary authorities accommodated a fiscal shock. In the language of macroeconomic theory, the (empirical) issue was whether the model's LM curve was so steep as to crowd out virtually all of any real impact resulting from a shift in the IS curve induced by a change in fiscal policy; and, along what dynamic path would the results occur. Little if any attention was paid to what difference it might make if domestic and foreign interest rate policies were coordinated or not, since the models had little to say about such matters.

During the 1980s U.S. models began to become more open in their specifications, and the openness began to be reflected in policy simulations. Hymans (1990) presented an early such study using MQEM, but warned that the quantitative results, especially for the later or "out-years" of the simulation, should not be taken literally because the relatively primitive rest-of-world sector of the model took too much, including foreign growth and inflation, to be exogenous and thus had no way to allow for proper rest-of-world impacts in response to changes in domestic policies. Such studies were more valuable on a technical than on a substantive level.

In the studies carried out in this paper, we take advantage of the greater degree of rest-of-world endogeneity that now exists in MQEM. The out-year results thus deserve far greater scrutiny than did those of the earlier paper. The reader is properly warned, however, that there remains an enormous difference in the quantity of information contained in the domestic and rest-of-world sectors of the model, and this may compromise the integrity of the simulation results in ways that are difficult to anticipate. That said, the main point is that we shall conduct simulation experiments intended to reveal the important dynamic characteristics of the responses of MQEM to fiscal and monetary policy perturbations, while explicitly accounting for the fact that U.S. economic policy necessarily operates in the context of a largely integrated world economy. The U.S. economy is large and domestic policy is powerful, but the U.S. alone is not all that counts.

Subsequent sections describe the specifics of the fiscal and monetary policy perturbations and the nature of the corresponding foreign responses. This section closes by describing the procedure that applies to all of the simulation experiments.

(1) We begin with a Control Run which is an "almost-pure" tracking solution of the 14 year period 1978:1 through 1991:4. In a pure tracking solution, the observed single-equation residuals would be added to the equation

intercepts (or "constant adjustments") so that a 56-quarter dynamic solution of the entire model would reproduce, quarter-by-quarter, the observed history of the period 1978-91.

Our Control Run contains one exception to this pure tracking procedure. As discussed earlier, the ratio of the U.S. to the rest-of-world short term interest rate, JUS.ROW, is an important determinant of the exchange rate in MQEM. In our simulations, much therefore depends on whether or not domestic policy changes are accompanied by international coordination of interest rates. But we did not think it reasonable to have the results depend on whether interest rates follow each other so as to maintain absolute or relative differentials. To avoid such a problem in a world in which national interest rates can and often do differ considerably, we changed the observed values of the rest-of-world short term interest rate, RROW3, so that it was identically equal to the U.S. 3-month T-Bill rate, RTB, in every quarter of the period 1978-91, and this meant that JUS.ROW was identically unity in the Control Run. Since no other variables were allowed to deviate from their historical values, this altered the single-equation residuals in those equations which contained RROW3 or JUS.ROW. It was these altered residuals that were used to generate the Control Run.

As used in this paper, therefore, the Control Run tracks the observed history of every variable in the model, except for RROW3, which is always equal to RTB, and JUS.ROW, which is always unity. As a result, when interest rates are coordinated internationally in our simulations, there is no ambiguity about relative versus absolute differences. Either RTB and RROW3 remain equal to each other in response to a change in U.S. policy (coordinated interest rates), or RROW3 remains at Control while RTB changes in response to U.S. policy (uncoordinated interest rates).

(2) An Alternate simulation is then derived from a specific set of perturbations applied to the inputs of the Control Run simulation.

(3) The effects of the particular perturbation are evaluated by comparing the Alternative relative to the Control, quarter-by-quarter, over the 14 year horizon.

IV. Fiscal Policy Simulations

The three fiscal policy experiments all involve shifting the *path* of real federal government purchases of goods and services up by $37 billion, measured in constant 1987 prices, beginning with 1978:1. The choice of a real path-shift of $37 billion derives from the fact that real GDP in 1978 is just under $3704 billion, so that the path-shift amounts to a fiscal-stimulus of one percent of the base level of real GDP. A convenient way to describe the outcome of the perturbation, then, is to focus on the *percent* difference between

Alternate and Control. Does a fiscal stimulus of one percent of real GDP, for example, raise real GDP by just about one percent, more than one percent, or less than one percent; and along what kind of dynamic path?

Before such a question can be answered, however, two kinds of concurrent conditions have to be specified; those dealing with the domestic economy, and those dealing with the rest-of-world sector. We deal first with the matter of what else happens in the domestic economy when the fiscal stimulus is applied; and the answer comes in two parts:

(i) *Nothing* happens to monetary policy in response to the change in fiscal policy. Specifically, the monetary aggregate M1 is held constant (i.e., on the path of the Control Run) in *nominal* dollars. This implies, of course, that domestic interest rates should be expected to change in response to the perturbation. In the language of macroeconomic theory, the LM curve (in interest rate/real GDP space, and given the price level) is being held fixed while the fiscal stimulus shifts the IS curve.

(ii) *Nothing* happens to fiscal policy at the state and local government level in response to the change in the federal budget. We assume that state and local *spending* is budgeted in *real* terms so that if the federal fiscal stimulus changes the price level, nominal state and local purchases are forced to change so as to keep real state and local purchases at Control. In addition, the state and local government sector is assumed to face a *nominal* budget constraint and we take that to mean that the state and local budget surplus must remain on the Control path. We use personal taxes collected by state and local governments as the instrument which keeps the state and local surplus fixed at Control.

These two concurrent conditions apply to all three fiscal experiments. It is therefore the treatment of the rest-of-world sector which distinguishes among the fiscal stimulus experiments. The first experiment -- labelled Experiment "JEXR" -- is that which we intend to be the baseline case approximating what would happen if the fiscal stimulus were applied to a closed U.S. economy. Since MQEM is structured to recognize that the U.S. economy operates in an open environment, our baseline experiment is necessarily artificial. Nonetheless, since much of our elementary policy intuition comes from the textbook cases dealing with the closed economy, it is useful to construct this baseline case.

(1) Experiment "JEXR" Since the economy is supposed to be closed in this simulation, we have to isolate the rest-of-world sector and not let it affect the domestic economy on the margin of the simulation. We accomplish this by forcing three variables to remain on their Control paths: the exchange value of the U.S. dollar (JEXR), the price of imported goods denominated in

foreign currency (PMROW), and real U.S. exports (X87). By this device, the rest-of-world sector is effectively eliminated from the model for the purpose of this simulation. Note, however, that U.S. imports should not be held at Control for such a closed economy simulation. Since parameters such as the marginal propensity to consume or invest implicitly include marginal purchases of imported goods and services, imports have to be allowed to respond to a change in GDP in order for GDP to continue to satisfy the national accounts definition of domestic output.

Experiment "JEXR" is thus the situation to which the neoclassical real/nominal dichotomy should apply. In the extreme, this implies that the fiscal stimulus should produce a combination of higher prices and interest rates sufficient to reduce other components of aggregate demand by exactly the increase in the federal government component and leave real GDP unchanged. In combination, financial and real crowding-out should reduce the long run "net multiplier" to zero. Such a result presupposes that full-employment real GDP is invariant to the fiscal stimulus.

(2) Experiment "JUS.ROW" This experiment opens the economy back up, and assumes that the rest of the world follows U.S. interest rates so that the ratio of U.S. to foreign interest rates (JUS.ROW) remains exactly on the path of the Control Run. Recalling Equation (3), this implies that JEXR can change in response to the fiscal stimulus, but *not* because of any interest rate differentials which might be generated by the fiscal stimulus. Rather, if the dominant effect of the fiscal stimulus in this experiment is to generate more domestic than foreign inflation and/or an increase in U.S. imports relative to exports, the result should be a decline in JEXR (dollar depreciation). And dollar depreciation, other things equal, would be expected to increase real net exports.

Experiment "JUS.ROW" is thus an open economy experiment characterized by international coordination of interest rates. If the dominant effect of the fiscal stimulus is therefore to depreciate the domestic currency, the fiscal stimulus in this experiment would be more effective in the short run in raising real GDP than in Experiment "JEXR".

(3) Experiment "RROW3" This too is an open economy simulation. In this case, however, the *level* of foreign interest rates (RROW3) is assumed to be held fixed on the Control path so that a stimulus-induced rise in domestic interest rates amounts to a rise in U.S. rates *relative* to foreign rates. This will clearly raise JEXR (appreciate the dollar) *relative to its value in the "JUS.ROW" experiment*, which implies a weakening of the fiscal stimulus effect relative to that in the "JUS.ROW" experiment. And it is clearly possible for the fiscal stimulus to raise domestic interest rates by enough to offset all or more of the dollar depreciation generated in the "JUS.ROW" experiment. This could leave

the dollar even stronger than it was in Experiment "JEXR" and thus in the Control Run. This is not, however, a logically necessary result; while the dollar must appreciate relative to its value in the interest rate coordination case ("JUS.ROW"), it need not rise above the Control value. It all depends on how much domestic interest rates rise in response to the fiscal stimulus. The more domestic interest rates rise, the greater the currency appreciation, and the greater the weakening of the fiscal stimulus.

It is worth pointing out that there is also a partial counter effect deriving from the reduction of foreign interest rates in this experiment relative to that of Experiment "JUS.ROW". The more the lowering of foreign interest rates stimulates rest-of-world growth, the greater the stimulation to U.S. exports, which directly strengthens the impact of the fiscal expansion in this experiment, but indirectly weakens it by further appreciating the domestic currency.

Experiment "RROW3" is thus an open economy experiment with a theoretically ambiguous result. The conventional wisdom is that in the absence of interest rate coordination, fiscal expansion will appreciate the currency by enough that we would observe the weakest fiscal stimulus effect in the open economy context of the "RROW3" experiment. Most textbooks simply assert that fiscal policy is less expansionary in an open economy than in an otherwise-the-same closed economy, often without even stating that this can only be true if foreign interest rates are fixed. In any case, all we can count on *a priori* is that fiscal stimulus is weaker in the "RROW3" than in the "JUS.ROW" experiment. Whether or not it is also weaker than in the closed economy ("JEXR") experiment is an empirical matter on which this simulation can shed some light.

Figure 2 provides a quick summary of the three fiscal-stimulus experiments by plotting the percent deviation of real GDP from Control for each experiment. The results can be described as follows relative to the issues raised in connection with each simulation.

(1) In the closed economy ("JEXR") experiment the one percent fiscal stimulus has its peak impact in the 4th quarter of the simulation when real GDP is nearly 1.6 percent above Control. The stimulus effect then declines over the next 3 years and reaches a deviation of less than 0.9 percent above Control in the 4th year of the run. The fiscal stimulus then begins to strengthen again, and the deviation attains a local peak about 1.2 percent above Control in the 6th year of the simulation. The fiscal impact then weakens again and the deviation falls to less than 0.7 percent above Control in the 9th year of the run. A very slight upswing occurs during the 10th year, after which the fiscal impact weakens for the next three years. The deviation hits zero in the 12th year, attains a negative of as much as -0.45 percent, and then heads back up toward zero in the 14th year of the run. Thus:

(a) In its closed economy version, MQEM does not obviously contradict the full crowding out, or zero long run multiplier, result of neoclassical theory. The long run, however, is very long indeed -- more than a decade.

(b) The response path is not monotonic, even after the build-up to the peak that occurs a year into the run. MQEM has significant accelerator effects which put a cyclical response pattern into the model's reaction to a perturbation. At least in the closed economy case, the cyclical response pattern is damped through time.

Figure 2: Fiscal Experiments
Real Federal Purchases Path UP by 1% of Real GDP

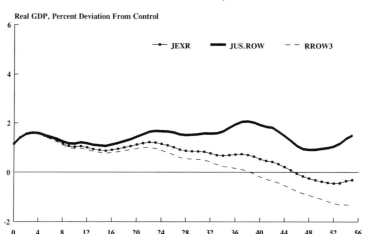

Further, as shown in Table 2, rates of interest and the price level both increase relative to Control in the "JEXR" experiment.   The interest-rate response, as measured by the 3-Month CD rate, is fairly rapid; within one year the CD rate is more than 110 basis points higher, about 150 basis points higher after two years, and then fluctuates between 140 and 270 basis points higher thereafter. The initial increase in interest rates has a fairly quick depressing effect on real residential construction which declines by almost 10 percent relative to Control by the end of the third year. The general price level, as measured by the private non-farm deflator, responds more sluggishly to the fiscal stimulus: after 2 years the price level has increased by less than one percent relative to Control, after six years the price level has increased by 3.4 percent, and after 12 years the price level is higher than Control by slightly more than 7 percent.

(2) The "JUS.ROW" experiment -- the open economy case with interest rate coordination -- is initially little different from the closed economy case in its aggregate impact. It takes a year and a half for the dollar to depreciate by 1.5 percent relative to Control, and it takes half a year longer to notice a stronger real GDP effect. Four years into the run, however, the dollar has fallen by more than 5 percent and by 6 years into the run it's down more than 7.5 percent relative to Control. By the 6th year, real GDP is nearly 1.7 percent above Control, well above the one percent deviation which characterizes the 6th year of the "JEXR" experiment, and a bit above the 1.6 percent first-year peak deviation common to both the "JEXR" experiment and this experiment.

One of the most interesting results of this experiment is that the opening of the economy, *with interest rate coordination*, appears to have neutralized the crowding out, for at least the first 10-12 years of the simulation horizon. Prices and interest rates in this experiment are indeed rising substantially above their deviations in Experiment "JEXR" (see Tables 2 and 3, below). But the current account balance of payments deteriorates for the first dozen years of the run and this keeps the dollar depreciating (Table 3), apparently by just about enough to offset the crowding out coming from higher prices and interest rates. By year 13, however, the currency depreciation has succeeded in turning the current account balance positive relative to Control, the dollar depreciation flattens out, and this may be the start of a period in which crowding out begins to dominate. However, to be definitive about this would require an even longer simulation than we carried out.

(3) The "RROW3" experiment -- the open economy case in which foreign interest rates are held at Control, and thus below their level in Experiment "JUS.ROW" -- adds a source of dollar appreciation to the simulation, which does indeed weaken the fiscal stimulus relative to the second experiment. And by the second year of the run, it is quite clear that there is enough dollar appreciation to weaken the fiscal stimulus relative to the closed economy ("JEXR") experiment as well. In MQEM, openness of the economy, *without interest rate coordination*, and with an accompanying constant-M1 monetary policy, reduces the efficacy of fiscal policy. Note the following, however. Even in this "least effective fiscal policy" case, a one percent fiscal stimulus provides an initial 3 years during which real GDP is no worse than one percent above Control, and another 3 years in which real GDP is on average 0.8-0.9 percent above Control. It takes 8 years before the appreciating dollar has had enough impact (along with the crowding out) to drive the deviation below 1/2 of one percentage point, and more than a decade before the fiscal expansion has turned perverse and driven real GDP below Control.

Tables 2-4 provide a more detailed look at the fiscal experiments by showing the path of deviations from control for eleven key variables in the

Table 2: Fiscal Experiment: "JEXR" Real Federal Purchases Path Up By 1% Of Real GDP

Quarters From Perturbation	Percent Deviation From Control					Deviation From Control					
	Real GDP	Private Non-Farm Deflator	Real Residential Construction	Real Business Fixed Investment	Exchange Value Of US $	Real Imports (bills. $87)	Real Exports (bills. $87)	Current Account Surplus (bills. $)	3-Month CD Rate (%-points)	Federal Purchases (bills. $)	Federal Gov't Surplus (bills. $)
0	1.15	0.00	0.51	0.05	0.00	1.77	0.00	-1.18	0.52	21.52	-14.02
1	1.40	0.02	0.39	1.18	0.00	3.92	0.00	-2.65	0.57	21.75	-11.86
2	1.55	0.06	-0.21	2.21	0.00	5.99	0.00	-4.10	0.86	22.29	-11.22
3	1.59	0.11	-1.14	2.98	0.00	7.65	0.00	-5.30	1.16	23.09	-11.67
4	1.57	0.21	-2.49	3.47	0.00	8.76	0.00	-6.17	1.18	23.66	-11.70
5	1.47	0.33	-3.88	3.56	0.00	9.26	0.00	-6.53	1.11	24.38	-13.76
6	1.38	0.48	-5.08	3.40	0.00	9.43	0.00	-6.52	1.35	25.03	-15.16
7	1.28	0.66	-6.20	3.20	0.00	9.43	0.00	-6.47	1.61	26.50	-17.10
8	1.16	0.87	-7.75	2.84	0.00	9.26	0.00	-6.24	1.49	27.54	-18.69
12	1.00	1.59	-9.89	1.69	0.00	9.27	0.00	-5.19	1.42	32.05	-22.59
16	0.90	2.18	-14.61	0.66	0.00	11.03	0.00	-5.91	1.75	36.40	-27.48
20	1.12	2.78	-9.57	0.51	0.00	14.66	0.00	-8.20	1.45	40.23	-26.87
24	1.14	3.40	-6.49	0.83	0.00	22.45	0.00	-13.71	1.83	43.66	-28.51
28	0.86	4.08	-7.66	0.05	0.00	27.72	0.00	-16.03	1.75	47.37	-33.97
32	0.76	4.76	-6.90	-0.38	0.00	34.05	0.00	-21.85	1.83	50.81	-36.55
36	0.71	5.34	-6.72	-0.66	0.00	36.15	0.00	-23.59	1.72	54.26	-37.08
40	0.53	5.94	-8.31	-0.66	0.00	40.74	0.00	-27.00	1.70	58.23	-37.89
44	0.21	6.49	-10.57	-1.44	0.00	42.58	0.00	-26.71	2.68	61.73	-41.59
48	-0.25	7.15	-14.59	-2.92	0.00	43.93	0.00	-23.88	2.29	67.02	-47.44
52	-0.45	7.60	-17.64	-4.10	0.00	43.99	0.00	-22.11	1.96	72.21	-50.08

Table 3: Fiscal Experiment: "JUS ROW" Real Federal Purchases Path Up By 1% Of Real GDP

Quarters From Perturbation	Percent Deviation From Control					Deviation From Control					
	Real GDP	Private Non-Farm Deflator	Real Residential Construction	Real Business Fixed Investment	Exchange Value Of US $	Real Imports (bills. $87)	Real Exports (bills. $87)	Current Account Surplus (bills. $)	3-Month CD Rate (%-points)	Federal Purchases (bills. $)	Federal Gov't Surplus (bills. $)
0	1.15	0.00	0.52	0.05	-0.05	1.77	0.07	-1.20	0.52	21.52	-14.02
1	1.40	0.02	0.39	1.18	-0.19	3.87	0.10	-2.76	0.57	21.75	-11.88
2	1.55	0.06	-0.22	2.22	-0.43	5.74	0.09	-4.32	0.86	22.29	-11.28
3	1.60	0.11	-1.14	2.99	-0.74	6.99	0.07	-5.53	1.17	23.10	-11.74
4	1.58	0.21	-2.50	3.47	-1.14	7.48	-0.04	-6.40	1.19	23.67	-11.80
5	1.50	0.33	-3.90	3.57	-1.52	7.17	-0.18	-6.59	1.14	24.39	-13.77
6	1.43	0.48	-5.11	3.42	-1.95	6.45	-0.29	-6.19	1.41	25.06	-14.99
7	1.35	0.67	-6.28	3.24	-2.49	5.52	-0.42	-5.96	1.71	26.54	-16.82
8	1.24	0.88	-7.92	2.89	-3.02	4.26	-0.68	-5.38	1.61	27.60	-18.27
12	1.17	1.66	-10.74	1.83	-4.48	0.47	-1.27	-1.70	1.61	32.25	-21.24
16	1.12	2.33	-16.94	0.78	-5.58	-0.46	-1.83	-1.42	2.04	36.81	-25.99
20	1.44	3.04	-11.57	0.76	-6.85	1.63	-1.37	-3.55	1.70	40.95	-25.00
24	1.66	3.80	-7.95	1.40	-7.60	5.78	-0.74	-8.87	2.24	44.79	-25.83
28	1.50	4.68	-9.67	0.84	-7.84	8.78	-1.01	-10.66	2.25	49.17	-30.86
32	1.56	5.64	-9.03	0.65	-11.42	10.16	0.04	-18.48	2.42	53.56	-34.18
36	1.94	6.55	-8.99	1.17	-13.90	3.42	0.84	-14.62	2.51	58.30	-30.89
40	1.91	7.64	-12.27	1.65	-15.19	5.14	0.64	-15.95	2.68	64.15	-30.45
44	1.46	8.81	-16.63	0.34	-15.45	4.47	-0.46	-11.07	4.36	69.73	-37.31
48	0.91	10.18	-22.14	-2.18	-17.67	0.80	-2.00	-4.29	3.88	78.21	-46.67
52	1.04	11.28	-25.86	-3.08	-20.74	-2.47	-0.30	4.54	3.56	86.59	-46.93

Table 4: Fiscal Experiment: "RROW3" Real Federal Purchases Path Up By 1% Of Real GDP

Quarters From Perturbation	Percent Deviation From Control					Deviation From Control					
	Real GDP	Private Non-Farm Deflator	Real Residential Construction	Real Business Fixed Investment	Exchange Value Of US $	Real Imports (bills. $87)	Real Exports (bills. $87)	Current Account Surplus (bills. $)	3-Month CD Rate (%-points)	Federal Purchases (bills. $)	Federal Gov't Surplus (bills. $)
0	1.15	0.00	0.52	0.05	0.23	1.77	0.07	-0.89	0.52	21.52	-13.93
1	1.41	0.02	0.40	1.18	0.71	4.19	0.15	-2.39	0.58	21.75	-11.74
2	1.55	0.06	-0.21	2.23	1.06	6.59	0.20	-3.83	0.86	22.28	-11.09
3	1.59	0.11	-1.14	3.00	1.46	8.72	0.28	-5.14	1.15	23.08	-11.58
4	1.55	0.20	-2.49	3.48	1.73	10.32	0.36	-6.26	1.17	23.65	-11.67
5	1.44	0.33	-3.88	3.57	1.79	11.25	0.41	-6.93	1.08	24.36	-13.90
6	1.34	0.47	-5.05	3.39	1.95	11.63	0.46	-7.05	1.31	25.01	-15.42
7	1.24	0.66	-6.13	3.17	2.19	11.79	0.43	-7.07	1.55	26.47	-17.43
8	1.10	0.85	-7.60	2.79	2.30	11.80	0.31	-7.14	1.42	27.50	-19.13
12	0.91	1.55	-9.48	1.60	2.91	12.33	-0.23	-7.13	1.31	31.96	-23.35
16	0.79	2.10	-13.55	0.56	3.49	14.74	-0.56	-8.25	1.61	36.20	-28.26
20	0.96	2.65	-8.68	0.35	4.58	19.25	-0.72	-10.58	1.34	39.89	-27.89
24	0.88	3.21	-5.82	0.50	5.15	28.98	-1.03	-16.70	1.63	43.12	-30.18
28	0.54	3.78	-6.65	-0.42	4.99	33.99	-1.32	-19.40	1.50	46.50	-36.01
32	0.42	4.32	-5.82	-0.84	7.74	41.30	-2.14	-23.84	1.57	49.47	-37.76
36	0.16	4.75	-5.70	-1.46	10.18	47.50	-3.28	-27.94	1.38	52.32	-40.11
40	-0.19	5.14	-6.65	-1.90	11.98	54.05	-4.68	-33.18	1.24	55.45	-42.72
44	-0.55	5.38	-7.77	-2.71	11.98	57.63	-5.37	-36.77	1.81	57.93	-46.68
48	-0.94	5.65	-10.05	-3.83	13.27	59.20	-6.17	-35.95	1.47	61.54	-50.98
52	-1.26	5.72	-11.85	-4.89	16.06	61.96	-8.48	-39.21	1.17	64.93	-54.28

model. Five are shown as percent deviations (real GDP, the core price level, real residential construction, real business fixed investment, and the exchange value of the dollar); six are shown as "Alternate minus Control" differences (real imports, real exports, the current account surplus, the 3-month CD rate, nominal federal purchases, and the federal budget surplus).

The results in these tables permit one to trace the impacts of each experiment in some detail, and we have already commented on some of the specifics in discussing the open economy simulations. Further detailed analysis is left for the reader, except to note that all three fiscal experiments produce a permanent deficit (relative to Control) in the federal budget. The increase in the deficit, however, is only about 50-60 percent of the (nominal) increase in federal purchases, which implies a 40-50 percent rate of endogenous revenue recovery.

Finally, we consider the results in Figures 3 and 4 which compare the real economic impacts of the fiscal stimulus on the U.S. and the rest-of-world economies for the two open economy simulations. For the U.S., economic activity is measured by real GDP; the closest analogue in our rest-of-world sector is our weighted average measure of industrial production, JIPROW. Figure 3 refers to Experiment "JUS.ROW" in which interest rates are coordinated internationally. The U.S. fiscal expansion is accompanied by an increase in domestic interest rates -- a non-accommodating monetary policy. Interest rate coordination thus means that foreign interest rates rise as well so that foreign economies are subjected to tighter credit conditions even though they've not been subjected to any direct fiscal stimulus. It is not surprising, therefore, that rest-of-world economic activity is negatively impacted, suffering a one to two percent loss of industrial production (relative to Control) during most of the simulation horizon. It should be noted that the rise of U.S. imports can be thought of as an indirect stimulus to foreign economies resulting from the positive impact of the fiscal stimulus on U.S. real GDP. Although the impact of this indirect stimulus is clearly dominated by the higher interest rates abroad, its existence is the reason that the size of the rest-of-world output loss is inversely related to the amount by which U.S. output is above Control.

Figure 4 refers to Experiment "RROW3" in which rest-of-world interest rates do not rise along with U.S. interest rates. In this case, the rest-of-world economies do not suffer from the tighter domestic credit conditions of the "JUS.ROW" experiment, and instead gain from the appreciating U.S. dollar which improves rest-of-world real net exports -- the other side of the deterioration in real net exports in the U.S. Industrial production is thus up by about one percent in the rest-of-world economies and remains positive even in the later years when U.S. output is below Control.

V. Monetary Policy Simulations

This section deals with simulations intended to reveal the model's response to a shift in monetary policy. There are three different monetary policy simulations defined according to the accompanying external conditions, precisely as in the case of the fiscal policy experiments.

Figure 3: Fiscal Experiment: "JUS. ROW"
Real Federal Purchases path Up by 1% of Real GDP

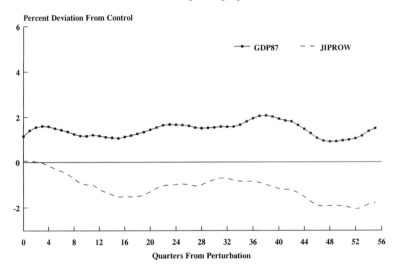

In all cases, the monetary policy perturbation lowers the *path* of the interest rate for 3-month Certificates of Deposit (CDs) by one percentage point (100 basis points) relative to the Control Run. We chose to impact the 3-month CD rate directly because it is the key short term interest rate in the model's basic money demand function. For purposes of perspective, we note that a 100 basis point drop in the path of the 3-month CD rate corresponds to a drop of about 85 basis points in the path of the 3-month Treasury Bill rate.

This monetary policy perturbation is not simply a one-time downward shift in the LM curve. The latter would be expected to lower the rate of interest which would lead to an increase in real GDP and the price level, whereupon the increase in the price level would then reduce the *real* money supply and start the rate of interest moving back up. In *this* monetary policy perturbation, the monetary authorities are required to provide whatever increases in liquidity are necessary to keep the rate of interest from moving back up; which amounts to a much stronger monetary stimulus policy.

The monetary stimulus is accompanied by *no change* in state and local government budgeting in precisely the same sense as discussed above in connection with the fiscal policy experiments. Analogously, we assume that federal government purchases are also budgeted in real terms, so that the federal government too is required to spend more to hold real purchases on the Control path when the monetary stimulus raises the price level.

Figure 4: Fiscal Experiment: ""RROW3"
Real Federal Purhcaes Path Up by 1% of Real GDP

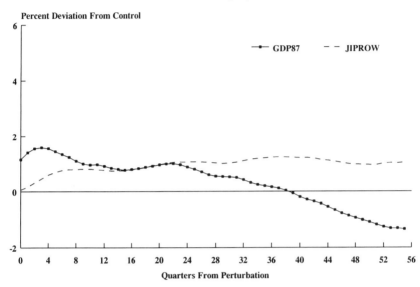

The three experiments can be characterized as follows:

(1)   Experiment "JEXR"   This experiment approximates monetary stimulus in a closed economy.   As in the corresponding fiscal policy experiment, we "close" the economy by forcing three variables to remain on their Control paths; namely, the exchange value of the U.S. dollar (JEXR), the price of imported goods denominated in foreign currency (PMROW), and real U.S. exports (X87). By this device, the rest-of-world sector is effectively eliminated from the model for the purpose of the simulation -- except for the fact that U.S. imports can change in response to the change both in domestic real income and in U.S. relative to foreign prices.

This closed economy experiment should be subject to significant long term real (price-induced) crowding out of the effect of the monetary stimulus on real GDP as a result of the negative impact of real-balance type effects on aggregate demand.   The crowding out may, however, be slowed down significantly as a result of the fact that the 3-month CD rate, as explained above, is held on a path fixed at 100 basis points below Control.

(2) Experiment "JUS.ROW" This is the open economy, coordinated interest rates experiment in which foreign economies match the decline in U.S. interest rates. This amounts to a matched monetary expansion, here and abroad. The outcome depends on whether the U.S. or the rest-of-world is more strongly impacted by a given path-shift in short term interest rates.

If the dominant effect of this symmetric monetary stimulus is to generate more domestic than foreign inflation and/or an increase in U.S. imports relative to exports, the result should be a depreciation of the U.S. dollar. And the weaker domestic currency will strengthen real net exports, hence real GDP, relative to Experiment "JEXR".   But the result would be quite the opposite if the dominant effect were reversed so that the rest-of-world currency did the depreciating.

In other words, "opening" the economy -- *with the rest of the world matching the U.S. interest rate policy* -- has an ambiguous result with respect to the efficacy of a stimulative monetary policy on the domestic economy.

(3) Experiment "RROW3" In this experiment, the *level* of foreign interest rates (RROW3) is held fixed so that the fall in U.S. interest rates *relative* to foreign interest rates provides a direct source of dollar depreciation. And this must of course enhance the efficacy of the monetary stimulus policy relative to that in the second experiment.

It might be thought that this presents another theoretical ambiguity, as follows. If the coordinated interest rates experiment ("JUS.ROW") reduces the domestic impact of an expansionary monetary policy, is it possible for this experiment ("RROW3") to bring the impact only part way back so that the expansionary impact remains weaker than in the closed economy ("JEXR") case? In fact, that cannot happen, as can be understood by comparing the current experiment directly with the closed economy simulation. If domestic interest rates decline and foreign interest rates do not, the domestic economy benefits from lower domestic interest rates, as in the closed economy simulation, and *additionally*, from a currency depreciation which further stimulates the domestic economy.   There is, in this case, no symmetric monetary stimulus abroad, hence no ambiguity.

Thus, in the case of monetary policy there is no ambiguity about Experiment "RROW3"; it is the experiment in which monetary policy should have its greatest impact on the domestic economy.

Figure 5 provides a quick summary of the three monetary-stimulus experiments by plotting the percent deviation of real GDP from Control for each experiment. The results can be described as follows relative to the issues raised in connection with each simulation.

Figure 5: Monetary Experiments
3-Month CD Rate Path Down by 100 Basis Points

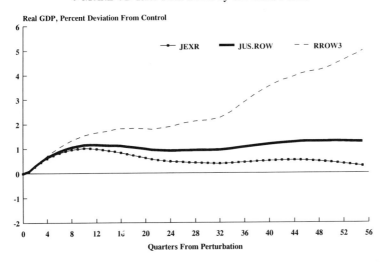

(i) In the closed economy ("JEXR") experiment the one percentage point drop in the rate of interest has virtually no effect on real GDP for half a year, raises real GDP by about 4/10 of one percent after one year, attains its maximum effect of a one percent increase in real GDP about a year later, and then sustains that effect for almost 2 more years before the real crowding out becomes significant. By the 6th year the stimulus effect is down to about 1/2 of one percent. A long, damped cycle then becomes evident and takes the stimulus down very slowly over the balance of the simulation horizon. Even in the 14th year, however, there is still a stimulus amounting to about 3/10 of one percent.

(ii) The result of the open economy, coordinated interest rates experiment ("JUS.ROW") is indistinguishable from that of the closed economy for about the first 2 1/2 years. By then, it is clear that the matched monetary expansion has larger impacts in the domestic economy, at least in the relevant dimensions (see Figure 6). Principally, the current account begins to deteriorate which puts downward pressure on the domestic currency and that begins to stimulate real net exports relative to the first experiment.

Apparently, the currency-depreciation effect is an approximate offset to the real crowding out that results from a higher domestic price level, so that the effect of the monetary stimulus stays roughly stable in the range of 1-1.3 percent over the balance of the simulation horizon.

The theoretical ambiguity is resolved in favor of the conclusion that openness of the economy enhances the efficacy of monetary policy even if the policy is matched abroad, but the degree of the enhancement is not quantitatively large.

Figure 6: Monetary Experiment: "JUS.ROW"
3-Month CD Rate Path Down by 100 Basis Points

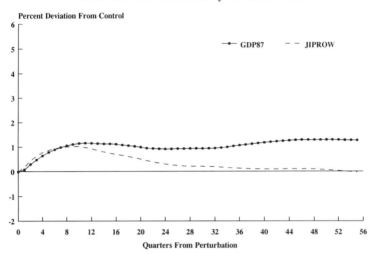

Percent Deviation From Control

Quarters From Perturbation

(iii) The third ("RROW3") experiment provides dramatically stronger results. Opening up a 100 basis point interest rate differential "against" the dollar provides a major economic stimulus for U.S. real GDP. The "fully open, go-it-alone" economy is subject to super-strong monetary policy. Indeed, compared to the most powerful open economy one-percent *fiscal stimulus* policy ("JUS.ROW"), this 100 basis point monetary stimulus policy is about equally powerful after 2 years, then becomes about 1/2 of one percentage point stronger in raising the level of real GDP for about the next 7 years. (Compare Figures 2 and 5.)

After the 9th year, the contrast with fiscal policy is even more startling. The dollar depreciation resulting from the sustained interest differential begins to provide an overwhelming and growing real net export advantage to the domestic economy (see the rest-of-world sector variables in Table 7), to the detriment of the rest-of-world economy. As discussed in greater detail below, this simulation assumes, quite unrealistically perhaps, that other countries permit the interest rate differential to persist despite the adverse consequences for their own economies.

Tables 5-7 provide information on nine key variables for each of the monetary simulations, and permit the economic effects of the simulations to be traced in considerable detail. The variables included are the same as in Tables 2-4, except for the exclusion of federal purchases and the federal government surplus.

Figures 6 and 7 are analogous to Figures 3 and 4 and compare the real economic impacts of the monetary stimulus on the U.S. and the rest-of-world economies for the two open-economy simulations. Figure 6 shows the case of a matched monetary expansion ("JUS.ROW"). While the domestic impact stabilizes so as to keep real GDP at about 1-1.3 percent above Control, the appreciation of foreign currencies adds to crowding out abroad, and rest-of-world output is back at Control in the closing years of the simulation.

Figure 7: Monetary Experiment: "RROW.ROW"
3-Month CD Rate Path Down by 100 Basis Points

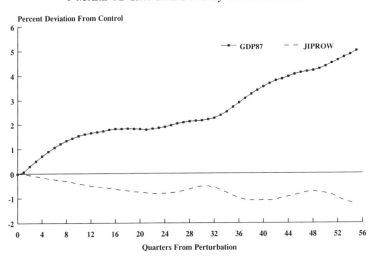

Figure 7 reveals a quite strong, and probably unsustainable, result. The "RROW3" monetary experiment in which domestic interest rates decline permanently relative to foreign interest rates, provides a permanent and increasingly strong domestic stimulus at the apparent cost of a permanent loss of output in the rest-of-world economies. As shown in Table 7, the permanent reduction of domestic relative to foreign interest rates in the "RROW3" monetary experiment produces the largest, sustained depreciation of the domestic currency of any of the experiments conducted (compare with Tables 6 and 3). The dollar depreciation amounts to 20 percent in the 10th year and 30 percent in the 14th year of the simulation. This has produced a huge positive swing of more than $65 billion (1987 dollars) in real net exports over the course of the simulation. But the corresponding strengthening of the current account balance is inadequate to stop the dollar depreciation in the face of the sustained differential in interest rates which works against the domestic

Table 5: Monetary Experiment: "JEXR" 3-Month CD Rate Path Down 100 Basis Points

Quarters From Perturbation	Percent Deviation From Control					Deviation From Control			
	Real GDP	Private Non-Farm Deflator	Real Residential Construction	Real Business Fixed Investment	Exchange Value Of US $	Real Imports (bills. $87)	Real Exports (bills. $87)	Current Account Surplus (bills. $)	3-Month CD Rate (%-points)
0	-0.01	0.00	-0.03	0.00	0.00	-0.02	0.00	0.01	-1.00
1	0.06	-0.04	0.74	0.00	0.00	0.04	0.00	-0.10	-1.00
2	0.26	-0.09	2.06	0.32	0.00	0.43	0.00	-0.43	-1.00
3	0.44	-0.10	3.29	0.65	0.00	1.10	0.00	-0.93	-1.00
4	0.59	-0.11	4.43	1.11	0.00	1.88	0.00	-1.55	-1.00
5	0.72	-0.12	5.20	1.57	0.00	2.65	0.00	-2.18	-1.00
6	0.82	-0.11	5.83	1.97	0.00	3.39	0.00	-2.78	-1.00
7	0.90	-0.09	6.39	2.34	0.00	4.03	0.00	-3.41	-1.00
8	0.96	-0.05	7.12	2.61	0.00	4.58	0.00	-3.95	-1.00
12	0.98	0.22	7.49	3.23	0.00	5.80	0.00	-4.75	-1.00
16	0.83	0.52	8.04	2.84	0.00	6.41	0.00	-4.80	-1.00
20	0.61	0.72	5.22	2.37	0.00	6.29	0.00	-4.33	-1.00
24	0.47	0.84	3.86	1.45	0.00	6.75	0.00	-4.46	-1.00
28	0.41	0.90	4.02	1.09	0.00	7.67	0.00	-4.88	-1.00
32	0.38	0.92	4.22	1.00	0.00	8.73	0.00	-6.24	-1.00
36	0.44	0.87	4.89	1.13	0.00	8.90	0.00	-6.77	-1.00
40	0.49	0.94	5.67	1.25	0.00	10.05	0.00	-8.04	-1.00
44	0.52	0.97	5.98	1.37	0.00	11.21	0.00	-9.28	-1.00
48	0.47	1.05	6.28	1.33	0.00	12.16	0.00	-9.76	-1.00
52	0.37	1.16	6.94	1.15	0.00	11.81	0.00	-9.06	-1.00

Table 6: Monetary Experiment: "!US ROW" 3-Month CD Path Down By 100 Basis Points

Quarters From Perturbation	Percent Deviation From Control					Deviation From Control			
	Real GDP	Private Non-Farm Deflator	Real Residential Construction	Real Business Fixed Investment	Exchange Value Of US $	Real Imports (bills. $87)	Real Exports (bills. $87)	Current Account Surplus (bills. $)	3-Month CD Rate (%-points)
0	-0.01	0.00	-0.03	0.00	0.00	-0.02	0.00	0.01	-1.00
1	0.07	-0.04	0.74	0.00	0.04	0.05	0.19	0.08	-1.00
2	0.28	-0.09	2.07	0.32	0.10	0.52	0.61	0.01	-1.00
3	0.47	-0.10	3.31	0.68	0.12	1.32	0.97	-0.35	-1.00
4	0.64	-0.11	4.45	1.16	0.14	2.19	1.26	-0.87	-1.00
5	0.78	-0.12	5.23	1.66	0.13	3.00	1.47	-1.41	-1.00
6	0.89	-0.11	5.86	2.09	0.13	3.68	1.61	-1.93	-1.00
7	0.98	-0.09	6.43	2.50	0.10	4.20	1.73	-2.47	-1.00
8	1.05	-0.04	7.16	2.79	0.04	4.53	1.80	-2.92	-1.00
12	1.15	0.26	7.56	3.55	-0.36	4.17	1.73	-2.90	-1.00
16	1.11	0.62	8.20	3.34	-0.75	2.74	1.31	-2.30	-1.00
20	0.99	0.91	5.37	3.17	-0.96	1.12	0.96	-1.69	-1.00
24	0.92	1.16	3.95	2.31	-1.08	-0.23	0.65	-2.23	-1.00
28	0.94	1.39	4.14	2.01	-1.17	-0.05	0.48	-2.54	-1.00
32	0.95	1.59	4.33	2.04	-1.28	0.09	0.36	-4.79	-1.00
36	1.07	1.74	5.02	2.34	-1.47	-0.79	0.24	-5.52	-1.00
40	1.18	2.03	5.82	2.45	-1.83	-0.62	0.17	-6.03	-1.00
44	1.26	2.30	6.17	2.60	-2.30	0.24	0.34	-5.96	-1.00
48	1.29	2.67	6.51	2.65	-2.90	0.13	0.41	-5.40	-1.00
52	1.29	3.09	7.25	2.67	-3.59	-2.05	0.23	-2.55	-1.00

Table 7: Monetary Experiment: "RROW3" 3-Month CD Path DOwn By 100 Basis Points

Quarters From Perturbation	Percent Deviation From Control					Deviation From Control			
	Real GDP	Private Non-Farm Deflator	Real Residential Construction	Real Business Fixed Investment	Exchange Value Of US $	Real Imports (bills. $87)	Real Exports (bills. $87)	Current Account Surplus (bills. $)	3-Month CD Rate (%-points)
0	-0.02	0.00	-0.03	0.00	-0.56	-0.02	0.00	-0.59	-1.00
1	0.07	-0.04	0.73	-0.01	-1.67	-0.57	0.11	-0.56	-1.00
2	0.29	-0.09	2.05	0.31	-2.41	-1.07	0.46	-0.53	-1.00
3	0.50	-0.10	3.30	0.66	-3.01	-1.57	0.72	-0.40	-1.00
4	0.70	-0.11	4.46	1.16	-3.51	-2.05	0.88	-0.37	-1.00
5	0.89	-0.11	5.26	1.70	-3.86	-2.56	0.99	-0.23	-1.00
6	1.06	-0.10	5.92	2.20	-4.35	-2.90	1.05	-0.01	-1.00
7	1.20	-0.07	6.52	2.70	-4.90	-3.24	1.16	-0.19	-1.00
8	1.33	-0.02	7.29	3.10	-5.41	-3.78	1.29	-0.01	-1.00
12	1.65	0.39	7.77	4.33	-7.22	-6.71	1.63	3.01	-1.00
16	1.82	0.92	8.56	4.56	-8.46	-8.94	1.52	4.01	-1.00
20	1.81	1.45	5.63	4.86	-10.26	-11.14	1.68	3.72	-1.00
24	1.90	1.98	4.16	3.99	-10.98	-16.56	2.03	3.86	-1.00
28	2.12	2.54	4.47	3.98	-10.67	-16.32	2.13	3.76	-1.00
32	2.25	3.14	4.64	4.28	-14.98	-19.53	3.35	-2.53	-1.00
36	2.89	3.72	5.55	5.55	-18.87	-31.18	4.65	2.73	-1.00
40	3.54	4.62	6.52	6.79	-21.20	-35.15	6.00	5.97	-1.00
44	3.95	5.69	6.92	7.67	-21.62	-35.36	6.81	12.35	-1.00
48	4.19	7.17	7.33	8.06	-24.41	-35.35	7.45	13.06	-1.00
52	4.61	8.85	8.40	8.72	-29.03	-44.60	10.43	28.55	-1.00

currency, and relative inflation is obviously working against the domestic currency as well.

It may be that a still longer simulation would eventually produceenough current account improvement to reverse the dollar depreciation; but one can't be certain. It may be that the permanent loss in foreign output is pointing to the fact that serious international policy problems could arise well before the domestic economy really starts to accelerate. For either economic or political reasons it may be impossible to maintain substantial interest differentials over such a long period of time.

APPENDIX

As mentioned in the text, an instrumental variable procedure was used to estimate the parameters in the exchange rate equation. The reason for this is that some of the goods that enter PMROW, the implicit price deflator denominated in foreign currencies for non-petroleum goods and services imported by the U.S., are priced by the U.S. Department of Commerce by multiplying the U.S. price of the comparable product by a component of the exchange rate itself. This procedure introduces a potentially troublesome errors-in-variable effect: a correlation between the explanatory variable Δ ln (PMROW/PX) and the disturbance term in the equation. The instrumental variables that were used for Δ ln PMROW are the predetermined explanatory variables in the Δ ln PMROW equation: Δ ln PCROW, Δ ln PCROW$_{-1}$, Δ ln JEXR$_{-2}$, Δ ln (RTB/RROW3)$_{-1}$, and Δ ln PMROW$_{-1}$. The remaining variables in the JEXR equation serve as valid instruments for themselves.

The main difference between the OLS and IV estimates of the parameters is that the OLS estimate of the coefficient of Δ ln (PMROW/PX) is 1.0328 with an estimated standard error of .0551 whereas the IV estimate is .8458 with a standard error of .1260. Although the OLS estimate comes closer to delivering purchasing power parity at the margin than does the IV estimate, there is good reason to believe that the OLS result is "too good to be true" because the OLS estimator is (upward) inconsistent.

E. Philip Howrey and Saul H. Hymans, Department of Economics, University of Michigan, Ann Arbor, MI 48109-1220.

ACKNOWLEDGEMENTS

Research Seminar in Quantative Economics, Department of Economics, The University of Michigan, Ann Arbor, MI 48109-1220; TEL:313-764-2567; FAX:313-763-1307; Howrey: Professor of Economics; Hymans: Professor of Economics and Director of the Research Seminar in Quantative Economics. The authors wish to express their deep appreciation to Pasquale (Pat) J. Rocco, who provided superb research assistance at all stages of this project.

REFERENCES

Green, R. Jeffery, Hickman, Bert G., Howrey, E. Philip, Hymans, Saul H. and Donihue, Michael R. (1991). The IS-LM core of three econometric models. Chapter 4 in Lawrence R. Klein (Ed.), *Comparative performance of U.S. econometric models*. (pp. 89-124). New York: Oxford University Press .

Hickman, Bert G., Huntington, Hillard G. and Sweeney, James L. (Eds.). (1987). *The macroeconomic impacts of energy shocks*. Amsterdam: North Holland.

Hymans, Saul H. (1990). The Michigan quarterly econometric model of the U.S. Economy: Structure and dynamic properties. Prepared for the U.S./U.S.S.R. Economic-Mathemtical Modelling Conference, Moscow/Sochi, U.S.S.R..

King, Robert G. and Watson, Mark W. (1992). Testing long run neutrality. mimeograph.

Klein, Lawrence R. and Goldberger, Arthur S. (1955). *An econometric model of the United States economy: 1929-1952*. Amsterdam: North Holland.

Klein, Lawrence R. and Burmeister, Edwin. (Eds.). (1976). *Econometric model performance: Comparative simulation studies of the U.S. economy*. University of Pennsylvania Press.

Klein, Lawrence R. (Ed.). (1991). *Comparative performance of U.S. econometric models*, New York: Oxford University Press.

Demand Management, Real Wage Rigidity, and Unemployment

Robert M. Coen and Bert G. Hickman

In this paper we investigate the relative importance of real wage rigidity and effective demand in determining excess unemployment in the United States. Unemployment is decomposed into its natural, classical, and Keynesian components during 1961-1990. This is partly an update of our earlier research (Coen and Hickman, 1987, 1988) which ranged only through 1984, but we have also modified our methodology in the process of respecifying the econometric model on which the study is based. The earlier research is also augmented by complete-model simulations of the impact of monetary and fiscal shocks on unemployment and its classical and Keynesian components.

I. Alternate Concepts of Classical and Keynesian Unemployment

In the standard fix-price model of price-taking competitive firms, Keynesian and classical unemployment are separate states according to whether notional product supply exceeds or falls short of market demand at the prevailing wage and price configuration, so that labor demand is either output constrained and determined by the inverted production function (Keynesian unemployment), or firms are on their notional product supply and labor demand functions but the real wage exceeds the Walrasian full-employment level (classical unemployment). Thus labor demand is independent of the real wage in the Keynesian state and depends only on the real wage in the classical state [Patinkin (1956), Clower (1965), Barro and Grossman (1971), Malinvaud (1977), and Benassy (1975, 1982)].

Our conceptual framework assumes instead that imperfectly-competitive firms price according to a markup rule and choose their inputs of labor and capital to minimize the cost of producing the output they expect to sell at the price they have set. The labor demand function is always conditional on both output and the real wage, and classical and Keynesian unemployment may therefore coexist. Keynesian unemployment exists when the economy is operating on an isoquant below that representing full-employment (F.E.) output, whereas classical unemployment occurs when the real wage exceeds its F.E. level either at or below F.E. output. For details on the theoretical specification see Hickman (1977) and Coen and Hickman (1987, 1988).

Notice that these concepts refer to unemployment states rather than the shocks which induce them. A supply shock, for example, may lead to classical unemployment by raising the real wage above the F.E. level, but it may also depress aggregate demand and create Keynesian unemployment.

In order to quantify the impacts of high real wages and deficient effective demand on unemployment in our framework, it is necessary first to estimate the F.E. rates of unemployment, real wages, and output for comparison with the actual levels of these variables. We begin with the F.E. unemployment rate.

II. Specification and Estimation of the Natural Rate of Unemployment[1]

Our specification of the natural rate of unemployment captures the influences of several factors. The natural rate is calculated as a weighted average of the natural rates for 16 age-sex groups, with weights proportional to the full-employment labor force in each group. Demographic changes can raise the natural rate insofar as they shift the composition of the full-employment labor force toward groups whose natural unemployment rates are normally relatively high. Demographic changes may also alter the natural unemployment rates of some groups; for example, disproportionate growth in the supply of younger workers might raise their unemployment rates insofar as it leads to a mismatch between the skills and work experience employers seek and those offered by younger job applicants.   Another set of factors includes those altering the real product wage or creating a wedge between the real wage paid by employers and the real wage received by workers.   These include indirect business taxes, employer contributions to social insurance, direct taxes on wage earnings, and changes in the relative price of consumption goods. The relative price of consumption goods in turn responds to changes in indirect business taxes and import price shocks.

One age-sex group, a prime-age male group, is selected as the "key labor force group," on the basis of its participation rate being relatively high and stable and its unemployment rate being relatively low and stable. Because of the key group's strong attachment to the labor force, its interests are likely to be dominant in wage negotiations. The natural unemployment rate of the key group is based in part on the notion that labor, as represented by the key group, and capital are engaged in a struggle over shares of aggregate income. Producers attempt to achieve a markup of price over unit labor costs to achieve their desired capital share, while workers seek a nominal wage that, in view of labor productivity and the price level, assures them of their desired income share. We assume that the bargaining strength of the key group is inversely related to its unemployment rate. Consistency of the desired income shares (they must sum to unity) is brought about by changes in the unemployment rate of the key group. Starting from a situation of equilibrium, a rise in taxes or import prices may create a distributional imbalance that would lead to spiraling inflation.   The imbalance is removed by an increase in the natural rate of unemployment.

The natural rates of the other age-sex groups are related to the natural rate of the key group, according to elasticities estimated from data on actual rates. The underlying determinants of the key group's natural rate thereby also influence the natural rates of the other groups. These other rates are also dependent on demographic shifts, according to estimated elasticities of each group's observed unemployment rate to its share of the population. The overall natural rate is a weighted average of the age-specific natural rates.

A. The Natural Rate of the Key Labor Force Group

Our model of the natural rate is a variant of the "Battle of the Markups" approach epitomized by Layard, Nickell and Jackman (1991). We assume that producers desire a capital share in income of γ and that the key labor force group (henceforth referred to simply as "labor" or "workers") desires a share of β. Suppose that the desired capital share is fixed, but that labor's desired share depends on the unemployment rate. A higher unemployment rate reduces workers' ability or willingness to bargain for higher wages, leading labor to settle for a smaller share.

If X is aggregate output, L labor input, W the nominal wage rate, and P the price level, then the desired labor and capital shares are, respectively:

$$\beta = \frac{WL}{PX} \tag{1}$$

$$\gamma = \frac{(PX-WL)}{PX} \tag{2}$$

To obtain the desired capital share, producers wish to set the price level as a markup on unit labor costs, that is,

$$P = \frac{1}{1-\gamma} \frac{W^e L}{X} \tag{3}$$

where the markup is on the expected wage and depends on the desired capital share.[2] The markup is assumed to be insensitive to changes in the level of activity, in view of the general finding in the literature that the elasticity of the markup with respect to changes in capacity utilization, orders, and similar variables is small. To obtain the desired labor share, workers wish to receive a wage that is a markup on the expected price level, namely:

$$W = \beta \frac{X}{L} P^e \tag{4}$$

The markup proportion here depends on the desired labor share and on the level of productivity. If, as we assume, the desired labor share itself is negatively related to the unemployment rate, U, then the markup proportion also depends on the unemployment rate. For example, we might have

$$\beta(U) = \beta_0 e^{-\beta_1 U} \tag{5}$$

in which case

$$W = \beta_0 \, e^{-\beta_1 U} \, \frac{X}{L} P^e \tag{6}$$

The desired factor shares will be consistent with one another only if they sum to unity. If capital and labor together desire shares that initially exceed unity, their interests are brought into balance by a rise in the unemployment rate, which reduces wage demands and makes labor content to settle for a smaller share. Given the specification of the desired labor share in (5), and imposing the condition that $\gamma + \beta = 1$, we can solve for the equilibrating, or natural, unemployment rate:

$$U^N = \frac{\ln \beta_0 - \ln (1 - \gamma)}{\beta_1} \tag{7}$$

Equation (7) is incomplete because is does not account for variations in tax rates or the relative price of consumer goods affecting the natural unemployment rate. A rise in the indirect tax rate t_i, for example, leads producers to increase the price level to maintain capital's share, and workers may respond with demands for offsetting wage increases to preserve their share. Similarly, the immediate impact of a rise in the employers' social insurance tax rate t_s is to raise labor costs and reduce capital's share. To restore capital's share, producers will attempt to recoup the added tax by reducing the wage rate or raising prices. Thus, if unemployment does not increase to induce labor to accept a lower share, the tax increases may set off an inflationary spiral as each group attempts to maintain its share by successive price and wage increases.

An increase in workers' combined personal income and social insurance tax rate t_x will reduce their disposable incomes and potential consumption and create upward pressure on wages, prices and unemployment.

A rise in the relative price of consumption goods P_c / P causes labor to demand higher wages in order to maintain potential consumption. If they are successful, capital's share declines, leading producers to raise prices, and so on. Once again a rise in the unemployment rate is required to coax labor into accepting a smaller share. There are two major factors affecting the real price of consumer goods. Differentially higher indirect taxes will tend to raise the relative price of consumer goods, since capital goods are not much affected by such taxes. Import price shocks may also drive up the real price of consumption goods.

The above analysis assumes that capital's share is invariant to changes in tax rates and relative prices, but this need not be so. To the extent that capital bears portions of these taxes and accepts a smaller share, the changes in the unemployment rate will be smaller. For example, in the extreme case where capital's share is reduced by the full amount of the taxes, the unemployment rate would be independent of the tax rates. An additional consideration arises in the case of social insurance taxes. To the extent that workers view these taxes, including the employer portions, as purchases of fairly-priced annuities which they would desire to purchase privately if they were not publicly provided, then tax increases should not lead to higher wage demands. Thus, the sensitivity of the natural unemployment rate to changes in taxes and relative prices is not clear a priori and needs to be studied empirically.

In our empirical implementation, we specify a log-linear relationship between the observed unemployment rate of the key group, which is 45-54 year old males, and its determinants:

$$\ln(U_{k,t}) = c_0 + c_1 \ln(t_{i,t}) + c_2 \ln(t_{s,t}) + c_3 \ln(t_{x,t}) + c_4 \ln s_m \left(\frac{P_{m,t}}{P_t}\right) + c_5 \ln(z_t) + v_t \quad (8)$$

where z_t is a capacity utilization variable and v_t is a stochastic disturbance.[3] We have substituted the real import price weighted by the import share for the real consumption price in order better to identify external supply shocks to the natural rate. The estimated equation omits the tax variables, which were insignificant. Finally, to estimate the natural rate of the key group, we purge the observed rate of its cyclical and stochastic components by setting z_t to unity and v_t to zero.

B. Natural Rates of Unemployment for Other Groups

Natural rates for other population groups are computed from regressions of their unemployment rates on the key unemployment rate and the population ratio of the group. For age-sex group i, the regression equation is:

$$\ln (U_{i,t}) = c_0 + c_1 \ln (U_{k,t}) + c_2 \ln (N_{i,t} / N_t) + w_t \tag{9}$$

where N is population and w_t is a stochastic disturbance. Thus, the effects of import-price shocks and cyclical variations in the age-sex specific rates are proxied by the corresponding variations in the key group's rate. The population ratios are included to allow for influences of labor market mismatches on the age-sex specific rates (Wachter, 1976). A rise in the proportion of young workers might, other things equal, increase their unemployment rates, insofar as employers do not regard them as good substitutes for older, experienced workers. Also, minimum wage legislation or other factors may prevent adjustments in demand favoring untrained and lower-skilled young workers.

To compute the natural rates from these regressions, we purge the observed rates of their cyclical and stochastic components by setting the unemployment rate of the key group on the right-hand side to its natural level and setting the disturbance to zero.

C. The Aggregate Natural Rate

The aggregate natural rate, U^N, combines the age-sex specific rates with weights equal to each group's share of the natural labor force:

$$U_t^N = \sum_{j=1}^{16} U_{j,t}^N \frac{L_{j,t}^N}{L_t^N} \tag{10}$$

where L^N refers to the natural labor force.

III. Prices and Wages

In this section we develop the theoretical specification of the price equation, discuss the empirical estimation of the equilibrium F.E. markup and relate it to the real wage gap, and describe the Phillips curve for nominal wage determination.

A. Theoretical Specification of the Price Equation

The distributional issues in wage and price determination were highlighted in the previous section by defining the desired price and wage markups in terms of the income shares of capital (γ) and labor (β), with the price markup equal to $1 / (1 - \gamma)$. What factors may determine γ and how does $1 / (1 - \gamma)$ relate to other markup specifications? These issues will be discussed in the context of our empirical model, which features a Cobb-Douglas technology with constant returns to scale and imperfect competition.

We assume a representative firm with value-added C-D function and neutral technical progress at rate ρ:

$$X = A\, e^{\rho t}\, K^{\alpha}\, L^{1-\alpha} \tag{11}$$

The market demand function is

$$X = BP^{-\phi_1} Y^{\phi_2} \tag{12}$$

where Y is aggregate income and $\phi_1 > 1$. Profit maximization yields the expression for optimal long-term price

$$P = [1-(1/\phi_1)]^{-1}\, AC \tag{13}$$

where

$$AC = A^{-1}\, \alpha^{-\alpha}\, (1-\alpha)^{-(1-\alpha)}\, e^{-\rho t}\, Q^{\alpha}\, W^{1-\alpha} \tag{14}$$

Equation (14) [the cost dual to equation (11)] is the expression for the minimum long-run average cost, AC, as a function of the production parameters, the wage rate, W, and the rental price of capital, Q. It is also the long-run marginal cost (MC) function under constant returns to scale. Thus the profit-maximizing condition (MR = MC) is represented in equation (13) as a markup on average cost, including the cost of capital as well as labor (Nordhaus, 1972). The markup is a function of ϕ_1, the elasticity of demand for the product, and its magnitude is determined by $[1 - (1/\phi_1)]^{-1}$, or the "degree of monopoly". Average cost may also be expressed, however, as[4] $AC = [1/(1-\alpha)]$ ULC, so that (13) may be transformed to

$$P = [(1-(1/\phi_1)]^{-1}\, [1/(1-\alpha)]\, ULC = \left(\frac{\theta}{1-\alpha}\right) ULC \tag{15}$$

where ULC = WL / X and $\theta = [1 - (1/\phi_1)]^{-1}$ is the degree of monopoly. Equation (15) shows that the markup over ULC basically depends on α, the relative ontribution of capital to production, and the elasticity of product demand ϕ_1.

Suppose that demand were perfectly elastic with $\phi_1 = \infty$. In that case labor and capital would be paid their marginal products, so that $\gamma = \alpha$. If $\infty > \phi_1 > 1$, however, labor and capital are paid their marginal revenue products and γ > α, as labor's share is reduced and "capital" receives a monopoly profit to augment its total interest return.[5] In this model, then, there is stability over the long-run in the income shares if the degree of monopoly and the marginal productivities are constant.[6]

B. Empirical Implementation of the Price Equation

Equations (13) - (15) refer to long-run equilibrium with firms on their production functions (no adjustment lags) and long-run cost curves. The actual markup may deviate from equilibrium because capital and labor inputs are not fully adjusted to the equilibrium levels and firms are not operating on their long-run cost curves or because of changes in the degree of monopoly associated with changes in demand pressure or capacity utilization.

For the master price equation we assume the existence of a constant, long-term equilibrium markup of price over unit labor cost and model the variations around the equilibrium markup by an error-adjustment mechanism. The adjustment variables include changes in the rate of capacity utilization, intended to capture the relatively small cyclical fluctuations in the markup, and changes in unit labor cost and lagged prices.

The equilibrium markup is estimated by the mean of the actual markup (P/ULC) during 1951-90 and is plotted against the actual markup in Figure 1. The reciprocal of the actual markup is, of course, the labor share (Figure 2). Evidently there are powerful forces tending to reverse deviations of the income distribution from its equilibrium, but the departures may sometimes persist for many years and attain considerable amplitudes.[7]

Figure 1: Actual and Equilibrium Price Markups

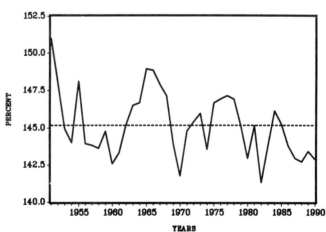

In addition to the master price equation for the aggregate price level, the model incorporates a set of auxiliary equations to explain the sector deflators as functions of the aggregate price index and other determinants.

C. The Real Wage Gap

We measure the real wage gap in relative terms as WR/WRF, where WR = W/P and WRF is the real wage that would prevail under full-employment conditions. In our model, the F.E. path implies an equilibrium income distribution, since otherwise the resulting wage/price pressures would be incompatible with maintaining a steady state growth path. The value of the long-run equilibrium markup over ULC compatible with optimal employment of labor and capital was defined as $\theta/(1-\alpha)$ in equation (15), and the equation can be rearranged to give WRF = $[(1-\alpha)/\theta]$ PRODF, where PRODF is the level of labor productivity at full-employment. Given a constant value for the equilibrium price markup, the real wage would grow at the same rate as productivity along the F.E. equilibrium path, and the equilibrium labor share would be constant at WRF/PRODF = $[(1-\alpha)/\theta]$, shown by the dotted line in Figure 2.

Figure 2: Actual and Equilibrium Wage Shares

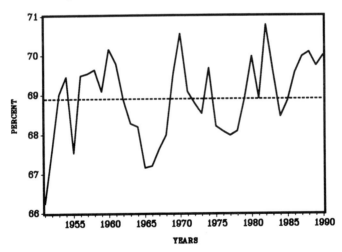

YEARS

The foregoing specification is an improvement over our earlier method of establishing the F.E. wage path in Hickman (1987) and Coen and Hickman (1987, 1988). In those studies the model was closed by assuming that the real wage grew at the same rate as labor productivity along the F.E. path, but this meant that the <u>level</u> of the real wage path was affected by the initial conditions at the beginning of the sample period. The new specification shares the same requirement of a constant income distribution along the growth path, but it establishes the equilibrium markup prior to the estimation of the F.E. path itself so that the level of the path is independent of the initial conditions.

D. Labor Market Phillips Curve

The theoretical money wage equation is a Phillips curve:

$$w = a - b\,(U_t - U_t^N) - c\,(U_t - U_{t-1}) + d\,(p^e) + e\,(p_{mw}) + f\,(D_iON) \quad (16)$$

where w is the rate of wage inflation, p^e is expected inflation, p_{mw} is the rate of change of import prices weighted by the import share, U and U^N are the actual and natural unemployment rates, and D_iON is a dummy variable which is zero before the ith time period and one thereafter. An import price shock may affect wage inflation directly through the p_{mw} term as well as indirectly by affecting U^N. The change in actual unemployment is included to test for persistence or hysteresis. Full hysteresis requires $b = 0$ and $c > 0$, so that wage inflation depends only on the change in U and there is no equilibrium NAIRU. If $b > 0$ and $c > 0$, there is persistence, implying loops around the Phillips curve of the Lipsey (1960) type, but not full hysteresis.

In the estimated equation p^e is an average of current and lagged inflation. The unemployment gap $(U - U^N)$ is highly significant, but the change of unemployment was statistically insignificant, implying at most a weak tendency toward persistence or hysteresis, and was omitted from the final regression. When actual and expected inflation are equal and $U = U^N$ the constant term equals the growth rate of the real wage. A fundamental determinant of real wage growth is the rate of technical progress, an estimate of which is obtained in our factor demand system. According to our findings, decelerations in technical progress occurred beginning in 1969 and again in 1974. By including appropriate dummy variables in the wage equation, we test whether a retardation in real wage growth accompanied these decelerations. A significant negative coefficient was found for the dummy beginning in 1969, but not for the one beginning in 1974.

IV. Determination of F.E. Output

We define F.E. output as the output that would be required to achieve full-employment of labor at the natural rate of unemployment, if the real wage were at its F.E. level. Estimates of F.E. output, the F.E. real wage, and the natural rate of unemployment are obtained by solving a simultaneous system of labor demand and supply equations that also determines the F.E. levels of the labor force and labor-hours. We first describe the specification of the equations on each side of the labor market and then indicate how the system is closed.

A. Labor Supply

Labor supply, measured in labor-hours, is the product of annual hours per worker and the number of employed persons, the latter depending on the unemployment rate and the size of the labor force. Labor force participation equations are estimated by sex for eight age groups. The principal determinants are the employment-population ratio (E/N), and the real after-tax consumption wage WRC. The employment-population ratio, an indicator of the probability of finding work, captures what are commonly referred to as discouraged-worker effects, while the after-tax real consumption wage measures the opportunity cost of leisure.

Denoting F.E. levels of variables by the suffix "F", the F.E. participation rate for the i-th group in year t is given implicitly by:

$$\ln\left[\frac{LPF_{i,t}}{1-LPF_{i,t}}\right] = c_{1,i} + c_{2,i}\,\frac{EF_t}{N_t} + c_{3,i}\,WRCF_t + c_{4,i}\,\frac{LA_t}{N_t}$$
$$+ c_{5,i}\,T49 + c_{6,i}\,\ln\left[\frac{LPF_{i,t-1}}{1-LPF_{i,t-1}}\right] \tag{17}$$

T49 is a time trend beginning in 1949, and N_i is the population in group i. LA is the armed forces, appearing only in the equations for male labor force participation. Additional trends and dummies are included for some groups. Given equations (10) and (17) and the volume of government employment, the aggregate labor force, aggregate employment, and private employment at full employment are determined by definitional identities.

The average hours equation is a hybrid relation combining both supply and demand factors. Hours per worker depend on the real after-tax consumption wage, which is presumed to affect workers' labor supply decisions, and on cyclical variations in labor demand as measured by the unemployment rate. Then F.E. average hours are given by:

$$HF_t = c_1 + c_2\,WRCF_t + c_3\,U_t^N + c_4\,LCW \tag{18}$$

where LCW is the proportion of women in the labor force, reflecting the fact that women tend to work fewer hours on the average than men. Finally, F.E. private labor-hours are then computed as the product of hours and employment.

The sub-system of labor supply equations can be solved for the F.E. values of the civilian labor force LF and civilian employment EF, conditional on the real wage and exogenous variables.

B. Labor Demand

The demand for labor-hours is derived on the assumption that firms choose labor and capital inputs to minimize the expected costs of producing expected output, subject to a Cobb-Douglas production function with constant returns to scale as in equation (11). Thus, desired labor-hour input depends on expected output, expected relative factor prices, and the level of total factor productivity. Because of adjustment costs, firms are assumed to close only a fraction of the gap between desired and actual labor-hours each period. Denoting expected values by the superscript "e", the disequilibrium demand function is:[8]

$$L_t = \left[\left(\frac{\alpha}{1-\alpha} \right)^{-\alpha} A^{-1} \left[\frac{(W^e)_t}{(Q^e)_t} \right]^{-\alpha} (X^e)_t \, e^{-\rho t} \right]^{\lambda} L_{t-1}^{1-\lambda} \tag{19}$$

where L is private labor-hours demanded, W is the private wage, Q is the implicit rental price of capital, X is private non-residential output, and λ is the speed of adjustment of labor-hours. The expected rental price is:

$$(Q^e)_t = (P_I^e)_t \, (r + \delta_t) \, TX_t \tag{20}$$

where P_I is the investment price deflator, r is the discount rate, δ is the depreciation rate, and TX captures the influence of business income taxation and investment subsidies. The discount rate is a constant equal to six percent per annum, which is a rough estimate of the average ex post real after-tax return on capital and can be thought of as the target return firms seek on new investment. We find that this specification gives superior fits for the labor input and investment functions to formulations using nominal or real long-term market interest rates before or after taxes. Expected values of wages, prices, and output are generally determined by predictions from autoregressions.

C. Full-Employment Output

Let XF be the F.E. level of X and WRIF be the F.E. real investment wage. Substituting LF for L on the left-side of (19), and XF for X^e and WRIF for W^e / P_I^e on the right, we can invert the labor demand equation to obtain the following expression for F.E. output:

$$XF_t = A \left(\frac{\alpha}{1-\alpha} \right)^{\alpha} \left[\frac{WRIF_t}{(r+\delta_t) TX_t} \right]^{\alpha} e^{\rho t} \, LF_t^{1/\lambda} \, L_{t-1}^{-(1-\lambda)/\lambda} \tag{21}$$

Our basic measure of aggregate output X is gross private nonresidential product, so F.E. GNP is obtained by adding to XF income originating in the housing and government sectors.

Note that XF is a function of both the real investment wage and the real consumption wage, the latter being a determinant of LF. To close the system, additional equations are needed to determine these F.E. real wages. We assume that the full-employment real <u>product</u> wage, WRXF, equals the reciprocal of the equilibrium F.E. markup times F.E. labor productivity:

$$ WRXF_t = \left[\frac{1}{MARKUPF_t} \right] PRODF_t = \left[\frac{1}{MARKUPF_t} \right] \frac{XF_t}{LF_t} \qquad (22) $$

Since we assume that the F.E. markup is constant, equation (22) implies that WRXF grows at the same rate as F.E. productivity and that the F.E. labor share is constant and equal to the reciprocal of the F.E. markup. Thus for reasons discussed earlier, our concept of the F.E. path imposes an equilibrium income distribution as well as a cleared labor market at the F.E. real wage.

Changes in real consumption and investment wages may differ from those in the real product wage for several reasons. First, there may be differential rates of technical progress in the two sectors. Second, consumption and investment goods prices may respond differently to external shocks, such as changes in energy prices. Finally, consumption and investment prices may display different patterns of cyclical behavior.

Changes in actual relative prices will, of course, be reflected in observed factor demands and supplies. To allow for their influence on the F.E. path, we express the F.E. real consumption wage as the F.E. real product wage multiplied by the actual ratio of the output deflator to the consumption deflator, and similarly for the F.E. real investment wage. Thus, we build into the F.E. path all observed changes in the relative prices of consumption and investment goods, although it would be preferable to abstract from purely cyclical changes. One result of this specification is that our empirical measures of real wage gaps are identical, whether they are expressed in terms of the output deflator, the consumption deflator, or the investment deflator.

The F.E. path generated by our specification is not the idealized smooth path of conventional measures of potential output. Instead, the path is disturbed by supply shocks to the real investment wage and by changes in taxation, both of which affect the rental-wage ratio in equation (21), by shocks to the real consumption wage, which affects LF, and by trend breaks in the rate of technical progress. The path is also sensitive to previous departures from full-employment, which impinge on current F.E. output in proportion to the gap between current F.E. labor input and lagged actual input. Finally, we use a stochastic concept of F.E. output, in which we add to the F.E. expressions for labor demand (as inverted in equation 21) and supply the current disturbances from the corresponding equations for actual demand and supply for labor.[9] In these respects our F.E. path specification resembles that of real business cycle theory, which emphasizes shocks of various sorts to the production technology and factor demands, although we do not assume continuous market clearing in the short-run and hence do allow for departures of actual inputs and outputs from their F.E. levels.

V. Theoretical Unemployment Decomposition

Our theoretical decomposition of employment is illustrated in Figure 3, where we abstract from variations in average hours, which are taken into account in the empirical estimates, and measure labor demand and supply in numbers of workers. The figure depicts a state of excess unemployment. LDF is the (log linear) labor demand schedule at F.E. output. The full-employment labor force is LF (drawn as wage-inelastic for simplicity), and the full-employment real wage WRF[10] would clear the labor market at the natural rate of unemployment (LF - EF) / LF. LD is the actual labor demand function whose position depends on current output and lagged employment. E persons are employed at the current real wage WR. There is a shortfall of (EF - E) between actual and full-employment employment. The actual or measured labor force, L, depends on the level of E, owing to the discouraged worker effect, which generates hidden unemployment of (LF - L). Actual or measured unemployment is (L - E), and the measured unemployment rate is (L-E)/L. The larger is the elasticity of labor force with respect to employment, the larger will be the increase in hidden unemployment and the smaller will be the increase in measured unemployment for a given decline in labor demand.

We adopt for our analysis the "adjusted unemployment rate," (LF-E)/LF, obtained by adding hidden unemployment to the numerator and denominator of the measured rate. The magnitude of discouraged-worker effects varies over time and among countries, and the adjusted unemployment rate facilitates intertemporal and international comparisons. The gap between the measured and natural unemployment rates shows the magnitude of excess unemployment, whereas the gap between the adjusted and measured rates shows the importance of hidden unemployment.

Given that WR > WRF, unemployment of (EF-EC) would exist even if actual and F.E. output, and hence LD and LDF, were equal. This is classical unemployment due to real wage rigidity and the failure of the labor market to clear. It may occur because of labor market bargaining on behalf of union members ("insiders") or because price setters are also efficiency-wage setters. It can be estimated empirically by substituting WR for WRF in the LDF function.

In past work we estimated the classical component by a counterfactual experiment in which WR was reduced to WRF with aggregate demand and output unchanged.[11] This increased employment to EA, so that of the original employment shortfall of (EF-E), (EA-E) was the classical component, attributable to the wage gap (WR-WRF), and (EF-EA) was the Keynesian element, assignable to deficient product demand. Given our constant-elasticit y log-linear labor demand function, it appears from Figure 3 that (EF-EC) = (EA-E). This is a misleading impression, however, because the labor demand function is actually specified for labor-hours rather than employees, and average hours are themselves an endogenous variable in the model. The partial wage elasticity of average hours is negative (the income effect dominates the substitution effect) so that one result of the hypothetical wage reduction is to increase average hours along with total labor-hours, thereby nullifying some of the stimulus to employment. Moreover, since the induced rise in employment from E to EA would encourage some workers to re-enter the labor force, the latter would increase from L to LA, further damping the estimated fall in unemployment from the hypothetical wage reduction. We now prefer to estimate the classical component directly at the disequilibrium wage - i.e. by (EF - EC) - so that the difference in labor-hour input due to real wage rigidity and the failure of the labor market to clear is attributed exclusively to a difference in employment at the existing levels of average hours and hidden unemployment.

Figure 3: Real Wage

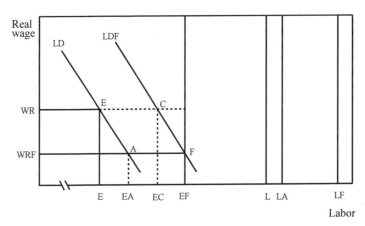

In the situation depicted in Figure 3, the adjusted unemployment rate is therefore decomposed into its natural, classical, and Keynesian components as follows:

$$(LF-E) / LF = (LF-EF) / LF + (EF-EC) / LF + (EC-E) / LF \qquad (23)$$

Figure 4 illustrates the five possible states of the decomposition. Point F shows the F.E. equilibrium, where there is neither Keynesian nor classical unemployment and measured unemployment is at the natural rate. Points E, B, H and J refer to the four disequilibrium employment states, and L, LB, LH and LJ are the corresponding actual labor supply functions reflecting the various degrees of employment-induced labor force change affecting hidden unemployment.

Two states are possible when the wage gap is positive, according to whether aggregate demand (AD) and hence LD fall short (point E) or exceed (point H) their F.E. levels. Classical unemployment (EF-EC) is the same in both cases, but excess Keynesian unemployment adds to total unemployment at point E, whereas at point H the classical unemployment arising from the wage gap is more than offset by the excess labor demand stemming from high aggregate demand. At point H the unemployment decomposition becomes

$$(LF-EH) / LF = (LF-EF) / LF + (EF-EC) / LF + (EC-EH) / LF \qquad (24)$$

The two remaining states involve a negative wage gap and either a negative (point B) or a positive (point J) AD and LD gap. In both cases excess demand for labor would exist at point G on LDF even if the AD and LD gaps were zero, and hence classical unemployment (EF-EG) is negative. The corresponding decompositions are

$$(LF-EB) / LF = (LF-EF) / LF + (EF-EG) / LF + (EG-EB) / LF \qquad (25)$$

where (EG-EB) is the employment shortfall from deficient aggregate demand, and

$$(LF-EJ) / LF = (LF-EF) / LF + (EF-EG) / LF + (EG-EJ) / LF \qquad (26)$$

where (EG - EJ) is the employment surplus or negative shortfall from excess aggregate demand.

VI. Partial Equilibrium Results

Our estimates of the natural unemployment rate and the unemployment gap during 1961-1990 are shown in Figure 5. The natural rate rose from 4.6 percent in 1961 to 6.6 in 1981 before declining to 5.7 in 1990. The deterioration in the eighties was due largely to shifts in the composition of the labor force, since the natural rate for the key labor force group was stable rather than declining. Actual unemployment was below the natural rate during the Vietnam war period and above it during the mid 70s and most of the 80s.

The gap between actual and F.E. or potential output is virtually the mirror image of the unemployment gap, as may be seen by comparing Figures 5 and 6. Substantial output losses were suffered in 1961-64, 1975-78, and especially in 1981-87, whereas large surpluses prevailed during 1966-70, and smaller ones in 1973-74, 1978-79 and 1988-89. The swings in the output gap primarily reflect the cyclical variations in actual output, but for reasons discussed in section IV.C, F.E. output does exhibit substantial variability in its growth rate, which actually turned negative in 1974 under the impact of the first oil shock.

Figure 4: Real Wages

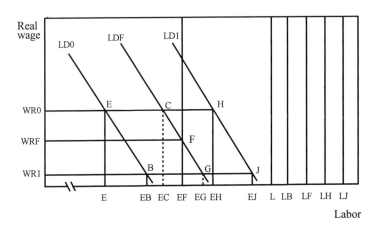

The Okun's Law relationship between the output and unemployment gaps is shown in Figure 7. The absolute slope of the regression line is 2.1, indicating that on average an increase of about two percent in aggregate output will be accompanied by a decline of one percent in unemployment. This is close to other current estimates of the Okun coefficient, most of which range between 2 and 2.5.

Both the actual and F.E real product wages trended irregularly upward during 1961-90 as the real wage gap alternated between positive and negative regions (Figure 8). The F.E. real wage declined absolutely during several years, reflecting corresponding declines in F.E. productivity induced by the oil shocks and other disturbances.

A positive relationship between the wage and output gaps is revealed by the scatter diagram in Figure 9. This positive correlation indicates that the real wage is procyclical, in the sense that fluctuations of actual output about F.E. output induce corresponding fluctuations of the real wage relative to the F.E. real wage. Market-clearing models require a reduction of the real wage to increase employment along a downward-sloping labor demand curve, but in our sticky-price model either actual or F.E. employment may increase at constant or rising real wages because output increases shift the labor demand function outward.

Figure 5: Actual and Natural Unemployment Rates

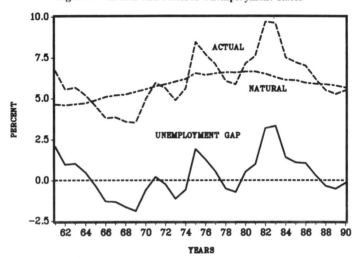

Thus the fluctuations in the wage gap are normally induced by corresponding fluctuations in aggregate demand. Supply shocks may, however, modify the procyclical pattern of the wage gap. Substantial wage gaps were opened during 1969 and 1974, for example, by absolute reductions in the F.E. real wage, and these reductions were induced by supply shocks affecting F.E. productivity. In 1969, the shocks were the discretionary elimination of the investment tax credit and an unexpected decline in the real investment wage, which together increased the rental-wage ratio and depressed the desired capital-labor ratio. In 1974, there was again an unexpected decline in the real investment wage under the inflationary impact of the first oil shock.[12] Since the F.E. wage is independent of the current state of aggregate demand, one may identify irregular movements of the F.E. wage with supply shocks broadly interpreted to include supply-side policies. When substantial shocks of this sort occur, they shift both F.E. output and F.E. productivity in the same direction, thereby augmenting the positive correlation of the output and wage gaps due to shifts in aggregate demand.

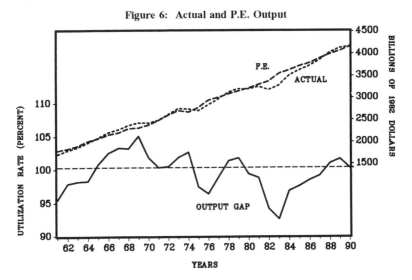

Figure 6: Actual and P.E. Output

Figure 7: Output Versus Unemployment Gap

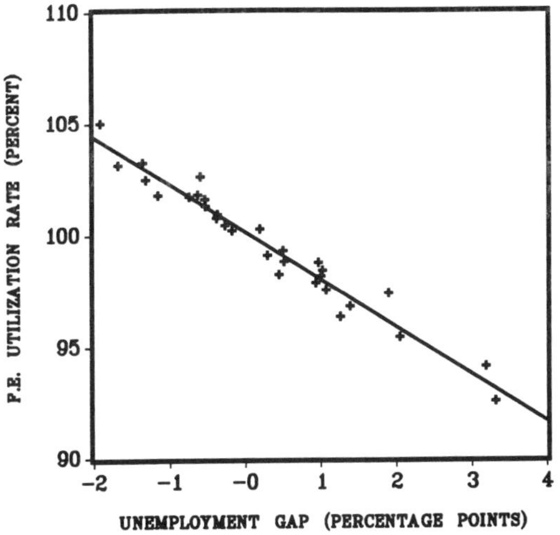

An interesting historical example of the foregoing combination of demand and supply shocks occurred in the early eighties. A large output gap developed in 1982 because of Volker's deflationary monetary policy, but in 1983 the subsequent deepening of the output gap and the development of a large negative wage gap reflected a marked increase in F.E. output and productivity. The productivity increment in turn is attributable to the effects of the computer revolution in reducing the investment goods deflator and rental price of capital relative to wage rates.

We turn now to the unemployment decompositions. The actual, adjusted and natural unemployment rates are shown in Figure 10. The gap between the actual and adjusted rates is a measure of the volume of hidden unemployment. Hidden unemployment - the difference between the F.E. and actual labor force - is usually positive when output falls short of its F.E. level and negative when the output gap is positive. The hidden unemployment rate reached 0.6 percent in 1975 and 0.9 in 1982-83, whereas it was - 0.5 during the Vietnam war years 1966-69.

Figure 8: Actual and P.E. Read Product Wages

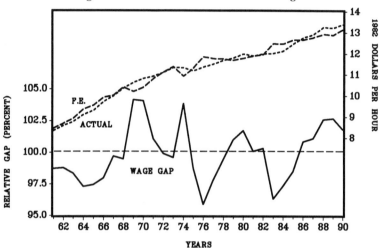

Excess unemployment - the difference between the actual and natural rates - also fluctuates cyclically but with much greater amplitude than hidden unemployment. The moderate rise in the natural unemployment rate in the mid seventies and early eighties accounts for little of the rise in actual unemployment in those periods.

Figure 11 decomposes the adjusted rate into its natural, demand and wage components as defined in equations (23)-(26). The demand component clearly dominated the fluctuations of excess unemployment throughout the period. The relative unimportance of the wage component, however, probably understates the role of supply shocks in the unemployment process. During the oil shocks, for example, there were adverse impacts on aggregate demand from terms-of-trade losses, and the inflationary consequences of the second oil shock helped to motivate the deflationary monetary policy of the early eighties.

The wage component is relatively unimportant in the U.S. for two reasons: both the wage gaps themselves and the wage-elasticity of labor demand are small. The short-run wage elasticity is the product of the capital coefficient α in the production function (11) and the adjustment speed λ in the labor

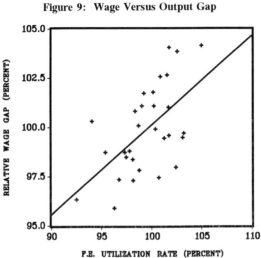

Figure 9: Wage Versus Output Gap

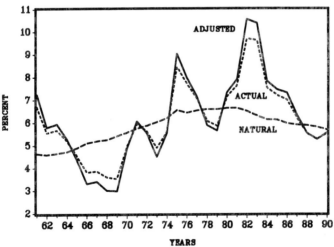

Figure 10: Adjusted, Actual and Natural Unemployment Rates

demand function (19), or (0.30)(0.66) = 0.20 in our estimates. The estimated wage gaps using our revised methodology range between plus and minus four percent, so the classical component ranges between about plus or minus 0.8 percent.

VII. Macroeconomic Policy Simulations

Given a state of Keynesian unemployment, should aggregate demand be stimulated by fiscal or monetary means? Would a positive stimulus be negated by financial crowding out? Would a positive response of the real wage increase classical unemployment and offset the reduction of Keynesian unemployment? Would the demand stimulus result in an unacceptably large tradeoff in increased inflation? In this section we investigate these issues by undertaking five-year simulations of fiscal and monetary policy shocks to the HC Annual Growth Model of the U.S. economy.

Figure 11: Adjusted Unemployment and Components

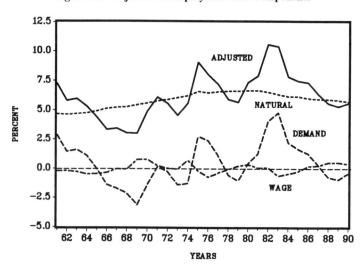

On the demand side the HC Model is a disaggregated, dynamic IS-LM system which explains six categories of consumption and three of investment, plus exports and non-oil imports and state and local government purchases. Aggregate supply is modeled as described above by the system of factor demand and production functions, labor force participation and average hours equations, the price markup equation, and the labor-market Phillips curve. In the updated system used in these simulations, the monetary sector has been reformulated as an error-correction model of money demand in velocity form with the option of treating either the money stock or the interest rate as the monetary policy instrument. The model has also been augmented to include endogenous exchange-rate determination embodying real interest rate parity, where the FRB multilateral exchange rate index is the measure of the effective exchange rate of the dollar.[13]

The first simulation is for a sustained shock to real federal spending equal to one percent of baseline GNP (see Table 1). During the first year, the shock induces an increase in real GNP of 1.8 percent, and since both the shock to government spending and the response of GNP are measured relative to baseline GNP, 1.8 is also the value of the impact multiplier on government expenditure (panel A). The dynamic income multiplier diminishes steadily in subsequent years as crowding out is induced by rising interest rates affecting consumer expenditures and residential construction and reinforced by multiplier-accelerator interactions (panels B and C).

The wage gap does increase as expected under the fiscal stimulus (panel B), but the increase is small and induces only a minor offset to the reduction in unemployment from increased demand (panel E). The reduction in actual unemployment is attenuated by a fall in hidden unemployment (panel D).

The unemployment-inflation tradeoff is favorable at first but becomes increasingly costly as inflation is sustained and employment gains are dissipated over the five-year horizon.

When the exchange rate is made endogenous under the fiscal shock, the entire multiplier stream is reduced and real GNP turns negative after three years (Table 2). This is because the induced rise in the interest rate appreciates the dollar, reduces international competitiveness, and crowds out net exports to a substantial degree (panels A and C). Inflation is lower because the appreciation puts downward pressure on import prices, but the tradeoff against unemployment is about the same as before.

The monetary scenario assumes a once-for-all increase in the level of M2 of four percent (Table 3). Interest rates fall accordingly and GNP rises from 0.5 percent of baseline in the first year to 1.9 in the second and tails off thereafter. The spurt in the second year is fueled by sharp jumps in the interest-sensitive areas of residential construction and consumer durables purchases. The pattern of unemployment change mirrors that of GNP and it is again the case that induced classical unemployment provides only a small offset to the reduction in Keynesian unemployment. Inflation is negligible at first but increases appreciably after about two years. The tradeoff between inflation and unemployment is favorable during the first two years but deteriorates thereafter.

Finally, the same monetary shock is imposed with the exchange rate endogenous in Table 4. As expected, induced dollar depreciation substantially augments net export demand and the GNP response to the monetary stimulus. The fluctuations in Keynesian and classical unemployment are amplified and so also is the inflation rate.

Table 1: Multiplier Restuls for Positive Fiscal Shock with Exogenous Exchange Rate

Variable and Units	1980	1981	1982	1983	1984
A. Percentage Deviations from Baseline					
GNP (billions of 1982 dollars)	1.8	.9	.5	.1	.0
GNP Deflator (1982=100)	.1	.6	1.3	1.8	2.4
M2 Money Stock (billions of dollars)	.0	.0	.0	.0	.0
Effective Exchange Rate (1973=100)	.0	.0	.0	.0	.0
B. Absolute Deviations from Baseline					
Inflation Rate (percent)	.1	.6	.7	.6	.6
Unemployment Rate (percent)	-.8	-.5	-.3	-.1	-.1
Short-term Interest Rate (percent)	1.5	2.3	2.5	2.1	3.0
Wage Gap (percent)	.6	.4	.2	.1	.0
C. Real Expenditure Components Absolute Deviations in Billions of 1982 Dollars					
GNP	57.6	30.0	14.4	4.2	.6
Consumer Expenditures	15.2	5.8	.0	-3.9	-5.4
Business Fixed Investment	9.7	7.8	3.2	.5	-1.0
Residential Construction	2.3	-6.7	-9.2	-11.7	-11.1
Inventory Investment	10.0	-1.3	-3.3	-3.0	-1.7
Government Purchases	30.9	31.5	30.0	29.3	29.5
Net Exports	-10.6	-7.2	-6.4	-7.0	-9.6
D. Adjusted Unemployment: Actual and Hidden Components Absolute Deviations in Percentage Points					
Adjusted Unemployment	-1.0	-.8	-.5	-.2	-.1
Actual Unemployment	-.8	-.5	-.3	-.1	-.1
Hidden Unemployment	-.2	-.2	-.2	-.1	.0
E. Adjusted Unemployment: Wage and Demand Components Absolute Deviations in Percentage Points					
Adjusted Unemployment	-1.0	-.8	-.5	-.2	-.1
Wage Unemployment	.1	.1	.1	.0	.0
Demand Unemployment	-1.1	-.9	-.5	-.2	-.1

Note: Details may not add to totals because of rounding.

Table 2: Multiplier Results for Positive Fiscal Shock with Endogenous Exchange Rate

Variable and Units	1980	1981	1982	1983	1984
A. Percentage Deviations from Baseline					
GNP (billions of 1982 dollars)	1.6	.7	.2	-.2	-.4
GNP Deflator (1982=100)	.1	.6	1.1	1.6	2.1
M2 Money Stock (billions of dollars)	.0	.0	.0	.0	.0
Effective Exchange Rate (1973=100)	2.3	2.6	5.8	7.8	10.7
B. Absolute Deviations from Baseline					
Inflation Rate (percent)	.1	.5	.6	.5	.5
Unemployment Rate (percent)	-.7	-.5	-.2	.0	.0
Short-term Interest Rate (percent)	1.4	2.0	2.0	1.5	2.1
Wage Gap (percent)	.5	.3	.1	-.1	-.2
C. Real Expenditure Components Absolute Deviations in Billions of 1982 Dollars					
GNP	52.5	23.4	6.5	-7.4	-15.0
Consumer Expenditures	15.6	7.1	4.8	4.7	9.8
Business Fixed Investment	8.9	6.3	1.4	-2.1	-4.4
Residential Construction	2.4	-5.7	-7.0	-7.9	-5.6
Inventory Investment	9.1	-1.9	-3.7	-3.8	-2.8
Government Purchases	31.1	32.1	31.3	31.4	33.1
Net Exports	-14.5	-14.5	-20.2	-29.6	-45.1
D. Adjusted Unemployment: Actual and Hidden Components Absolute Deviations in Percentage Points					
Adjusted Unemployment	-.9	-.6	-.3	.0	.2
Actual Unemployment	-.7	-.5	-.2	.0	.2
Hidden Unemployment	-.2	-.2	-.1	.0	.0
E. Adjusted Unemployment: Wage and Demand Components Absolute Deviations in Percentage Points					
Adjusted Unemployment	-.9	-.6	-.3	.0	.2
Wage Unemployment	.1	.1	.0	.0	-.1
Demand Unemployment	-1.0	-.7	-.3	.1	.3

Note: Details may not add to totals because of rounding.

Table 3: Multiplier Results for Positive Monetary Shock with Exogenous Exchange Rate

Variable and Units	1980	1981	1982	1983	1984
A. Percentage Deviations from Baseline					
GNP (billions of 1982 dollars)	.5	1.9	.8	.9	.4
GNP Deflator (1982=100)	.0	.3	.9	1.6	2.5
M2 Money Stock (billions of dollars)	4.0	4.0	4.0	4.0	4.0
Effective Exchange Rate (1973=100)	.0	.0	.0	.0	.0
B. Absolute Deviations from Baseline					
Inflation Rate (percent)	.0	.2	.7	.7	.9
Unemployment Rate (percent)	-.2	-.9	-.5	-.6	-.3
Short-term Interest Rate (percent)	-3.5	-2.2	-2.5	-1.6	-1.6
Wage Gap (percent)	.2	.6	.4	.4	.2
C. Real Expenditure Components Absolute Deviations in Billions of 1982 Dollars					
GNP	15.7	60.2	26.5	30.6	13.0
Consumer Expenditures	10.3	23.5	14.9	18.4	14.4
Business Fixed Investment	2.7	11.0	7.4	5.9	2.8
Residential Construction	.6	25.7	9.0	13.2	7.6
Inventory Investment	2.7	8.7	-2.8	-.4	-3.2
Government Purchases	2.3	3.3	5.4	5.2	4.0
Net Exports	-2.9	-12.0	-7.4	-11.7	-12.5
D. Adjusted Unemployment: Actual and Hidden Components Absolute Deviations in Percentage Points					
Adjusted Unemployment	-.3	-1.1	-.7	-.8	-.5
Actual Unemployment	-.2	-.9	-.5	-.6	-.3
Hidden Unemployment	-.1	-.3	-.2	-.2	-.1
E. Adjusted Unemployment: Wage and Demand Components Absolute Deviations in Percentage Points					
Adjusted Unemployment	-.3	-1.1	-.7	-.8	-.5
Wage Unemployment	.0	.1	.1	.1	.0
Demand Unemployment	-.3	-1.2	-.8	-.8	-.5

Note: Details may not add to totals because of rounding.

Table 4: Multiplier Results for Positive Monetary Shock with Endogenous Exchange Rate

Variable and Units	1980	1981	1982	1983	1984
A. Percentage Deviations from Baseline					
GNP (billions of 1982 dollars)	1.1	2.8	1.9	1.8	.7
GNP Deflator (1982=100)	.0	.5	1.6	3.0	4.5
M2 Money Stock (billions of dollars)	4.0	4.0	4.0	4.0	4.0
Effective Exchange Rate (1973=100)	-7.8	-11.6	-19.7	-20.4	-19.3
B. Absolute Deviations from Baseline					
Inflation Rate (percent)	.0	.5	1.2	1.4	1.5
Unemployment Rate (percent)	-.5	-1.3	-1.1	-1.1	-.6
Short-term Interest Rate (percent)	-3.2	-1.1	-.9	.1	.7
Wage Gap (percent)	.4	1.1	.9	.9	.6
C. Real Expenditure Components Absolute Deviations in Billions of 1982 Dollars					
GNP	33.7	91.7	59.8	60.3	26.1
Consumer Expenditures	9.1	18.8	-3.4	-10.3	-29.0
Business Fixed Investment	5.8	17.6	15.4	13.3	6.6
Residential Construction	.3	22.2	.1	-1.8	-10.3
Inventory Investment	5.9	12.3	-1.0	-.3	-5.9
Government Purchases	1.7	1.1	.6	-2.5	-7.1
Net Exports	11.0	19.7	48.2	62.0	71.7
D. Adjusted Unemployment: Actual and Hidden Components Absolute Deviations in Percentage Points					
Adjusted Unemployment	-.6	-1.7	-1.5	-1.5	-.9
Actual Unemployment	-.5	-1.3	-1.1	-1.1	-.6
Hidden Unemployment	-.1	-.4	-.4	-.4	-.3
E. Adjusted Unemployment: Wage and Demand Components Absolute Deviations in Percentage Points					
Adjusted Unemployment	-.6	-1.7	-1.5	-1.5	-.9
Wage Unemployment	.1	.2	.2	.2	.0
Demand Unemployment	-.7	-1.9	-1.7	-1.6	-1.0

Note: Details may not add to totals because of rounding.

VIII. Conclusions

Our concepts of classical and Keynesian unemployment refer to states of the economy rather than to the nature of the shocks which induced them. The decompositions for 1961-90 show that fluctuations in excess unemployment dominated the movements of actual unemployment - the changes in the natural rate were gradual throughout the period - and that classical unemployment was a minor part of excess unemployment. The importance of supply shocks in affecting aggregate demand directly and as a catalyst for restrictive policies is not captured, however, in the measures of the classical component.

The measures of unemployment states are incomplete guides to policy because both aggregate demand and real wages are endogenous variables. Supply-side measures to reduce a wage gap may also dampen aggregate demand, and a monetary or fiscal stimulus will induce a countervailing change in the real wage. Supply-side policies were not investigated in this paper, but use was made of the complete HC model to quantify the real wage offset to demand stimulation and the tradeoff between unemployment and inflation. The induced wage offset was found to be small. The tradeoff was favorable at first, but became less favorable over time as inflation increased and employment gains were eroded by crowding out.

Robert M. Coen, Departmet of Economics, Northwestern University, Evanston, IL, 60208, and Bert G. Hickman, Department of Economics, Stanford University, Stanford, CA, 94305.

ACKNOWLEDGEMENTS

The generous financial support of the Jubiläumsfonds of the Austrian National Bank is gratefully acknowledged.

NOTES

1. The topics in sections II-IV are more fully developed in Coen and Hickman (1992).

2. For simplicity, we specify the markup as determined by actual rather than smoothed productivity despite the fact that most econometric models assume normal cost pricing on the basis of trend productivity. Similarly, in equation (4) the desired wage depends on actual productivity.

3. For the derivation of this reduced form for $U_{k,t}$ from the expressions for the desired shares including the effects of the wedge variables, see Coen and Hickman (1992). Capacity utilization is an endogenous variable in our model. We assume that if firms are increasing their net capital stock, this indicates that they are using their existing stock more intensively than is optimal. Given our adjustment hypothesis for capital, which is analogous to that shown for labor in equation (19), capacity utilization can thus be measured as the ratio of the current to the lagged capital stock, raised to the reciprocal of the speed of adjustment of capital, which is estimated to be 0.156. (See Hickman and Coen, 1976, pp. 12-17, for the theoretical specification of the short-run production function and the rate of capacity utilization). In order to isolate the cyclical component of capacity utilization, the original series is normalized by its sample-period average.

4. See Hymans (1972). The transformation involves multiplying the r.h.s. of equation (14) by $(Ae^{\rho t}K^\alpha L^{1-\alpha}) / (X^\alpha X^{1-\alpha})$, collecting terms to yield $AC = \alpha^{-\alpha}(1-\alpha)^{-(1-\alpha)}(QK / X)^\alpha (WL / X)^{1-\alpha}$, making use of the marginal revenue productivity conditions $Q / P = [1 - (1/\phi_1)]\,\alpha(X / K)$ to eliminate (QK / X) and $W / P = [1 - (1 / \phi_1)]\,(1 - \alpha)\,(X / L)$ to eliminate $(WL / X)^{-\alpha}$, and collecting terms. Note that this derivation shows that in the long-term equilibrium of our system, the configuration of inputs, output, and prices will maximize profits as well as minimize costs, even though our factor demand specification is based only on cost minimization.

5. See Coen and Hickman (1992) for a specific algebraic imputation of the monopoly rent to (the owners of) the capital stock. It should be noted that our measure of monopoly power $P / MC = [1 - (1 / \phi_1)]^{-1}$ ranges upward from one without bound. The more common expression is $(P - MC) / P$, with range zero to one. In either metric, monopoly power is greater and the markup larger, the less elastic is demand (as long as it exceeds unity).

6. The foregoing development is closely related to target-return pricing formulas, which assume that price is set to cover unit labor cost and a target rate of return on unit capital cost, both at a standard rate of output, and where the target rate includes profit as well as interest. See Eckstein (1964), Eckstein and Fromm (1968), and Ball and Duffy (1972).

7. The hypothesis that the markup is stationary is accepted at the 5% level in an Augmented Dickey-Fuller test. The corresponding hypothesis that P and ULC are cointegrated is also accepted at the 5% level in an ADF test.

8. There is a similar disequilibrium demand function for capital input, which shares common parameters from the underlying production function. To honor restrictions arising from the production function, we estimate the input demand functions jointly using the seemingly unrelated regressions procedure.

9. The reason for adopting the stochastic specification of F.E. output is as follows. Suppose that the actual value Y (e.g., of labor input) depends on X and the residual e, so that (a) $Y = a + bX + e$. If the corresponding F.E. value YF were computed as (b) $YF = a + b\,XF$, and if $XF = X$, YF would equal Y only if e were zero. To ensure that $YF = Y$ whenever $XF = X$, the disturbance e must be added to the r.h.s of equation (b).

10. In this theoretical discussion we use WRF as the (single) real wage variable in the labor market, although our empirical model uses the real investment wage in the demand for labor and the real consumption wage in the supply of labor.

11. This procedure was followed because LD would not necessarily coincide with LDF under the potential output concept which we employed in the earlier studies. In those studies potential output was defined along a natural growth path of continuous equilibrium, so that the labor demand function at potential output depended on EF(-1), whereas the actual labor demand function depended on E(-1), and LD would not coincide with LDF unless the lagged employment terms were equal. See Hickman (1987) and Coen and Hickman (1987) for this distinction and the related concept of carryover unemployment.

12. Equation (21) may be recast to solve for F.E. labor productivity by dividing both sides by LF_t. One may then calculate the contributions of the various r.h.s. determinants to movements in F.E. productivity. The statements in the text are based on a calculation of the effect of changes in the rental-wage ratio on productivity together with a numerical decomposition of the separate determinants of the rental-wage ratio.

13. Space limitations preclude the presentation of the model equations, but a complete listing and data description is available upon request to the authors.

REFERENCES

Ball, R. J. and Duffy, Martyn. (1972). Price formation in European countries. *The econometrics of price determination*, Board of Governors of Federal Reserve System and Social Science Research Council, 347-68.

Barro, Robert J. and Grossman, Herschel I. (1971). A general disequilibrium model of income and employment. *American Economic Review*, March, 82-93.

Benassy, Jean-Pascal. (1975). Neo-Keynesian disequilibrium theory in a monetary economy. *Review of Economic Studies*. *42*, 503-523.

_____ (1982). *The economics of disequilibrium*, New York: Academic Press.

Clower, Robert W. (1965). The Keynesian counterrevolution: A theoretical appraisal. In F. H. Hahn and F.P.R. Brechling (Eds.), *The theory of interest rates*. London: Macmillan.

Coen, Robert M. and Hickman, Bert G. (1987). Keynesian and classical unemployment in four countries. *Brookings Papers on Economic Activity*. *1*, 123-193.

_____ (1988). Is European unemployment classical or Keynesian? *American Economic Review*. *78*, 2, 188-93.

_____ (1992). U.K. unemployment revisited. CEPR Publication No. 337, Center for Economic Policy Research, Stanford University, December, 1-30.

Eckstein, Otto. (1964). A theory of the wage-price process in modern industry. *Review of Economics and Statistics*, October, 267-86.

Eckstein, Otto and Fromm, Gary. (1968). The price equation. *American Economic Review*, 58, 5, Part 1, 1159-1183.

Hickman, Bert G. (1987). Real wages, aggregate demand, and unemployment. *European Economic Review*, *31*, 1531-1560.

Hickman, Bert G. and Coen, Robert M. (1976), *An annual growth model of the U.S. economy.* Amsterdam: North-Holland.

Hymans, Saul H. (1972). Prices and price behavior in three U.S. econometric models. *The econometrics of price determination*. Board of Governors of Federal Reserve System and Social Science Research Council, 309-24.

Layard, Richard and Nickell, Stephen. (1986). Unemployment in Britain. *Economica*. 53, S121-S169.

Layard, Richard, Nickell, Stephen and Jackman, Richard. (1991). *Unemployment*, Oxford University Press.

Lipsey, Richard G. (1960). The relation between unemployment and the rate of change of money wage rates in the United Kingdom, 1862-1957: A further analysis. *Economica*. February, 1-31.

Malinvaud, Edmond. (1977). *The theory of unemployment reconsidered*. Oxford: Blackwell.

Nordhaus, William D. (1972). Recent developments in price dynamics. *The econometrics of price determination*. Board of Governors of Federal Reserve System and Social Science Research Council, 16-49.

Patinkin, Don. (1956), *Money, interest and prices*. 2ed., New York: Harper and Row.

Poterba, J. M., and Summers, L. H. (1983). Dividend taxes, corporate investment and 'Q', *Journal of Public Economics*. 22, 135-167.

Wachter, M. L. (1976). The changing cyclical responsiveness of wage inflation. *Brookings Papers on Economic Activity. 1,* 115-159.

Macroeconometrics and Growth in the Global Economy

John F. Helliwell

Macroeconometric modeling, perhaps the longest-standing branch of applied macroeconomics, has faced an increasing number of challenges and opportunities over the past two decades. The increasing opportunities have arisen from the remarkable growth in availability of reasonably long samples of comparable macroeconomic data from an increasing number of interestingly different national economies, and the even larger growth of the power, flexibility and sophistication of the computer hardware and software available to support research. The real world challenges have come in part from big surprises, among which the 1973-74 and 1979-80 oil price increases loom large, but which probably include post-1973 exchange rate volatility and varying combinations of slower growth, higher fiscal deficits and debts, and increasing unemployment. The intellectual challenges have come in part from the need to adapt and specify models that come to grips in a reasonable way with new circumstances, including the doubling of trade shares over the past thirty years, and in part from academic attacks on the legitimacy of structural models of national economies.

At different times, and in different ways, proponents of rational expectations, vector autoregression, real business cycles and other new waves have made their case for primacy on the basis of what were said to be intellectual or empirical vacuums in the existing stock of macroeconomic wisdom, at least to the extent that it was embodied in structural econometric models. It is perhaps a reflection of the nascent state of scientific standards in empirical work in the social sciences, but I may be not the only one who was surprised at the extent to which there have been several so-called revolutions in macroeconomics conducted on the basis of so little empirical work testing the relative abilities of the competing frameworks to explain macroeconomic data. In the absence of comparable tests, it is not even clear if each revolution offered a framework which was to be interpreted as a reasonable response to changing economic structure and events, or if the fashion cycle for ideas is causally unconnected with the changing economic environment. In any event, the combination of challenges was large enough, when coupled with the increasing demands for researchers to produce early published results from their own individual research, to sharply reduce, or even eliminate, structural macroeconometric modeling in U.S. universities. There were notable exceptions, including Lawrence Klein at Pennsylvania, Ray Fair at Yale, John Taylor at Stanford and Saul Hymans at the University of Michigan, among others, but generally the teaching of macroeconomics became more theoretical than empirical, and the empirical work was seldom based on structural representations of a general equilibrium system.

How has macroeconometric modeling responded to these challenges and opportunities, and what are the most promising future opportunities? In addressing this main question, I shall deal in sequence with three other questions. First, to what extent are the impressions of decline in interest and activity, especially in the United States, verifiable, and to what extent has applied macroeconomics been more actively pursued outside the United States? Second, has applied macroeconomics in the United Kingdom been more active than elsewhere, and to what extent can this be attributed to Economic and Social Research Council (ESRC) funding and direction? And finally, what are the main theoretical and policy challenges facing applied macroeconomists in the future? These three questions form the basis for the rest of the paper, with one section devoted to each. The concluding section will then try to pull the disparate pieces together.

By way of introduction, and fittingly so in this Festschrift volume, it is appropriate to note that Lawrence Klein has been a leader in each of the three major ways in which modeling activity is changing in response to globalization. In the first section, I shall be noting first that macroeconometric modeling has become of increasing importance as a share of total published research, principally through the spread of this activity outside North America. Lawrence Klein was not only himself the leader of the first U.K. model-building team (Klein, Ball, Hazelwood and Vandome 1961), but over the years has trained many of those researchers who later undertook modeling activity in other countries. Section II pays special attention to modeling in the United Kingdom, where recent international leadership in several areas of model specification, estimation and comparison has been encouraged by a system of competitive granted-funded research.   One of the most successful parts of the U.K. experience has been the ESRC Macroeconomic Modeling Bureau at the University of Warwick, started at instigation of the ESRC funding consortium, and doing much to make models better understood and available, and in the course of this encouraging more widespread interest in model-related research. The initial model comparison activities of the Bureau were in part based on the U.S. model comparison exercises of the 1970s (as reported in Klein and Burmeister, eds., 1974), in which Lawrence Klein played a leading role. The Warwick Bureau raised the levels of model comparability and access, in part because of the research leverage provided by the funding consortium. In an example of positive feedback noted by Lawrence Klein (in Klein, ed., 1991, v), the success of the Warwick model comparison exercises were in part responsible for the existence and breadth of a new round of U.S. model comparisons, recently reported in Klein, ed. (1991) and reviewed by Ken Wallis (1992), the director of the Warwick Bureau. After reviewing the U.K experience, I shall review a number of the globalizing trends in modeling activity, including especially the increasing use of multicountry models. Here too, the leadership of Lawrence Klein is apparent, as he was the chief instigator and remains the key figure in Project LINK, which was the first (as reported in Ball, ed. 1973 and many subsequent publications) and remains the largest and most comprehensive of the multicountry models.

I. What Happened to Macroeconomic Modeling?

This question, which was also the title of a session at the American Economic Association meetings in January 1993, suggests, as did some of my introductory comments, that macroeconomic modeling has disappeared, either out of the range of professional interest or at least out of the United States. Is the presumption true, and if so, has the activity moved outside professional interest or just migrated outside the United States? There are two sorts of evidence that might help to answer these questions, the first being the records of some 220,000 professional publications between 1969 and 1990, as recorded by the *Journal of Economic Literature*, and now made available in CD-ROM form, and the second being a bibliography of macroeconometric models prepared on a continuing basis by Gotz Uebe (1991) of the University of Munich. Uebe records about 4300 macroeconometric models published in some form between Tinbergen's first model of the Netherlands in 1937 and mid-1991, and shows increasing relative activity in models of developing countries as time has passed. For our purposes, the *JEL* records may be the more useful source, since they contain comparable information for all research topics, and hence help to place macroeconometric research in a comparative context.

Figure 1 shows total annual publications, in the journals and books recorded on the *JEL* data base, in four fields chosen to cover most papers dealing with the construction and use of macroeconometric models, shown as a percent of all publications recorded in the *JEL* database.[1] The total shows fluctuations around an average of 2% for the period from 1969 through 1983, and thereafter a jump to a new post-1983 average of about 3.3%. The difference between the two means is highly significant, and regressions show that a regime-shift model with two different means dominates a model explaining the series either as time-dependent or as a mixture of time-dependence and regime shift.[2] Thus there would appear to have been no secular decline in macroeconometric research, but rather a significant increase during the 1980s. To delve further, Figure 2 shows the publications separately for each of the topic areas, showing the largest part of the growth to have been in the construction, analysis and use of econometric models (*JEL* code 2120).

Before using some sub-samples of the data to learn more about the patterns of research, it is necessary to consider some changes in data collection and reporting procedures that might be influencing the results in Figures 1 and 2. Figure 3 shows the selected econometric modeling publications by type of publication, while Figure 4 shows the same information for total publications. In both cases, there is a sharp upsurge in the total number of reported publications from 1984 on, reflecting first the inclusion of chapters in collective volumes and later books and PhD dissertations. These reporting changes are very large, as in 1988 the number of chapters in collective volumes was more than 8000, about two-thirds as large as the number of journal articles. In 1989 and 1990, the number of these publications drops away sharply, reflecting the lags in getting these publications recorded on the database.[3] This change in reporting raises the possibility that the upsurge in macroeconometric publications may simply be due to their greater relative importance in collective volumes, rather than to any increases in their frequency of appearance in academic journals. Figure 3 shows that this is not the case, and that the major part of the increase in macroeconometric publications took place within the category of journal articles.[4]

Figure 1: Macroeconometric Publications

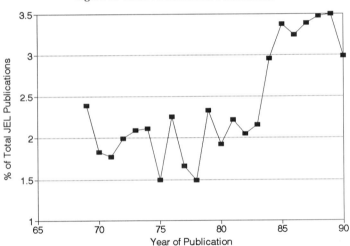

Year of Publication

To get some idea of the source and nature of the increase in publications in macroeconometrics, I selected a subset including all publications containing the words 'macroeconometric' or 'macro-econometric' in the title or abstract and also being in one of the five selected *JEL* topic areas.   Of these 119 publications, 42 were in the 1969 through 1983 period, and 77 were from 1984 to 1991. This represents a post-1984 increase similar to that evident for the larger group of all publications in the topic area.[5] Attempts to divide the publications by their country of focus showed a remarkable migration of interest and activity from North America to Europe and the developing countries in the 1980s. Of the total of 119 papers, 26 had no specific country focus, and half of

Figure 2: Publications by JEL Field

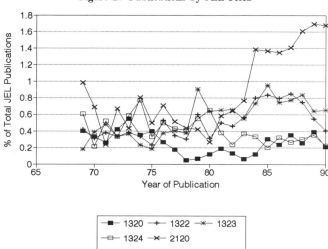

Figure 3: Macroeconometric Publications by Type

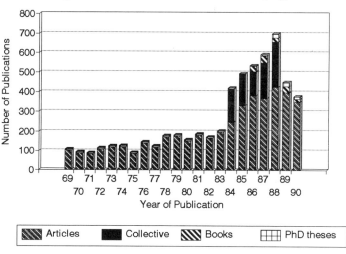

these were published before 1984. Of the 93 papers with a country focus, 21 related to the United States, 12 of which were published before 1984. Thus the United States was the focus of more than 40% (12/29) of the country-specific modeling papers before 1984, but less than 15% thereafter (9/63). The number of publications dealing with European countries, by contrast, including the United Kingdom as well as Central and Eastern Europe, grew from 6 before 1984 to 35, or more than half the total, from 1984 on. U.K modeling papers

Figure 4: All Publications by Type

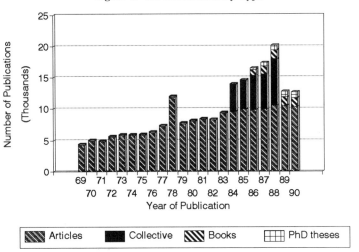

grew the most remarkably, from only 1 before 1984 to 18 in the later period. Publications related to developing countries also grew rapidly, from 5 in the earlier period to 19 in the later period. The small numbers of papers reflects the fact that I have chosen only a small subset of the 5823 papers in the five subject categories, but I suspect that the sample chosen is likely to be fairly representative, and that the pattern of migration of modeling activity from the United States to other countries would be replicated if alternative samples were chosen. Thus it would appear that the surge of post-1983 publications in macroeconometric modeling is principally due to research outside the United States. Macroeconometric modeling may have largely disappeared from view in the United States, but this was apparently due to its emigration rather than its demise.

II. Macroeconometric Modeling in the United Kingdom

In the previous section it was shown that there has been a sharp increase in macroeconometric model-building research and publication in Europe, and especially in the United Kingdom. Perhaps even more noteworthy is that of the 77 selected post-1983 publications in all countries, 18 related to the United Kingdom, and 15 of the 18 were clearly identifiable as flowing from research supported by the ESRC macroeconomic modeling consortium. On a per capita basis, the post-1983 published U.K. contributions have been ten times those relating to the United States. Given the strong preponderance of consortium-sponsored research among the published papers, and especially the strong representation of papers emanating from the Warwick Bureau (formed in 1983), it is reasonable to conclude that the substantial funding by the consortium has been largely responsible for, or at least supportive of, the surge of modeling activity and publication in the United Kingdom. To find the research productivity of the grants themselves, it would be necessary to obtain a much larger data-set that would permit other variables to capture the human capital and research proclivities of the funded and unfunded researchers.[6]

Given the 1980s preponderance of U.K modeling activities, what can be said about its likely payoff in terms of increased knowledge about macroeconomic structure and forecasting? I think it is possible to conclude that the combination of several factors, including the number and diversity of the modeling groups, the availability of sustained financial support, the careful and competitive research monitoring provided by the ESRC/Treasury/Bank of England Consortium, and the support and mandate for model comparison by the Warwick Bureau, has clearly put the United Kingdom in the vanguard of scientific macroeconomic research in the 1980s. The consistency, comprehensiveness, and appropriateness of the model evaluation and comparison techniques used in the United Kingdom are clearly superior to those in general use elsewhere, and the models themselves are correspondingly more transparent, better understood, and carefully tended.[7]

Have there been offsetting tendencies for the imposition of particular versions of 'best practice' imposed by the greater peer pressure likely to accompany the UK model evaluation processes? The first point to be made here is that nowhere else has there been such a continuing and extensive survey of model properties, so that it is not easy to see whether convergence has been greater in the United Kingdom, and, if it has been greater, whether this has increased the likelihood that the resulting set of models has become mistakenly single-minded. I shall return to consider some of the costs and benefits of model diversity in the next section. I have noticed two possible areas where there may have been more convergence than is so far supported by the data. Beyond these examples, there may be a general case for broadening the scope of the model types under review so as to guard against inbreeding, and to provide more guidance for those making modeling decisions in other countries.

One area where there has been especially notable progress in the United Kingdom, relative to other countries, is in the study of the specification and comparative structure of the supply side, generally following the leadership of Layard and Nickell, and as surveyed by Steve Nickell (1988). The chief characteristic and advantage of their approach is to recognize and ensure that the specification and estimation of the wage, price and factor demand equations are consistent with a coherent underlying specification of the production constraints and characteristics of the markets for labour and output. Such an approach would seem to invite a more explicit treatment of the underlying production structure, and hence a greater focus on the rate and longer-term determinants of technical progress. Unfortunately, although there has been much welcome analysis of short-term price and wage determination, and considerable emphasis on labour supply and demand, and on the determination of unemployment rates, there has been relatively little analysis of overall factor productivity, either in terms of its integration within the production structure or its growth over a medium-term forecasting horizon. There has been emphasis on the assessment of long-term properties of the models, but not yet as much as attention as would seem to me appropriate on longer-term technical progress, so often treated as an unexplained residual in the past and as an extrapolated trend for the future.

Another feature of UK modeling activity has been admirably great attention to econometric testing of alternative specifications, using a variety of powerful techniques. If there has been excessive convergence, without adequate continuing testing of the alternatives, it may lie in the use of estimation techniques that tie down the longer-term properties of a relationship but leave the intermediate dynamics more-or-less unconstrained, with the potential consequence of sharply differing dynamic properties from model to model, or from version to version. There is probably much more testing of this sort than I am familiar with, but however much there is, there probably should be more, given the increasing attention being paid to path dependence and the perennial importance for model users of having dynamics that are both plausible and comprehensible.

My final and more general point in the realm of model comparison strategy, before I turn to some specific issues of model specification and use, is to advocate extension of the sets of models being evaluated and compared. The battery of tests and the range of experiments already being applied by UK researchers, at the Warwick Bureau and elsewhere, is without doubt the most sustained and consistent set of procedures used anywhere to evaluate parallel research projects. So far so good. But think how valuable it would be to extend these tests on a more regular and uniform basis to include a wider range of alternative model structures, including ones that are often advocated as being superior to structural models without the application of methodical tests of the claims. Perhaps the weakest feature of applied macroeconomics is that dramatically diverging conceptions of macroeconomics can be propounded, and even become tools or assumptions widely used in teaching and research, without methodical testing against the alternatives. For the last period of almost twenty years, it has been widely claimed, at least in the United States, that structural macroeconometric models are inadequate for either forecasting or policy analysis, in the former case because they were simply no good, and in the latter case because of the 'Lucas Critique' arguing that the failure of existing models to derive restrictions on expectations from any first principles grounded in economic theory made them inadequate for predicting the effects of policy changes. The combination of these two inadequacies was said to underlie "the spectacular failure of the Keynesian models in the 1970s" (Lucas and Sargent,

1978). What was needed then, and can be more easily done now that history has unfolded for so many intervening years, is to derive and test comparably the competing simplifications of macroeconomic behaviour. Instead, there have been series of VAR models, real business cycle models, monetary models (of real GDP, prices, the balance of payments, prices and the exchange rate), the 'supply side' models of the early 1980s, computable general equilibrium models, and a wide variety of econometrically estimated structural models of the sort supported by the ESRC, without adequate monitoring of their relative predictive power either in or outside their estimation periods.

It is possible to do some of the testing among the competing approaches in terms of the different restrictions they impose on a more general specification in which they can all be nested, in which case the comparison can be done as a separate exercise.[8] For many of the bigger questions, the interest is in which of a variety of models fits the past and explains the future in a more adequate way. The activities of the Warwick Bureau, and the collaborative activities of the major modeling groups, both in the United Kingdom and elsewhere, do this to a greater or lesser extent for the existing mainstream models. The usefulness of these exercises for the broader group of macroeconomists, and for policy-makers trying to decide which horse or horses to back when making policy decisions, would be much greater if the range of models to be assessed were expanded to include a number of the simplified or structurally unrestricted forecasting devices with which econometric models have often been informally compared. This could be done for any country, of course, but the payoff would be especially great in the United Kingdom, where the range of available models is already quite wide, and where there are already well-established procedures for comparison and evaluation.

III. Trends and Prospects for National Modeling in a Global Context

I would like to concentrate in this section on three more specific questions of model structure and use. I shall deal first with the implications that recent work on policy coordination may have for the design of national and multicountry models. Then I shall consider the implications that recent work in growth theory and cross-country studies may have for the structure of national models, and finally consider other ways in which globalization is likely to influence the desirable structure for national and international macroeconometric models.

A. Policy Optimization and Policy Coordination When Model Structure is Uncertain

One of the most long-standing uses of national macroeconometric models has been to provide methodical guidance to the selection of macroeconomic policies. Qualifications to the simple application of the Tinbergen techniques were required to deal with cases where the number of targets exceeds the number of instruments, where account needs to be taken of external constraints, and where the economy is appropriately modelled by a non-linear system subject to stochastic shocks. Some of the early applications of the use of optimal control procedures to inform policy choices took place in the United Kingdom[9], and the U.K. has remained a centre for such research ever since. One of the issues that has arisen in more recent years relates to uncertainty about model structure, and how to embody such uncertainty in the choice of optimal policies. In the case of early research based on the use of a single model, the issue was sometimes couched as a worry that a full-scale search for optimal policies over some feasible range of values for the available policy instruments could produce choices that either exploited some previously unobserved bit of nonsense in the structure of the model, or took advantage of the lack of terminal conditions to leave some fiscal or inflationary time bombs set to go off after the end of the time period chosen for assessment. The more frequent use of models with forward-looking expectations, which require terminal feasibility constraints to be set and respected, and which mirror future problems in current bond prices and exchange rates, has helped to establish realistic terminal conditions. The issue of model uncertainty has increasingly been addressed by recognizing that there are several alternative models available, and that they may all have some information that could be used in the choice of policies.

There are several possible strategies for using a stable of models to reduce the risks that the result of a model-specific search for optimal policies is being driven by some feature that is both implausible and peculiar to the model in question. One element common to most strategies is to use model diversity as a means of testing the robustness of policy choices to uncertain features of model specification or estimation. In the recent Brookings tests of the optimal choice of monetary strategies in a stochastic simulation environment (Bryant et al, eds., 1993), the model-builders ranked the various policy rules by using several alternative loss functions applied to each of the alternative econometric models, and then looked for policy rules that appeared to be robust across different models. If the models are all taken to be independent drawings from some distribution of equally probable models, then a ranking of policy rules that is uniform across alternative models is more credible than a ranking based on results from a single model.

The same issue arose in the aftermath of a paper by Frankel and Rockett (1988), in which it was shown that international agreements on policy coordination based on one particular multicountry model often turned out to be disadvantageous if one of the other multicountry models were in fact true. The paper hence tended to be skeptical of coordination efforts. However, the subsequent literature (Holtham and Hughes Hallett 1992; Frankel, Ervin and

Rockett 1992) pointed out that if the choice of policy bargains is limited to those that are advantageous under a variety of models, then the outcomes are more likely to be favourable, even if the models used to assess the outcomes *ex post* do not include any of the models used to inform the policy selection. In this context, model diversity is an asset rather than a liability, since the diversity provides a check against policy choices that are based on peculiarities of the structure or parameters of a particular model. On the other hand, as Wallis (1992) points out, the range of model properties may be very wide, and this may mean (as pointed out by Holtham and Hughes Hallett 1992) that there are no policies robust enough to be preferred by all models. This suggests that some preliminary tests of model plausibility should precede their inclusion in sets of models used for checks of policy robustness. Similarly, guard should be taken against untested convergent swings of fashion in model-specification that may either artificially narrow or shift the agreed range of policy outcomes.

Although much of the recent activity using diverse models to represent policy choices has related to multicountry models, and to the choice of internationally agreed policies, precisely the same issues occur in the context of domestic policy choices. Thus there would appear to be a continuing payoff to model diversity, and to parallel research paths that may give rise to quite distinct model types. However, this diversity will be of most use if Bureau-like standards can be maintained for model transparency, availability, and comparative testing.

B. Modeling Growth

I have already argued above that most existing national models, including those for the United Kingdom, do not attach sufficient importance to the representation of technical progress. In the past, when modellers were mainly concerned with short-term forecasting, and with the analysis of the short-term effects of monetary and budget policies, it might have seemed natural to adopt a demand-side determination of aggregate output, with supply-side constraints flowing back through prices, wages and interest rates as determinants of effective demand. This was turned on its head by the real business cycle modellers, who assumed an underlying production function to hold, subject to autocorrelated technology shocks. But neither type of model spent much effort explaining the economic determination of technical progress or total factor productivity, in the one case because it was implicit in some of the trends and constant terms in the equations determining prices and wages, and in real business cycle models because it was taken to follow a stochastic trend.

New theoretical and empirical work in comparative growth suggests that over a longer-term horizon than the business cycle, the really interesting questions require the isolation and attempted explanation of national rates of technical progress. This is not the place to survey that literature in any depth, but some stylized facts are starting to emerge to suggest that treating technical progress as either constant or subject to a stochastic trend is not sufficient. For one thing, construction of time series of Solow residuals for 19 OECD countries from 1960 through 1989 shows that there is no chance that all countries share the same production function or that the rate of technical progress within any country can be treated as a constant (Helliwell 1992). For the industrial countries as a group, for the period since 1960, there is systematic evidence of convergence in the rates of growth (but not of the levels) of technical progress. Furthermore, gradual convergence appears to dominate a post-1973 break as an explanation of the decline in the average rate of technical progress that has characterized the industrial countries over the past three decades (Helliwell and Chung 1991). As emphasized by Mankiw, Romer and Weil (1992) and Barro (1991), the convergence of rates of growth of real output per capita is consistent with an underlying Solow model with equal constant rates of technical progress in all countries, but only as long as the convergence is fuelled by higher investment rates in the initially poorer countries. However, analysis based on the Solow residuals shows that convergence is taking place there, thus rejecting any model of constant technical progress, and providing some support for models (e.g. such as those of Grossman and Helpman 1991) that account explicitly for international transfers of technical progress. What can be ruled out, however, are models that rely only on national returns to scale fuelled by spillovers from the domestic accumulation of either physical or human capital (Romer 1986), since there is evidence of international convergence of rates of technical progress even after allowing for the estimated economies of scale and allowance for investment in human capital. Still unresolved, I would judge, is whether physical investment has any external effects; experiments with data for the industrial countries show no evidence that countries with higher investment rates have faster rates of growth of their Solow residuals (Helliwell and Chung 1991, Helliwell 1992), but De Long and Summers (1991, 1992), using a global sample of countries, find some evidence that countries with higher rates of machinery and equipment investment have higher rates of growth of real output per capita GDP, to a greater extent than would be expected given usual estimates of the marginal product of capital.

The facts that technical progress has been shown to have systematic variance over time, and that productivity remains the largest contributor to per capita income growth over the longer term, suggest to me that treatment of technical progress as part of an explicit production function, accompanied by more systematic attempts to explain its variations, should be part of the modeling agenda for national models. It may be that efficient estimates of the rates of technical progress in national economies will have to rely on the use of pooled data from a number of comparable economies, given the difficulties of tying down the determinants of slowly-moving trended series. For the construction of national models, this may represent efficient use of extraneous information to identify some of the key features of domestic economic structure. It may also increase the likelihood that model-builders will start to question the autonomy of some of the structure of their national models in the face of increasing globalization, a topic to which I now turn.

C. Modeling When Globalization is Increasing

I have already argued that technical progress is not constant, either through time or across countries, and that there appears to be international convergence in its rate of growth. There is also some evidence that productivity growth is faster for countries that trade more (Harrison 1991, Helliwell 1992) and for countries that have recently reduced their trade barriers (Ben David 1992), thus providing a further potential linkage from standard macroeconomic structure to the underlying trends of technology. Increasing globalization has been marked by much greater mobility of goods, capital, people and communications, each of which is likely to have changed the degree of domestic autonomy, and to have led to at least parameter shifts in some key relationships in national models. Without attempting to be exhaustive, two examples may help to illustrate some of the problems.

Trade shares have doubled for most of the world's economies over the past twenty-five years, giving rise, among other things, to very high income elasticities for imports estimated in equations containing only income and relative prices. Whether models are being used for cyclical or longer-term purposes, it is probably important to distinguish trend changes in imports from cyclical ones, preferably in some way that also explains trends in openness. For the richer industrial countries, which are destined to have smaller shares of world exports to the extent that poorer countries have faster growth rates, as convergence would imply, and open themselves to trade, it is necessary to be careful in choosing the scale variables used to determine both imports and exports. It is also likely to be important to treat globalization-induced increases in trading activity separately from the cyclical variance. The changing pace of globalization is likely to influence both, but in different ways.

Another key part of national macroeconometric models explains the rate of domestic investment, usually on the basis of factor demands derived, often consistently in the U.K. context, from the expected growth of domestic output and relative factor costs. However, from the point of view of globally oriented enterprises, the location of new plant and equipment depends only partly on the growth of the national market in which the plant is to be located. Increasingly important are likely to be the relative costs of producing and assembling in a variety of places, with global products (ones comprising bits and pieces from all over the globe) being increasingly common. Is it possible to conceive of an augmented version of the Tobin Q model, wherein the ratio of expected future revenues (based on sales to any number of countries) from investment in the domestic economy is compared to the present value of the capital, labour, environmental and tax costs over the life of the facility? What would matter for the pace of domestic investment would be not only whether that ratio exceeded 1.0, in the usual way, but also how it related to comparable calculations for the same investment located elsewhere. Thinking in portfolio terms would also suggest taking account of the marginal riskiness of domestic investment when added to an existing portfolio of plant and equipment. The value of diversification may help relatively high-cost jurisdictions to continue to draw substantial investment, particularly if domestic economic and political risks are low, or uncorrelated with risks elsewhere.

The examples I have chosen to illustrate the impact of globalization on the future relationships in national models are only two among many, but even they have implications for each other. If physical investment decisions are becoming increasingly globalized, then it will become increasingly important to model trade flows as depending on the size, and perhaps the recency, of the economy's productive capacity. This has always been obvious in the case of export-oriented investments for resource developments in locations that would otherwise not be heavily involved in trade, but the logic is just as strong in the case of footloose manufacturing investment. To the extent that the doubling of trade shares over the past thirty years has really represented a decline in the perceived costs of international trade, then there must have been a corresponding increase in the importance of relative international costs in decisions about the location of production facilities, and an increase in the linkage between these investments and subsequent trade flows.

Perhaps a word about labour and financial markets would be in order, by way of conclusion to this section. In the context of Europe after 1992, it might be expected that there should be a shift as well as a trend in the tightness of international labour market linkages, but even the trends have had little direct representation in national models. For capital market linkages, it is more common now to model them by means of some modified form of uncovered interest parity, which may well overstate the tightness of the linkages to the extent that foreign exchange risks are not duly accounted for. However, to overstate the linkage in this way is probably better than to assume autonomy of domestic interest rates, as was frequently done (for better reason then) a decade or two ago. There are also implications of globalization for tax policy and even the modeling of tax yields, since as Razin and Sadka (1991) and others have pointed out, tax competition, and the benefits of international tax harmonization, increase with factor mobility.

Overall, the modeling implications of increasing globalization are too great to require emphasis; but the difficulties of capturing gradual changes in the relative importance of domestic and foreign prices and markets should not be underestimated.

IV. Conclusions

The first section presented evidence that macroeconometric modeling is alive and living in Europe, with a leading presence in the United Kingdom. The widespread view (at least in the United States) that empirical macroeconomics, especially that based on structural models, is now a professional backwater, or a matter only of concern to consultants and business forecasters, is clearly not supported by the global data on published research over the past twenty years. There has certainly been an increase in the variety of models, in terms of theory, estimation techniques and country coverage. The country coverage has been extended to include many more developing countries, and there has been much more comparative and multicountry modeling over the past decade.

The second section, which looked especially at modeling in the United Kingdom, argued that in all of these areas the United Kingdom has been a leader in terms of the quality and quantity of applied macroeconometric research. The resources devoted to this research have been great, and the coordination among macroeconomic researchers at a level unmatched in other countries. Although there may be occasional risks of loss of model diversity as a result, the overall benefits of having models widely available, frequently compared, and extensively tested have been impressive. The resulting published research has been impressive in its quality and quantity, and will surely have substantial spinoffs to users of models and forecasts, as well as to modellers in other countries.

The third section of the paper listed a few of the likely priorities for future research. A discussion of the use of differing models to test the robustness of policy advice was advanced as additional support for model diversity. I then advocated more attention to the determination of longer-term technical progress, and argued for a more explicit, and more internationalized, treatment of the production sector. Finally I listed some areas where existing model structures were likely to be unable to reflect the consequences of increasing globalization, including, among many others, internationally oriented plant and equipment investment decisions, and the interplay of trade and technical progress.

The advice at the end was not intended to diminish the credit due for the progress made so far. When it comes to macroeconometric modeling, there is little doubt that in the aggregate, and especially on a per capita basis, the United Kingdom is a world leader, with a strongly positive balance in the associated flows of inspiration and ideas. In this, as in so much else in the history of macroeconometric modeling, Lawrence Klein has contributed much by way of model innovations, training, research collaboration and intellectual leadership.

John F. Helliwell, Department of Economics, University of British Columbia, Vancouver, B.C., Canada V6T1W5.

ACKNOWLEDGEMENTS

Harvard University and the University of British Columbia. The first version of this paper was given at ESRC/CEPR Conference on the Future of Macroeconometric Modeling in the United Kingdom, London, November 30-December 1, 1992, and I am grateful for the suggestions of many participants. In preparing a revised version for this volume, I have received helpful advice from Jan Dutta. I am grateful for the research support of the Social Sciences and Humanities Research Council of Canada.

NOTES

1.   The five topic areas are 1320 (Forecasting-Econometric Models-General), 1322 (General Forecasts and Models), 1323 (Specific Forecasts and Models), 1324 (Forecasting and Econometric Models-theory) and 2120 (Construction, Analysis and Use of Econometric Models). These categories were chosen by first searching for key words in titles, e.g. 'macroeconometric', and then seeing what subset of *JEL* codes would include all of the relevant papers in the selected group. To provide a cross-check, the CD-ROM was asked to list all of the papers by several researchers publishing extensively in macroeconometric modeling, e.g Currie, Fair, Klein, Taylor, and Wallis to see what subject codes were used for their papers dealing with macroeconometric modeling. The chosen groups tend to cover from one-quarter to one-half of the publications by each, with the more modeling-related papers usually included.

2.   An equation containing only a constant and a second constant starting in 1984 explains 85% of the variance of the series, with a primary constant of 1.98 and a coefficient of 1.29 (s.e.=.121) on the 1984-and-after variable. An alternative equation using a linear time trend explains only 54% of the variance, while an equation using both variables shows a highly significant shift and no time trend.

3.   This research makes use of the CD-ROM for December 1991, released in January 1992.

4.   From 1984 on, the macroeconometric publications (as represented by the five selected topic areas) comprise 3.5% of the 74,500 journal articles, 3.1% of the 28,000 articles in collective volumes, 2% of the 7000 books and 1.9% of the 4000 dissertations.

5.   They increased from 44/106166 to 77/109500 (including all listed publications except theses), an increase of about 70% in the proportion of total listed publications, almost exactly the same increase as for the larger group, which went from 2% to 3.3% of total articles.

6.   I was able to do this (in Helliwell 1993) for the population of Canadian research economists, making use of their publication records, age, educational background and research grants, and was able to identify a strong linkage between research grants and subsequent publications even after allowing for the influence of the other variables.

7.   The differences between UK and US model comparison procedures and pressures are documented by Ken Wallis (1992) in his review of the U.S. model comparison exercises.

8.   Thus, in Helliwell (1986) I attempted to specify a general equation for output determination in which a number of the Keynesian, neo-classical, new classical and monetarist specifications could be nested as special cases, and found a fairly clear ranking, with preference for a model involving influences from both the supply and demand sides, and hence being more general than either the simple Keynesian or new classical specifications.

9.   For example, by Jeremy Bray (1974, 1975). For parallel work in North America, see

Holbrook (1974) and Cooper and Fischer (1974).

REFERENCES

Ball, R. J., (Ed.) (1973). *The international linkage of national economic models*. Amsterdam: North Holland.

Barro, Robert J. (1991). Economic growth in a cross section of countries. *Quarterly Journal of Economics. 106*, 407-44.

Ben-David, D. (1992). Income disparity among countries and the effects of freer trade. Paper prepared for the IEA Conference on Economic Growth and the Structure of Long-Term Development, Lake Como, October 1992.

Bray, Jeremy. (1974). Predictive control of a stochastic model of the U.K. Economy simulating present policy making practice by the U.K. government. *Annals of Social and Economic Measurement. 3*, 239-56.

_____ (1975). Optimal control of a noisy economy with the UK as an example. *Journal of the Royal Statistical Society Series A. 138*, 339-73.

Bryant, Ralph, Henderson, Dale, Holtham, Gerald, Hooper, Peter, and Symansky, Steven, (Eds.) (1988). *Empirical macroeconomics for interdependent economies*, Washington: Brookings Institution.

Bryant, Ralph, Hooper, Peter, Mann, Catherine, and Tryon, Ralph, (Eds.). (1993). *Evaluating policy regimes: New research in empirical macroeconomics* Washington: Brookings Institution.

Cooper, Phillip, and Fischer, Stanley. (1974). A method for stochastic control of nonlinear econometric models and an application. *Annals of Economic and Social Measurement. 3*, 205-6.

De Long, J. B. and Summers, L. H. (1991). Equipment investment and economic growth. *Quarterly Journal of Economics 106*, 445-502.

_____ (1992). Equipment investment and economic growth: How strong is the nexus? *Brookings Papers on Economic Activity. 2*(September), 157-211.

Frankel, Jeffrey, and Rockett, Katherine. (1988). International macroeconomic policy coordination when policymakers do not agree on the true model. *American Economic Review. 78*(June), 318-40.

Frankel, Jeffrey, Erwin, Scott, and Rockett, Katherine. (1992). International macroeconomic policy coordination when policymakers do not agree on the true model: Reply. *American Economic Review. 82*(September), 1052-56.

Grossman, Gene, and Helpman, Elhanan. (1991). *Innovation and growth in the global economy*. Cambridge: MIT Press.

Harrison, Ann. (1991). Openness and growth: A time-series, cross-country analysis for developing countries. Policy Research Working Paper WPS809. Washington: World Bank.

Helliwell, John F. (1986). Supply-side macro-economics. *Canadian Journal of Economics. 19*, 597-625.

Helliwell, John F. (1992). Trade and technical progress. *NBER Working Paper* No. 4226. Cambridge: National Bureau of Economic Research.

_____ (1993). What have Canadian economists been doing for the past twenty-five years? *Canadian Journal of Economics. 25*, 39-54..

Helliwell, John F., and Chung, Alan. (1991). Macroeconomic convergence: international transmission of growth and technical progress. In P. Hooper and J.D. Richardson, (Eds.), *International economic transactions: issues in measurement and empirical research.* (pp. 388-436). Chicago: University of Chicago Press.

Holbrook, Robert. (1974). A practical method for controlling a large nonlinear stochastic system. *Annals of Economic and Social Measurement. 3*, 155-75.

Holtham, Gerald, and Hughes Hallett, Andrew. (1992). International macroeconomic policy coordination when policymakers do not agree on the true model: Comment. *American Economic Review.* September, 82, 1043-51.

Klein, Lawrence R. (Ed.) (1991). *Comparative performance of U.S. econometric models.* New York: Oxford.

Klein, Lawrence R., Ball, R. J., Hazelwood, A. and Vandome, P. (1961). *An econometric model of the United Kingdom.* Oxford: Basil Blackwell.

Klein, Lawrence R, and Burmeister, E. (Eds.) (1974). *Econometric model performance.* Philadelphia: University of Pennsylvania Press.

Lucas, Robert E. and Sargent, Thomas J. (1978). After Keynesian macroeconomics. In Federal Reserve Bank of Boston *After the Phillips Curve: Persistence of high inflation and high unemployment.* (pp. 49-72). Boston: Federal Reserve Bank.

Mankiw, Greg, Romer, David, and Weil, D. (1992). A contribution to the empirics of economic growth. *Quarterly Journal of Economics. 107*, 407-37.

Nickell, Stephen. (1988). The supply side and macroeconomic modelling. In Bryant et al, (Eds.). (pp. 202-21).

Razin, Asaf and Sadka, Efraim. (1991). International tax competition and gains from tax competition. *Economics Letters. 37*, 69-76.

Romer, Paul M. (1976). Increasing returns and long-run growth. *Journal of Political Economy. 94*, 1002-37

Uebe, Gotz. (1991). *Macroeconometric models: An international bibliography.* Brookfield: Gower.

Wallis, Kenneth F. (1992). Comparing macroeconometric models. ESRC Macroeconomic Modelling Bureau Discussion Paper No. 30, July 1992. Warwick: University of Warwick.

Structural Versus VAR Modeling: An Empirical Comparison with a Small Model of the United States

F. Gerard Adams and Ronald C. Ratcliffe

In the past ten years, the literature of macro-econometrics has taken an abrupt turn away from the tradition of structural econometric models toward empirical vector autoregressive systems, VARs. Some would call this shift, from using theory to impose restrictions on the functional relations to empirical equation fitting, a move from "theory with measurement" toward "measurement without theory". But the VAR practitioners respond that many theories can fit the data, that traditional macroeconometricians have paid too little attention to the time series properties of the data, and that VAR models produce better forecasts.

It is not likely that this controversy can be resolved on theoretical grounds. But there is ample room to investigate the relative merits of various approaches on an empirical basis. This paper presents a "horse race" between the structural econometric approach, the VAR approach, and the cointegrated error correction method. Using a common data set for the United States, we estimated three models:

-- A conventional structural econometric model (SEM).
-- A VAR model (VAR-I) recognizing the identities that link the variables in the national accounting system.
-- A cointegrated error correction model (ECM).

To provide a comparison for pedagogic as well as empirical use, we have chosen a simple but, hopefully, not simplistic specification for our model and we have estimated it over a relatively short (20 year) annual time horizon.[1]

In Section 2, we contrast the alternative methodologies and consider their likely implications. In Section 3, we describe the structural model. In Section 4, we present the VAR-I model. In Section 5, we consider the ECM model. Sections 6 and 7 evaluate the simulation properties of the three systems. A final section presents our conclusions.

II. Alternative Approaches to Model Construction

The typical dynamic linear simultaneous equation model has the structural form,

$$\Gamma y_t = \Phi_1 y_{t-1} + \cdots + \Phi_r y_{t-r} + B_0 x_t + \cdots + B_n x_{t-n} + u_t \qquad (1)$$

where Γ is an NxN matrix of unknown parameters, $\Phi_1, \ldots, \Phi_r$ are NxN matrices of autoregressive parameters, $B_0, \ldots, B_n$ are NxK matrices of parameters associated with the Kx1 vector of exogenous variables, x_t, and their lagged values, and u_t is the Nx1 disturbance vector distributed $NID(0,\Omega)$. Pre-multiplying through by Γ^{-1} gives the reduced form

$$y_t = \Pi_1 y_{t-1} + \cdots + \Pi_r y_{t-r} + \Pi_0 x_t + \cdots + \Pi_s x_{t-n} + v_t \tag{2}$$

Restrictions imposed on the structural parameters affect the reduced form parameters indirectly.

Alternative Models

The essence of the structural econometric model is the use of theory to determine the variables to be included in each structural equation, in other words to impose zero restrictions on some of the possible coefficients. In fact, model building proceeds experimentally, testing whether the structural coefficients obtained are statistically significant and have the correct sign and value based on theoretical expectation and past experience. We note, incidentally, that some of the equations are identities. Their coefficients are known and there is no error term. They take this form not only because of the theory of the underlying structure, but also because of the data generation process from which the information is derived. The national statisticians who prepare the historical statistics assume that certain well-defined relationships must hold, additive ones in the accounts and multiplicative ones between real values and price indexes.[2]

In contrast, the VAR approach introduced by Sims (1980) emphasizes the empirical interrelationships between variables, arguing that it is not possible to identify the structural coefficients and to evaluate their individual meaning or significance. No zero restrictions on any of the relationships are imposed. A general multivariate vector autoregressive model has the form,

$$Y_t = \Phi_1 y_{t-1} + \cdots + \Phi p y_{t-p} + z_t \tag{3}$$

where Φ_i, $i = 1, \ldots, p$ are NxN matrices of parameters and z is an Nx1 disturbance vector. The Φ matrices are unrestricted. Generally in VAR models, no distinctions are made between endogenous and exogenous variables. In this system of equations, every variable is a function of its lag and the lag of all of the other variables. Only the lag length needs to be determined. Typically, information criteria based on the residuals, not on economic theory, are used for this purpose.

The unrestricted VAR could serve as a multivariate naive model. The scarcity of the data points available on which to estimate the parameters means that even a pure VAR model cannot be entirely unrestricted. Choices must be made with respect to the number of right-hand side variables and/or the number of lags to be included, or there will be more variables than observations.

To improve on the naive model requires introducing information in the form of restrictions. Information can be from economic theory or from the data in the form of known data-generating relationships or econometric analysis. The pure VAR model is more naive than is appropriate since more is known about the data generation process than is assumed. Consequently, we propose an alternate VAR model, VAR-I, that preserves the relationships used in constructing the data. In particular, we include in the VAR-I system the national accounting identity, the capital stock equation and the definition of the price level. We include this information by making these identities part of the system and imposing restrictions on the VAR. Variables appearing on the right-hand side of the identities are excluded from the right-hand side of the VAR equations. In its most general form, the model is now the same as in equation (3) with restrictions on the parameters imposed indirectly on the reduced form coefficient matrix. Some coefficients are forced to be one for the identities. We refer to this model as the VAR-I model since identities are added to the naive VAR.

An alternative approach, cointegration with error correction, which seeks to combine the short term time series characteristics of the data with the long run relationships of the economic structure, has gained considerable attention. The origins of this work lie in the concern of Granger and Newbold (1974) with the spurious regressions that can be obtained by working with non-stationary time series. In most cases, stationarity can be obtained by differencing the series, but at the cost of a loss of information on the underlying long term relationships. When a cointegrating relationship is present, i.e., if non-stationary series can be combined to produce a stationary or equilibrium variable, they can be modeled by an error correction mechanism (Engle and Granger 1987).

Statistical tests of the stationarity of time series (Dickey and Fuller, 1979) and procedures for estimating the cointegrating vector (Johansen, 1988) have been proposed. The Johansen method is based on a vector autoregressive approach and emphasizes statistical as compared to economic theoretical considerations in selecting the appropriate variables. The basic equation specification obtained is:

$$\Gamma \Delta Y = \varphi_1 \Delta Y_{t-1} + \ldots + \varphi \Delta Y_{t-\pi} + B_0 \Delta X_t + \ldots + B_n \Delta X_{t-n}$$

$$+ A \ ECM_{t-1} + w_t \tag{4}$$

The notation is the same as in equation (1) except that the variables, other than ECM, are in differences of logs. ECM is the vector of error correction terms, A is the corresponding coefficient matrix. These are the short term adjustment equations.

$$ECM_t = Y_t - ZY_t + w_t \tag{5}$$

where Y is the matrix of the cointegrated variables and Z represents the corresponding coefficients. These are the cointegration equations, reflecting long term equilibrium relationships between the variables.

It is possible to substitute more statistically ad hoc procedures with emphasis on the theoretical specification of the model for the strict Johansen approach. We found it necessary to do so in the ECM version of our model. We discuss these issues further below when we consider the characteristics of the models.

Implications

The structural econometric model (SEM) maximizes reliance on traditional theoretical views. While it does not altogether impose these theories on the data, it seeks to maintain a theoretically-based structure and relies on the data principally to provide suitable structural parameter estimates. In contrast, the pure VAR eschews all reliance on prior theory. Our combination of a VAR with identities (VAR-I) imposes only the identities of the national accounting structure. The ECM model adds the constraints imposed by the cointegrating relationships.

Our priors are that:

--   The SEM and the ECM models will have behavioral responses to shocks (and/or policy changes) that are reasonably in accord with experience.[3]

--   The behavioral responses to shocks of the VAR-I model depend on the parameter estimates which may or may not catch structural aspects of the economy.[4]

What can we expect with respect to forecasting ability? In the short run, in-sample period, forecasting:

--   The VAR-I probably makes better use of the available data. Even though it suffers from the lack of current exogenous variable information (all the variables are entered as lags), it makes maximum use of the serial properties of the data.

--   The SEM does not make as much use of the time series properties of the data and is more constrained by theoretical assumptions. Whether this means that it will forecast more or less accurately depends in part

on the role of current values of exogenous variables in the forecast.[5]
-- The ECM model is likely to catch some of the VAR's advantages for short term forecasting.

With respect to longer run dynamic simulations, the order is likely to be reversed:

-- The SEM may well be able to catch the dynamic cyclical response of the system in a more systematic manner than the VARs.

-- The ECM takes additional advantage of the long run cointegrating constraints.

-- On the other hand, the VAR system may not contain sufficient structure to simulate well.

Unfortunately, at the time this work was carried out, there was not sufficient post sample data to do substantial "out of sample" period forecast tests.

III. Structure of the SEM Econometric Model

Even though our minimum econometric model of the United States economy includes only 5 behavioral equations and 3 identities, it deals comprehensively with demand, supply, and the monetary sector.[6] The equations are summarized in Table 1, and explained briefly here.

-- **Equation 1** is the GNP demand identity. Total output (GNP) is demand determined as the sum of consumption (CONSUMP), investment (INVEST), government purchasing (GOVT), exports of goods and services (EXPORTS) less imports (IMPORTS).

-- **Equation 2** explains consumption as a function of GNP after income taxes (TR*GNP), where TR is an exogenously determined tax and transfer policy parameter, the interest rate (INT), the inflation rate, and lagged consumption. A DUMMY variable in 1974 and 1975 allows for consumer response to the oil shock which is not well captured by the model.

-- **Equation 3** explains investment in terms of the interest rate (INTEREST), changes in GNP, the rate of inflation (INFLAT) and the lagged capital stock (CAPITAL). This equation combines a flexible accelerator and a cost of capital model. The inflation rate here has a positive effect reflecting the reduced real interest cost of inflation. The DUMMY variable also enters this equation.[7]

-- **Equation 4** is the behavioral equation for the interest rate. INTEREST is explained by the rate of inflation, the nominal GNP (GNP*Price/100) and by monetary policy variables, the money supply (MONEY) and the discount rate (RD).

--   **Equation 5** explains inflation (INFLAT) in terms of a GNP gap, a measure of the shortfall between Potential GNP (YPOT) and actual GNP. The latter is computed as an exogenous growth trend of 2.75% per year starting in 1970. The relationship is nonlinear: the lower the available capacity the faster the rate of inflation. World prices (PRICEW) also influence domestic inflation. The lagged price term accounts for expectations and adjustment processes.

--   **Equation 6** explains imports in relation to GNP, and import prices relative to domestic prices (PRICEW/PRICE). The inclusion of the lagged dependent variable IMPORTS[-1] means that the adjustment to changes in GNP and prices is a gradual one.

--   **Equation 7** is the identity that accumulates capital stock (CAPITAL) with investment and an assumed rate of depreciation of 5% per year.

--   **Equation 8** is the identity translating the inflation rate (INFLAT) into the price level (PRICE).

Such a set of equations makes a complete simultaneous system, using as exogenous variables only GOVT and TR as fiscal policy variables, MONEY and RD as monetary policy variables, EXPORTS and PRICEW as variables describing the foreign environment, and potential output (YPOT) which is, as we have noted, an exogenous trend. Variables were initially considered for inclusion in the equations on the basis of our theoretical priors. The estimation procedure involved tests of numerous alternative formulations going from the "specific to the general", i.e., from the simplest versions to equations including a larger number of variables. The coefficients estimated by ordinary least squares[8] meet theoretical expectations as to sign and value.

IV.   Structure of the VAR-I Model

We began with an unrestricted VAR where every variable is a function of its own lag and that of all the other variables. Then we replaced the GNP, CAPITAL and PRICE equations with the identities. Further, in order to reduce the number of regressors, we exclude from the right-hand side of the remaining equations all variables appearing on the right-hand side of the identities since their impact will show up through the inclusion of the identities. CONSUMP, INVEST, GOVT, EXPORTS, and IMPORTS are excluded based on the GNP identity and INFLAT and PRICEW are excluded based on the PRICE equation.

To maintain sufficient degrees of freedom with twenty observations, no more than one lag of the variables can be included. Consequently, we do not test for the appropriate lag length with information criteria. However, a lag of one year should capture the economic relations of interest for comparison with the structural model.[9]

    The resulting VAR-I model, shown in Table 2, contains three identities and nine equations containing lagged values of all the variables, subject to the restrictions above, as shown in Table 2. On the right-hand side of each equation are lagged values of GNP, INTEREST, TR, RD, PRICE, MONEY and CAPITAL.

Table 1: Structural Model
1: GNP = CONSUMP + INVEST + GOVT + EXPORTS − IMPORTS

2: CONSUMP = − 46.6051 + .5309 * CONSUMP (−1) + .3840 * (TR * GNP)
       (70.4609)   (.1525)   (.1247)
 	
        − 1.9194 * INTEREST - 5.3882 * INFLAT-13.8403 * DUMMY
         (2.9547)   (3.8096)   (21.5022)
                                                  $\overline{R^2}$ = .9976

3: INVEST = 165.6698 * - 14.6074 * INTEREST(-1) + .3533 * (GNP(-1) − GNP(-2))
      (47.5704)   (3.2673)   (.0945)
 	
    + .0803 * CAPITAL(−2) + 2.9918 * INFLAT-32.5253 * DUMMY
    (.0077)   (3.8704)   (23.9449)
                                  $\overline{R^2}$ = .9489

4: INTEREST = 1.4968 + .8064 * (GNP * (PRICE/100)) -
        (.7417) (.0023)
 	
      .1251 * MONEY + .7519 * RD(− 1)
      (.0037)   (.0938)
                               $\overline{R^2}$ = .8946

5: INFLAT = -.4688 + .7603 * INFLAT(− 1) +
      (.9534) (.0852)
 	
      .0859 * (1/((YPOT(− 1) − GNP(− 1)) / YPOT(− 1)) +
      (.0587)
 	
      5.5359 * (PRICEW − PRICEW(− 1)) / PRICEW(− 1)
      (2.8535)
                                  $\overline{R^2}$ = .8777

6: IMPORTS = -268.9444 + .3099 * IMPORTS(− 1) + .2033 * GNP -
       (41.0175) (.0983)   (.0255)
 	
       125.1334 * (PRICEW/PRICE)
       (19.2269)
                                $\overline{R^2}$ = .9945

7: CAPITAL = 0.95 * CAPITAL(− 1) + INVEST(− 1)
8: PRICE = (PRICE(− 1)) * (1 + (INFLAT / 100))
standard errors in parentheses

Unlike the case with standard VAR analysis, the ordering of the equations is not important in the VAR-I setting since the shocks are introduced directly to the estimated equations or the identities. Since the coefficients of the VAR equations represent reduced forms, they are not evaluated from a theoretical perspective.

V.   Structure of the ECM Model

We began our ECM model calculations with the intent of using the Johansen technique (Johansen 1988) for estimating all the equations. Even though this approach is data-oriented, it is necessary to select the appropriate variables entering each equation--in each case we used those corresponding to the SEM equation--and to choose one of multiple cointegrating vectors. It turned out, however, that some of the equations estimated in this way resulted in an unstable model when they were combined with other equations into a simultaneous system. In particular, it was not possible to use a current income change variable (ΔGNP) in the investment equation, nor a current inflation change (ΔINFLAT) in the inflation equation even though the coefficients of these variables were significant with the right sign in the single equation estimates. Consequently, as in the SEM, we imposed our theoretical and practical priors and we estimated most of the equations by ordinary least squares in two stages, first on the levels, and then in first differences with an error correction term. This provided a long run equilibrium relationship, and a short term adjustment mechanism. Choices among the possible regressors were guided by theoretical considerations and by the performance of the equation in simulating the entire system. The equations used are summarized in Table 3. The aim was to include the same variables in this model as in the others, and, specifically to treat as exogenous the same variables as in the SEM. The ECM variables (ECMC, etc.) are the error correction terms for each structural equation.[10]

VI.   Sample Period Simulation

Simulations of the models have been run within sample. Our initial comparison considers the dynamic simulation of the models over the entire sample (Table 4). We note that the SEM makes explicit use of the year-by-year values of the exogenous variables. The VAR-I system does not do so since some of the exogenous policy variables are themselves explained by equations.[11]

Table 4 summarizes the results, but a great deal more is apparent visually from Figure 1 that show the path of actual and simulated values. The mean absolute percentage error (MAPE) of the SEM for GNP is 1.13 percent, quite close to the performance of much larger structural econometric models (Klein, 1991), and it provides a good explanation of cyclical movement of GNP

Figure 1: Dynamic Simulations 1971-1990 GNP 1982 $

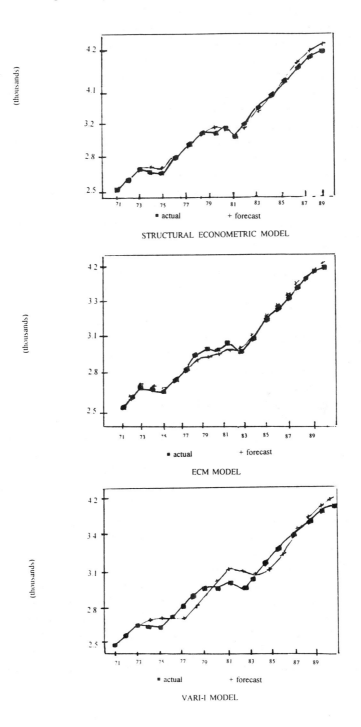

Table 2. VAR-I Model

1: GNP = CONSUMP + INVEST + GOVT + EXPORTS - IMPORTS

2: CONSUMP = -388.614 + .500*GNP(-1) - 24.592*INTEREST(-1)
              (670.153) (.066)   (3.366)

          + 550.023*TR(-1) - 7.326*RD(-1) + 13.445*PRICE(-1)
            (753.481)   (3.826)   (1.755)

          -.301*MONEY(-1) + .009*CAPITAL(-1) $\overline{R}^2$ = .999
          (.069)   (.023)

3: INVEST = 100.473 + .288*GNP(-1) - 36.106*INTEREST(-1)
            (1675.438) (.165)   (8.416)

          -146.575*TR(-1) - 10.382*RD(-1) + 11.882*PRICE(-1)
          (1883.765)   (0.565)   (4.389)

          -.289*MONEY(-1) - .104*CAPITAL(-1) $\overline{R}^2$ = .947
          (.174)   (.056)

4: INTEREST = 97.907 - .001*GNP(- 1) - .666*INTEREST(- 1)
              (54.988)   (.005)   (.276)

          - 105.228*TR(- 1) - 1.397*RD(- 1) + 0.38*PRICE(- 1)
            (61.825)   (.314)   (.144)

          + .0001*MONEY(- 1) - .0006*CAPITAL(- 1) $\overline{R}^2$ = .886
          (.006)   (.002)

5: IMPORTS = 152.383 - .087*GNP(- 1) - 8.681*INTEREST(- 1)
             (573.020)   (.056)   (2.878)

          - 185.624*TR(- 1) + .476*RD(- 1) - 1.551*PRICE(- 1)
            (644.270)   (.476)   (1.501)

          + .173*MONEY(- 1) - .001*CAPITAL(- 1) $\overline{R}^2$ = .996
          (.059)   (.019)

6: INFLAT = - 37.007 - .005*GNP(- 1) - .047*INTEREST(- 1)
             (34.133)   (.003)   (.171)

          - 22.696*TR(- 1) + .522*RD(- 1) - .251*PRICE(- 1)
            (38.378)   (.195)   (.089)

          - .003*MONEY(- 1) - .005*CAPITAL(- 1) $\overline{R}^2$ = .950
          (.004)   (.001)

7: TR = .912   - .0000001*GNP(- 1) - .00009*INTEREST(- 1)
         (.245)   (.00002)   (.001)

          -. 032*TR(- 1) - .002*RD(- 1) - .0003*PRICE(- 1)
          (.275)   (.001)   (.0006)

          - .00002*MONEY(- 1) - .000003*CAPITAL(- 1) $\overline{R}^2$ = .699
          (.00002)   (.000008)

Table 2:   VAR - I Model (cont.)

8: RD = 77.149   - .113*GNP(- 1) - 1.112*INTEREST(- 1)
    (56.324)   (.011)   (.283)

    - 108.427*TR(- 1) + 1.073*RD(- 1) - .330*PRICE(- 1)
    (63.328)   (.322)   (.148)

    - .015*MONEY(- 1) - .002*CAPITAL(- 1) $\overline{R^2}$ = .872
    (.006)   (.001)

9: MONEY = - 1210.245   -   .013*GNP(- 1) - 3.689*INTEREST(- 1)
    (1636.050)   (0.160)   (8.218)

    - 1194*TR(- 1) - 3.737*RD(- 1) - 10.460*PRICE(- 1)
    (1839.478)   (9.340)   (4.286)

    - .687*MONEY(- 1) - .012*CAPITAL(- 1) $\overline{R^2}$ = .999
    (.169)   (.055)

10: GOVT = 217.527   -   .029*GNP(- 1) + 3.819*INTEREST(- 1)
    (472.303)   (.046)   (2.372)

    - 585.557*TR(- 1) + .800*RD(- 1) - 2.427*PRICE(- 1)
    (531.030)   (2.696)   (1.237)

    + .262*MONEY(- 1) - .047*CAPITAL(- 1) $\overline{R^2}$ = .993
    (.048)   (.015)

11: CAPITAL = 0.95*CAPITAL(- 1) + INVEST(- 1)

12: PRICE = (PRICE(- 1))*(1 + (INFLAT / 100))

standard errors in parentheses

---

except in 1974 and 1989 and 1990.   For other variables, the model fits somewhat less well, but still well within acceptable levels.

The ECM model achieves a better dynamic simulation result for GNP, MAPE of 0.87%, but the difference between the SEM and ECM results may not be significant.[12] Curiously the simulation for other variables is not as good, indeed, after some years the simulation for INTEREST and INFLAT drifts away from the actual statistics.   This drift is consistent between the two series, as a result of the cointegration link between them.   Since the real interest rate operates in the rest of the system, the real results are not greatly affected.   We tried a number of specifications to eliminate the drift but did not succeed.   It seems to be related to the fact that we use the second difference of prices for the short term inflation equation and the first difference for the cointegrating relation of the ECM model.   This works well enough on the single equation level but difficulties with error cumulation are encountered when the price differences are translated to levels of prices.

In sharp contrast, the dynamic simulation of the VAR-I model yields poor results, a MAPE on GNP of 3.47%.   The inability of this model to catch

Table 3: ECM Model

1: GNP = CONSUMP + INVEST + GOVT + EXPORTS - IMPORTS

2a: CONSUMP = EXP(CONSUMP(- 1) + (- 0.4457 + 0.6925*Δln(GNP*TR)
$$\hspace{6cm}(0.068)$$
$$\hspace{1.5cm}- 0.1507*\Delta\text{INTEREST} - 0.6332*\Delta\text{INFLAT} + 0.4154*\text{ECMC}(- 1))$$
$$\hspace{1cm}(0.1126)\hspace{2.5cm}(0.1912)\hspace{2.5cm}(0.1508)$$

$$\overline{R}^2 = 0.9038$$

2b*: ECMC = ln CONSUMP - (1.1006*ln(GDP*TR)
$$\hspace{3cm}- 0.0586*\text{INTEREST}/100 - 0.0348*\ln\Delta\text{PRICE})$$

3a: INVEST = EXP(lnINVEST(- 1) + (0.0018 + 2.0903*ΔlnGNP(- 1)
$$\hspace{6cm}(0.5638)$$
$$\hspace{1.5cm}- 1.7423*\Delta\text{lnGNP}(- 2) - 0.0477*\Delta\text{INTEREST}(- 1)$$
$$\hspace{1cm}(0.5604)\hspace{3cm}(0.0082)$$
$$\hspace{1.5cm}+ 0.0985*\text{DUMMY78} - 81 - 1.0965*\text{ECMI}(- 1)$$
$$\hspace{1cm}(0.0338)\hspace{3.5cm}(0.3115)$$

$$\overline{R}^2 = 0.8604$$

3b: ECMI = lnINVEST - (- 5.037 + 2.7057*lnGNP
$$\hspace{4cm}(0.2611)$$
$$\hspace{1.5cm}- 0.0095*(\text{INTEREST} - \text{INFLAT}(- 1)) - 1.7122*\text{lnCAPITAL}(- 1))$$
$$\hspace{1cm}(0.0047)\hspace{4.5cm}(0.2045)$$

$$\overline{R}^2 = 0.9525$$

4a: IMPORTS = EXP(lnIMPORTS(- 1) - 6.8845 + 2.552 ΔlnGDP
$$\hspace{6cm}(0.3498)$$
$$\hspace{1.5cm}+ 0.1375*\Delta\text{ln}(\text{PRICE} - \text{PRICEW}(- 1)) - 0.5272\text{ECMM}(- 1))$$
$$\hspace{1cm}(0.1040)\hspace{4cm}(0.2283)$$

$$\overline{R}^2 = 0.8992$$

4b*: ECMM = lnIMPORTS - (0.1519*ln(PRICE/PRICEW)
$$\hspace{3cm}- 2.3456*\text{lnGNP})$$

5a: INTEREST = INTEREST(- 1) + 100*(- 0.0130 - 0.4059*ΔlnMONEY
$$\hspace{6cm}(0.0551)$$
$$\hspace{1.5cm}+ 0.5465*\Delta\text{ln}(\text{PRICE}*\text{GNP}) + 0.0039*\Delta\text{RD}(- 1)$$
$$\hspace{1cm}(0.0665)\hspace{3cm}(0.0010)$$
$$\hspace{1.5cm}- 0.0067*\text{ECMINT}(- 1))\hspace{3cm}\overline{R}^2 = 0.9282$$
$$\hspace{1cm}(0.0025)$$

5b: ECMINT = INTEREST - (- 166.7315 - 32.5030*lnMONEY
$$\hspace{4cm}(6.8117)$$
$$\hspace{1.5cm}+ 0.6216*\text{RD}(- 1) - 32.915*\text{ln}(\text{PRICE}*\text{GNP}))$$
$$\hspace{1cm}(0.0930)\hspace{3cm}(6.8107)$$

$$\overline{R}^2 = 0.9277$$

Table 3: ECM Model (cont'd)

6a: PDOT = PDOT(- 1) + EXP(0.0288 + 0.0648 + ln(GNP/YPOT)
 (0.0996)
      + 0.3615*ln(GNP/YPOT(- 1)) + 0.0391ΔlnPRICEW(- 1)
      (0.1504)
      - PRICEW(- 2)) - 0.4273*ΔlnINFLAT
          (0.2203)
      - 0.3081*ECMPDOT(- 1))
      (0.1704)
 	
 $\overline{R^2}$ = 0.7653

6b: ECMPDOT = lnINFLAT - (0.5025 + 0.1443*ln(GDP/YPOT(- 1))
          (0.0911)
      + 0.0760ΔlnPRICEW + 0.1419ΔlnPRICEW(- 1))
      (0.0293)   (0.0302)
 	
 $\overline{R^2}$ = .7537

7: PRICE = PRICE(- 1)*(1 + INFLAT/100)
8: CAPITAL = 0.95*CAPITAL(- 1) + INVEST(- 1)
*Estimated by the Johansen method, no R^2 or standard errors are available for the cointegrating relationship estimated by the Johansen method.

business cycle fluctuations is clearly apparent. The results show that the VAR model fails to represent the dynamic path of the economy over the sample period. Simulation properties for projections one and two periods ahead are presented in Tables 5 and 6. We remind the reader that these are in-sample forecasts from a fixed parameter model. Exogenous values have not been introduced into the basic VAR-I, so that the VAR-based forecasts are unconditional whereas the structural forecasts are conditional on the exogenous values assumed.

Tables 5 and 6 suggest that for one year projections the VAR-I achieves a higher degree of accuracy than the SEM or the ECM. This is particularly remarkable in view of the fact that current values of exogenous variables are excluded for the VAR-I model.

The ECM model does only marginally better than SEM for GNP and a little worse for INVEST and IMPORTS.

In the two period forecasts, the VAR-I model does considerably better than the others for GNP and CONSUMP, though not for the other variables. The ECM does almost as well as the VAR-I for GNP. For the nominal variables, INTEREST and INFLAT the results are mixed, none of the versions produce consistently the best results.

VII. Multiplier Properties Compared

Multiplier simulations have been run with all three models to compare the dynamic behavioral properties of the three systems. These tests show the impact of external "shocks" or policy changes, on the entire system. Their primary relevance is to the application of the models as tools for policy analysis. Do they realistically depict the impact of alternative policies on the economy? There is no objective standard by which to gauge the results since the underlying properties of the economy, the data-generating system, are not known. On the other hand, there is much prior information, from theory, history and policy-making practice that can be brought to establish some priors.

The dynamic multiplier results for an increase in government spending (GOVT) at 1% of GNP are shown on Table 7 and 8. The SEM has typical multiplier results. The government expenditure multiplier reaches a peak of 1.6 after one year. The remainder of the system responds as anticipated: the increase in interest rates ultimately crowds out investment and leads to a decline in GNP below its original path, a somewhat greater negative swing than might have been expected.

The ECM model shows the same initial multiplier effect, but this effect is somewhat more sustained during the fifth and sixth year of the simulation largely because investment does not show as much decline even though the effect on the nominal interest rate is somewhat greater.

Typically, policy analysis with VAR's involves shocks of one standard deviation to an impulse response function which is a linear combination of the innovation terms from all of the equations. Since the identities contain no innovation terms and for comparison with the results from the structural model, we do not follow this approach with the VAR-I model. Instead, we shock the equations directly by the same economically-meaningful amounts as in the structural policy analysis. Since we have equations for the policy variables in the VAR-I model, shocking the equations is equivalent to changing the exogenous variables in the structural model.

The VAR-I model shows a similar but somewhat greater effect of fiscal stimulus on GNP reflecting a stronger positive investment response. This is not surprising in view of the fact that the interest rate and inflation responses are somewhat muted compared to those in the SEM and ECM simulations. We note that in a pure VAR model lacking the identities (not shown), the simulation responses to a government expenditure shock were not realistic. The national income identity included in VAR-I appears to be essential to generate the typical multiplier effect.

A monetary policy stimulus has been introduced by increasing money supply by 1 percent and by reducing the discount rate, the other monetary policy variable, by 1 percent.

In the SEM, the result is as typically found in macro model multiplier studies (Klein (1991)). Interest rates decline with a lag of one year by a little less than one percent. The resulting impact particularly on investment, but also somewhat on consumption, mounts gradually and shows a little decline toward the end of the six year simulation period. GNP increases to a peak of 0.86 percent and then shows some decline, following a moderate increase in interest rates. To summarize in Period 4, the monetary stimulus has reduced interest rates by 0.7, increased inflation by 0.5 percent and caused an increase of 3.8 percent in investment and 0.9 percent in GNP.

Table 4: Sample Period Simulation Errors: Dynamic Simulations			1971-90
VARIABLE	MAPE (%)	RMSE (%)	THEIL U
DYNAMIC SIMULATIONS-STRUCTURAL MODEL			
GNP	1.128	1.374	0.007
CONSUMP	1.635	1.855	0.009
INVEST	5.745	6.685	0.031
INTEREST	15.888	20.996	0.087
INFLAT	11.526	15.177	0.059
IMPORTS	3.200	3.822	0.016
DYNAMIC SIMULATIONS-ECM MODEL			
GNP	0.847	1.059	0.010
CONSUMP	0.798	1.052	0.009
INVEST	4.829	5.538	0.054
INTEREST	20.079	39.675	0.332
INFLAT	23.124	24.666	0.182
IMPORTS	2.536	3.202	0.026
DYNAMIC SIMULATIONS-VAR-I MODEL			
GNP	5.029	8.120	0.048
CONSUMP	5.111	8.408	0.050
INVEST	12.512	17.855	0.099
INTEREST	13.939	17.074	0.069
INFLAT	27.792	43.588	0.118
IMPORTS	2.622	3.083	0.015

Table 5: Sample Period Simulation Errors: One Step Ahead Simulations			
VARIABLE	MAPE (%)	RMSE (%)	THEIL U
ONE-STEP AHEAD SIMULATIONS-STRUCTURAL MODEL			
GNP	0.817	1.009	0.005
CONSUMP	1.006	1.229	0.006
INVEST	3.659	4.936	0.021
INTEREST	10.172	13.304	0.059
INFLAT	8.678	11.219	0.062
IMPORTS	1.909	2.653	0.011
ONE-STEP AHEAD SIMULATIONS-ECM MODEL			
GNP	0.731	0.902	0.004
CONSUMP	0.611	0.813	0.004
INVEST	4.492	5.364	0.025
INTEREST	8.706	10.838	0.053
INFLAT	7.755	10.099	0.040
IMPORTS	1.773	2.298	0.009
ONE-STEP AHEAD SIMULATIONS-VAR-I MODEL			
GNP	0.478	0.611	0.003
CONSUMP	0.367	0.493	0.002
INVEST	3.979	4.747	0.021
INTEREST	9.434	12.525	0.051
INFLAT	7.755	10.099	0.040
IMPORTS	1.773	2.298	0.009

Table 6. Sample Period Simulation Errors: Two-step Ahead Simulations			
VARIABLE	MAPE (%)	RMSE (%)	THEIL U
TWO-STEP AHEAD SIMULATIONS-STRUCTURAL MODEL			
GNP	0.913	1.136	0.006
CONSUMP	1.157	1.523	0.008
INVEST	3.814	5.055	0.022
INTEREST	11.127	14.284	0.060
INFLAT	9.015	13.558	0.062
IMPORTS	2.326	2.945	0.012
TWO-STEP AHEAD SIMULATIONS-ECM MODEL			
GNP	0.663	0.838	0.004
CONSUMP	0.707	0.8743	0.009
INVEST	3.410	4.522	0.040
INTEREST	10.178	11.968	0.097
INFLAT	12.386	14.801	0.115
IMPORTS	2.052	2.4855	0.020
TWO-STEP AHEAD SIMULATIONS-VAR-I MODEL			
GNP	0.629	0.762	0.004
CONSUMP	0.389	0.493	0.003
INVEST	4.551	5.666	0.025
INTEREST	13.195	17.451	0.071
INFLAT	9.393	11.782	0.045
IMPORTS	2.287	2.977	0.012

Table 7: Simulations: Response to Govt Spending Increase					
PERIOD	GNP(%)	CONSUMP(%)	INVEST(%)	INT(*)	INFLAT(*)	IMPORTS%
STRUCTURAL MODEL						
1	1.15	0.61	0.00	0.10	0.00	2.55
2	1.66	1.10	2.13	0.18	0.24	4.24
3	1.45	1.12	0.88	0.25	0.63	4.13
4	0.99	0.54	0.09	0.42	1.75	3.49
5	0.57	-0.09	- .81	0.64	1.98	2.37
6	0.30	-0.46	-1.70	0.89	1.59	1.33
ECM MODEL						
1	1.16	0.65	0.00	0.69	0.08	2.98
2	1.63	0.85	2.55	1.07	0.57	4.10
3	1.30	0.67	1.02	1.16	0.96	3.24
4	1.13	0.54	0.36	1.51	1.19	2.95
5	1.15	0.49	0.73	2.01	1.32	3.22
6	0.99	0.35	0.24	2.40	1.42	3.02
VAR-I MODEL						
1	1.01	0.00	0.00	0.00	0.00	0.00
2	1.68	0.78	1.69	- 0.02	0.14	0.94
3	2.25	1.26	3.72	0.37	0.37	1.58
4	2.07	1.05	3.14	0.93	0.71	0.64
5	1.33	0.29	0.57	1.34	0.89	-1.58
6	0.65	-0.43	-2.11	1.33	0.7453	-3.94

*in original units

PERIOD	GNP(%)	CONSUMP(%)	INVEST(%)	INT(*)	INFLAT(*)	IMPORTS (%)
STRUCTURAL MODEL						
1	0.01	0.02	0.00	-0.09	0.00	0.02
2	0.14	0.18	0.31	-0.84	0.001	0.29
3	0.69	0.55	2.78	-0.78	0.03	1.48
4	0.86	0.68	3.76	-0.70	0.49	2.32
5	0.65	0.50	3.38	-0.62	0.90	2.18
ECM MODEL						
1	0.04	0.09	0.00	-0.38	0.00	0.11
2	0.49	0.37	2.04	-0.44	0.051	1.26
3	0.65	0.44	2.86	-0.44	0.25	1.63
4	0.29	0.22	1.36	-0.57	0.39	0.71
5	0.05	0.08	0.17	-0.56	0.36	0.19
6	0.05	0.05	0.31	-0.47	0.30	0.28
VAR-I MODEL						
1	0.00	0.00	0.00	0.00	0.00	0.00
2	-0.27	0.32	-2.96	-1.40	-0.54	0.33
3	1.84	2.02	6.72	-0.83	-0.48	5.49
4	1.96	2.00	6.79	0.14	0.29	5.95
5	0.44	0.46	1.56	0.97	0.82	2.56
6	-1.63	-1.52	-5.75	0.95	0.72	-2.24

Table 8: Simulations: Response to Monetary Policy, Increase Money by 1% & Reduce RD by 1

In the ECM, the interest rate decline is somewhat smoother and more moderate to a maximum of 0.6. The effect on investment follows a patternsimilar to that in the SEM solution, but the investment stimulus declines to close to zero in period 5 and 6. Not surprisingly, the effect on GNP fades away after period 4.

The VAR model shows much stronger and more variable effects. The positive impact on GNP is large and increasing until period 4, but becomes sharply negative in Period 6. This reflects wide cyclical movements in investment. While there is no objective way to evaluate the multiplier path, the results of the VAR-I simulation appear to exaggerate the cyclical effect of the monetary stimulus assumed.

VIII.   Conclusions

What are the conclusions that can be obtained from such a comparison? There are few objective criteria by which to compare model performance (Wallis (1985)) but comparisons of the behavior of model performance over the sample period can give some insights into their respective merits.

A few results are clear:

--   In comparisons of short term in-sample forecasting, the VAR-I system is unequivocally better than the SEM or ECM, though the ECM system does almost as well for some variables.

--   In terms of dynamic simulation over the sample period, which reflects the model's ability to mimic the longer term dynamic properties of the economy, the VAR model comes out badly. The ECM model does best for the real variables, but interest rate and inflation drift away from the base solution during the simulation. The SEM model provides a fairly consistent picture of the economy over a long period simulation with errors not significantly different from those encountered in large disaggregated models. The VAR does not provide satisfactory long term simulation patterns particularly over the business cycle.

--   With respect to policy shocks the SEM, understandably, shows patterns similar to those observed in other macro model simulations. The ECM shows similar multiplier properties although with somewhat different dimensions. The VAR-I shows a reasonable real response to a government expenditure shock, but the results with respect to a monetary shock do not seem realistic.

The simulations suggest that if only unconditional forecasts are required, VAR type models may be the instrument of choice. On the other hand, if more analytical support is needed to explain the forecast or for policy analysis, the SEM model seems more appropriate. The ECM system does not yet seem to be the ideal instrument that some of its supporters have suggested.

F. Gerard Adams and Ronald C. Ratcliffe, Department of Economics, University of Pennsylvania, 3718 Locust Walk, Philadelphia, PA 19104-6297.

ACKNOWLEDGEMENTS

We thank Zafer Yavan, Massimo Tivegna, and Raymond Courbis for their contributions at various stages of this project.

NOTES

1. The original version of the macro model was prepared as a teaching tool and is presented in Adams and Ratcliffe (1994). The data set and the solution (in a Lotus 1-2-3

spreadsheet) are available from the authors on request.

2. In some cases national statisticians will include a statistical discrepancy between two related statistics. This is typically not a random error term (Adams and de Janosi (1966)).

3. One difficulty is that we do not really know what past experience represents. We cannot rely too heavily here on the results of other SEMs since their results vary considerably, Klein (1991).

4. While we cannot be sure about the structural reliability of the coefficients of the SEM, in that case we observe the structural coefficients and we can evaluate them.

5. For true ex ante forecasts, the contribution of the current exogenous variables may be offset by the fact that they are subject to forecast error.

6. Klein's classic model I, which dates from 1950, had 4 behavioral equations.

7. The short term interest rate works considerably better than a long term rate, probably because it catches other aspects of financial markets, "tight money" for example, better than the long term rate.

8. This conforms to practice with many large econometric models used for forecasting and policy analysis. However, the rationale here was to be able to show both estimation and solution of our basic model in the framework of one Lotus 1-2-3 worksheet.

9. Sims (1980) also uses information dating back one year; quarterly variables are lagged by four in his unrestricted VAR.

10. DUMMY 78-81 is a dummy variable with a value of one for the 1978-1981 period.

11. We also tested an alternate version that assumes policy variables as exogenous.

12. The errors of the simulations are very sensitive to the specification of the model. Small changes in one equation produce clearly apparent differences in the error statistics.

REFERENCES

Adams, F. G., and Ratcliffe, R. (1994). A small structural model of the us economy. *Social Science Computer Review. 12*(1), 83-99.

Adams, F. G., and deJanosi, P. E. (1966). On the statistical discrepancy in the revised U.S. national accounts. *Journal of American Statistical Association.* December, 1219-1229.

Dickey, David A. and Fuller, W. A. (1979). Distribution of estimates for auto regressive time series with unit root. *Journal of American Statistical Association,* 427-431.

Engle, R. F. and Granger, C. W. J. (1987). Cointegration and error connection models: Representation, estimation, and testing. *Econometrica. 55,* 251-276.

Fair, R. C. (1979). An analysis of the accuracy of four macroeconometric models. *Journal of Political Economy. 87,* 701-18.

Fair, R. C. (1984). Specification, estimation and analysis of macroeconometric models. Cambridge University Press.

Granger, C. W. J., and Newbold, P. (1974). Spurious regressions in econometrics. *Journal of Econometrics. 2*(July), 111-120.

Johansen, S. (1988). Statistical analysis of cointegration vectors. *Journal of Economic Dynamics and Control. 12,* 231-254.

Klein, Lawrence R. (1991). *Comparative performance of U.S. econometric models.* New York: Oxford University Press.

Sims, C. A. (1980). Macroeconomics and reality. *Econometrica. 48,* 1-48.

Wallis, K. J. (1985). *Models of the U.K. economy.* New York: Oxford University Press.

Manufacturing Industry and Balance of Payments Adjustment: The United Kingdom Case

Sir James Ball and Donald Robertson

It is over thirty years ago, that Pamela Drake and one of the present authors published an article that tried to estimate the potential rate of economic growth of the United Kingdom consistent with equilibrium in the balance of payments (Ball and Drake, 1962). For this purpose equilibrium was defined as a current account of zero. The analysis was based on a simple expenditure model of the economy. It explicitly accepted that in the absence of capital mobility, the current account of the balance of payments represented a 'constraint' on the level of output and employment and in a dynamic context on the rate of growth. It was supposed that the balance of payments equilibrium requirement might lead to the actual rate of growth falling below the rate of growth of potential output that was feasible from a supply side point of view. Later in the decade it was suggested that in the last analysis it might well be supply side factors that would limit the rate of economic growth rather than the balance of payments. (Ball and Burns, 1968).

This preoccupation with the current account of the balance of payments took place against the background of the Bretton Woods system. The models used for analysis generally ignored the effect of price changes and assumed zero capital mobility, partly as a matter of description but also because of the early post-war belief based on experience of the 1930s that capital mobility, particularly of a short term nature, resulted in welfare losses. Even seminal models, such as that of Mundell (1962), which introduced capital mobility into the adjustment process, lacked a comprehensive treatment of the behaviour of nominal prices. It is of course very easy to build models of the economy that exhibit balance of payments constraints when all the potential adjustment mechanisms that tend to restore equilibrium are firmly nailed down.

Concern about the balance of payments as a constraint on economic growth in the United Kingdom has, in recent years, been associated with the relative decline in manufacturing industry. This linkage was graphically set out in the Report of the House of Lords in 1985. Their Lordships' prime concern was with the balance of trade in manufactures (House of Lords, 1985).

"But trade in manufactures and the output of manufacturing industry are indissolubly linked." (p.5.) and subsequently

"Manufacturing even now represents over a fifth of all activity in the United Kingdom and it provides over 40% of our earnings. Its performance is therefore crucial in an economy which depends upon imports of food and raw materials ... (p. 4)

Furthermore,

"The Committee fully recognises that growth of GDP is the important objective. But sustainable growth *has not been possible and will not be possible without a favourable trade balance in manufactures* (our italics) and manufacturing output would be expected to grow faster with a favourable trade balance in manufactures than without." (p. 42)

The thesis of this essay is that it is unlikely that economic growth in the United Kingdom during the rest of the 1990s will be balance-of-payments constrained in any significant sense. Our conclusion is that manufacturing industry does not and will not play a *unique* role in Britain's economic future. These propositions are not to be confused with two distinct and related ones. The first is that the fact that the balance of payments need not be seen as a constraint does not rule out the possibility of other problems associated with the balance of payments such as we have seen in the recent past. Secondly, the fact that manufacturing industry is not in any sense a unique part of economic activity in no way lessens its importance as a major and significant sector. This paper is in no sense a critique of United Kingdom manufacturing industry.

I. The Balance of Payments as a Constraint

Before we turn to the United Kingdom as a specific case, it is worth considering what we may learn from the current state of economic analysis with regard to the balance-of-payments adjustment process. In particular, we should consider what we mean by the term 'constraint'. A general question relates to the possible persistence of large imbalances between nations and the extent to which they may be intractable without major changes in international relationships. It is a stylised fact of the 1980s that imbalances between major countries in the OECD have been larger and have persisted for longer than in previous decades. This raises questions as to whether the disequilibria we observe are in the main the result of specific shocks to the economic system over time, or whether they also reflect an inability of economic systems to adjust.

In a significant contribution that deals with the 'external constraint' in the United Kingdom, Bean (1991) states that:

"The phrase 'external constraint' is open to a variety of interpretations. It is here construed narrowly as reflecting the country's ability to borrow to cover a current account deficit. If the opportunities for borrowing at the 'world' interest rate are unlimited then this would amount (at most) to an intertemporal solvency condition. A wider interpretation is certainly possible: for instance, expectations of future monetary expansion or the need for future real depreciation to satisfy intertemporal solvency may precipitate an attack on the currency now, and make it difficult or even impossible to maintain the current level of the exchange rate. However, we may prefer to think of this as a 'credibility constraint'!" (p.202)

This definition leads to a number of important questions with regard to the conduct of stabilisation policy, national liquidity and solvency and the significance of the external debt ratio. However, this approach to the 'external constraint' does not relate to what many are concerned with. It is possible that developing country X would have a higher rate of economic growth if it could borrow more abroad to finance the import of capital goods that it cannot manufacture. Unfortunately its exports of primary products and the resulting flow of foreign exchange may be insufficient to raise its growth to the level of which it is capable. On Bean's definition it might be said that country X has an 'external constraint'. Standing the illustration on its head, we might envisage country Y which exports manufactures but imports raw materials to produce them. If it can borrow more abroad at a steady rate, it might be able to grow faster than in the absence of borrowing. If it would like to borrow more, it too might be said to suffer an 'external constraint'. This certainly is getting closer to the concern of the House of Lords (1985). But in the light of what follows, the two cases might be better described not as suffering from 'external constraints' or 'balance-of-payments constraints', but simply as 'import constrained'. Neither of the illustrations remotely describes the position of the United Kingdom today.

The focus on the ability to borrow externally as a possible macroeconomic constraint has been prompted by the emergence of the so-called Burns' doctrine, as described by Muellbauer and Murphy (1990). But the concerns of those who regarded the economy in the 1960s as 'balance-of-payments constrained' were not couched in these terms. In modern parlance, in the absence of capital mobility, the external constraint of the 1960s was seen as a liquidity constraint, the other end of the spectrum from the solvency constraint that is of importance in a world of mobile capital. The 1960s concern in modern dress has recently been set out by Thirlwall (1992):

"Demand, determined by export performance and the balance-of-payments position governs output growth but supply-side policies such as investment technology, research and development effort, education and training in skills etc. determine the income elasticity of exports and therefore how fast exports grow as world income grows. The view that

I cannot accept is that a mere augmentation of the supply of resources *will necessarily improve the growth performance of a country if it does not at the same time improve the long run balance of payments position* (our italics). If exports remain static and imports rise, the deficit on the balance of payments will be unsustainable and therefore demand will have to be retracted and resources will remain under-utilised. It is in this sense that the balance of payments becomes the ultimate constraint on growth." (p. 141)

This echoes the earlier statement made by the House of Lords Committee to the effect that sustainable growth is not possible without a favourable trade balance in manufactures. Here it is alleged that growth cannot be improved unless there is an improvement in the balance of payments. It is tempting to dismiss this out of hand by pointing out that since trade deficits and surpluses must add up for the world as a whole, these propositions cannot be general, without asserting that improvements in growth rates cannot be shared. Economic growth would become a zero-sum game, which must be nonsense. However, Thirlwall's detailed analysis does not in fact require the italicised statement to be true.

The implication of Thirlwall's analysis and the concerns expressed by the House of Lords Committee seems to be that the structural relationships that drive the level of domestic activity and the components of the balance of payments may be such as to make it *impossible* simultaneously to achieve internal and external equilibrium. External equilibrium is a constraint that must be observed. This possibility is arrived at by Thirlwall essentially by 'fixing' the relationship between income and imports so that there is a *unique* level of imports at the full employment level of income - a constant income elasticity of demand for imports. Price effects are eliminated by the assumption that in the medium term the real exchange rate follows close to its purchasing power parity value.

The idea that income elasticities are exogenous rather than endogenous is criticised by Krugman (1989).

"I am simply going to dismiss *a priori* the argument that income elasticities determine growth rather than the other way round. It just seems fundamentally implausible that over stretches of decades balance of payments problems could be preventing long term growth, especially for relatively closed economies like the US in the 1950s and 1960s." (p. 1037)

The existence of a balance-of-payments constraint is not implied by a world of less than 'perfect' adjustment, if by that is meant a rapid return to equilibrium following some kind of shock to the economic system. The assertion implicit in the argument of Thirlwall seems to be that no such equilibrium entailing both internal and external balance may exist. The consequence of this may be a growth rate below that of productive potential and permanently unemployed resources.

If we abandon for a moment the assumption that there is a unique relationship between the rate of growth of income and the rate of growth of imports, are there any other reasons for supposing that there is in a significant sense the serious possibility that a country like the United Kingdom may be balance-of-payments constrained?

It is appropriate to remind ourselves of the accounting identities that hold between the current account of the balance of payments, domestic savings and investment, the capital account of the balance of payments and the change in reserves. In general we have:

$$CUR \equiv S - I \equiv CAP + \Delta R$$

where

CUR	=	the current account
S	=	domestic savings (both public and private)
I	=	domestic investment (both public and private)
CAP	=	the capital account
ΔR	=	change in foreign exchange reserves

Expressed in this general form, it is clear that changes in the current account can emanate from a number of different sources. Typically, those who have been concerned with the relationship between manufacturing exports and the current account have focused on real changes in the structure of traded goods relationships. The monetary theory of the balance of payments has emphasised the importance of changes in the money stock and the demand of money in inducing changes in the current account. The role of saving and investment behaviour has long been recognised in the absorption approach to current account analysis. While some theorists have focused on the question of exchange rate behaviour in affecting trade volumes, others have asserted that balance of payments behaviour is essentially a monetary phenomenon (e.g. Frenkel & Johnson, 1976) that is independent of 'real trade' theory.

It is difficult not to agree with Williamson (1983):

"Several of these approaches are presented by their creators as representing a conflict with the preceding approaches which have indeed been derided by some as erroneous orthodoxy. My view is that such exclusiveness is unmerited; that any adequate understanding of the macroeconomics of an open economy demands an integration of all the various approaches within the context of a general equilibrium model." (p.37)

It is arguable that 'real' changes in the structure of trading relationships involving relative changes in factor productivity, technology etc. will require changes in the real exchange rate, recognising shifts in underlying competitiveness. But the same cannot be said necessarily if the current account disturbance is the result of a shock emanating from the expenditure side of the system or policy shocks derived from the conduct of fiscal and monetary policy.

This of course lies at the heart of the so-called Burns' doctrine to which reference was made earlier. In over-simplified terms the plunge of the United Kingdom trade balance into the red after 1987 has been attributed primarily to the collapse of net (not gross) private saving. The resulting current account deficit was not therefore attributable to a sudden and dramatic decline in the structural competitiveness of UK industry (even allowing for the fact that some decline in the real exchange rate might have been justified by the fall in oil prices). It resulted, as in other post-war UK episodes, from excess demand. Most previous episodes had been the result of a shock due to a fall in public sector saving rather than private. The so-called privatisation of the deficit leads to the idea that eventually the deficit will control itself.

This proposition was unfortunately muddled in the public view with another proposition altogether, namely that in general, current account deficits do not matter, or in particular they do not matter when they are 'private'. This is wrong on two counts. The first is that the shock was largely due to the conduct of public policy, and did not arise from a spontaneous shift in demand. Secondly, widescale departures from equilibrium subject the economy to major welfare costs and therefore do matter. The analytical point in this debate is whether a return to equilibrium requires a permanent fall in the real exchange rate.

Against this background, and in the absence of any assumption of a constant relationship between the rate of growth of income and the rate of growth of imports, how can the balance of payments be a constraint on economic growth? Provided monetary and fiscal policy are conducted on a prudent basis, there is little evidence that economic fluctuations on the whole have reflected *endogenous* shifts of any magnitude in private spending behaviour, so that it should be possible to limit and contain disturbances arising from the savings and investment account - at least in principle. This leaves us with the problem of dealing with real secular change, which requires shifts in the equilibrium real exchange rate. In the United Kingdom the focus has been on the need for such a shift as the result of the fall in the contribution of North Sea oil both to the current account of the balance of payments and to the growth/level of national output.

To summarise, a balance-of-payments constraint may be interpreted in terms of the ability or inability of an individual country to finance its current account deficit. This is not what has traditionally been meant by a balance-of-payments constraint - nor is it the problem that some have seen in modern dress. The problem, as seen by those currently who focus not only on the balance of payments as a possible constraint on economic growth but also on the important connection with manufacturing industry, can be interpreted in two ways. Firstly, that there is some unique relationship between imports and income at 'full employment' that constrains the economy in its mix between foreign and domestic output, irrespective of the real exchange rate. Secondly, that conventional adjustment processes that affect the real trade side of the system 'will not work', either in the sense that factor price adjustments are so inflexible that adjustments will not take place at all or over a sufficiently long period as to be socially unacceptable.

Recent literature on this subject has raised a number of problems that relate to the nature of, and the dynamics of, the adjustment process. There are many disagreements. But the nature of these disagreements does not suggest to us any widely canvassed view that theory points to a strong probability that a balance of payments constraint for the United Kingdom is likely. For example, recent debate about United Kingdom membership of the ERM, as illustrated by the contributions of Minford (1992), Currie and Dicks, and Wren-Lewis (also 1992), reflects very different views about the appropriateness of both exchange rate regimes and the value of the pound from time to time. But none of them challenges the basic proposition that there exists a real exchange rate that is consistent with both internal and external balance.

II. The Balance of Payments

In the 1950s and 1960s the view that the current account of the balance of payments as a constraint was simply derived. Over short periods of time, growth was slow and unemployment rose modestly by the standards of the day. Fiscal expansion was put in place and eventually imports rose faster than exports and the current account deteriorated. Having stepped on the accelerator, the policy-makers then had to step on the brake. The result was described as stop-go. It was also described as demonstrating the existence of a balance of payments (current account) constraint on economic growth, in the sense that the authorities were limited in their ability to pump up the rate of growth of demand.

There were those who believed that the constraint manifested itself in the maintenance of the pegged exchange rate, in which case a major devaluation of the currency, or better still, perhaps a floating of the currency would remove the constraint. This view unfortunately was based on rudimentary thinking about the determinants of the inflation rate. There was a failure to see that current account deficits under pegged exchange rates and rapid inflation under floating ones would simply be two sides of the same coin. They both reflected policy shocks emanating from the conduct of fiscal and monetary policy that pushed the rate of growth of demand beyond the ability of the supply side of the economy to respond. The current account constraint theory turned a supply side/labour market problem into a so-called balance-of-payments problem. The intellectual pressure for such a development is obvious. The former problem is micro-economic in character, difficult to deal with precisely, and without doubt calling for changes that were painful and long in coming. The latter suggested quick fixes, such as subsidy, protectionism and exchange rate manipulation. It is not surprising that the balance of payments constraint hypothesis found so many adherents - as to some extent it does today.

The story of the current account of the balance of payments since the second world war falls into two parts - the post-war years of Bretton Woods and the period since the collapse of Bretton Woods and the first oil price shock (Figure 1). The concept of a balance-of-payments constraint on the rate of

Figure 1: The Current Account % Nominal GDP

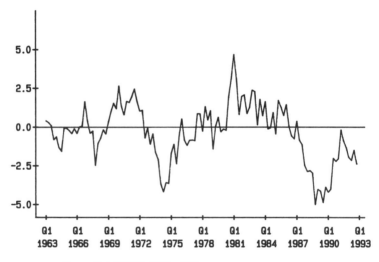

Source: Central Statistical Office, 1992

economic growth emerged from the Bretton Woods era as described above. Over this period with unemployment averaging under 3%, the constraint on growth was not related to the current account of the balance of payments *per se*. Relatively slow growth stemmed from a lack of competitive ability to prevent the United Kingdom's share of world demand from falling, i.e. it was essentially a supply-side problem.

The second period can be sub-divided into three significant periods as illustrated in Figure 2 which looks at the current account as the sum of net

Figure 2: Sector Balances % Nominal GDP

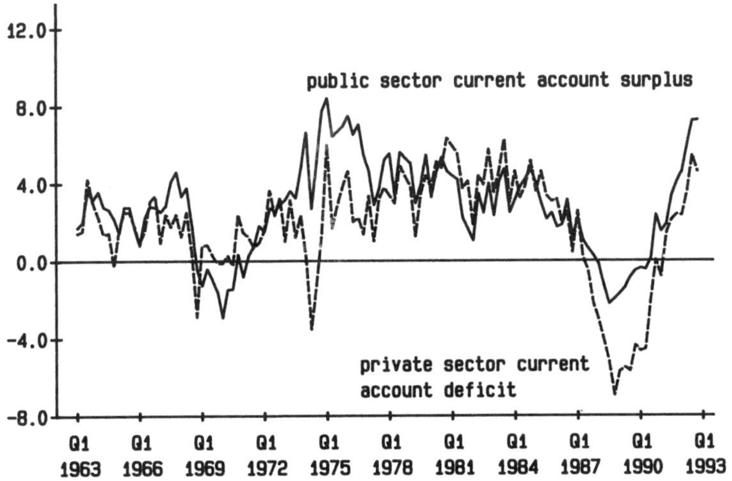

Source: Central Statistical Office, 1992

private saving and the general government deficit. Looked at from this point of view, the large current account deficits of the 1970s are clearly associated with the behaviour of public sector saving. The experience of the period suggested support for the so-called New Cambridge economics of the period. In its simplest form, this suggested that net private saving was generally close to balance or at least bore a stable relation to income. Fluctuations in the current account were therefore largely to be attributed to fluctuations in the public sector deficit.

Clearly, however, while the behaviour of the public sector deficit may be associated with current account disequilibrium, it bears little or no relationship to the surplus period of the 1980s, which was materially influenced by revenues from oil. Even less was public sector savings responsible for the major deterioration of the current account after 1986. As can be clearly seen from Figure 2, while for the last years of the 1980s the public finances were in surplus, net private saving collapsed. The current account deficit that it matched was produced by the behaviour of the private as opposed to the public sector.

Current account behaviour in the United Kingdom since the breakdown of Bretton Woods has been dominated by the behaviour of supply-side shocks in the form of oil price changes and the conduct of fiscal and monetary policy. Over much of the time, particularly in the 1970s, economic performance in terms of economic growth was disappointing. However, there is little evidence that performance was constrained by, or influenced by the balance of payments as such.

The Balance of Payments and Manufacturing Industry

Since the early 1970s, overall disappointing economic performance has been associated by many with the disappointing performance of manufacturing industry, and particularly the behaviour of manufactured exports. The share of United Kingdom manufactured exports in total world exports of manufactures is shown in Figure 3, which depicts the dismal story of the decline in the United Kingdom market share from 1960 (and before) up to the middle 1980s. Since then the share has risen modestly.

Figure 3:   U.K. Share of World Trade in Manufactures (1985=100)

Source: Central Statistical Office, 1992

It seems to us indisputable that a better performance with regard to export growth and import substitution for manufactures would have improved the United Kingdom's overall economic performance in the 1970s and early 1980s, other things being equal. The question that concerns us here is the special significance that manufacturing industry holds for some with regard to the behaviour of the current account of the balance of payments *per se*, which is thought to feed back on to the rate of growth of total output. Some have argued that manufacturing exports can provide the only means of securing what might be described as Britain's 'import needs' at full employment. To this was added the prediction that this 'need' would increase *pari passu* with the decline in revenues from North Sea oil. Most recently, Davis, Flanders and Star (1992) comment:

"It is commonly observed that the economy requires a strong tradeable goods sector to support our consumption of imports. Correspondingly the UK's balance of payments deficit is said to reflect our growing inability to do this" (Davis, Flanders and Star, 1992, p.46).

Statements of this kind seem to ignore completely the major change in the structure of the United Kingdom current account of the balance of payments over the last forty years. This is ironic for, as we shall see below, looked at from another point of view, major changes in the share of manufactures in total imports have been a focus of attention. The flavour of the observations made, and the concerns expressed, still characterise the United Kingdom as a country which exports manufactured goods in order to import food to eat and raw materials to keep the wheels of industry turning.

The structure of the credits and debits for the last thirty years that determine the current account are set out in Table 1. Starting in 1960, we see that on the credit side, exports of goods and services, the export of goods, principally manufactures accounted for 63% of all receipts, and services 37%. Manufactures were 52% of the total and 77% of receipts from goods alone.

On the debit side, imports of goods and services, imports of goods amounted to 67% of total receipts, and services 33%. Manufactures were 23% of the total and 34% of goods imported. The category 'other goods' includes food and raw materials. As can be seen from the table, if we take the value of 'other goods' plus 'oil' as reflecting the economy's basic 'needs', we see that they were roughly equal to the value of manufactured exports in 1960. Under these circumstances it might have been reasonable to conclude in 1960 that the United Kingdom exported manufactured goods to pay for food and imports into the economic process.

Thirty years later the situation on the debit side of the account has changed dramatically. On the credit side, following the fall in oil revenues after 1986, the situation as far as the goods account above was concerned is almost identical to that of 1960. 80% of the revenue from the sale of goods is attributable to manufactures. However, the export of goods relative to receipts from services has fallen materially. Putting matters the other way round, receipts from the invisible account have risen from 37% to over half of the total. It is puzzling why there is a contemporary view that contributions from services can never substitute for manufacturing earnings when, in a significant sense, they already have.

But the major change on the visible account has been the rise in imports of manufactures. For total manufactures, i.e. both finished and semi-finished goods, the proportion of manufactured imports in total imports of goods rose from 34% in 1960 to nearly 79% in 1990. Imports of 'other goods' plus 'oil' fell to 12% of the value of total imports of goods. Reflecting a similar development on the credit side of the current account, the ratio of payments on goods account also fell relative to receipts from services, from 67% to 52%. Payments of interest, profits and dividends rose from 7% in 1960 to a third of total debits on current account by 1990.

Table 1: U.K. Exports and Imports of Goods and Services 1960-1990, £m (%)

	1960	%	1970	%	1980	%	1990	%
Visible Exports								
Semi-manufactures	1131	22	2782	21	14152	16	28875	13
Finished manufactures	1756	30	4100	31	20727	23	53879	25
Oil	104	2	180	1	6118	7	7451	3
Other goods	566	10	1088	8	6392	7	11833	5
Total Visibles	3737	63	8150	62	47389	54	102038	47
Invisible exports	2207	37	5103	38	41059	46	117350	53
TOTAL EXPORTS	5944	100	13253	100	88448	100	219388	100
Visible imports								
Semi-manufactures	922	15	2323	19	12561	15	31565	14
Finished manufactures	474	8	1997	16	16871	20	62629	27
Oil	480	8	676	5	5818	7	5933	3
Other goods	2262	37	3188	26	10811	13	20586	9
Total visibles	4138	67	8184	66	46061	54	120713	52
Invisible imports	2034	33	4273	34	39504	46	113055	48
TOTAL IMPORTS	6172	100	12457	100	85565	100	233768	100

Sources: UK Balance of Payments, Pink Book

These dramatic changes in the structure of the United Kingdom current account have received insufficient attention in the discussion relating to the performance of the current account. Financial liberalisation has fundamentally altered the impact of capital flows on current account behaviour. On the credit side, receipts from profits, interest and dividends have risen from 11% in 1960 to 37% in 1990. On the goods side, it is simply untrue to suggest that the United Kingdom exchanges manufactured goods for imports of food and raw materials. The statement that the United Kingdom needs to export manufactured goods to 'pay its way in the world' is totally meaningless. What the United Kingdom now does, and has for a long time, is to export manufactured goods in order to import manufactured goods. The next question is why has this come about and should we be concerned about it as many clearly are?

Figure 4 illustrates the dramatic change in the balance between imports and exports of manufactures since 1970. As the Figure shows, the surplus in manufactures in nominal sterling terms grew over the period 1970-1980. In itself, this is interesting since as already suggested, the period of the 1970s was one in which the overall economic performance of the United Kingdom was disappointing. Indeed, the period 1973-1979, which bridged two peaks of the 1970s from the point of view of real output behaviour, was one in which the volume of output of manufactures declined absolutely. While the early 1980s saw a major collapse in manufacturing output of some 15% (the implications of which are discussed further below), it began to recover after the trough of 1981, and in the last five years of the decade grew more strongly than over any other comparable period since 1970. Yet as Figure 4 shows, it was during this period of recovery in manufacturing output that the manufacturing trade balance in the United Kingdom declined most rapidly.

There are three principal reasons why the share of manufactures in imports rose. The first is specific to the United Kingdom, namely the impact on the current account balance of the coming on stream of North Sea oil. The second is a worldwide change in the structure and nature of trade as a result of imperfect competition and economies of scale. The third and most obvious is

Figure 4: U.K. Current Balance in Manufactures £ bn

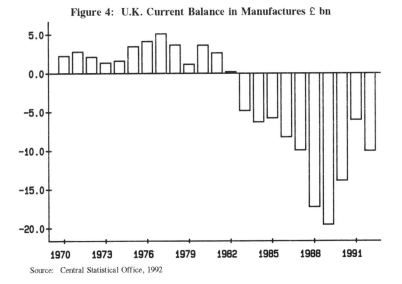

Source: Central Statistical Office, 1992

the deterioration of the competitive position of United Kingdom producers of manufactures in both domestic and international markets. There has been a remarkable unwillingness to accept the idea that the impact of North Sea oil in itself could either in theory or practice exert a major influence on both the current account of the balance of payments and the economics of manufacturing industry. A typical media response of the mid-1980s was to observe the deficit in manufacturing trade and to comment that it was 'fortunate' that the favourable oil balance existed to offset it. The idea that the two statistics were related was hard to understand!

To summarize familiar conclusions - the effects of North Sea oil on the economy appeared in various forms. They all derive from the lack of a need to export in order to pay for the oil that the revenues from North Sea oil made unnecessary. The consequences of this were reflected in a combination of the foreign exchange saved (as a result of our not having to spend it) and the real resource cost of not having to produce the exports of goods previously required. The latter was emphasised in the familiar paper by Forsyth and Kay (1980), which suggested a need for an expansion in the output of service industries to make up for the decline in manufacturing output. The foreign exchange implications of North Sea oil revenues focused on the ability of the United Kingdom economy to enjoy a higher volume of imports than would otherwise be the case, and to acquire overseas assets by investing the foreign exchange savings overseas.

Under the circumstances it was to be expected that both imports and capital exports should therefore rise sharply in the first half of the 1980s. Much ink has been spilt on the question of how much the rapid rise in sterling and the deterioration in competitiveness of exports in the early 1980s was attributable to North Sea oil or the fiscal and monetary policies pursued by the incoming Conservative administration in 1979.

There is little evidence that the adjustment process was of a kind suggested by Forsyth and Kay. The most visible consequences were those of an increase in imports of manufactures and a rise in capital exports. In our view no reliable estimates exist of the impact of North Sea oil on manufactured imports. Nonetheless *a priori* considerations suggest that it was likely to have been considerable. Other things equal, one would have expected a major substitution of manufactured imports for oil in the absence either of a significant diversion of real resources into service industries or a major expansion of capital exports. In view of the latter, it is clear that the rise in manufactured imports into the United Kingdom in the first half of the 1980s could not entirely be attributed to the effects of North Sea oil, although the impact of oil was *prima facie* material.

A second influence on the level of manufactured imports and the ratio of manufactured imports to exports might have been expected over the last twenty years as a result of changes in the overall structure of trade worldwide. The concept of the United Kingdom as a supplier of manufactures in exchange for goods and raw materials exemplifies the traditional theory of comparative advantage that accounts for *inter-trade* between nations. But it is now generally recognised that the principal driving force behind the growth of trade within the OECD over the last two decades has been the development of *intra-trade* deriving from the twin forces of economies of scale and imperfect competition. This has led to much discussion of the benefits of free trade in a world where the classical assumptions that underlie Ricardo's seminal statement of free trade principles, constant returns to scale and perfect competition no longer hold.

To illustrate the point we may consider the growth of trade between members of the European Community.   The increase in trade between Community members has been substantially due to a rise in intra-trade based on manufactured goods.   In itself this is surely unsurprising on *a priori* grounds.

Given a higher income elasticity of demand for manufactured goods relative to foodstuffs and a declining technical ratio between raw materials and the level of economic activity, countries that began the period as major sellers of manufactured goods might have been expected to spend an increasing proportion of their export receipts on manufactured imports. This would result in a secular trend upward in the ratio of imports of manufactured goods relative to total visible imports. As Table 2 demonstrates, the behaviour of the United Kingdom in this respect is not significantly different from the experience of its major competitors.

The discussion suggests that secular trends in the consumption of raw materials, and the coming on stream of oil in the first half of the 1980s, could

Table 2: Imports of Manufactured Goods as % of Total Visible Imports by Value

(Manufactured Goods defined as SITC groups 5-9 except for Italy & US
where their own categories have been aggregated)

	1960	1965	1970	1975	1980	1985	1990
Germany	45	54	61	58	59	62	77
France	-	51	63	63	58	62	61
Italy	-	51	54	46	50	53	68
Japan	-	23	30	20	23	29	50
USA	49	56	68	56	62	81	82
UK	34	40	53	53	64	69	78

Source: CD Economic Surveys for France, Germany, Italy & Japan, Statistical Abstract of the United States for the
US, UK Pink Books (various)

have themselves explained a rise in the ratio of the imports of manufactured goods relative to exports. Over that period it is also possible that the ratio was affected by a decline in price competitiveness brought about by the conduct of economic policy by the Conservative government as part of the contra-inflation stance it adopted. A crucial question that has been posed by Muellbauer and Murphy (1990) is whether the deteriorating balance in manufactures reflects a long-term secular decline in competitiveness that can be expected to continue into the 1990s.

As pointed out there are both secular factors and fluctuations in short-term competitiveness that suggest that imported manufactures may have risen substantially relative to exports in the *first half* of the 1980s, irrespective of any longer-term secular decline in manufacturing competitiveness. Moreover, it is puzzling that such a rapid rise in the ratio should occur in the *second half* of the 1980s, when, as will be argued in the next section, looked at from the export side, manufacturing industry had become more rather than less competitive. Given, for the sake of argument, that that was the case, this leaves the major deterioration in the manufacturing trade balance to be accounted for by the conduct of economic policy in the mid-1980s that led to the boom conditions that followed. This is not to say that such a rapid expansion of domestic demand as has been seen on numerous occasions since the Second World War may not have had some permanent effects on the trade balance. Hysteresis may set in when people change suppliers. However, taking all these factors together, we arrive at a somewhat agnostic conclusion with regard to the influence of a structural secular decline in manufacturing competitiveness that might be expected to continue into the next decade.

III. The State of Manufacturing Industry

The development of the argument started with the possibility that from a theoretical point of view the balance of payments *per se* could exercise a serious long- term 'constraint' on the economic growth of a country like the United Kingdom.   We concluded that that was unlikely to be the case. However, leaving that on one side, we examined the extent to which the case might be made empirically that taking the period since the Second World War, such a constraint was operative. This led us to look specifically at the role and significance of manufacturing industry in the context of the balance of payments, and in particular to look at the importance of the change in the ratio of manufactured imports to manufactured exports. We concluded that changes in this ratio had a number of possible explanations and we remained agnostic with regard to the hypothesis that the observed changes led inevitably to the conclusion that secular decline in the competitiveness of United Kingdom manufacturing industry would continue over the next decade.

However, for the sake of argument let us concede the proposition - that there is a significant case on theoretical grounds for believing that a 'balance-of-payments constraint' is not only possible but likely; moreover, that over the period since 1945 we have witnessed a long-term secular decline in the competitiveness of British manufacturing industry that potentially has boded ill for United Kingdom economic performance - particularly in relation to the balance of payments. The final question therefore relates to the state of manufacturing industry in the United Kingdom today. In particular, will its competitive decline continue into the decade of the 1990s? And will this result in the re-imposition of what is seen by many as 'the balance-of-payments constraint'?

It is both convenient and appropriate to look at these issues in the context of what have come to be known as the 'Thatcher years', covering the period from the advent of Thatcher's first government to the end of her decade in 1989. For the most part, professional economists, both of a macroeconomic and a microeconomic persuasion, were not particularly disposed toward the economics of Thatcher's administration. But there is little doubt that the one phenomenon that intrigued them all was the rise in manufacturing productivity. The central question to which they addressed themselves was to what extent had the Thatcher years produced a real change in the underlying competitiveness of British manufacturing industry.

In an impressionistic sense, three versions of the story emerged which are not necessarily mutually exclusive. The first is that the collective changes in both macro- and micro-economic policy, including major changes in taxation, trade union legislation and privatisation had fundamentally changed the supply side of the economy and had led to the Thatcher revolution on the micro-economic side of the economy. The second, that was diametrically opposed to this view, was that the so-called Thatcher revolution had destroyed large sections of British manufacturing industry, over-emphasised the growth of financial services and wasted the potential benefits of North Sea oil. The third view struck a balance between these two views. It recognised that material changes had taken place during the Thatcher regime. It was difficult to deny it. The peak of manufacturing output in the pre-Thatcher cycle occurred in 1979. Manufacturing output did not regain that level until 1987. But at that stage the same level of output was being produced with nearly 2.5m fewer people.

None of these analyses fully took on board the longer run implications of the cost behaviour of the 1970s. The seeds of the problem of British manufacturing industry were sown long before Margaret Thatcher was on the scene. They are related to the appalling behaviour of real wages and unit labour costs that underlies the performance of manufacturing. The facts of the 1980s as opposed to their interpretation were simple enough. The growth of manufacturing productivity during the 1980s was exceptional. It had to be - how, in less than a decade, could you produce the same level of manufacturing output with one-third fewer people? Even historical comparisons with the 1960s were of no interest. The only serious question was how was this phenomenon to be explained.

Comprehensive accounts of the alternatives have been set out by

Muellbauer (1986, 1991). These accounts explored both the cyclical and secular possibilities of why productivity in the 1980s increased so much both absolutely and from an historical perspective. Consideration was given to the phenomena of accounting increases - the so-called batting average phenomenon - measurement effects, the spread of innovation, particularly in information technology, the change in industrial relations, and changes in the ratio of capital to labour.

The consensus that clearly emerged was that there was no doubt over the change in productivity. The key issue was the extent to which the change was once and for all, without affecting the dynamic ability of the economy to carry forward the progress made during the 1980s.

Some have their doubts about the underlying change in the dynamics of productivity increase, and believe that the improved performance of the 1980s was of an essentially one-off nature. Metcalf (1989) and Bean and Symons (1989) have focused on the significance of changes in the nature of labour relations and in particular the structure of bargaining relationships to account for the change in productivity, particularly in heavily unionised industries. But all these analyses are essentially of a static nature. They all reflect a considerable scepticism with regard to the dynamics of economic change.

Such a conclusion is derived from the material evidence of the lack of investment in the United Kingdom in education and human capital, and in research and development. But the lack of human capital investment, which is crucial both on the management side and the input of direct labour, remains a central issue for the United Kingdom. The issue of technology transfer that is central to the dynamic progression of productivity remains an open question. A recent Confederation of British Industry study of manufacturing industry which, despite expressing many conventional views of the importance of manufacturing industry that are not shared in this paper, nevertheless took an optimistic view of the future of manufacturing.

"The 1980s have seen more dramatic changes in manufacturing industry than any other decade since the Second World War. Specifically:

(i)   The key indicators of output, productivity, and share of world trade all show an improving trend relative to our competitors.
(ii)   There has been a considerable restructuring of industry, with the help of inward investment, increasing the presence of the UK in high-technology sectors.
(iii)   Some long-standing weaknesses in terms of quality, customer service, innovation and skills are being addressed.
(iv)   Employer communications and industrial relations have been transformed."

(CBI, The Report of the CBI Manufacturing Advisory Group, Autumn 1991, p.11)

The bottom line, however, reflects the perennial difficulty of assessing the dynamics as opposed to the statics of productivity change. It is easy to describe the key long term factors that are critical to the process of economic change in manufacturing. The stylised facts are well known. The United Kingdom has been deficient in investing in human resource skills, and its record on research and development has been much criticised.

The unpredictability of future productivity growth and competitiveness, however, lies critically in what happens to the performance of management and labour. The studies referred to above have identified changes in the relationship between the two in the 1980s as a major part of the story of productivity, at least when looked at from a static point of view. The central question is whether such changes in both industrial relations and managerial performance will have carry-over effects on to the quality of inputs into the manufacturing process.

At this stage, the only conclusion that we can draw is an agnostic one. While United Kingdom manufacturing productivity in the 1980s grew faster than any of the major industrial countries, the absolute level is still far behind that of its principal competitors. There is potential opportunity for catch-up and relative improvement. But what evidence there is makes it unclear whether that opportunity will be taken or not.

IV. Summary and Conclusions

The idea that a mature industrial economy such as the United Kingdom may be 'balance-of-payments' constrained can be interpreted in a number of different ways. Most recently, emphasis has been placed on the financing of current deficits, and the possible constraints on domestic policy making. But in the United Kingdom context we have suggested that a 'constraint' may be placed on the realisation of full employment and a socially acceptable rate of economic growth as a result of the problem of reconciling internal and external equilibrium. In particular, we have suggested that this problem might arise if external equilibrium is consistent with only one rate of growth of output that can be below the potential growth rate of output of the economy. The focus of this paper has been on the latter interpretation of the 'balance-of-payments constraint'.

This concern has been identified in more recent years with the performance of manufacturing industry. The argument is straightforward enough. Imports are related to the level of domestic activity, in particular the supply of food and raw materials for which historically Britain has exchanged its manufactured goods. The economy has to 'pay its way' in the world. That it is said has been a problem for the United Kingdom since the Second World War. During the 1980s, Britain benefited from the windfall gains that accrued as the result of the coming on stream of North Sea oil. But as the contribution of oil to the current account of the balance of payments and the level of national income declines, it is crucial that these contributions are replaced by a superior manufacturing industry performance. The prospects of this are not seen to be good, largely as a result of the dramatic increase in the 1980s of imported manufactures. The manufacturing trade balance has changed materially since 1980, as has been shown in Figure 4. What has happened to the manufacturing trade balance has been seen by some (Muellbauer and Murphy, 1990, are an excellent example) as representative of a long- term decline in the competitiveness of British manufacturing industry. Muellbauer and Murphy are somewhat ambiguous about what the implications of this are. But others (e.g. Thirlwall, 1992) are clear that the change in the manufacturing trade balance is related to the 'balance-of-payments constraint' idea that we have already set out. The problem resided clearly in the behaviour of real wages and unit labour costs.

We have argued that from a theoretical point of view there is no strong reason to suppose that an industrial economy as mature as the United Kingdom would be 'balance-of-payments constrained' in the sense in which we have focused on the issue. The co-existence and feasibility of joint internal and external equilibrium at different rates of economic growth can always be made impossible by 'fixing' the adjustment mechanisms. We have emphasised the welfare costs of adjustment if and when shocks to equilibrium occur. In this sense balance-of-payments disequilibrium matters. But the lesson it teaches us, to which we return below, has nothing to do with the balance of payments *per se* but relates to the conduct of stabilisation policy.

The dramatic deterioration in the current account in the late 1980s can be assigned again to oil price changes combined with the disastrous conduct of stabilisation policy. While it is true that during the 1980s the balance in manufacturing trade declined, we have suggested reasons why that might have been the case, irrespective of any secular decline in the competitiveness of British manufacturing industry. While we would not dispute the fact that some such decline took place over the period 1970-1990, that cannot have been the major reason for the decline in the trade balance in the second half of the 1980s, when it is widely accepted that the competitiveness of British manufacturing had in fact improved.

We accept the fact that the evidence relating to the performance of British manufacturing industry during the 1980s leads to agnostic conclusions with regard to its future. But in balance-of-payments terms, we have emphasised the major changes that have taken place in the structure of the current account over the last thirty years. On the goods side, the United Kingdom is primarily engaged in producing and selling manufactures in order to buy manufactures. We reject the idea that this change in structure is some sort of national disaster. We are also less concerned about the problem of dealing with the decline in oil receipts and the contribution of oil output to total output since the most dramatic changes have already taken place.

We have two central concerns that shape our attitude to the issues discussed. The first concerns the importance of export behaviour in determining economic growth. The second concerns the conduct of fiscal and monetary policy in a balance of payments context.

The significance of international competitiveness is that it determines an individual country's ability to determine its share of *total* world demand. As a matter of simple arithmetic, to grow at the world average rate requires a country at least to maintain its share of world demand. Almost inevitably those countries with relatively strong export performance have tended to have strong current account performances. But that is not the point. If all current account balances were zero, the phenomenon would still exist. Export performance is a supply side phenomenon that rests on supply side fundamentals. In our view, it is a mistake to confuse the supply side problems with problems of the balance of payments *per se*. It is the idea that there is a balance-of-payments problem *per se* that leads us down the road to protectionism, subsidies, and intervention in trade affairs. The realisation that export performance and international competitiveness are crucial for *economic growth* leads one down a different path to issues of productivity, labour market behaviour, and investment in human capital.

In this context the role of fiscal policy would no longer be assigned to the problem of stimulating output, employment and economic growth. It has a major role to play in the determination of the current account balance at a given nominal exchange rate.

There are two important issues here that are familiar in present day literature with regard to monetary analysis and the balance of payments. The first is the nature of the exchange rate regime under which the system operates, which alters the relative powers of fiscal and monetary policies in a familiar textbook way. The second is the distinction between the real and nominal behaviour of the economy, and the extent to which one is prepared to distinguish clearly between the real implications of trade and the behaviour of the balance of payments in a nominal sense.

Our view is that the significance of real trade performance, whether in manufacturing or otherwise, *as far as the balance of payments is concerned*, is not a central issue. The conduct of fiscal and monetary policy is. The importance of real trade performance may be of crucial significance, for employment, output and economic growth. To mix up the problem of balance of payments adjustment with economic growth and real competitiveness is to mix up apples with pears, the result of which predictably, is often rhubarb!

In the kind of model that underlies such an analysis, there is no interaction between the balance of payments and the real economy in the long run, although clearly interaction is not only possible but likely in the event of shocks that cause a major disequilibrium to occur. The proposition that there is no such thing as a balance of payments constraint has the same logical status as the proposition that the long run Phillips Curve is vertical. It denies the existence of a long-term trade-off that makes a balance of payments constraint possible, in the sense in which we have interpreted such a constraint in this paper.

Sir James Ball and Donald Robertson, London Business School, Sussex Place, Regent's Park, London NW1 4SA, United Kingdom.

ACKNOWLEDGEMENTS

We are grateful to David Currie and Harold Rose for their comments on an earlier draft of this paper, and to Geoffrey Dicks for editing the final product. All errors and omissions remaining are entirely ours.

REFERENCES

Ball, R. J. and Drake, P. (1962). Export growth and the balance of payments. *Manchester School. XXX*(2), 105-119.

Ball, R. J. and Burns, T. (1968). The prospects of faster growth in Britain. *National Westminster Bank Review*. pp. 3-22.

Bean, C. and Symons, J.. (1989). Ten years of Mrs. T. *NBER Macroeconomics Annual*, pp. 13-72 (with discussion)

Bean, C.. (1991). The external constraint in the U.K. In G. Alogoskoufis, L. Papademos and R. Portes (Eds.), *External constraints on macroeconomic policy: The European experience*. (pp.193-218) Cambridge: Cambridge University Press.

CBI Report. (1991). Manufacturing Advisory Committee

Currie D.A. and Dicks, G.R. (1992). Policy options for the U.K. *Economic Outlook. 16*(5), 39-45.

Davis E., Flanders S., and Star, J. (1992). British industry in the 1980s. *Business Strategy Review. 2*(1), Spring, 45-69.

Forsyth, P.J. and Kay, J. A. (1980). The economic implications of North Sea oil revenues. *Fiscal Studies. 1*(3), 1-28.

Frenkel, J.A., and Johnson, H. G., (Eds.). (1976). Essential concepts and historical origins. *The monetary approach to the balance of payments*. London: Allen and Unwin, pp. 20-45.

House of Lords. (1985). *Report from the Select Committee on Overseas Trade*. Volume I - London: HMSO.

Krugman, P. R. (1989). Differences in income elasticities and trends in real exchange rates. *European Economic Review. 33*, 1031-1054.

Metcalf, D. (1989). Water notes dry up: The impact of Donovan reform proposals and Thatcherism at work on labour productivity in British manufacturing industry. *British Journal of Industrial Relations 27*, 1-31

Minford, P. (1992). Why we should leave the ERM. *Economic Outlook. 16*(5), 31-34.

Mundell, R. A. (1962). The appropriate use of monetary and fiscal policy for internal and external stability. *IMF Staff Papers. IX*(1), 70-79.

Muellbauer, J. (1986). Productivity and Competitiveness in British Manufacturing. *Oxford Review of Economic Policy. 2*(3), 1-25.

Muellbauer, J. and Murphy, A. (1990). The UK current account deficit: Is the UK balance of payments sustainable? *Economic Policy*. October, 347-395,

Muellbauer, J. (1991), Productivity and competitiveness. *Oxford Review of Economic Policy*. 7(3), 99-117,

Thirlwall, A. P. (1992). The balance of payments as the wealth of nations. *The economics of wealth creation*. Aldershot: Edward Elgar Publishing Ltd, Chapter 9, pp. 134-170

Williamson, J. (1983). *The open economy and the world economy*. New York: Basic Books, Inc.

Wren-Lewis, S. (1992). Why the pound should be devalued inside the ERM. *Economic Outlook*. 16(5), 35-38.

The J-curve and the Indian Trade Balance:
A Quantitative Perspective

Kanta Marwah

...by the basic teaching of Lawrence R. Klein,
"Because of time lags involved in ordering
goods at great distances and required periods
of production for many goods, exports of a
given period are likely to be associated with
prices and production of an earlier period." *An
Introduction to Econometrics*, Prentice-Hall,
1962, p.35.

This paper broadly examines the impact on trade from the recently
liberalized exchange rate management system of the Indian rupee. Ideally any
trade analysis should be conducted in the context of a multinational model of
world trade such as in Project Link pioneered by Lawrence R. Klein [See
Hickman (1991)]. This study draws its motivation from Project Link, but
focuses on a particular case of the Indian trade on the assumption that it
belongs to small open economy.

To understand fully the trade perspective of the new exchange rate
system, it is important to briefly trace the economic circumstances and the
policy background it emanated from. Faced with a dangerously low level of
foreign exchange reserves and critically worsening balance of payments
position, the newly elected Government of India, early in the summer of 1991,
announced some rapid-fire changes in its economic policies. First, it devalued
the Indian rupee in two sharp bursts on July 1 and 3, by as much as 20 percent
against the U.S. dollar, the pound sterling, the Deutsche mark and the Japanese
yen. These currencies supposedly constitute the primary components of India's
official currency basket by which the rupee is evaluated.[1] Within a couple of
days of devaluation the government also suspended the cash compensatory
support to the exporters and introduced instead an *Exim* (export-import) *scrip*
scheme. *Exim scrip* was a freely tradeable replenishment license given to
exporters. Under this scheme exporters were entitled to use 30 percent of their
foreign exchange earnings in the form of a replenishment (REP) license for
importing other goods. More importantly, however, exporters who neither
needed nor planned to import any new goods were allowed to freely sell their
REP licenses in the foreign exchange market at a huge premium. Thus, for the
first time merchandise imports were pegged directly to exports.

Soon there were a series of other watershed announcements aimed at
liberalizing the industrial structure. The first budget which the new government

presented on July 24,1991 for the fiscal 1991-1992 year included a large scale deregulation of industry, a cut in subsidies, a reduction in defense expenditure, in real terms, changes in tax structure, and more incentives for non-resident Indians (NRI's) to invest in India.[2] The Bombay stock market responded buoyantly to these budgetary changes and the prices of stocks surged high.

Exim scrip, however, was short lived. On February 29, 1992, Finance Minister Manmohan Singh presented his second budget for the fiscal 1992-1993. The most dramatic thrust of this budget related to the exchange rate and import policy. The rupee was made partially convertible under a new Liberalized Exchange Rate Management System (LERMS), while the newly instituted *Exim scrip* was scrapped, custom duties were sharply cut and import licensing was virtually abolished.

Under this new system of partial convertibility, first, the foreign exchange market is expected to function with dual exchange rates of the rupee, an official rate and a free market rate. The Reserve Bank of India (RBI) would set the official rate based on the fixed value of a basket of currencies, and let the free market rate be determined by the demand and supply forces. All incoming foreign exchange earnings would be legally converted into rupees with a split of 40 percent at the official rate and 60 percent at the market rate. The RBI expects that the difference between the official rate and the market rate would not exceed 15 percent.[3] In any case the RBI retains the power to intervene if the foreign exchange market becomes extremely volatile. The Foreign Exchange Dealers Association of India (FEDAI) is expected to announce the value of the free market rate on every business day as the RBI does the value of the official rate. Second, there is a new element of flexibility. Whereas under the old system (also managed according to the fixed value of a given basket of currencies), the exchange rate of the rupee was only stated in terms of the pound sterling, the RBI now has the option to set the official rate in terms of either the pound sterling or the US dollar. Thus, from time to time the RBI would have the flexibility of using either of the two as intervention currency.

And finally, the LERMS is intended to initiate a gradual movement towards full convertibility of the rupee for all external transactions in the current account. In small future moves, the RBI also intends to liberalize exchange control regulations pertaining to these transactions. For the time being, however, all transactions in official as well as in the market rate will be governed by the existing exchange control rules.[4] The government has declared its commitment to make the rupee fully convertible within a period of three to five years.

The main thrust of this paper lies in assessing the 'expected' or potential effectiveness of the basic change in the exchange rate system in restoring India's balance of trade equilibrium. In practical terms, the partial convertibility has virtually amounted to another devaluation of the rupee. Thus, specifically for methodological purpose, we may draw on a similar mechanism that has been developed to assess the effectiveness of devaluation in restoring trade equilibrium.[5] Devaluation changes the terms of trade by changing the relative prices of domestic versus foreign produced goods when both are measured in one common currency. And, in real terms, the new terms of trade are expected to increase the value of exports and reduce the value of imports of the home country provided certain elasticity conditions are met. For example, one well known condition recognised long as the Marshall-Lerner condition simply states that the sum of absolute values of a home country's import and export price elasticities must exceed unity. Otherwise, if this sum was less than unity, devaluation would have perverse effects on the trade balance. It has been widely accepted that the Marshall-Lerner condition generally holds, and devaluation does improve a country's balance of trade.

In juxtaposition, however, another proposition which has come to be known as the J-Curve hypothesis, and which is supported by sporadic evidence, suggests that following the devaluation the trade balance of the home country may first become worse before getting better. In other words, the dynamic adjustments in the trade account may trace a J-shaped pattern. The J-curve phenomenon is basically ascribed to overlapping of pre- and post-devaluation export-import contractual transactions. It is generally believed that due to delivery lags, the completion of contracts initiated prior to devaluation (at the pre-devaluation real exchange rates) may dominate the short-term changes in the balance of trade account.[See Krueger(1983).] The short-term initial impact may further get magnified by uncertainty about the new situation, by the decision lag, and by the lags in the completion of new contractual post-devaluation transactions.

There are, therefore, at least three interrelated aspects of trade adjustment which need to be quantitatively analyzed for each country in assessing the effectiveness of devaluation or changes in the real exchange rate. The three issues are :

(a)   What is the expected time profile of the trade adjustment in response to any change in the exchange rate of the currency?

(b)   How long does it take for the entire adjustment process to be completed?

(c)   When the adjustment is completed, would the trade balance equilibrium be restored?

This paper addresses the first two issues directly and draws some broad implications with respect to the third in the context of India. A very special, and a new, feature of the analysis is that the time profile of the trade balance is placed individually on bilateral tracks in terms of the exchange rate of the Indian rupee vis-a-vis the Canadian dollar, the French franc, the Deutsche mark, the Japanese yen, the pound sterling and the U.S. dollar. The empirical analysis is based on the quarterly data for 1974-1986. The methodology is described in section 1 and the results are discussed in section 2. Some overall observations and implications are stated in section 3. The final section 4 contains brief concluding remarks.

I. Empirical Design

To track the Indian J-curve phenomenon for assessing the effectiveness of most recent changes in the exchange rate policy, I use the basic trade model in which the volume of trade responds to both income and relative prices. The relative prices are measured in *real* terms which are commonly interpreted as real exchange rates. That is, if ϵ is the nominal exchange rate defined as rupee per unit of a foreign currency, p is the export price and p_w is the world price level, then the *real* relative price h, defined as

$$h = \frac{p_w \epsilon}{p} \quad , \text{ is the real exchange rate.}$$

Furthermore, denoting India's total volume of exports by X, imports by M, income by Y, and world income by Y_w, the volume of Indian trade may simply be explained by equations (1)-(2):

$$X = A_x Y_w^{\alpha_x} \left(\frac{p_w \epsilon}{p} \right)^{\beta_x} \mu_x \quad , \tag{1}$$

$$M = A_m Y^{\alpha_m} \left(\frac{p_w \epsilon}{p} \right)^{\beta_m} \mu_m \quad , \tag{2}$$

where μ_x and μ_m are stochastic errors. In a world of gross substitutes, relative price elasticities for exports and imports respectively are, a priori, $\beta_x > 0$ and $\beta_m < 0$. But the income elasticities, α_x and α_m, are both positive.

I analyze the trade balance in the form of a ratio that is independent of any unit of measurement, be it domestic currency or foreign. This ratio, denoted by T, is simply defined as the ratio of the value of exports over the value of imports, and is mathematically expressed as

$$T = \frac{pX}{p_w \epsilon M} = \frac{X}{hM} \quad \begin{matrix} > \\ < \end{matrix} \; 1 \quad . \tag{3}$$

The trade balance would be in equilibrium when T = 1, in surplus when T > 1, and in deficit when T < 1. And, if T stays persistently much below unity (T << 1), it would signalize building of a trade deficit crisis.

The dynamic adjustments in the trade balance can be studied by differentiating T with respect to time, t. As has been shown in the Appendix,

$$\frac{dT}{dtT} = (\eta_{X/Y_w} r_{Y_w} - \eta_{M/y} r_y) + (\eta_{X/h} - \eta_{M/h} - 1)r_h , \qquad (4)$$

where $\eta_{i/j}$ represents the elasticity of i with respect to j and r_i the rate of change in i.

Substituting export-import elasticity parameters from trade equations (1)-(2) into (4), we obtain

$$\frac{dT}{dtT} = (\alpha_x r_{y_w} - \alpha_m r_y) + (\beta_x - \beta_m - 1)r_h . \qquad (5)$$

The first term on the right hand side, enclosed in the parentheses, measures the effect of relative growth rates of incomes, and the second of movements in the real exchange rate. If we impose the restriction that income elasticities of export and import trade are equal in magnitudes, that is $\alpha_x = \alpha_m = \alpha$, and also, if we 'net' out the price elasticities to an elasticity of net exports as $\beta_x - \beta_m = \beta$, equation (5) gets simplified to

$$\frac{dT}{dtT} = \alpha(r_{y_w} - r_y) + (\beta - 1)r_h . \qquad (6)$$

Although under normal conditions both α and β are expected to be positive, $(\beta - 1)$ may be $\gtrless 0$. The Marshall – Lerner condition of stability of the trade balance adjustment is satisfied when $(\beta - 1) > 0$. For otherwise, for $r_h > 0$, the trade balance will remain unchanged if $(\beta - 1) = 0$, and will certainly deteriorate or become adverse if $(\beta - 1) < 0$, that is if $(\beta_x - \beta_m) < 1$.[6] Furthermore, since $r_h = r_{p_w} + r_\epsilon - r_p$, we may explicitly state that for a given depreciation or devaluation in the nominal exchange rate $(r_\epsilon > 0)$, *ceteris paribus*, the trade balance will improve $\left(\frac{dT}{dt} > 0\right)$ if and only if $(\beta - 1) > 0$.

We may also note that if the adjustment in the nominal exchange rate is kept strictly in accord with the purchasing power parity relationship, $r_\epsilon \neq 0$ would imply that $r_h = 0$ and the trade balance will remain unaffected.

Trade balance adjustment by market shares:

We may view the real exchange rate h as a geometric weighted average of the real exchange rates of the rupee against individual currencies of India's n trading partners. That is

$$h = \frac{p_w \epsilon}{p} = \prod_i \left(\frac{p_i \epsilon_{Ri}}{p} \right)^{w_i} = \prod_i h_i^{w_i}, \; \sum_i w_i = 1 \quad , \tag{7}$$

where p_i is the price of country i exporting to India and ϵ_{Ri} is the exchange rate defined as rupees per unit of currency of country i. Thus, $p_i \epsilon_{Ri}$ is simply the import price of the Indian imports coming from country i. Relying on my earlier study of the international trade flows (1976) and on my follow-up study on analytical properties of the associated price elasticities (1978), I assume that India's export flow elasticities are invariant across markets of destination by virtue of a world-wide common competition, but in her import market, the share elasticities of substitution do vary across her major suppliers. Using this assumption her overall trade balance adjustment response can be analytically ascribed to and differentiated by her individual trading partners. As has been shown in the Appendix, the trade balance equation in terms of bilaterally differentiated response is reduced to

$$\frac{dT}{dtT} = \alpha \left(r_{y_w} - r_y \right) + \sum_i \beta_i r_{h_i} \tag{8}$$

where $\beta_i = (\beta - 1)w_i + (1 - w_i)\beta_{m_i}$.

Even when the overall stability condition for trade balance adjustment is satisfied and $(\beta - 1) > 0$, $\beta_i \gtrless 0$ depending upon the behaviour of β_{mi}. β_i will be positive as long as

(a)   stability condition $[(\beta - 1) > 0]$ is satisfied, and
(b)   (i) $\beta_{mi} \geq 0$, or
          (ii) when $\beta_{mi} < 0$, $|(\beta - 1)w_i| > |(1 - w_i)\beta_{mi}|$.

A positive value of β_i would imply that any depreciation of the rupee against currency i would improve India's trade balance.

β_{mi} is the parameter that measures country i's share elasticity of substitution in India's import market. Defining M_i as India's imports coming from country i, $\sum_i M_i = M$ is total imports of India. The share of country i of the Indian import market is described by the market share coefficient α_i defined as $\alpha_i = \frac{M_i}{M}$, $\alpha_i \geq 0$, $\sum_i \alpha_i = 1$, and the corresponding share elasticity of substitution of country i in the Indian market is defined as

$$\beta_{m_i} = \frac{\partial \log a_i}{\partial \log \left(p_w \epsilon / p_i \epsilon_{Ri}\right)} \overset{>}{\underset{<}{}} 0 .$$

In my (1978) paper I showed that β_{mi} is a composite entity whose properties depend upon the relative parametric forces of import demand in the importing country and the import supply provided by country i in competition with other countries. The nature of the trade balance adjustments response to fluctuations in the real exchange rate of the Indian rupee against currency i in equation (8) will be determined by the value of β_{mi}.

The trade ratio analogue of the trade balance adjustment equation (8) including stochastic error,

$$T = A \left(\frac{Y_w}{Y}\right)^\alpha h_1^{\beta_1} h_2^{\beta_2} \ldots h_n^{\beta_n} u , \qquad (9)$$

is the base equation that provides the focal point to our empirical investigation. Equation (9) implies that India's trade balance would improve (deteriorate) with general depreciation (appreciation) in the real exchange rate of the Indian rupee

as long as $\sum_i \beta_i > 0 \left(\sum_i \beta_i < 0\right)$.

Since the primary purpose of this paper is to assess the J-Curve phenomenon, a distributed lag structure of *a priori* unspecified length is imposed on each of the real exchange rate variables. Furthermore, it is assumed that lag coefficients lie on a polynomial of a certain degree as suggested in the Almon distributed lag scheme. Both the lag length k and the degree of polynomial r are determined empirically.[7] With the lag structure thus imposed, equation (9) can be expressed as

$$T_t = A \left(\frac{Y_w}{Y}\right)_t^\alpha \prod_{j=1}^k h_{1t-j}^{\beta_{1j}} \prod_{j=1}^k h_{2t-j}^{\beta_{2j}} \ldots \prod_{j=1}^k h_{nt-j}^{\beta_{nj}} \mu_t . \qquad (10)$$

For the J-Curve phenomenon to exist, the distributed lag coefficients summed over i's, that is $\sum_i \beta_{ij}$, should initially be negative and then become positive over time. The steady state long run adjustment can then be measured by the base equation (9) where any β_i can be interpreted as the sum of lag coefficients of i over time, β_{ij} , $\left(\beta_i = \sum_j \beta_{ij}\right)$. The overall stability condition

would now require $\sum_j \sum_i \beta_{ij} > 0$.

Equation (10) forms the focus for our econometric analysis as described and presented in the next section.

II. Econometric Estimation

A. Quantifiable Entities

The econometric analysis of equation (10) is based on the quarterly data for 1974-1986. The first eight quarters enter only through the lags in the real exchange rate variables. The equation is estimated by ordinary least squares using the remaining 44 observations for the sample period 1976:1-1986:4.

Two separate measures of trade ratio based primarily on f.o.b. and c.i.f. evaluation of merchandise exports and imports are used to obtain the estimates. Canada, France, West Germany, Japan, U.K. and the U.S.A. are six of India's major trading partners recognised explicitly in the equation. Thus the real exchange rate variables, h_i (i=1,2,...,6), correspond to the real exchange rate of the Indian rupee per Canadian dollar ($h_{Rc\$}$), French franc ($h_{Rf}$), Deutsche mark ($h_{Rm}$), Japanese yen ($h_{Ry}$), U.K.'s pound sterling ($h_{R\pounds}$) and the U.S. dollar ($h_{R\$}$). Interestingly, initial experiments revealed a strong structural similarity between the impacts of the real exchange rates of the rupee against the Canadian dollar and the pound sterling on one side and between the exchange rates against the French franc and the German mark on the other.[8] Therefore, it was judged appropriate to substitute each pair of exchange rates by their corresponding geometric means. Thus, the two new variables, $h_{Rc\$\pounds}$ and h_{Rfm} represented respectively the geometric means of $h_{Rc\$}$ and $h_{R\pounds}$, and of h_{Rf} and h_{Rm}.[9] And also to eschew from potential pitfalls of aggregation across national incomes, the world income variable was substituted by total world trade index (WXT); and similarly, the index of Indian industrial production (PRODS) was used as a proxy for income in India. Equation (10), for the J-Curve, in estimable form was expressed as

$$\ln T_t = A' + \alpha \ln\left(\frac{WXT}{PRODS}\right)_t + \sum_{j=0}^{k} \beta_{1j} \ln(h_{RC\$\pounds})_{t-j}$$

$$+ \sum_{j=0}^{k} \beta_{2j} \ln(h_{Rfm})_{t-j} + \sum_{j=0}^{k} \beta_{3j} \ln(h_{Ry})_{t-j} \qquad (11)$$

$$+ \sum_{j=0}^{k} \beta_{4j} \ln(h_{R\$})_{t-j} + error \ .$$

where (with the trade ratio T, measured as the ratio of the value of India's exports to India's imports, used in two alternative forms):

$$T \quad = \quad \left[\begin{array}{ll} \text{TF} & (\text{ trade ratio measured f. o. b. }) \\ \text{TC} & (\text{ trade ratio measured c. i. f. }) . \end{array} \right.$$

WXT	=	index of total world trade.
PRODS	=	index of industrial production.

h_{Ri} = real exchange rate defined as $\dfrac{p_{ei} \, \epsilon_{Ri}}{p}$.

h_{Rij} = $(h_{Ri} \cdot h_{Rj})^{1/2.}$

p_{ei} = unit value of exports of country i.

ϵ_{Ri} = exchange rate, Indian rupee per unit of currency of country i.

p = wholesale price index of India.

i = countries: Canada, France, West Germany, Japan, U.K. and the U.S.A. currencies: Canadian dollar (C\$), French franc (f), West German mark (m), Japanese yen (y), pound sterling (£), and the U.S.\$.

f.o.b. = free on board.

c.i.f. = including cost, insurance, and freight.

t = time period.

ln = natural logrithm.

^ = estimate.

All indices are based on 1980 = 100, the values of exports and imports used in the trade ratio are measured in millions of SDRS. The ratio TF is computed by using merchandise trade entries in the balance of current account. TC is derived by adding, for other goods, services and income, credit items to merchandise exports and debit items to merchandise imports.[10] All data are obtained from *International Financial Statistics*, and *Balance of Payment Yearbook* of the IMF.

B. Estimates

Since the true lag length k and the degree of polynomial r are *a priori* unknown, we first determine the optimal size of k by using both Akaike information and Schwarz-Bayes information criteria. Both these statistics reach minima over time at k = 8 as is shown by their respective curves presented in Figures 1 and 2.[11] Having determined the lag length k = 8, the optimal degree of polynomial r = 4 was then located using a sequential testing procedure.[12] No endpoint constraints were imposed. Table 1 contains estimates of trade ratio TC for k = 8 and r = 4 and Table 2, corresponding estimate for TF. Their steady-state long run relations together with relevant diagnostic statistics are summarized in Table 3. All estimates, as stated above, are obtained by ordinary least squares using quarterly data for 1976:1-1986:4.

Figure 1: Akalke 1C and Lag Size (k)

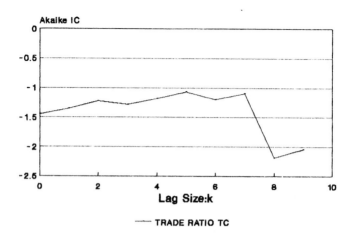

Figure 2: Sch-Bayes 1C and Lag Size (k)

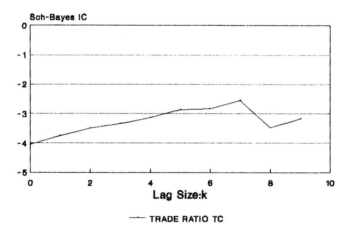

Table 1: Trade Ratio (c.i.f.)
($|t|$ - Statistics in Parentheses)
r = 4, k = 8

$$\ln(TC)_t = \underset{(0.477)}{4.9786} + \underset{(1.53)}{0.4705} \ln\left(\frac{WXT}{PRODS}\right)_t + \sum_{i}^{4}\sum_{j=0}^{8}\beta_{ij}\ln(h_{ri})_{t-j} \quad \text{that is} \qquad (12)$$

h_{Ri}:	$\ln h_{RCS\mathcal{E}}$	$\ln h_{Rfm}$	$\ln h_{Ry}$	$\ln h_{RS}$	SUM
$\hat{B}_{ij}$ →					
j = 0	-2.1919 (2.36)	0.5737 (0.99)	0.1289 (0.22)	2.3097 (2.66)	0.8204
1	0.2907 (0.51)	-0.9905 (1.96)	0.7178 (1.67)	-0.7477 (1.54)	-0.7297
2	-0.1519 (0.34)	-0.6075 (1.46)	0.4731 (1.41)	-0.5146 (1.28)	-0.8009
3	-1.1997 (3.84)	0.3718 (1.11)	-0.0169 (0.07)	0.7268 (1.93)	-0.1180
4	-1.4882 (3.50)	1.0705 (2.35)	-0.3686 (1.20)	1.5062 (3.30)	0.7199
5	-0.6076 (1.93)	1.0858 (2.42)	-0.4036 (1.33)	1.1648 (2.84)	1.2394
6	0.8963 (2.35)	0.4889 (1.31)	-0.1482 (0.47)	-0.1446 (0.29)	1.0924
7	1.5227 (2.84)	-0.1747 (0.54)	0.1663 (0.49)	-1.4574 (2.69)	0.0569
k = 8	-1.1842 (1.79)	0.1144 (0.24)	0.1039 (0.20)	-0.9976 (1.14)	-1.9635

$\bar{R}^2 = 0.795$ $d = 3.230$ $SE = 0.09486$ Mean $\ln TC = -0.23803$ $N = 44$

Table 2: Trade Ratio (f.o.b.)
(|t| - Statistics in parentheses)
r = 4, k = 8

$$\ln(TF)_t = 6.2609 + 0.4186 \ln\left(\frac{WXT}{\prod}\right)_t + \sum_i \sum_{j=0}^{8} \overset{n}{\beta}_{ij} \ln(h_{ri})_{t-j}, \quad \textit{that is} \qquad (13)$$
$$_{(0.483)} _{(1.09)}$$

h_{Ri}:	lnh_{RCSE}	lnh_{Rfm}	lnh_{RY}	lnh_{RS}	SUM
$\hat{B}_{ij} \rightarrow$					
j = 0	-2.8290	0.4668	0.4055	2.9945	1.0378
	(2.46)	(0.65)	(0.55)	(2.78)	
1	0.1510	1.3223	1.0152	-0.6762	-0.8323
	(2.11))	(2.11)	(1.907)	(1.12)	
2	-0.2871	-0.6455	0.5739	-0.4805	-0.8392
	(0.51)	(1.25)	(1.38)	(0.96)	
3	-1.460	0.6880	-0.1190	0.9352	-0.0442
	(3.77)	(1.66)	(0.38)	(2.00)	
4	-1.794	1.5318	-0.5593	1.8508	1.0288
	(3.40)	(2.71)	(1.47)	(3.27)	
5	-0.7878	1.4025	-0.5388	1.4723	1.5482
	(2.01)	(2.51)	(1.43)	(2.90)	
6	0.9528	0.4794	-0.1446	-0.0681	1.2195
	(2.01)	(1.03)	(0.37)	(0.11)	
7	1.7293	-0.3952	0.2407	-1.7121	-0.1373
	(2.60)	(0.98)	(0.57)	(2.55)	
k = 8	-1.2504	0.2835	-0.0612	-1.4751	-2.5032
	(1.52)	(0.48)	(0.09)	(1.36)	

$\bar{R}^2 = 0.803$ d = 3.273   SE = 0.11769   Mean lnTF = -0.2795   N = 44

Table 3: Implied Steady-state Trade Ratios and Diagnostic Statistics
($|T|$ - Statistics in Parentheses)

	lnTC (12)	lnTF (13)
Constant	4.9786 (0.48)	6.2609 (0.48)
ln (WXT/PRODS)	0.4705 (1.53)	0.4186 (1.09)
$lnh_{RC\$\epsilon}$	-4.1138 (2.89)	-5.575 (3.16)
lnh_{Rfm}	1.9324 (2.00)	2.4890 (2.08)
lnh_{RY}	0.6527 (0.39)	0.8125 (0.39)
$lnh_{R\$}$	1.8456 (1.47)	2.841 (1.82)
Mean Lags:		
$h_{RC\$\epsilon}$	1.47	1.45
h_{Rfm}	5.82	6.01
h_{RY}	1.18	3.44
$h_{R\$}$	3.68	2.91
Diagnostics:		
Normality (JB)	0.72	0.83
Homoscedasticity (Arch)	0.94	0.86
Serial Independence:		
Breusch/Godfrey LM (2)	0.80	0.77
ADF (Unit root)	0.00	0.00

Note: These entries represent the level of significance needed to reject the associated null hypothesis.
(The larger the number the smaller the level of confidence in rejecting the associated hypothesis.)

Among the major findings of interest are: First, the trade balance responds to fluctuations in the real exchange rate of the rupee with dramatic variations over time and across currencies. Moreover, these variations, with all their general characteristics, remain intact irrespective of whether the trade ratio is measured c.i.f. or f.o.b. Second, the trade balance adjusts in slow convulsions rather than in a well-behaved and well defined systematic pattern. For example, the impact reaction to a fall in the value of the rupee *vis-à-vis* the Canadian dollar and the pound sterling is perverse, and the trade balance deteriorates, but in response to a fall vis-à-vis the US dollar, it improves. As for a fall against all other currencies, the trade balance does adjust in the desirable direction but its magnitude is not statistically significant. Third, sharp reversals of response follow immediately after initial convulsions in three cases. The effect of $h_{Rc\$\pounds}$ turns positive from negative, but of h_{Rfm} and $h_{R\$}$ reverses from positive to perverse. Fourth, in so far as the time distribution of adjustment is concerned, the trade balance response to fluctuations in $h_{Rc\$\pounds}$ is front-loaded while convulsive, the mean lag is rather short (around two quarters). But the same response to fluctuations in h_{Rfm} and $h_{R\$}$ is more or less slow and steady with a longer mean lag (around 4-6 quarters). Fifth, with the exception of the Canadian dollar and the pound sterling, a fall in the real value of the rupee in terms of all other currencies improves the trade balance over time; the net aggregate effect is positive. The effect of h_{Ry} is statistically weak and relatively less robust. And finally, if we further analyse the net aggregate effect across currencies - the SUM column in Tables 1 and 2 - we clearly discern a pattern of delayed J-curve. After some initial improvement in the trade ratio, there is a marked deterioration for two periods which is later followed by slow improvement once again.

Having made these broad overall observations on the estimates, we next, in section 3, examine the latent J-Curve and draw some policy implications.

III. Implications

The implications of special interest relate broadly to the questions raised in the introductory discussion. For easy understanding and simplification of the analysis the relevant information from equations (12) and (13) has been synthesized in Table 4.

First, in Table 4, β_i-SUM represents the summation of β_{ij} coefficients across $i = 1, \ldots, 4$ exchange rates for each distributed lag size j, from 0 to 8,

$$\left(\sum_i \hat{\beta}_{ij}, \quad j = 0, 1, \ldots, 8\right).$$ The J-SUM represents the cumulative effect of the exchange rate variables over time by the end of period j

$$\left(\sum_0^j \sum_{i=1}^4 \hat{\beta}_{ij}, \quad j = 0, 1, \ldots, 8\right).$$ The J-SUM thus provides an overall time profile of the cumulative trade adjustment in response to shocks in the real exchange rates. A close look at the J-SUM leads to an inescapable conclusion that it tracks a J-Curve pattern over time. Following an initial positive convulsion immediately after a general devaluation of the rupee, the trade balance deteriorates for as long as three-quarters of a year and then starts improving

Table 4: Conjectural J-curve and Implied Optimal Level of Devaluation (IOD)

	TC: EQ. (12)			TF: EQ. (13)		
	B_i-SUM	J-SUM	IOD% 1986:4	B_i-SUM	J-SUM	IOD% 1986:4
j = 0	0.8204	0.8204		1.0378	1.0378	
1	-0.7297	0.0907		-0.8323	0.2055	
2	-0.8009	-0.7102		-0.8392	-0.6337	
3	-0.1180	-0.8282		0.0442	-0.5895	
4	0.7199	-0.1083		1.0288	0.4393	
5	1.2394	1.1311	36.5	1.5482	1.9875	21.2
6	1.0924	2.2235	17.1	1.2195	3.2070	12.6
7	0.0569	2.2804	16.7	-0.1373	3.0697	13.2
8	-1.9635	0.3169		-2.5032	0.5665	

Note: TC (1986:4) = 0.7035
         TF (1986:4) = 0.6824

slowly. The J-curves latent in the adjustment pattern of TC and TF are presented in Figure 3. They are drawn by plotting the J-SUM against j periods (j = 0, 1, . . . , 8).

Second, although the estimates provide clear evidence that devaluation would improve the trade balance over the long run [as $(\beta - 1)$ = 0.3169 for TC, and 0.5665 for TF is positive], the process of adjustment is expected to spread over eight to nine quarters. Moreover, since no far-end constraints are imposed on the distributed lag coefficients, a residue weak effect may continue to exist somewhat longer.

Third, it is noteworthy that the positive cumulative effect peaks out by the end of eight quarters and thereafter some negative pressures reemerge. These pressures may presumably reflect conjectural feedback effect from the rest of the domestic economy. Nonetheless, the estimates give a signal of a real danger that the gains made early in the trade account could dissipate subsequently. Perhaps at this point of time some corrective policy action is needed.

Fourth, the minimum amount of waiting time required for devaluation to start producing positive gains in trade is close to five quarters. It is only when more than one full year has passed that positive gains become visible, β_i-SUM turns > 0, and trade balance starts improving.

Fifth, the trade ratio TF measured as f.o.b. adjusts somewhat faster and registers higher gains than the trade ratio TC measured broadly in c.i.f. terms.

Finally, on the important question of optimal level of devaluation or whether a given amount of devaluation is sufficient to correct the trade imbalance, the answer must depend upon two things: (a) the size of the initial imbalance, and (b) how soon the trade imbalance must be corrected.

Figure 3: J - Curve J-SUM Trade Ratio

The trade ratio function estimated and presented above is homogeneous of degree $\sum_i \beta_{ij}$ in real exchange rates. That is scaling all h_i by a positive factor λ, from $(h_{Rc\$£}, h_{Rfm}, h_{Ry}, h_{R\$})$ to $(\lambda h_{Rc\$£}, \lambda h_{Rfm}, \lambda h_{Ry}, \lambda h_{R\$})$, scales trade ratios by $\lambda^{\sum_i \beta_{ij}}$, from TC to $TC\lambda^{\sum_i \beta_{ij}}$ and from TF to $TF\lambda^{\sum_i \beta_{ij}}$. Thus for trade balance to be restored and the trade ratio T to become unity after j number of periods ($T_{+j} \approx 1$), the required amount of devaluation would be equal to ($\lambda - 1$) such that

$$T_0 \lambda^{\sum_0^j \sum_i^4 \beta_{ij}} \approx 1 \qquad (14)$$

where T_0 is the trade ratio at the (initial) time of devaluation (presumably < 1, indicating trade deficit). Solving (14) for λ

$$\lambda = (T_0)^{-\dfrac{1}{\sum_0^j \sum_{i=1}^4 \beta_{ij}}} . \qquad (15)$$

Thus λ depends upon (a) T_0, and (b) the choice of j which would determine

$$\sum_{0}^{j} \sum_{i=1}^{4} \beta_{ij} \text{ to be substituted in (15).}$$

Columns IOD (implied optimal devaluation) in Table 4 present an optimal degree of devaluation that might have been required at the end of 1986 to correct the then prevailing trade deficit. It may be noted that in the 4th quarter of 1986 the value of Indian exports was around 70 percent of the total value of its imports (TC = 0.7035, TF = 0.6824). The scenarios derived are conjectural, but the estimates obtained are implied by J-SUM coefficients for each j length of time. For example, if the goal at the year end of 1986 had been to eliminate the trade deficit and equate the trade ratio TF to unity by the end of j = 5 periods, that is within one and one-half year, the requisite amount of devaluation should have been around 21 percent. Similarly, for balancing TC within the same period the rupee would have needed to fall in value by a much larger amount of nearly 36 percent. Furthermore, if the country could afford an additional six months to let the process work through, a much lower amount of devaluation would have sufficed, that is 12-13 percent for TF and 16-17 percent for TC.

At this juncture, we may venture to infer parenthetically that around 20 percent devaluation incurred by the Indian rupee in the summer of 1991 ought to start showing some gains in the trade account in another three months or so. In accepting these numbers we must also underscore an implicit caveat that these estimates are conjectural and are conditioned by two underlying assumptions. First, that the Indian economy would grow at the same rate as the rest of the world, and second, that any feedback effect of devaluation from the rest of the economy is neutral.

IV. Conclusions

On the three issues raised in the introduction, we may conclude that the latent time profile of the trade balance adjustment process following the partial convertibility of the rupee indeed tracks a J-CURVE pattern. The entire aggregate adjustment would most likely be completed not sooner than eight quarters or so. And barring any major new crisis, the level of trade balance at the completion of the process should become healthy even if not fully restored. The amount of devaluation experienced by the rupee indeed falls close to the optimal level of devaluation implied by quantitative results that is needed to correct the trade balance disequilibrium within a couple of years. These conclusions conditioned by the two underlying assumptions are somewhat conjectural, but emerge from the distributed lag analysis of the effects of real exchange rates of the rupee across six major currencies on the balance of Indian trade.

In closing it may also be pointed out that, although the discussion in the paper has been couched in terms of devaluation and with reference to persistent trade deficit, a similar analysis of the reverse order remains valid if the country was faced instead with persistent trade surplus.

A postscript: After this study was completed in September 1992, the Indian rupee was made fully convertible on the trade account on February 27, 1993. And, in accordance with the J-Curve predictions of the timing and the pattern of adjustment computed in the paper, India's trade gap has narrowed substantially.

APPENDIX

The trade ratio,

$$T = \frac{px}{p_w \epsilon M} = \frac{x}{hM} \quad given \ in \ (3)$$

when differentiated over time yields

$$\frac{dT}{dt} = \frac{hM \dfrac{dx}{dt} - X \left[h \dfrac{dM}{dt} + M \dfrac{dh}{dt}\right]}{h^2 M^2}. \tag{A1}$$

Similarly from (1)

$$\frac{dx}{dt} = \frac{\partial x}{\partial Y_w} \frac{dY_w}{dt} + \frac{\partial X}{\partial h} \frac{dh}{dt} \tag{A2}$$

$$= X \left[\eta_{x/Y_w} r_{Y_w} + \eta x_{/h} r_h\right], \quad and \ from \ (2)$$

$$\frac{dM}{dt} = \frac{\partial M}{\partial Y} \frac{dY}{dt} + \frac{\partial M}{\partial h} \frac{dh}{dt} \tag{A3}$$

$$= M \left[\eta_{M/y} r_y + \eta_{M/h} r_h\right] ,$$

where r_i denotes the rate of change in (i) and $\eta_{i/j}$ the elasticity of (i) with respect to j, that is

$$r_i = \frac{d(i)}{dt(i)},$$

$$\eta_{i/j} = \frac{\partial(i)}{\partial(j)} \frac{(j)}{(i)} = \frac{\partial \log(i)}{\partial \log(j)}$$

Substituting (A2) and (A3) into (A1), and simplifying yields

$$\frac{dT}{dtT} = \left[\eta_{x/Y_w} r_{y_w} - \eta_{M/Y} r_y\right] + \left[\eta_{x/h} - \eta_{M/h} - 1\right] r_h \tag{A4}$$

where, in terms of the parameters of equations (1) and (2), $\eta_{x/yw}$ and $\eta_{M/y}$ are respectively α_x and α_m and $\eta_{x/h}$ and $\eta_{M/y}$ are β_x and β_m. Thus we get equation (5).

<u>Decomposition of relative price elasticities</u>: h defined by (7) is the weighted geometric average of h_i (i = 1, 2...n).

<u>Decomposing β_x</u>: let export flow X_i be defined as volume of India's exports going to country i, and functionally expressed as

$$X_i = A_{xi} \left(\frac{p_w \epsilon}{p} \right)^{\beta_{xi}}, \quad \beta_{xi} > 0, \quad i = 1, 2 \ldots n \qquad (A5)$$

where, for simplicity, the income effect is subsumed in the constant term A_{xi}. Taking logarithms of (A5) and summing over n,

$$\sum_i \log X_i = \sum_i A_{xi} + \sum_i \beta_{xi} \log \left(\frac{p_w \epsilon}{p} \right). \qquad (A6)$$

Assuming X_i follows log-normal distribution such that

$$AM_x = GM_x \, e^{\sigma_{x/2}^2}\,,$$

(A6) is reduced to

$$\log \sum_i X_i = \log A_x' + \sum_i \beta_{x_i} \log \left(\frac{p_w \epsilon}{p} \right), \qquad (A7)$$

where $\log A_x' = \sum_i \log A_{xi}' - n^{\sigma_{x/2}^2}$. And since $\sum_i X_i = X$, (A7) with the stochastic term

added implies equation (1) where

$$\sum_i \beta_{xi} = \beta_x \quad and \quad A_x' = A_x \, Y_w^{\alpha_x}\,.$$

Disintegrating further the real exchange rate term by individual real relative prices in (7), and simplifying, we get

$$\log X = \log A_x' + \sum_i B_{x_i} \log [h_1^{w_1} \, h_2^{w_2} \ldots h_n^{w_n}]$$

$$or \, \log X = log \, A_x + w_1 \sum_i \beta_{x_i} \log h_1 + w_2 \sum_i \beta_{x_i} \log h_2 \ldots + w_n \sum_i \beta_{xi} \log h_n\,,$$

$$(A8)$$

where $w_i \sum_i \beta_{xi} = \beta_{xi}$ and $\sum_i w_i \sum_i \beta_{x_i} = \sum_i \beta_{xi} = \beta_x$.

Thus the aggregate elasticity of exports with respect to real exchange rate is simply a sum of individual elasticities, or the individual elasticities may be interpreted as fractions of the total as ascribed to individual markets by appropriate weights w_i (i = 1, 2..., n and $\sum_i w_i = 1$). That

is

$$\eta_{x/hi} = \frac{\partial \log X}{\partial \log h_i} = w_i \, \beta_x = \beta_{x_i}\,.$$

Decomposing β_m:

Denoting M_i as the volume of India's imports coming from country i, and $\Sigma_i\, M_i = M$, country i's share in the Indian import market is functionally described as

$$\frac{M_i}{M} = A_{m_i} \left(\frac{P_w \epsilon}{P_i \epsilon_i}\right)^{\beta_{m_i}}, \quad \beta_{m_i} \gtrless 0$$

$$= A_{m_i} \left(\frac{P_w \epsilon}{p}\right)^{\beta_{m_i}} \left(\frac{P_i \epsilon_i}{p}\right)^{-\beta_{m_i}} \tag{A9}$$

Next, substituting equation (2) and definition (7) into (A9), and taking logarithim we get

$$\log M_i = \log (A_{m_i}\, A_m') + (\beta_{m_i} + \beta_m) \log \left[h_1^{w_1}\, h_2^{w_2}..h_i^{w_i}..h_n^{w_n}\right] - \beta_{m_i} \log h_i, \tag{A10}$$

where $A_m' = A_m\, Y^{\alpha_m}$ with the error suppressed. Collecting $\log h_i$ terms, (A10) is written as

$$\log M_i = \log (A_{m_i}\, A_m) + (\beta_{m_i} + \beta_m) \sum_{j \neq i} w_j \log h_j + [w_i(\beta_{m_i} + \beta_m) - \beta_{m_i}] \log h_i \tag{A10'}$$

The own price elasticity σ_{ii} of bilateral imports from country i can then be defined and obtained as

$$\sigma_{ii} = \frac{\partial \log M_i}{\partial \log h_i} = w_i \beta_m - (1 - w_i)\beta_{m_i} \tag{11}$$

and cross price elasticity σ_{ij} as

$$\sigma_{ij} = \frac{\partial \log M_i}{\partial \log h_j} = w_j (\beta_m + \beta_{m_i}),$$

$$j = 1, 2, .. (i - 1),(i + 1) .. n. \tag{12}$$

Aggregate of cross effects on imports from country i is $\sum_{j \neq i} \sigma_{ij} = (\beta_m + \beta_{m_i}) \sum_{i \neq j} w_j$, where $\sum_{j \neq i} w_j = (1 - w_i)$ *remembering* $\sum_i w_i = 1$.

It can be easily seen that for a change in all prices across all h_i $(i = 1, 2 \ldots n)$ the elasticity of import flow M_i would then be the sum $\sigma_i = \sigma_{ii} + \sum_{j \neq i} \sigma_{ij} = \beta_m$, that is M_i will change by the same percentage as total imports such that market share a_i would remain unchanged and the zero-degree homogeneity property would be satisfied.

Aggregating (A10)' over n countries,

$$\sum_i \log M_i = \sum_i \log (A_{mi}\, A_m') + \sum_i (\beta_{mi} + \beta_m) \sum_{j \neq i} w_j \log h_j + \sum_i [w_i(\beta_m + \beta_{m_i}) - \beta_{m_i}] \log h_i. \tag{A11}$$

The second term on the right hand side measures all the cross-price effects and the third term own-price effects on M_i flows. Correspondingly,

$$\sum_i \sigma_i = \sum_i \sum_{j \neq i} \sigma_{ij} + \sum_i \sigma_{ii} = n\, \beta_m, \tag{A12}$$

that is β_m is simply a mean of σ_i elasticities ($\bar{\sigma}_i = \beta_m$). Next, by invoking that M_i follows lognormal distribution (A11) is reduced to,

$$\log \sum_i M_i = \log A_m + \sum_i (\beta_{mi} + \beta_m) \sum_{j \neq i} w_j \log h_j + \sum_i [w_i(\beta_m + \beta_{mi}) - \beta_{mi}] \log h_i,$$

(A13)

where $\log A_m = \sum_i \log (A_{m_i} A'_m) + n \dfrac{\sigma_m^2}{2}.$

On the assumption that for across the board (equal) changes in h_i, all interactive cross country substitution effects will either mutually cancel or converge to some constant quantity, all the price terms in the import adjustment can be summarized and indentified as $\sum_i w_i \eta_{m_{hi}} r_{h_i} = \sum_i \sigma_{ii} r_{h_i}$. Next, by substituting the disintegrated export and import price terms in the trade adjustment equation (6) we get

$$\frac{dT}{dt\,T} = \alpha(r_{y_w} - r_y) + \sum_i w_i \beta_x\, r_{h_i} - \sum_i \sigma_{ii}\, r_{h_i} - r_h$$

(A14)

where $r_h = \sum_i w_i r_{h_i}$, *and* r_{h_i}, as may be recalled, is the rate of change in h_i. . Using the expression for σ_{ii} and r_h and simplifying (A14),

$$\frac{dT}{dt\,T} = \alpha\,(r_{y_w} - r_y) + \sum_i (w_i \beta_x - \sigma_{ii} - w_i) r_{h_i},$$

that is

$$\frac{dT}{dt\,T} = \alpha\,(r_{y_w} - r_y) + \sum_i [w_i \beta_x - w_i(\beta_m + \beta_{mi}) + \beta_{mi} - w_i]\, r_{hi}$$

and rearranging

$$\frac{dT}{dt\,T} = \alpha\,(r_{y_w} - r_y) + \sum_i [w_i(\beta_x - \beta_m - 1) + (1 - w_i)\beta_{m_i}] r_{h_i}$$

(A15)

which is our equation (8) $\dfrac{dT}{dt\,T} = \alpha\,(r_{y_w} - r_y) + \sum_i \beta_i\, r_{h_i}$;

here $\beta_i = w_i(\beta_x - \beta_m - 1) + (1 - w_i)\beta_{m_i}$. It is this equation with its analogue, (eq.(9)), which forms the core in the analysis.

Kanta Marwah, Department of Economics, Carleton University, Ottawa, Ontario, Canada K1S 5B6.

ACKNOWLEDGEMENTS

I am thankful to M. Bahmani-Oskooee, Ronald G. Bodkin and Jaime Marquez for helpful comments, and to Kamal Sharan and Ayyad brothers, Hatem and Wael for assistance in sorting out the data.

NOTES

1. For mechanism of `basket-link' management, see Verghese (1984), and Marwah and Palsson (1988).

2. The professed objective of the finance minister was to meet the challenge created by the balance of payments crisis in a manner that would stabilize economic situation, bring the budget deficit under control, reverse inflationary expectations, and yet not hurt economic growth or put undue burden on the poor; that is the policy which would generate growth- oriented adjustment, an adjustment with a `human face.' The clarion call of his first federal budget could best be summed up as `Reform or be dammed'.

3. In practical terms the partial convertibility of the rupee amounted to devaluation of the rupee despite a strong denial by the Reserve Bank Governor, S. Venkitaramanan. The RBI initially fixed the official exchange rate of the rupee about Rs 25.35 to the U.S. dollar and expected the free market parity to settle around Rs 29 to the U.S. dollar. The government also maintained that partial convertibility was simply an extension and refinement of the *Exim scrip* scheme.

4. Although the new system does not imply complete abandonment of the exchange control regulations, there has been some relaxation on that score. For example, under the new system the recipient does not have to convert the entire remittances representing current account transactions into rupee. The recipient can legally retain 15 percent of the receipt in a foreign currency account with a bank in India. Moreover, the authorized dealers in the foreign exchange market are allowed to retain 60 percent of their receipt from current account transactions with themselves for future sale in compliance with the trade and exchange control regulations. Under the previous system the RBI was the sole custodian of the foreign exchange inflows except for the working balance the authorized dealers were permitted to hold in their overseas accounts. There has also been some relaxation in the import of gold.

5. See, for example, the methodology suggested by Haynes and Stones (1982).

6. Recall that $\beta_x > 0$ and $\beta_m < 0$. Thus the stability condition $(\beta_x - \beta_m) > 1$ also implies $|\beta_x| + |\beta_m| > 1$, that is the sum of absolute values of export and import price elasticities must be greater than unity. Also see Stern (1973) for detail analysis of the elasticities condition in evaluating the effect of devaluation on trade. Marquez (1992) surveys econometric estimates of trade elasticities.

7. Among other studies which have used similar methodology to assess the J-Curve phenomenon are Bahmani-Oskooee (1985), Bahmani-Oskooee and Pourheydarian (1991), Haynes and Stone (1982), Himarios (1985, 1989), Miles (1979), Moffet (1989), and Rosensweig and Koch (1988). My equation is a further extension across various trading partners.

As far as the methodology is concerned, there still remain two potentially valid issues. (a) One may argue that some allowance should also be made for possible lags in the relative income variable. However, our main purpose in this paper is to concentrate only on the dynamic structure of the real exchange rate variabe. (b) The statistical estimate of equation (10) may suffer from the `spurious regression' problem, and thus alternative modeling procedures should be used. But, prior investigation of the data, and the diagnostic test reported later in the paper do not indicate any real need for cointegration or error-correction modeling.

8. This was not entirely unexpected in view of the traditional trade linkages among the Commonwealth countries and India's traditional geo-economic relationship with other European countries.

9. The geometric mean $h_{Rc\$\pounds} = (h_{Rc\$} \cdot h_{R\pounds})^{1/2}$, for example, could be interpreted as rupee per one-half c$ plus one-half £.

10. Specifically, from India Table 1A in the Balance of Payments Yearbook, exports and imports f.o.b. are entries against merchandise trade under item A. Their c.i.f. values include credit for exports and debit for imports against other goods, services and income items.

11. Although Figures 1 and 2 are based on Akaike and Schwarz-Bayes' information criteria statistics related to only trade ratio TC, a thorough search to locate the optimal lag length was conducted using some other measures of trade account. These measures included: (a) the total current account balance (value of exports of goods and services minus the value of imports of goods and services plus net unrequited transfers), (b) the balance on merchandise trade f.o.b., and (c) the balance on merchandise trade c.i.f. In all cases both statistics indicated a global minima at k = 8.

12. See any graduate textbook of econometrics. For example, see Judge et al (1988).

REFERENCES

Bahmani-Oskooee, Mohsen. (1985). Devaluation and the J-Curve: Some evidence from LDCs. *The Review of Economics and Statistics*. August, 500-504.

Bahmani-Oskooee, Mohsen, and Pourheydarian, Mohammad. (1991). The Australian J-Curve: A reexamination. *International Economic Journal*. 5(Autumn), 49-58.

Haynes, Stephen E. and Stone, Joe A. (1982). Impact of the terms of trade on the U.S. trade balance: A reexamination. *The Review of Economics and Statistics*. November, 702-706.

Hickman, Bert G. (1991). 'Project LINK and multi-country modelling. In R. G. Bodkin, L. R. Klein, and K. Marwah, *A history of macroeconomatric model-building*. Edward Elgar.

Himarios, Daniel. (1985). The effects of devaluation on the trade balance: A critical view and re-examination of Miles's new results. *Journal of International Money and Finance*. December, 553-563.

Himarios, Daniel. (1989). Do devaluations improve the trade balance? The evidence revisited. *Economic Inquiry*. January, 143-168.

Judge, G. G., Griffiths, W. E., Hill, R. Carter, Lutkepohl, H., and Lee, T.-C. (1984). *The theory and practice of econometrics*. New York: John Wiley.

Klein, Lawrence R. (1962). *An Introduction to Econometrics*. Englewood Cliffs, N.J.: Prentice-Hall.

Krueger, Aanne O. (1983). *Exchange-Rate Determination*. Cambridge: Cambridge University Press.

Marquez, Jaime. (1992). The autonomy of trade elasticities: Choice and consequences. International Finance Discussion Paper, No. 422, Washington, D.C.: Board of Governors of the Federal Reserve System.

Marwah, Kanta. (1976). A world model of international trade: Forecasting market shares and trade flows. *Empirical Economics*. *1*, 1-39.

Marwah, Kanta. (1978). On the variability of share elasticity of substitution of international trade flows. *Carleton Economic Papers*, No. 78-10, 17pp.

Marwah, Kanta and Palsson, Halldor P. (1988). The tracks of the managed exchange rate of the Indian rupee with monetary shocks and external disturbances. In M. Dutta (Ed.), *Asian industrialization: Changing economic structures*, 1, Part B, JAI Press, 1988, 139-172.

Miles, Marc A. (1979). The effects of devaluation on the trade balance and the balance of payments: Some new results," *Journal of Political Economy*. June, 600-620.

Moffet, M. (1989). The J-curve Revisited: An Empirical Examination for the United States. *Journal of Money and Finance*. *8*, 425-444.

Rosensweig, Jeffrey A. and Koch, Paul D.(1988). The U.S. Dollar and the delayed J-Curve. *Economic Review*. Federal Reserve Bank of Atlanta, July-August, 2-15.

Stern, Robert M. (1973). *The Balance of Payments: Theory and Economic Policy*, New York: Aldine Publishing Co..

Verghese, S. K. (1984). Management of exchange rate of the rupee since its basket link. *Economic and Political Weekly*, 1096-1105 and 1151-1158.

Models of Exchange Rate Behavior:
Application to the Yen and the Mark

Sung Yeung Kwack

Since the industrial countries adopted floating exchange rate systems in early 1973, nominal and real exchange rates of the main industrial countries have greatly varied.   Greater variability in the exchange rates have greatly affected the economies of industrial nations and the rest of the world. Consequently, the behaviors of exchange rates of major currencies and the U.S. dollar in particular have become focal points of many international economic policy issues in the 1980s.   In consequence, many detailed studies have addressed the exchange rate behaviors and determination.   However, many theoretical models of exchange rate determination yield poor explanation of the actual behaviors of the exchange rates in the 1980s. The use of such models to analyze issues of economic policy involves some embarrassment and reflects an element of faith rather than economic logic.

As part of the effort to enhance our understanding of the actual movements of exchange rates under the floating exchange rate system, the purpose of this paper is to review the development of major exchange rate modeling in the past two decades and to build a better empirical model of exchange rate determination. This paper presents an empirical model of yen-dollar and mark-dollar exchange rates tested over the period from January 1976 to June 1990.

In section one, we will discuss major theories on exchange rate determination for industrial countries generated since the early 1970s to cast some light on exchange rate modeling. In section two, we discuss some of the empirical results on the theories of exchange rate determination applied to yen-dollar and mark-dollar. Section three concludes this paper.

I. Selected Exchange Rate Theories

We consider the flow model and the asset models in determining the exchange rate of a small open economy. The country in question is small compared to the rest of the world, so all the conditions of the foreign country, including prices, interest rates and outputs, are assumed to be predetermined. For analytical simplicity, we assume that economic agents have perfect foresight on all the future events; perfect foresight assumption is consistent with the rational expectations hypothesis.

I.A Balance of Payments Flow Approach

The balance of payments is the net inflows of foreign exchange resulting from the current and private capital transactions. Traditionally the current account balance is assumed to depend negatively (positively) on domestic (foreign) real income and positively on the real exchange rate and net foreign assets. The real exchange rate is the nominal exchange rate multiplied by the ratio of foreign goods prices to domestic goods prices. It is assumed that foreigners do not invest in the assets of the other country including money, so that capital account transactions are made only by domestic residents. The capital account balance is affected positively by the excess of domestic interest rate over foreign interest rate. Hence, the flow equilibrium condition in the balance of payments is

$$\overset{(-)\;(+)\quad\;(+)\quad\;(+)\quad\quad(+)\;(+)}{C\,(\,y\;,\,y^{*},\,e+p^{*}-e,\;f\,)+K\,(\,i-i^{*},\,f\,)=0} \tag{1}$$

where C = the current account balance, y = the logarithm of domestic real income, log(domestic income Y), p = the logarithm of domestic prices, log(domestic prices P), f = the logarithm of net foreign asset holdings, log(foreign assets holding F), i = nominal interest rate at home. * indicates foreign variables, e.g y* is the logarithm of foreign real income.

Equation (1) is solved for e:

$$\overset{(+)\;\;(-)\;\;(+)\;(-)\;\;(-)\;(+)\;(-)}{e=H\,(\,y,\;y^{*},\;p,\;p^{*},\;i,\;i^{*},\,f\,)} \tag{2}$$

The exchange rate in the balance of payments flow equilibrium is related positively to domestic real income and prices and negatively to domestic interest rate and holdings of net foreign assets. It is also affected positively by foreign interest rate and negatively by foreign real income and prices. If the domestic and foreign produced goods are perfect substitutes in the consumer demand preferences, the price elasticity of goods is infinite. In this case, equation (2) indicates the exchange rate to be determined exclusively by the relative price of goods, namely e = p - p*. In the case where the domestic and foreign assets are perfect substitutes in the investor's portfolio preferences, the exchange rate is determined by the differential between domestic interest rate and foreign interest rate, that is to say e = i - i*.

I.B Asset Approach

The asset approach regards the exchange rate as the relative price of two assets, domestic and foreign. Among the models of the asset approach, we will discuss the flex-price monetary model (FPM), sticky-price monetary model (SPM), and portfolio balance model (PB). The differences among these models are apparent in the degree of flexibility in prices and substitutability in assets.[1]

As a starting point, we consider that the country trades only commodities with the rest of the world. If we assume perfect mobility of goods across the countries and perfect substitutability in demand preference, the domestic price level is equal in equilibrium to the foreign price multiplied by the exchange rate, and therefore purchasing power parity (PPP) holds. From the PPP condition, the exchange rate is determined by the relative price of goods:

$$e = p - p^* \qquad (3)$$

I.B.a Flex-price Monetary Model (FPM)

The flex-price monetary model (FPM) is based on the maintained assumption that the prices are perfectly flexible, so that any disequilibrium in the goods market is instantaneously cleared by changes in the prices. Because of the complete flexibility in prices, the output (real income) is always at the full employment level determined by the supply condition of the economy. The model regards the exchange rate to be the relative price of two monies. For the determination of exchange rate, therefore, the FPM model relies upon the supply of and demand for the money.

The flex-price monetary model by Frenkel (1976) and Bilson (1978) is summarized by equations (3) and (4):

$$e = p - p^* \qquad (3)$$

$$m - p = \phi y - \lambda i \qquad \phi > 0, \ \lambda > 0 \qquad (4)$$

where m is the logarithm of the nominal money stock supplied by the monetary authority, log(nominal money stock M). It is assumed in (4) that a higher real income causes the demand for real money balance to rise, whereas a higher domestic interest rate reduces the demand for real money balance.

The money supply affects the price level, which in turn determines the exchange rate. We solve recursively equations (3) and (4) for e

$$e = m - \phi y + \lambda i - p^* \qquad (5)$$

An increase in domestic output or a decrease in the domestic interest rate creates an excess demand for the money balance. The excess money demand is satisfied by reducing the domestic absorption, thereby causing a fall in the prices. Consequently, a rise in domestic real income or a decrease in the domestic interest rate appreciates the exchange rate. Variations in the nominal interest rate reflects both the real interest rate, r, and the rate of change in the prices, π. Thus, a rise in the nominal interest rate induced by a rise in the expected inflation depreciates the exchange rate.

In addition to the trade of goods, we introduce the transactions of capital across the countries. We assume that the domestic and foreign assets are perfect substitutes, implying that domestic interest rate equals foreign interest rate plus the expected rate of depreciation; this equality is called uncovered interest parity (UIRP). In the world of the only one commodity and the only one bond, the FPM model consists of three equations (3), (4), and (5):

$$e = p - p^* \tag{3}$$

$$m - p = \phi y - \lambda i \tag{4}$$

$$\dot{e} = i - i^* \tag{5}$$

where $\dot{e} = e_{t+1} - e_t$ for any t is the rate of expected depreciation from time t to t + 1.

To examine the factor determining the exchange rate we solve for e

$$e = m - \phi y + \lambda i^* + \lambda \dot{e} - p^* \tag{6}$$

At the given expected exchange rate, e_{t+1}, initially, the exchange rate is

$$e = (1 + \lambda)^{-1} (m - \phi y + \lambda i^* - p^*) + \lambda e_{+1} \tag{7}$$

At the full equilibrium, $\dot{e} = 0$ and hence $i = i^*$, the equilibrium price $\bar{p}$, and equilibrium exchange rate, $\bar{e}$, are

$$\bar{p} = m - \phi y + \lambda i^* \tag{8}$$

$$\bar{e} = \bar{p} - p^* = m - \phi y + \lambda i^* - p^*$$

The subtraction of (8) from (6) yields $\dot{e} = -\lambda^{-1} (\bar{e} - e)$, which describes the time path toward the equilibrium exchange rate. The PPP condition implies that the rate of expected depreciation equals the expected inflation differential. Using this condition $\dot{e} = \pi - \pi*$ and the dynamic path equation above, we obtain [2]

$$e = m - \phi y + \lambda i^* + \lambda(\pi - \pi^*). \tag{9}$$

The exchange rate is determined by the money supply, real income, foreign interest rate, foreign prices, and the inflation differential. When there are exogenously given shocks, the motion of the exchange rate towards the equilibrium value is indicated by an arrow in Figure 1.

For instance, a one percent rise in the nominal money stock immediately increases the price level and the exchange rate by one percent, thereby keeping the constancy of the real exchange rate.

Figure 1: Flex-price Monetary Model

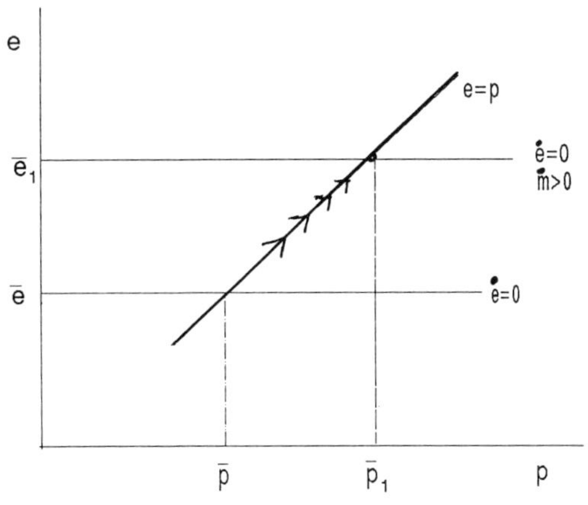

I.B.b Sticky-Price Monetary Model (SPM)

The flex-price monetary model does not allow for changes in the real exchange rate and output in the short-run. In order to relax this property of the FPM model, Dornbusch (1976) introduces an additional assumption that the rate of inflation slowly responds to excess demand in the goods market. As a result, the domestic price moves slowly to an equilibrium level. Consequently, changes in the money market conditions instantaneously affect domestic interest rate, which determine the exchange rate. Therefore, the exchange rate is determined by the UIRP condition, $e = i^* + e_{t+1} - i$, implying that the capital account conditions influence the exchange rate.

A simplified version of the sticky price monetary model by Dornbusch is

$$m - p = \phi y - \lambda i \tag{4}$$

$$\dot{e} = i - i^* \tag{5}$$

$$\dot{p} = \sigma [AD - y] = \sigma \{\delta(e + p^* - p) + x\}, \quad 0 < \sigma < 1 \tag{9}$$

where AD is aggregate demand for goods, and x denotes exogenous components of aggregate demand such as real government expenditures, g, and foreign real income, y^*; $x = g + y^*$. Implicit is the assumption in (9) that private total absorption equals the full employment level of output, and thus

$$AD = y + \delta(e + p^* - p) + x \, .^3$$

The operation of the sticky price monetary model (SPM) is as follows.

The relationship between the expected rate of depreciation and the money market conditions is obtained by substituting the money market equation (4) for i in the UIRP relation (5):

$$\dot{e} = \frac{1}{\lambda} (p + \phi y - m) - i^*$$ (10)

Consequently, the SPM model is reduced to equations (9) and (10), which jointly determine the time path of e and p.   In the long run, $\dot{e} = 0$ and $\dot{p} = 0$, and

$$\bar{p} = m - \phi y + \lambda i^*$$ (11)

$$\bar{e} = \bar{p} - p^* - \frac{1}{\lambda}x = m - \phi y + \lambda i^* - \frac{1}{\lambda} x - p^*$$ (12)

The time path of domestic price and exchange rate is described by the following two differential equations:

$$\dot{p} = \sigma\delta [(e - \bar{e}) - (p - \bar{p})]$$ (13)

$$\dot{e} = \frac{1}{\lambda} (p - \bar{p}).$$ (14)

The above system has a unique convergent saddle path, since $- \sigma\delta / \lambda < 0$ holds. The saddle path is

$$e = \bar{e} - (\lambda\theta)^{-1} (p - \bar{p}).$$ (15)

The substitution of p and e by (11) and (12) yields the reduced form exchange rate equation:

$$e = (1 + \frac{1}{\lambda\theta}) \{ m - \phi y + \lambda i^* \} - \frac{1}{\lambda}x - p^* - \frac{1}{\lambda\theta} p$$ (16)

In the SPM model, the price is constant in the short run, relatively speaking, and becomes perfectly flexible in the long-run. As a consequence, the exchange rate overly responds in the short run to shocks generating a shift in the saddle path and moves gradually to equilibrium, as indicated by arrows of motion in Figure 2. From (16), it is obvious that de / dm > 1 when p is fixed and de / dm = 1 in the long run. The short run effect on the exchange rate of a rise in the money supply, for example, exceeds the long run effect [see the distance $e_1 - \bar{e}_1$]; this phenomenon is called "overshooting" in the exchange rate. The resulting excessive depreciation in the real exchange rate increases aggregate demand which pushes up domestic prices and interest rate. This causes the exchange rate to appreciate, until the price and the exchange rate rise at the same rate as the rate of monetary expansion. As a result, the PPP condition holds in the long run, and the interest rate moves back to the level which existed before the monetary injection. In an implicit form, the SPM model is

$$
\begin{array}{ccccccc}
(+) & (-) & (+) & (-) & (-) & (-) & (-) \\
e = H & (m , & y , & i^* , p , & p^* , g , & y^*)
\end{array}
\tag{17}
$$

Figure 2: Sticky Price Monetary Model

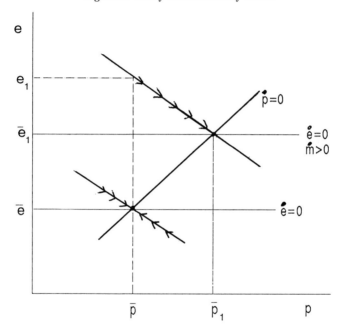

The SPM model of Dornbusch does not permit a role for the inflation rate differential and produces "overshooting." Frankel (1979) postulates that the expected rate of depreciation is affected not only by the deviation of the current exchange rate from the long run exchange rate value but also by the differentials in expected inflation rates. Frankel's expectation scheme is

$$
\dot{e} = \theta \ (\bar{e} - e) + (\pi - \pi^*).
\tag{18}
$$

The replacement of the Dornbusch's regressive expectation scheme

$\dot{e} = \theta(\bar{e} - e)$ by (18) yields

$$e = \bar{e} - \frac{1}{\lambda\theta} (p - \bar{p}) + \frac{1}{\theta} (\pi - \pi^*)$$

$$= (1 + \frac{1}{\lambda\theta}) \{ m - \phi y + \lambda i^* \} - \frac{1}{\lambda} \times - \frac{1}{\lambda\theta} p - p^* + \frac{1}{\theta} (\pi - \pi^*) \tag{19}$$

We substitute p in (19) by the money market equation (4) and get

$$e = m - \phi y + \lambda i^* - \frac{1}{\lambda} x - \frac{1}{\theta} ((i - \pi) - (i^* - \pi^*)) - p^* \tag{20}$$

implicitly,

$$
\begin{array}{cccccccccc}
(+) & (-) & (-) & (+) & (+) & (+) & (-) & (-) & (-) \\
\end{array}
$$
$$e = H \quad (m, y, p, p, i^*, \pi, \pi^*, g, y^*).$$

Considering the role of real interest rate differential, the version of the SPM model by Frankel (SPMRID) is referred to as the real interest rate model.

Hooper and Morton (1982) argue that a change in the equilibrium exchange rate affects the expected rate of depreciation, but it is not incorporated in the expectation scheme of Frankel. Further, the long run exchange rate is assumed to be associated with the net foreign assets.[4] Hence, the expectation scheme of Hooper and Morton is approximated by

$$\dot{e} = \theta(\bar{e} - e) + (\pi - \pi^*) - \theta_1(f - \bar{f}) \tag{21}$$

$\bar{f}$ and f are the logarithm of equilibrium and actual net foreign assets, log (foreign assets F), respectively. The substitution of Dornbursch's regressive expectation by (21) yields the SPM model of Hooper and Morton (SPMHM):

$$e = (1 + \frac{1}{\lambda\theta})(m - \phi y + \lambda i^*) - \frac{1}{\lambda} x - \frac{1}{\lambda\theta} p - p^* + \frac{1}{\theta}(\Pi - \Pi^*) - \frac{\theta_1}{\theta}(f - \bar{f}), \tag{22}$$

$$
\begin{array}{cccccccccc}
(+) & (-) & (-) & (-) & (+) & (+) & (-) & (-) & (-) & (-) \\
\end{array}
$$
implicitly, $e = H \quad (m, y, p, p^*, i^*, \pi, \pi^*, f, g, y^*).$

Frankel and Rodriguez (1982), and Driskell (1981) examined whether the overshooting in the exchange rate is a general phenomenon or a particular case associated with the characteristics of the SPM's specifications. They argued that the overshooting results largely from the high (perfect) substitutability in assets. To condense their arguments, we assume that the domestic and foreign assets are imperfect substitutes. The capital account equilibrium does not necessarily assure the balance of payments equilibrium. Therefore, the balance of payments equilibrium needs to be introduced as an substitute for the UIRP condition and is given by

$$\delta(e + p^* - p) + \beta(i - i^* - e_{+1} + e) = 0,\ 0 \le \delta \le \infty,\ 0 \le \beta \le \infty \qquad (23)$$

Real incomes and the foreign asset position are ignored in (23). As $\beta \to \infty$, the UIRP condition holds. However, as $\beta \to 0$, the payments equilibrium depends exclusively on the current account conditions. In the short run where e_{t+1} and p are predetermined, the effect of a rise in the money supply via its decreasing the interest rate is

$$\frac{de}{dm} = \frac{\partial e}{\partial i}\frac{\partial i}{\partial m} = \frac{\beta}{\lambda(\delta + \beta)} \qquad (24)$$

The effect lies within the limit of zero to one. When the elasticity of trade flows with respect to a change in the real exchange rate is infinite, i.e. $\delta = \infty$, no change in the exchange rate occurs. If the elasticity of the capital account with respect to a change in the interest rate differential is closer to zero, i.e. $\beta \to 0$, the initial exchange rate depreciation is closer to zero. Hence, it is possible that the monetary expansion produces the "undershooting" in the exchange rate.

To investigate the possibility of undershooting in the SPM model, we generalize the SPM model by extending price equation (9) to include the holdings of net foreign assets as an additional explanatory variable and by introducing the balance of payments equilibrium condition:[5]

$$m - p = \phi y + \lambda i \qquad (4)$$

$$\dot{p} = \sigma \{ \delta_1 (e + p^* - p) - \delta_2 f + x \} \qquad (25)$$

$$\delta_1(e + p^* - p) - (\delta_2 - 1)f + x - \delta_3 g - b + e - \beta(\dot{e} + i^* - i) = 0 \qquad (26)$$

The imperfect capital outflows are assumed to take place in order to close the gap existing between the desired and actual holdings of foreign assets. The desired stock is determined on the basis of portfolio diversification [see equation (35) below]. b is the logarithm of domestic non-tradable bonds, log (domestic bond B), and x= g + y*. Given the value of net foreign assets f = $\bar{f}$, the long run equilibrium values of domestic price and the exchange rate are

$$
\begin{aligned}
\bar{e} &= \bar{p} - p^* - \frac{1}{\delta_1}x - \frac{\delta_2}{\delta_1}\bar{f} \\
&= [\beta \{-p^* + m - \phi y + \lambda i^* + ((\delta_2 - (\lambda/\beta))/\delta_1)\bar{f}\} \\
&\quad -\lambda(x - \delta_3 g - b)] (\beta - \lambda (1 - \delta_1 - \beta))^{-1}
\end{aligned}
\qquad (27)
$$

$$
\begin{aligned}
\bar{p} &= [\beta (m - \phi y + \lambda i^*) + \lambda \{ (\delta_2 - 1 -(\delta_1 + \beta)\delta_2)\bar{f} \\
&\quad - (x - \delta_3 g - b) \}] (\beta - \lambda(1 - \delta_1 - \beta))^{-1}
\end{aligned}
\qquad (28)
$$

The dynamic path of p and e are

$$\dot{p} = \sigma\delta_1 \left((e - \bar{e}) - (p - \bar{p}) \right) \tag{29}$$

$$\dot{e} = \frac{\beta - \lambda}{\beta}(p - \bar{p}) + \frac{\delta_1 + \beta}{\beta}(e - \bar{e}) \tag{30}$$

If $\beta \to \infty$, these two equations are the same as (13) and (14) of Dornbusch's SPM model. The saddle path is

$$e = \bar{e} - \frac{\beta - 1}{\lambda(\beta\theta + \delta_1 + \beta)}(p - \bar{p}) \tag{31}$$

Equation (30) indicates that the slope of $\dot{e} = 0$ is negative when $\beta > 1$ [Dornbusch's case], as shown in Figure 3. For the value of β to be smaller than one [low degree of asset substitutability], the slope is positive. For the case of low substitutability, therefore, the slope of the saddle path is positive, and a monetary expansion leads to an undershooting in the exchange rate [see the distance $e_1 - \bar{e}_1$ in Figure 3. The magnitude of the undershooting is larger when the real exchange rate elasticity, δ_1, and interest rate elasticity of money demand, λ , are larger. Domestic price and the exchange rate move in the same direction. The model in low asset substitutability case is written as follows:

$$\begin{array}{cccccccccc}
(+) & (-) & (+) & (+) & (-) & (-) & (+) & (-) & (-) \\
e = H\,(& m\,, & y\,, & i^*, & p\,, & p^*, & f\,, & b\,, & g\,, & y^*).
\end{array} \tag{32}$$

I.C Portfolio Balance Model

In the monetary model we have discussed above, money is valued as the distinctive asset, whereas the domestic and foreign non-money assets are treated as a single identical asset. The portfolio balance model recognizes the exchange rate as the relative price of assets (non-money assets) and takes an explicit account of the wealth holder's diversification among the available different assets in determining the exchange rate.

Following Branson (1977), we assume that there are three types of assets in portfolios: money, bonds, and foreign assets. In the case of three assets, the portfolio balance model consists of the demand equations for three assets and the identity that the sum of the values of three assets held is equal to the value of wealth. Equilibrium takes place when the demand for each asset equals its supply. We postulate that the log of the share of an asset demanded to wealth is a linear function of domestic interest rate, foreign interest rate, and the expected rate of currency depreciation. The linear form of the wealth identity is replaced by the log form of the identity that the log of wealth is the average of the logarithmic values of three assets weighted by their relative value shares.[6] After the substitution of the log of the wealth in the three demand functions by the weighted average of the log values of three assets being made, the three asset demand equations are

$$m = m_0 - m_1 i - m_2 i^* - m_2 \dot{e} + b_{m0} b + f_{m0} e + f_{m0} f \tag{33}$$
$$b = b_0 + b_1 i - b_2 i^* - b_2 \dot{e} + m_{b0} m + f_{b0} e + f_{b0} f \tag{34}$$
$$e + f = f_0 - f_1 i + f_2 i^* + f_2 \dot{e} + m_{f0} m + b_{f0} b \tag{35}$$

where all the parameter values are positive.

Figure 3: Sticky Price Monetary Model (Low Substitutability Case)

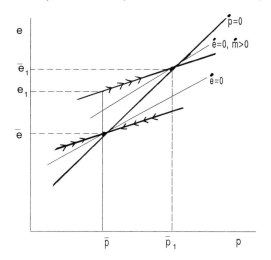

When the values of m, b, f, i, i$_*$ and e$_{+1}$ are taken as predetermined, the exchange rate can be determined from equation (35), regarding that the exchange rate is the relative price of foreign assets:

$$e = (1 + f2)^{-1} (f_0 - f_1 i + f_2 i^* + f_2 e_{+1} + m_{f0} m + b_{f0} b - f) \tag{36}$$

A rise in the domestic interest rate and foreign assets supplied tends to appreciate the exchange rate, whereas a rise in the foreign interest rate and the expected exchange rate tends to depreciate it.

In the short run, the portfolio balance model determines both the interest rate and the exchange rate, with the given values of m, b, f, i*, and e$_{+1}$. We use of equations (33) and (35) to determine the two variables i and e. The domestic interest rate is solved for from the money market condition (33), and the resulting equation substitutes for i in the demand function for foreign assets (36). The reduced form exchange rate equation is

$$e = k_0 i^* + k_1 m + k_2 b - k_3 f + k_4 e_{+1} \tag{37}$$

where , $k_0 = (m_1 f_2 + f_1 m_2) k_5 > 0$, $k_1 = (m_1 m_{f0} + f_1) k_5 > 0$, $k_2 = (m_1 b_{f0} - f_1 b_{m0}) k_5 \gtrless 0$, $k_3 = (m_1 + f_1 f_{m0}) k_5 > 0$, $k_4 = (m_1 f_2 + f_1 m_2) k_5 > 0$, $k_5 = (f_1 f_{m0} + m_1 + m_1 f_2 + f_1 m_2)^{-1} > 0$.

Thus, the exchange rate is determined by the foreign interest rate, the money stock, bonds, foreign assets, and the expected exchange rate. An increase in the domestic bond supplied either depreciates or appreciates the exchange rate, depending largely upon the degree of substitutability between domestic and foreign bonds. To the extend that the expected future exchange rate e_{t1} is formed on the information on expected relative prices, current account and government policies, the public perception on future monetary and fiscal policies will have an effect on the exchange rate determination in the current period.

The exchange rate changes arising with the given levels of the money stock, bonds, and foreign assets induce changes in the current account. The changes in the current account in turn feeds back into the asset market and the exchange rate. The evolution of the exchange rate over time involves the dynamic stock-flow interactions of changes in the exchange rate and the current account. Hence, the long run equilibrium is established when the dynamic process is complete.

The substitutions of i in equations (34) and (35) by (33) yield equations for domestic bonds and foreign assets:

$$b = z_0 - z_1 i^* - z_1 \dot{e} + z_2 m + z_3 (e + f), \quad z_1 > 0, z_2 \gtrless 0 , z_3 > 0 \tag{38}$$

$$e + f = v_0 + v_1 i^* + v_1 \dot{e} + v_2 m + v_3 b, \quad v_1 > 0, v_2 > 0, v_3 \gtrless 0 \tag{39}$$

Equations (38) and (39) solve for e + f, and in addition, we introduce the equation describing the current account behavior:

$$e + f = q_0 + q_1 i^* + q_1 \dot{e} + q_2 m, \quad q_1 > 0, q_2 > 0 \tag{40}$$

$$\dot{f} = \delta_1 (e + p^* - p) - \delta_2 f + x - \delta_3 g \tag{41}$$

where direct effects are assumed to outweigh indirect effects in determining the signs of q_1 and q_2.[7] Equations (40) and (41) describe the behaviors of e and f over time. At the long run equilibrium,

$$\bar{e} = p - p^* + \frac{\delta_2}{\delta_1}\bar{f} - \frac{x - \delta_3 g}{\delta_1}$$

$$= \frac{1}{\delta_1 + \delta_2}\{-(x - \delta_3 g) + \delta_1(p - p^*) + \delta_2(q_0 + q_1 i^* + q_2 m)\} \tag{42}$$

$$\bar{f} = q_0 + q_1 i^* + q_2 m - \bar{e}$$

$$= \frac{1}{\delta_1 + \delta_2}\{(x - \delta_3 g) - \delta_1(p - p^*) + \delta_1(q_0 + q_1 i^* + q_2 m)\} \tag{43}$$

The dynamic paths obtainThe dynamic paths obtained from (40) and (41) are

$$\dot{e} = \frac{1}{q_1}((e - \bar{e}) + (f - \bar{f})) \tag{44}$$

$$\dot{f} = \delta_1(e - \bar{e}) - \delta_2(f - \bar{f}) \tag{45}$$

Recalling that $x = g + y^*$, the exchange rate is determined by

$$e = \bar{e} - \frac{1}{1 + q_1\theta}(f - \bar{f}) = \frac{1}{(1 + q_1\theta)(\delta_1 + \delta_2)}\{q_1\theta(-y^* + \delta_1 (p - p^*)) - (\delta_1 + \delta_2)f$$

$$+ (\delta_1 + \delta_2 + \delta_2 q_2\theta)(q_0 + q_1 i^* + q_2 m) - (1 - \delta_2)q_1\theta g\} \tag{46}$$

The exchange rate of the portfolio balance model is

$$\begin{array}{cccccc} (+) & (+) & (+) & (-) & (-) & (-) \end{array}$$
$$e = H (p - p^*, m , i^*, f , g , y^*) \tag{47}$$

since $(1 - \delta_2) > 0$.

As shown in Figure 4, a rise in the money stock yields the overshooting in the exchange rate. The initial overshooting causes the current account to be a surplus. The resulting rise in the foreign assets holding leads the exchange rate to appreciate at a decreasing rate as the foreign assets accumulate. At the same time, a rise in the domestic prices by the monetary expansion generates additional appreciation of the real exchange rate. Since the increased foreign asset holdings produce added interest income, zero balance on the current account at the long run equilibrium requires the depreciation of the nominal exchange rate and the appreciation of the real exchange rate. Therefore, the PPP condition does not hold in the long run.

Figure 4: Portoflio Balance Model

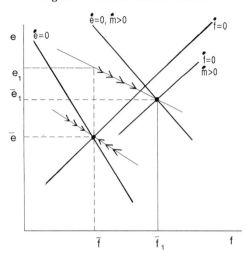

When the investment income from the net foreign assets is used to purchase foreign goods, its effect on the current account is zero, and $\delta_2 = 0$ into (46) gives

$$ e = \frac{1}{1 + q_1\theta} \{ q_1\theta \left(-\frac{x - \delta_3 g}{\delta_1} + (p - p^*) \right) - f + q_0 + q_1 i^* + q_2 m \} \qquad (48) $$

We have presented the essence of selected models to determine exchange rates. Although most of the models are originally reported for the two countries of an equal size, our presentation of the models is made in the context of a small open economy. Table 1 summarizes the coefficient signs on major determinants implied in the models. Because each of the models emphasizes the role of particular markets in the economy, these models appear to differ substantially. Surprisingly, however, they agree the coefficient signs on the majority of determinants to each other.[8] The differences arise with respect to domestic prices, domestic interest rate, and domestic real income. In the SPM model by Dornbusch and Frankel, negative sign on domestic price holds only in the short run and becomes zero in the long run. Therefore, the coefficient sign of the domestic price variable is positive, implying that a rise in the domestic price leads to a currency depreciation. The sign on domestic interest rate is positive for the FPM model, because the nominal interest rate reflects mainly the expected inflation expectation which is formed on the basis of the rate of growth in the money stock, as it is a likely case when monetary expansions are excessive over a period of longer period. In the case of a modest monetary growth, the liquidity effect of a monetary growth is greater than its price expectation effect. Thus, the nominal interest rate represents real interest rate, and the coefficient sign of the domestic interest rate is expected to be negative, which is consistent with positive sign on the foreign interest rate variable predicted in all the models we have included in Table 1. The coefficient sign of the domestic income variable in monetary models is negative. This is attributed to the real income as representing the output supplied. A larger supply of goods relative to demand implies a current account surplus, which leads to currency appreciation. Positive sign on domestic real income is the result of real income generated from aggregate demand for goods and services. Therefore, sign on real income is not uniquely determined, depending on which of the supply and demand conditions primarily affects the output and thus real income.

Table 1: Coefficient Signs of Exchange Rate Equations

	p	p*	i	i*	π	π*	y	y*	e_{t+1}	g	m	b	f
BOP Flow Model (2)	+	-	-	+			+	-					
PPP Basis (3)	+	-											
FPM (5)		-	+				-				+		
+UIRP (9)		-		+	+	-	-				+		
UIRP Basis			-	+					+				
SPM (17)	-	-		+			-	-		-	+		
SPMRID (20)	-	-		+	+	-	-	-		-	+		
SPMHM (22)	-	-		+	+	-	-	-		-	+		-
Portfolio Basis													
Foreign Assets (26)			-	+					+		+	+	-
Short-term (37)				+					+		+	?	-
Short & Long-Run (47)	+	-		+				-		-	+		-

Figure in parentheses indicates equation number in the text. * denotes foreign country.
p = domestic goods prices, i = nominal interest rate at home, π = domestic expected inflation rate, y = real income at home,
e_{t+1} = expected exchange rate, g = real governemnt expenditure, m = nominal money stock, b = bonds, f = foreign assets.

II. Empirical Tests of the Models

A large number of empirical studies on exchange rate models have appeared since 1976. These studies are largely confined to variants of monetary and portfolio balance models, because a balance of payments flow model is thought to possess theoretical shortcomings. The portfolio balance model has been tested less frequently than monetary models, perhaps due to the difficulty in obtaining proper data required for the portfolio balance model.

The models work well with data for the period of 1973 - 1978. The coefficients of key explanatory variables are statistically significant and have expected signs. However, when the coverage of data is extended beyond 1978 to the late 1980s, the regressions of models produce unexpected coefficient signs and insignificant coefficients for key explanatory variables. Meese and Rogoff (1983, 1984, 1988) compared the out-of-the sample performance of reduced form monetary models with the random walk model and concluded that the performance of monetary models is no better than those of the random walk model.

This conclusion is confirmed by studies of Backus (1984) and Boughton (1984), although Boughton reported that the portfolio balance model in general performed better in the case of the dollar-DM exchange rate. Doubt on the ability of theoretical models to explain what has behaved is expressed clearly by Dornbusch (1989, p.401): "After 20 or 30 years of exchange rate modeling, from the work of Meade and Mundell to the New Classical Economics, we are left with an uncomfortable recognition that our understanding of exchange rate movements is less than satisfactory."

Why have these theoretical models performed unsatisfactorily in recent years? Among the many possibilities, we may present two reasons.[9] First, the equations employed for empirical studies are reduced-form equations. Reduced-form equations are easily subject to mis-specifications. The reduced form coefficients vary substantially, especially when the underlying economic structures of countries vary. This is confirmed indirectly by the work of Schinasi and Swamy (1986), in which the equations estimated with varying coefficients for selected monetary models performed better than those with fixed coefficients and the random walk models. Secondly, the interdependence among not only financial and capital sectors within a country, but also industrial and developing countries, has been growing at a rapid pace during the 1980s. The growing interdependence causes unexpected events in a country which in turn affects economic conditions of other countries in the world and thus exchange rates. News and political risks accompanied by the growing interdependence of nations are hard to adequately capture in exchange rate modeling.

Earlier empirical studies found the presence of the first-order serial correlations between the residuals. This suggests that the form of change in the levels of an exchange rate is preferable to the form of the levels of the exchange rate. In order to examine this appearance, the unit root tests are performed on the basis of the following regression equations:[10]

$$\log(e(t)) = a1\ \log(e(t-1)) \tag{49}$$

$$\log(e(t)) = a0 + a1\ \log(e(t-1)) \tag{50}$$

$$\log(e(t)) = a0 + a1 \ \log(e(t-1)) + a2 \ t \qquad (51)$$

$$\log(e(t)p^*(t)/p(t)) = a0 + a1 \ \log(e(t-1)p^*(t-1)/p(t-1)) + a2 \ t \quad (52)$$

where e is an exchange rate, t is time trend, p^* and p are foreign and domestic price levels, respectively. Equations (49)-(52) are tested with the monthly data over January 1976 to June 1990 on the yen-dollar and the mark-dollar exchange rate. As shown in Table 2, the results confirm statistically that a0 = 0, a2 = 0, and a1 = 1, indicating that the specification of an exchange rate equation in the difference form is preferable.

Before we begin our regression run, we test whether a change in an exchange rate follows the movement of relative prices which are the maintained hypothesis of flexible price models or the movement of interest rate differentials, using the following difference equations:

$$\Delta \ 100 \ \log(e) = a0 + a1 \ \Delta \ 100 \ \log(p^*/p) \qquad (53)$$

$$\Delta \ 100 \ \log(e) = a0 + a2 \ \Delta \ (i^* - i) \qquad (54)$$

$$\Delta \ 100 \ \log(e) = a0 + a3 \ \Delta \ (i^* - \pi^* - i + \pi) \qquad (55)$$

As presented in Table 3, the estimates of a2 and a3 for Japanese yen and the Deutsche mark are significantly different from zero, and the estimated values of a1 do not differ from zero. These simple tests suggest that changes in the exchange rates are largely affected by changes in interest rates.

Variations of the nominal exchange rate have been similar to variations of the real exchange rate since the adoption of the flexible exchange rate system, as shown in Table 4, which reports the co-variance of monthly percent changes in the logarithms of nominal, real exchange rates and national price levels from March 1973 to June 1990. The monthly percent changes in the logarithm of the national price levels are small and less correlated to monthly percent changes in the nominal exchange rate. This is evidence of the slow adjustments in the exchange rates to changes in the price levels. As a result of the sluggish response, a change in the real exchange rate brings about approximately equal change in the nominal exchange rate. Thus, the specification for the real exchange rate determination is relevant to the specification for the nominal exchange rate behavior.

Now we utilize the variety of the monetary and portfolio balance model equations to explain the behavior of the yen-dollar and the mark-dollar exchange rates. The typical equations are as follows:

$$\Delta 100 \ \log(e) = ao + a1 \ \Delta \log(M1^*/M1) - a2 \ \Delta \log(y^*/y) - a3 \ \Delta(i -i^*) \qquad (56)$$

$$\Delta 100 \ \log(ep^*/p) = a0 - a1 \ \Delta(i - i^* - \pi + \pi^*) + a2 \ (USST/(p^* \ y^*)) - a3 \ FEI/FER(-1) \qquad (57)$$

$$\Delta\,100\,\log(ep^*/p) = a0 + a1\,\Delta\,(i-i^*-\pi+\pi^*) + a2\,\Delta\,(USST/p)$$
$$- a3\,\Delta\,(M2^*/p^*) + a4\,\Delta\,(M2/p) \tag{58}$$

where FEI and FER represent the amount of foreign exchange intervention during a unit period of time and foreign exchange reserves. USST is the net foreign assets of the United States, which are the accumulated sum of the U.S. trade balance; M1 and M2 are the narrowly and broadly defined money stock.

	a_0	a_1	a_2	R^2 (adj.)	DW
Table 2: Unit Root Tests, 1976.3 - 1990. 6					
Yen - Dollar					
(2.1) Nominal (JAE)		0.99		0.98	1.36
		(2383)			
(2.2)	0.05	0.99		0.98	1.36
	(1.0)	(114)			
(2.3)	0.10	0.98	0.00	0.98	1.35
	(1.0)	(87)	(0.6)		
(2.4) Real (JAE.USCPI/JACPI)	0.09	0.98	0.00	0.97	1.40
	(1.0)	(70)	(0.0)		
Mark - Dollar					
(2.5) Nominal (WGE)		0.99		0.97	1.45
		(384)			
(2.6)	0.01	0.99		0.97	1.45
	(0.5)	(85)			
(2.7)	0.01	0.99	0.00	0.97	1.45
	(0.7)	(82)	(0.5)		
(2.8) Real (WGE.USCPI/WGCPI)	0.01	0.99	0.00	0.97	1.45
	(0.9)	(83)	(0.2)		

1. JAE = Yen per dollar; WGE = mark per dollar; USCPI= U.S. consumer price; JACPI = Japapnese consumer price; WGCPI = West German consumer price.
2. R^2(adj.) = coefficient of determination adjusted for degree of freedom; DW = Durbin-Watson statistic. Figures in parentheses are t-statistic.

Equation (56) represents the monetary model, and (57) and (58) represent the portfolio balance model. The ratio of the U.S. foreign assets relative to U.S. income in (57) represents the risk associated with the holding of U.S. dollar-denominated assets. When the asset to income ratio is higher, the risk is lower, leading to a depreciation of foreign currency. The intervention data of monetary authorities are not available to the public, making it impossible to know the nature of official intervention. However, a proxy of the amount of intervention can be assumed to be changes in official foreign exchange reserves after subtracting estimated interest earnings on the reserves during the period. Since Japan and West Germany have maintained a surplus on their current account balance, whereas the United States has had a deficit in its current account, it is likely that the monetary authorities desire to have an appreciation of the yen and the mark against the dollar. Therefore, the exchange interventions are likely to be made to smooth foreign exchange operations and to prevent the depreciation of the yen and the mark.

Table 3: Simple Test						
	a_0	a_1	a_2	a3	R^2 (adj.)	DW
Yen - Dollar (JAE)	-0.48 (2.1)	0.42 (1.3)			0.01	1.36
	-0.41 (1.8)		0.57 (2.2)		0.02	1.32
	-0.38 (1.7)			0.45 (2.3)	0.02	1.32
Mark - Dollar (WGE)	-0.25 (1.0)				0.00	1.45
	-0.25 (1.2)		0.55 (2.2)		0.02	1.40
	-0.25 (1.2)			0.45 (2.0)	0.02	1.42

Data used as follows: JARCM = Japanese call money rate, WGRS = West German interbank deposit rate, USRTB = U. S. Treasury bill rate, JAPHI = Japanese expected inflation, WGPHI = West German expected inflation, and USPHI = U. S. expected inflation.

Table 4: Exchange Rate Variability				
	Yen - Dollar		Mark - Dollar	
	1973.3 -1990.6	1985.10 - 1990.6	1973.3 - 1990.6	1985.10 - 1990.6
Variance of				
100 Δ log(e)	8.01	10.8	15.4	7.74
100 Δ log(ep*/p)	8.76	11.5	15.7	7.79
100 Δ log(p*/p)	0.65	0.34	0.13	0.07
Covariance of				
100 Δ log(e), 100 Δ log(p*/p)	0.05	0.18	0.05	-0.01

Industrial production and consumer price indices were used in place of income and national prices, respectively. Two different measures of the money stock, M1 and M2, are alternatively tried. With regard to interest rates, short-term rates and long-term rates are used, although short-term rates turn out to be preferable. The proxy of expected inflation is the actual inflation in one year ahead from the present time. In the absence of monthly data on the current account of the United States, net foreign asset values are the accumulated sum of the trade balance of the United States. Alternatively we tried in the place of U.S. net foreign assets the accumulated sum of Japanese and German trade balance in domestic currency and U.S. dollars, with the bench mark values in December 1971 which are sums of the trade balance from January 1950 to December 1971. The sources of monthly data are *International Financial Statistics of the International Monetary Fund* and the *Federal Reserve Bulletin*.

The regression period is from January 1976 to June 1990. The estimation techniques used are ordinary least squares. Whenever the past movements of an economic variable affect the present levels of an exchange rate, the time lag effects are captured through a polynomial lag distribution scheme. When regressions indicate the presence of serial autocorrelations between the residuals, the equations were re-estimated with a first-order correction of auto-correlations between the residuals. The variety of the specifications (56) - (58) was tried.

Table 5 reports the estimates of equations (56), which represents a monetary approach to determine an exchange rate. The variables of relative money stock, industrial production, and expected inflation have insignificant coefficients, even though the coefficients of the relative money stock and expected inflation variables are of theoretically expected signs. The interest rate variables in the equations of the yen-dollar as well as the mark exchange rate have significant coefficients. The results shows that interest rates are important determinants of the exchange rates.

Table 6 contains the estimates of equations (57) and (58) for the yen-dollar exchange rate. The coefficient estimates for all the determinants except for the real money stock of the United States are significant and correctly signed in the case of real yen exchange rate. The Durbin-Watson statistics reject the presence of significant serial correlations between the residuals at the five percent level of significance. Real interest rate differentials and the U.S. foreign assets relative to the U.S. income are significant in affecting the nominal yen-dollar exchange rate, although the relative money stock, which is assumed to determine the relative goods prices, is found to be insignificant. On the whole, our results show that the portfolio balance model can explain a substantial portion of the yen-dollar movements. The real interest rate differentials affected the exchange rate over a period of one year. Based on the coefficient estimates of the real interest rate variable, a one percentage point rise in the real interest rate of the United States leads to a depreciation of the yen by 0.2 percent in the Table 5 short-run and approximately by 2 percent in the long-run. When we evaluate at the mean value, 7585, of U.S. nominal income, USCPI.USIP, a one billion dollar rise in the U.S. trade balance is estimated to depreciate the yen against the U.S. dollar by 1.7 to 2.6 percent.[11] The effect of an improvement in the U.S. trade balance is conditioned to the size of U.S. nominal income. Thus, the effect of an equal rise in the U.S. trade balance is smaller in recent years, since U.S. income is higher in recent years than the mean value. The coefficient estimate, -25.3, of the exchange intervention implies that a one billion dollar-worth intervention tends to appreciate the yen by about 0.8 percent, at the mean value of foreign exchange reserves. Although the effect is small, the intervention helps a great deal to explain the actual changes in the yen rate. This indicates that foreign exchange interventions smoothed the changes in the exchange rate.

	Table 5: Monetary Model Specifications			
	Yen - Dollar 100 Δ log (JAE)		Mark - Dollar 100 Δ log (WGE)	
Equation No.	(5.1)	(5.2)	(5.3)	(5.4)
Constant	-0.40	-0.39	-0.24	-0.24
	(1.8)	(1.7)	(1.1)	(1.1)
Δ log(USM1 / FM1)	-6.05	-5.86	-6.90	-6.91
	(1.2)	(1.2)	(1.0)	(1.0)
Δ log(USIP / FIP)	0.11	0.14	4.24	4.22
	(0.2)	(0.3)	(0.4)	(0.4)
Δ (FR - USRTB)	-0.57		-0.55	
	(2.3)		(2.2)	
Δ (FR - USRTB - FPHI + USPHI)		-0.57		-0.56
		(2.3)		(2.2)
Δ (FPHI - USPHI)		-0.32		-0.53
		(0.8)		(0.9)
R²(adj.)	0.02	0.02	0.02	0.01
SEE	2.01	2.92	2.71	2.71
DW	1.3	1.3	1.3	1.3

FM1 = foreign money stock M1,(F = JA, WG) i.e. Japanese M1 (JAM1) and West German M1 (WGM1); FR = foreign short-term interest rate, JARCM, WGRS; FPHI = foreign expcted inflation, JAPHI, WGPHI; USIP = U.S. indusrtrial production index; FIP = foreign production index, (JAIP, WGIP) . SEE = standard error of estimate.

Table 7 reports the estimates of equations (57) and (58) for the Deutsche mark. The results are, more or less, similar to what we obtained for the Japanese yen; namely, all the determinants, except for the real money stock and relative money stock, are significant. Again, the real interest rate differentials are found to be important.[12] A change in the real interest rate differentials has an effect over one and half years and yields greater effect than the equal change in the real interest rate differential affects the yen exchange rate. A one percentage point rise in the U.S. interest rate depreciates the mark-dollar rate by about 0.3 percent in the short-run and 4.5 percent in the long-run. An increase of 10 billion dollars in the U.S. trade balance depreciates the mark by 2 to 2.3 percent, at the mean value of U.S. nominal income. While the effect of a rise in the U.S. trade balance on the German mark rate is similar to the effect on the Japanese yen, a rise in the real interest rate differential has greater impact on the mark exchange rate than on the yen rate. The greater impact on the German mark rate may result from higher degree of capital mobility between the United States and Germany or absence of capital controls. This implies that as capital controls lessen, interest rates become more important in the determination of exchange rates. The foreign exchange interventions affect the mark rate: a one billion dollar-worth intervention leads to about 0.2 percent appreciation of the mark.

Table 6: Portfolio Balance Specifications for the Yen					
	Real Yen 100 Δ log(JAE.USCPI/JACPI)			Nominal Yen 100 Δ1 og(JAE)	
Equation No.	(6.1)	(6.2)	(6.3)	(6.4)	(6.5)
Constant	0.71	0.96	0.75	0.48	0.72
	(2.0)	(2.5)	(1.7)	(1.4)	(1.9)
Δ(JARCM - USRTB - JAPHI + USPHI)	-1.9	-2.1	-2.1	-1.7	-2.0
	(3.4)	(3.5)	(3.3)	(3.2)	(3.3)
Δ(USST /(USCPI.USIP))	1361	2028		1327	1983
	(2.7)	(3.6)		(2.7)	(3.7)
Δ(USST / USCPI)			15.9		
			(2.6)		
Δ(USM2 / USCPI)			-2.1		
			(1.4)		
Δ(JAM2 / JACPI)			0.01		
			(2.3)		
JAFEI / JAFER(-1)	-25.3			-25.8	
	(6.5)			(7.0)	
Δlog(USM1 / JAM1)				-1.57	-1.6
				(0.1)	(0.3)
R^2 (adj.)	0.32	0.14	0.14	0.33	0.13
SEE	2.5	2.8	2.8	2.4	2.7
DW	1.8	1.6	1.6	1.7	1.5

JAM2 (U.S. M2) = Japanese (U.S.) money stock M2; JAFEI = foreign exchange intervention; JAFER = Japanese official exchange reserves. The coefficient of real interest rate differntials is the sum of current and past eleven month lag coefficients, estimated by PDL with 2nd degree and far end constraint.

Finally, one questions how much the portfolio balance exchange rate model, consisting of equations (6.1) and (7.1), traces the actual movements of the yen and the mark rates. Alternatively, is the model's performance better than the performance of a random walk model, equations (2.4) and (2.8) of Table 2? In order to compare the performance within the sample period, the portfolio balance model and the random walk model are dynamically simulated from January 1976 to June 1990 in addition to the simulation from January 1988 to June 1990. The root mean squared percentage errors obtained from the simulations are summarized in Table 8. Further, the predicted yen and mark rates in June 1990 by the model simulation from January 1988 are 130 and 1.7, respectively. On the other hand, the predicted values for the yen and the mark

Table 7: Portfolio Balance Specifications for the Mark					
	Real Mark 100 Δ log(WGE.USCPI/WGCPI)			Nominal Mark 100 Δ log(WGE)	
Equation No.	(7.1)	(7.2)	(7.3)	(7.4)	(7.5)
Constant	0.92	0.99	0.75	0.62	0.72
	(2.8)	(3.1)	(1.9)	(1.9)	(1.9)
Δ(WGRS - USRTB - WGPHI + USPHI)	-4.5	-4.8	-4.8	-4.7	-5.1
	(3.6)	(3.8)	(3.7)	(3.7)	(4.1)
Δ(USST /(USCPI.USIP))	1592	1761		1528	1682
	(3.3)	(3.6)		(3.2)	(3.4)
Δ(USST / USCPI)			10.0		
			(1.8)		
Δ(USM2 / USCPI)			-2.1		
			(1.6)		
Δ(WGM2 / WGCPI)			14.9		
			(0.9)		
WGFEI / WGFER(-1)	-5.1			-5.1	
	(2.9)			(3.0)	
Δlog(USM1 / WGM1)				-12.9	-5.8
				(0.9)	(0.9)
R^2 (adj.)	0.22	0.18	0.18	0.21	0.18
SEE	2.4	2.5	2.5	2.4	2.4
DW	1.8	1.7	1.7	1.8	1.6

WGM2 = West German money stock M2; WGFEI = foreign exchange intervention amount; WGFER = West German official holdings of foreign exchange reserves. The coefficient of real interest rate differntials is the sum of current and past eleven month lag coefficients, estimated by PDL with 2nd degree and far end constraint.

by the random walk model are 132 and 1.56, respectively. On the basis of the root means squared percentage errors and the predicted values, the model is clearly better in predicting the mark rate than the random walk model. The model's performance in predicting the yen rate is no worse than the random walk model's performance. However, the model's predictive ability over a longer period is better than the random walk model. Since the forecasts by the random walk model depend on the initial condition and are monotonic, the forecasts would not capture a long cycle of exchange rateswings generated by stock-flow interactions; the accumulated current account surplus tends to keep the overvalued currency for a long period, until the accumulated foreign assets are consumed by a flow of current account deficit. It is followed by the undervalued currency for a long period, until the accumulated deficits are eliminated by a flow of current account surplus.

Table 8: Root-Mean Squared Percentage Errors

	1976.1 - 1990.6	1988.1 - 1990.6
Yen - Dollar		
Model	10.6	8.5
Random Walk	18.1	7.3
Mark - Dollar		
Model	8.6	3.6
Random Walk	17.9	11.7

III. Concluding Remarks

This paper discusses major theories about flexible exchange rate behaviors over the past twenty years and tests the variety of the monetary and portfolio balance models on the yen-dollar and mark-dollar exchange rates. The real interest rate differentials, the net external assets to income ratio and the official foreign exchange interventions are major determinants of the actual movements of the yen and the mark rates since 1976. Further, the in-sample forecasting tests indicate that the estimated portfolio balance model performs better than or no worse than the random walk model.

Because the external value of a currency is an asset price, the exchange rate is affected by the past, present and future courses of economic factors including economic policies. In the development of our empirical models, we did not fully incorporate forward-looking aspects of exchange rate determinations. Further effort should take forward-looking mechanism into account to bring about a better understanding of exchange rate behaviors.

Sung Y. Kwack, Department of Economics, Howard University, Washington, D.C. 20059.

ACKNOWLEDGEMENTS

I am grateful to Peter Clark and the participants in seminar at the Korea International Economic Policy Institute for valuable suggestions. This research is supported partially by the University-Sponsored Faculty Research Program in the Social Scineces, Humanities, and Education of Howard University.

NOTES

1. For the discussions on the theories of exchange rate determination, see Dornbusch (1980, 1989), Frankel (1983) and Isard (1987).

2. The real interest rate is regarded as constant at the full employment level of output. The nominal interest rate reflects the inflation expectation. If the nominal interest rate in (5) is substituted by the inflation expectation, the resulting equation is similar to (9).

3. Dornbusch assumes that the interest rate affects aggregate demand and thus the inflation. We do not include the interest rate. But the results from our simplified version do not differ qualitatively from those of Dornbusch.

4. Hooper and Morton include a change in the cumulative current account balance as a variable determining an unexpected change in the long run equilibrium real exchange rate.

5. The stability of an open macroeconomic system requires that the effect on the current account of a rise in the foreign assets is negative, other things being equal. See Branson and Buiter (1983). This requirement implies that the effect on domestic absorption of a change in the income and the wealth resulting from a rise in the net foreign assets is substantial.

6. The approximation holds from the fact that arithmetic average of the values of different assets is approximated by the geometric average.

7. A high government spending reduces the current account balance by increasing the domestic income or price. Since x includes g, the coefficient value of g in (44) is greater than 1.

8. For persuasive discussions on this point, see Gylfason and Helliwell (1983).

9. For discussions, see Isard (1987) and Dornbusch (1989).

10. See Phillips (1987) for a unit root test.

11. Slightly different estimates are reported in Fukao (1989). The real interest rate effect is 2.5-3.7 percent depreciation on the average, whereas the trade balance improvement effect is 1.8-2.7 percent depreciation.

12. The importance of real interest rate differentials is reported by Marston (1989) and Stein (1989), whereas Meese and Rogoff (1988) cast some doubt on the relationship between real interest rates and exchange rates.

REFERENCES

Artus, J. (1976). Exchange rate stability and managed floating: the experience of the Federal Republic of Germany. *IMF Staff Papers. 23*(July), 312-333.

Ayanian, R. (1988). Political risk, national defense and the dollar. *Economic Inquiry. 26*(April), 345-351.

Backus, D. (1984). Empirical models of the exchange rate: Separating the wheat from the chaff. *Canadian Journal of Economics. 17*(4), 824-846.

Bilson, J. F. O. (1978). The monetary approach to the exchange rate: Some evidence. *IMF Staff Papers. 25*(March), 1-15.

Bisignano, J. and Hoover, K. (1982). Some suggested improvements to a simple portfolio balance model of exchange rate determination with special reference to the U.S. dollar, Canadian dollar rate. *Weltwirtschaffliches Archiv. 118*(Heft1), 20-37

Boughton, J. M. (1987). Tests of the performance of reduced-form exchange rate models. *Journal of International Economics. 23*(September), 41-56.

Branson, W. H. (1977). Asset markets and relative prices in exchange rate determination.

Sozialwisseschaftliche Annalen. 1, 69-89.

Branson, W. H., Halttunen, H. and Masson, P. (1979). Exchange rates in the short-run: Some further results. *European Economic Review. 12* (October), 395-402.

Branson, W. H. and Buiter, W. H. (1983). Monetary and fiscal policy with flexible exchange rates. In J.S. Bhandari and B. H. Puttnam (Eds.), *Economic interdependence and flexible exchange rates.* (pp. 251-85). Cambridge, MA: MIT Press.

Dickey, D. A. and Fuller, W. A. (1981). The likelihood ratio statistics for autoregressive time series with a unit root. *Econometrica. 49*(July), 1057-1072.

Dornbusch, R. (1976). Expectations and exchange rate dynamics, *Journal of Political Economy. 84*(December), 1161-1176.

Dornbusch, R. (1980). Exchange rate economics: Where do we stand? *Brookings Papers on Economic Activity.* (1), 143-155.

Dornbusch, R. (1989). Real exchange rates and macroeconomics: a selective survey, *Scandinavian Journal of Economics. 91*, 399-432.

Dornbusch, R. and Fischer, S. (1980). Exchange rates and the current account, *American Economic Review. 70*(December), 960-971.

Driskell, R. A. (1981). Exchange rate dynamics: an empirical investigation, *Journal of Political Economy. 89*(April), 357-371.

Dooley, M. and Isard, P. (1980). Capital controls, political risks, and deviations from interest-rate parity. *Journal of Political Economy. 88*(April), 370-384.

Frankel, J. A. (1979). On the mark: a theory of floating exchange rates based on real interest differentials. *American Economic Review. 69*(September), 610-622.

Frankel, J. A. (1983). Monetary and portfolio balance models of exchange rate determination. In Bhandari and Putnam (Eds.), *Economic interdependence and flexible exchange rates.* (pp. 84-115). Cambridge, MA: MIT Press.

Frenkel, J. A. (1976). A monetary approach to the exchange rate: Doctrinal aspects and empirical evidence. *Scandinavian Journal of Economics, 78*, 200-224.

Frenkel, J. A. and Rodriguez, C. (1982). Exchange rate dynamics and the overshooting hypothesis. *IMF Staff Papers. 29*(March), 1-30.

Fukao, M. (1989). Exchange rate fluctuations, balance of payments balances and internationalization of financial markets. *Bank of Japan Monetary and Economic Studies. 7*(August), 25-70.

Granger, C. W. J. (1981). Some properties of time series data and their use in econometric model specification. *Journal of Econometrics. 16*, 121-130.

Gylfason, T. and Helliwell, J. F. (1983). A synthesis of Keynsian,monetary, and portfolio approached to flexible exchange rates. *Economic Journal. 93*(December), 820-831.

Gros, D. (1989). On the volatility of exchange rates: Tests of monetary and portfolio balance of models of exchange rate determination. *Weltwirtschaffliches Archiv. 125*(Heft2), 273-295.

Hooper, O. and Morton, J. (1982) Fluctuations in the dollar: A model of nominal and real exchange rate determination. *Journal of International Money and Finance. 1*(April), 39-56

Isard, P. (1987). Lessons from empirical models of exchange rates. *IMF Staff Papers. 34*(March), 1-28.

Kouri, P. J. K. (1976). The exchange rate and the balance of payments in the short run and in the long run: A monetary approach. *Scandinavian Journal of Economics. 78*, 280-304.

Marston, Richard C. (1989). Systematic movements in real exchange rates in the G-5: evidence on the integration of internal and external markets, NBER Working Paper No. 3332.

Meese, R. A. and Rogoff, K. (1983). Empirical exchange rate models of the seventies: Do they fit out of sample. *Journal of International Economics. 14*(February), 3-24.

Meese, R. A. and Rogoff, K. (1984). The out of sample failure of empirical exchange rate models: sampling error or misspecification? In J.A. Frenkel(Ed.), *Exchange rates and international macroeconomics.* (pp. 67-112). Chicago: University of Chicago Press.

Meese, R. A. and Rogoff, K. (1988). Was it real? The exchange rate-interest differential relation over the modern floating-rate period. *Journal of Finance*. *18*(September), 933-948.

Phillips, P. C. B. (1987). Time series regression with a unit root. *Econometrica*. *55*(March), 277-301.

Schinasi, G. J. and Swamy, P. A. V. B. (1989). The out-of-sample forecasting performance of exchange rate models when coefficients are allowed to change. *Journal of International Money and Finance*. *8*(September), 375-390.

Stein, J. E. (1989). The real exchange rate, 1989, manuscript.

Tobin, J. (1969). A general equilibrium approach to monetary theory. *Journal of Money, Credit and Banking*. *1*(February), 15-29.

Exchange Rate Unions Versus Flexible Exchange Rates: An Empirical Investigation

Michael G. Papaioannou

This paper attempts to assess empirically the impact of various real, monetary and bond market disturbances on output and exchange rates under a flexible exchange rate system and an exchange rate union setting.[1] Traditional macroeconomic theory has been used extensively to analyze such disturbances under flexible rates but not in a union environment. The participating currencies in an exchange rate union (like the European Monetary System (EMS)) are tied together in a joint float against non-member currencies and union countries are often committed to exchange market interventions in order to achieve fixity of the cross exchange rates of their currencies. For intervention to be successful, it is customarily assumed that member countries' economic policies ought to follow some scheme of international coordination. In theory, disturbances would generally affect the behavior of output and other macroeconomic variables in union countries in a different way than would in a flexible exchange rates regime.

Our analysis is conducted in the framework of a three-country model, with two of the countries entering an exchange rate union. For each regime, a structural model is estimated where the behavioral parameters are assumed to be variant with respect to the two alternative exchange rate systems. Therefore, Lucas' critique that changes in policy regimes may alter structural relationships under rational expectations is assumed to be directly applicable. In particular, we presume that the choice of the exchange rate regime affects the economic and financial behavior of the two union countries (such as investment activity and substitutability between assets) during the span of our simulation horizon. When various types of disturbances are applied on the models representing the two regimes, macroeconomic responses differ in the two countries depending on the extent of transformation that their respective economies are undergoing when they switch from flexible rates to a union setting. Our results show further that an exchange rate union is preferred to flexible exchange rates when bond market disturbances are prevailing in the two union countries, while a flexible exchange regime tends to be favored when real shocks are present.

This study is organized as follows: Section I describes briefly the theoretical model employed in our analysis. Section II presents the methodology and data, and provides an overview of the estimated models for a flexible exchange rates regime and an exchange rate union. Section III introduces various disturbances and presents the results in terms of variances for the dollar exchange rates and real outputs of the two union countries--France and Germany. Section IV concludes by summarizing the main findings.

I. The Theoretical Model

The model employed to study the effects of various disturbances under the two alternatives regimes, flexible rates and currency unions, derives from Marston's work on the subject.[2] It consists of a three-country world: two potential members of an exchange rate union, countries 1 and 2, along with a third country representing the rest of the world.[3] Three exchange rates are thus defined: x_t^1, x_t^2 representing the price of countries' 1 and 2 currencies, respectively, in terms of the third country's currency and x_t^{12}, representing the price of country 1's currency in terms of the other union country's currency. The cross exchange rate between currencies of countries 1 and 2 is determined by triangular arbitrage conditions, i.e., . $x_t^{12} = x_t^1 / x_t^2$

Real, monetary, and financial structures in the two union countries are assumed to be identical.[4] The real sector includes the aggregate supply and demand equations, the monetary sector incorporates a transactions demand for money equation, and the financial sector consists of a demand for domestic bonds equation. Financial market integration is examined in the context of financial ties between the respective countries' bond markets through interest rates. Under flexible rates, these four equations will determine output, prices, interest rates, and the dollar exchange rate for each union country against the third country's currency. Under a union setting, the demand for a union country's bonds can be additionally determined, since only one union country's exchange rate needs to be specified. The two countries of the union are assumed to be too small to influence the output, prices and interest rates of country 3. All disturbances considered are assumed to have zero mean, to be serially uncorrelated and to be serially uncorrelated with each other. Lower case letters denote variables expressed as a percent of change from their corresponding stationary values, with the exception of interest rates, r_t^i, and the demand for bonds, $H_t^i - H_o^i$, which express absolute changes from their stationary values.[5] For the sake of expositional convenience, the initial values of all prices and exchange rates are set equal to unity.

Aggregate Supply Equation

Supply behavior may vary depending on the time response of wages to prices and the scheme of wage indexation to current changes in the general price level.[6] The general type of a supply equation is introduced as

$$y_t^i = c(p_t^i - E_{t-1} p_t^i) - cb(l_t^i - E_{t-1} l_t^i) \tag{1.1}$$

where y_t^i is country i's output, p_t^i is the corresponding price in local currency and l_t^i is the (percentage change of the) general price level. $E_{t-1}(.)_t$ denotes the expected value of variable (.) at time t based on information available at t − 1. The parameter b is the indexation factor ranging from zero (no indexation) to one (full indexation). In case of no indexation, supply behavior depends only on deviations of domestic prices from their expected levels. When full indexation prevails, supply depends on the difference between the prediction errors of country i's price of output and the general price level. With less than complete indexation, supply of output will generally increase if p_t^i and l_t^i increase equally as a result of a fall in real wages.

If there is no wage indexation (i.e., b = 0) and nominal wages in country i, w_t^i, are fixed in one-period labor contracts, formed according to price expectations at time t − 1, i.e., $w_t^i = E_{t-1} p_t^i$, then the previously specified supply equation (I.1) can be rewritten as

$$y_t^i = c \, (p_t^i - w_t^i) \tag{1.1a}$$

This last relation indicates that only unexpected changes in prices at time t may alter output since contract wages are assumed to be fixed on one-period lag of expected prices. In other words, output increases only if real wages are lowered. Finally, note that in the absence of contract lags output is unaffected by either monetary or aggregate demand shocks.

Aggregate Demand Equation

The aggregate demand equation for each union country is expressed as a function of the terms of trade and output of the two other countries with which it trades, the real rate of interest, and a stochastic factor. That is,

$$y_t^1 = d_1 \, (p_t^2 + x_t^1 - x_t^2 - p_t^1) + d_2 \, (p_t^3 + x_t^1 - p_t^1) +$$

$$d_3 \, y_t^2 + d_4 \, y_t^3 - d_5 [r_t^I + (l_t^1 - E_t \, l_{t+1}^1)] + u_{d1t} \tag{1.2}$$

$$y_t^2 = d_1 \, (p_t^1 + x_t^2 - x_t^1 - p_t^2) + d_2 \, (p_t^3 + x_t^2 - p_t^2) +$$

$$d_3 \, y_t^1 + d_4 \, y_t^3 - d_5 [r_t^2 + (1_t^2 - E_t \, 1_{t+1}^2)] + u_{d2t} \tag{1.3}$$

where $E_t l_{t+1}^i$ is the expected value of variable l^i at time t + 1 based on information available at t and u_{dit} is the aggregate demand disturbance of country i. All d coefficients bear the conventional elasticity interpretation. The price coefficients, d_1 and d_2, depend on export and import price elasticities. The income coefficients, d_3 and d_4, depend on the income elasticities of demand for the respective union country's good in the two foreign countries.

A rise in foreign prices relative to the price of the country in question is assumed to increase the aggregate demand of that country as the output in either foreign country increases. A rise in the real interest rate is assumed to reduce aggregate demand. When the demand for country i's output does not respond to its real interest rates, i.e., $d_5 = 0$, the interest rate term drops from the aggregate demand equations. In the case of perfect substitution between domestic and foreign goods, i.e., when d_1 and d_2 become infinite in size, the aggregate demand equations reduce to simple PPP conditions.[7] Finally, it should be noted that the disturbances u_{d1t} and u_{d2t} represent either fiscal policy shocks or changes in private sector demands.

The Demand for Money Equation

The transactions demand for money in each union member country is assumed to depend on the return only of its own bond, nominal transactions[8] and a stochastic term. That is, we assume that the effects of foreign returns on the demand for money are negligible. If the income elasticity of the demand for money is further assumed to be one, we can express the transactions demand for money, m^i_t , as

$$m^i_t = p^i_t + y^i_t - kr^i_t + u_{mit} / M^i_o \qquad (1.4)$$

where u_{mit} is the disturbance term and M^i_o is the stationary value of money demand for country i.

The Structure of the Financial Sector

Each union member country i is assumed to have, besides money, M^i_t, a second financial asset, a domestic bond, H^i_t, with each domestic bond bearing interest rate R^i_t. Further, the private sector of country i is assumed to hold its wealth in the form of four different assets, money, M^i_t , and bonds denominated in its currency, H^i_{it} , bonds denominated in the other member country's currency, H^i_{jt} , and in the third country's currency, F^i_t. The later two assets are translated into domestic currency units at the current exchange rates. Each bond demand in country i is a function of the three expected returns on the three bond assets, while money demand in each union country depends only on its own bond return.

Since there are no dynamics in the model and all disturbances in the model are assumed unanticipated and temporary, the expected exchange rate for the next period is equal to the stationary value of that exchange rate. That is, we assume rational expectations and therefore realignments are not expected to occur or, equivalently, economic fundamentals are anticipated to remain unaltered. Thus, the return S^i_{qt} , $q = 1, 2, 3$ on each of the three bonds expressed as a percentage change from a stationary value is approximately equal to

$$S^1_{1t} = r^1_t \tag{1.5}$$

$$S^1_{2t} = r^2_t - x^{12}_t \tag{1.6}$$

$$S^1_{3t} = r^3_t - x^1_t \tag{1.7}$$

The monetary authorities of country i hold as assets domestic currency bonds, H^{im}_{it} and the third country's bonds, F^{im}_t, as well as bonds, H^{im}_{jt}, and money, M^{im}_{jt} of the other member country j. Again, all foreign assets are converted to country i's currency by the appropriate exchange rates. Liabilities consist of money held by the private sector, M^i_t, and by the monetary authorities of the other member country j, $M^{j,m}_{jt}$, plus a balancing item which offsets changes in exchange rates in the monetary authorities' balance sheet, A^{im}_t. The government of country i issues domestic currency bonds, $\bar{H}^i_t$, with the net supply available to the private sector, H^i_t, equal to

$$\bar{H}^i_t - H^{im}_{it} - H^{jm}_{it}, \quad i \neq j = 1,2.$$

The Demand for Bonds Equation

The demand for each country's bond is expressed as a function of transactions in both union countries (since residents of both member countries hold each of the bonds), interest returns on the union bonds, the exchange rates of the union currencies against the third country's currency and a disturbance term.[9] That is,

$$H^1_t - H^1_0 = -b_1(p^1_t + y^1_t) - b_2(p^2_t + y^2_t) + b_3 r^1_t - b_4 r^2_t + b_5 x^1_t - b_6 x^2_t + u_{h1t} \tag{1.8}$$

$$H_t^2 - H_0^2 = -b_1(p_t^2 + y_t^2) - b_2(p_t^1 + y_t^1) + b_3 r_t^2 - b_4 r_t^1 + b_5 x_t^2 - b_6 x_t^1 + u_{h2t} \quad (1.9)$$

where u_{hit} is the demand for bonds disturbance term. These specifications assume that the demand for each bond is a negative function of transactions since a rise in nominal transactions causes a shift from all assets into money. If $b_1 > b_2$, a rise in the income of union country i (holding the other union country's income constant) leads to a greater shift out of its own bond than out of the bond of the other union country. Further, it is assumed that all assets are gross substitutes, that is, asset demands are positively related to their own returns and negatively related to the returns on any other asset.[10] The assumptions of serially uncorrelated disturbances, with mean zero, and rational formation of expectations about the exchange rates lead to a positive relationship between depreciation of each union country's currency and the demand for its own bonds.[11] (It is implicitly assumed that the current price of the country's exchange rate is higher than its stationary value, i.e., in a sense, we may observe an overshooting phenomenon.) The parameters b_3, b_4, b_5, and b_6 measure the degree of substitutability between union members' bonds. Perfect substitution between these bonds implies that all of these coefficients acquire infinite values. The differences $(b_3 - b_4)$ and $(b_5 - b_6)$ indicate the substitution between the countries 1 and 2 bonds together and money and the third country's bonds, respectively. If all three bonds are perfect substitutes, the demand for bond equations reduce to their expected returns which can be expressed (in the same money) approximately by

$$r_t^1 = r_t^3 - x_t^1 \qquad\qquad (1.10)$$

$$r_t^2 = r_t^3 - x_t^2. \qquad\qquad (1.11)$$

Description of an Exchange Rate Union

When country 1 joins in an exchange rate union with country 2, their monetary authorities undertake the obligation to intervene in the foreign exchange markets so that $x_t^{12} = x_t^1 / x_t^2$ remains constant. By assuming that x_t^{12} is initially equal to one, intervention must ensure that the cross exchange rate in percentage changes is kept equal to zero at all times, i.e., $x_t^{12} = x_t^1 - x_t^2 = 0$. Thus, only one exchange rate needs to be determined in a union.

Models designed to capture the workings of countries 1 and 2 in an exchange rate union framework introduce an additional equation to those appearing under flexible exchange rates. Thus, the model that determines the changes in the demand for country 1's bonds incorporates the equilibrium condition $H_t^2 - H_0^2 = - \dfrac{1}{X_0^{12}}(H_t^1 - H_0^1)$ which corresponds to an exchange

market intervention being sterilized. Consequently, both demand for bonds variables are determined endogenously in this model. Note that, by having the demand for country 2's bonds explained endogenously in the model, we maintain the effects of changes in country 1's bond market on its output through changes in its exchange rate (X_t^1). In a more realistic course of events, bond market changes should appear in the country's interest rate equation and changes in interest rates should, in turn, influence its real output. In our model, however, output is not directly affected by interest rate changes. Hence, we have to assume that an exchange intervention (which keeps the movements of both union currencies against the third country's currency equal) is perfectly sterilized in order to keep an effective channel of response between country 1's bond market disturbances and its output.

An exchange market intervention resulting in an increase of the foreign exchange reserves may be sterilized by selling official holdings of domestic bonds in equal value, so as to keep bank reserves constant. This is the case when monetary authorities are exclusively committed to the attainment of their domestic policy goals and no offsetting capital flows are initiated by investors in response to the initial open market sale. The condition of opposite changes in the demand for bonds of the two union countries, that emerges from the assumption of sterilized intervention, can effectively be thought of as a swap of country 2's bonds for country 1's bonds in private portfolios. Thus, from the bonds market equilibrium condition for country 1,

$$H_t^1 = H_{1t}^1 + H_{1t}^2 = \bar{H}_t^1 - H_{1t}^{1\,m} - H_{1t}^{2\,m}$$

we can derive the following expression in deviations from stationary values:

$$(H_{1t}^{1\,m} - H_{10}^{1\,m}) = - (H_t^1 - H_0^1) \tag{1.12}$$

For an intervention increasing country 1's official holdings of country 2 assets by $(H_{2t}^{1m} - H_{20}^{1m})$, the change in the money supply of country 1, ceteris paribus, becomes,

$$(M_t^1 - M_0^1) = X_0^{12}(H_{2t}^{1\,m} - H_{20}^{1\,m}) = - X_0^{12}(H_t^2 - H_0^2) \tag{1.13}$$

When the intervention carried out by country 1 is fully sterilized, then

$$(H_{1t}^{1\,m} - H_{10}^{1\,m}) = - X_0^{12}(H_{2t}^{1\,m} - H_{20}^{1\,m}) \tag{1.14}$$

Hence, utilizing equations (I.12) and (I.13), we obtain,

$$(H_t^1 - H_0^1) = - X_0^{12}(H_t^2 - H_0^2) \qquad (1.15)$$

This latter equation solved for $H_t^2 - H_0^2$ appears as the addendum to the model representing the flexible rates case.

II. Methodology and Estimation

For assessing empirically the macroeconomic implications of various disturbances, we consider France and Germany as the two union countries, before and after the formation of the EMS, and the United States as an approximation for the rest of the world. The model that we attempt to estimate for each regime consists of four equations for each of the two union countries: an aggregate supply equation, in which wages are not indexed but rather fixed to prices expected at the previous period (equation (I.1a)); an aggregate demand equation, in which no real interest rate effects and imperfect substitution between domestic and foreign goods are assumed (equations (I.2), (I.3) with $d_5=0$); and two demand equations for assets (money and domestic bonds) (equations (I.4) and (I.8), (I.9)). In addition, we utilize equation (I.15) for the exchange rate union case.

Methodology

The variables of the theoretical model are expressed in percentage changes, except for interest rates and the demand for bonds that are expressed in differences, from their stationary values. Hence, to estimate such a model through classical (multistage) regression methods would require data that are expressed in deviations from stationary values. Data in this form, however, are not readily available or easy to construct. Furthermore, structural estimation requires that a model is subject to identifying and possible overidentifying restrictions, which in our case cannot always be warranted. In addition, problems of interdependencies and multicollinearities that invariably arise among the right-hand-side variables of the equations of the system are not easy to overcome.

To avoid such shortcomings of conventional methods of structural estimation and data limitations, a two-step procedure is utilized instead. In the first step, all structural coefficients are estimated through forward simulations with the world system of The WEFA Group (Wharton Econometric Forecasting Associates) as explained below. The coefficients obtained in the first step are then used to simulate dynamically the models representing the flexible rates and the union regimes over the historical periods of flexible rates (1973-78), and the EMS (1979-90), respectively. In the second step, the final coefficient-estimates are derived as the parameters that minimize the (squared) deviations between the actual and (dynamically) estimated values of the endogenous variables over the relevant sample periods. This method is the OLS analog of the extension of Jorgenson's multi-stage least squares approach suggested by L.R. Klein.[12]

The procedure for obtaining estimates of coefficients suggested in step one involves performance of dynamic simulations with WEFA's large-scale global model for the period 1991-94. By perturbing a set of appropriately selected policy instruments in the relevant country-models of the WEFA world system and using a baseline solution as a proxy for the stationary values of the relevant variables, we attempt to establish a causal effect between two variables (representing the dependent variable and a determinant) in a certain relationship of our theoretical models. Through these simulations, we effectively "extract" economic properties embodied in the individual models and the interlinkage relationships of the WEFA model. These properties, in essence, are implied reduced form parameters and constitute first-round estimates of the coefficients of our models. Such estimates are derived as arithmetic averages of annual impulse effects, stemming from shocks in the relevant policy instruments.

Each parameter in a behavioral multivariate relationship indicates the influence of the corresponding determinant on the dependent variable, provided that the rest of the determinants in the relationship remain unaltered. This conventional interpretation of coefficients in a multivariate regression is followed in the set-up of our dynamic simulations. Isolation of the effects of each determinant has been accomplished by fixing certain policy instruments at a predetermined level through optimal control methods. Note that the estimates of the coefficients are not totally invariant to the choice of policy instruments. However, since most economic functions in the WEFA model are linear, the size of the shock for each instrument is irrelevant. Furthermore, the prospect for getting the correct sign and magnitude of coefficients becomes higher not only because the obtained coefficients are reduced forms of established models, but also because collinearities among predetermined variables, as well as identification and exogeneity restrictions, have been eliminated by appropriately designing the simulation exercises. Finally, the structure of the simulation for each regime is formed by introducing simple dynamics (additive lagged dependent variables) to the theoretical model and solving for certain specified variables.

An outline of the second step of our estimation procedure is in order. To obtain estimates of the parameters that minimize residuals over the simulation period, the multiperiod solutions of the individual dependent variables are computed dynamically from fixed initial conditions and the actual independent variables of the sample period. The lagged dependent variables are *not* assumed predetermined at period t. Solution of the first-order conditions of the minimization of the (squared) deviations between the actual and (dynamically) estimated values of the endogenous variables produces the values for the coefficient-estimates of the model.

The second step of the parameter-estimation procedure can be summarized as follows. Let the i-th equation of our linear system be expressed in final solution form as $z_{it} = k_i \lambda^t - \dfrac{\alpha_i(L)}{\Delta(L)} \theta v_t + \dfrac{\alpha_i(L)}{\Delta(L)} e_t$ $i = 1, 2, \ldots, ms$

where z_t, v_t, θ and e_t are the vector of endogenous variables at time t, current exogenous variables, parameters and disturbances, respectively. And k_i is the i-th row of the k matrix (m x m) of constants determined by the initial conditions and parameter values of the system, λ is an m-element vector of roots of the characteristic polynomial associated with the autoregressive structure, $\alpha_i(L)$ denotes the i-th row of the adjoint of A(L) with A(L) being a polynomial expression in the lag operator where

$$L^\mu z_t = z_{t-\mu} \qquad \mu = 0, 1: \qquad\qquad A(L) = A_0 + A_1 L$$

and $\Delta(L) = |A(L)|$ denotes the determinantal polynomial of A(L). The estimation of the parameters θ can be derived from the final solution form by minimizing the mean square error over the sample period t_1 to t_2. That is,

$$S = \sum_{t=t_1}^{t_2} \epsilon^2_{it} = \min \text{ where } \epsilon_{it} = \frac{\alpha_i(L)}{\Delta(L)} e_t \, .$$

Equivalently, we choose those parameter values that satisfy $S = \sum_{t=t_1}^{t_2} (z^A_{it} - k_i \lambda^t - \dfrac{\alpha_i(L)}{\Delta(L)} \theta v_t)^2 = \min$. Recall that λ is a function of θ and Z^A_i denotes the actual values of the endogenous variables.

The minimization process seeks structural parameter estimates and not simply estimates of the parameters in the solution form. Through a search procedure (algorithm) combinations of values of the parameters in an interval (determined by economic theory) may be chosen sequentially until a value with minimal squared error is reached. This approach establishes estimates that minimize the squared error along the sample-period solution path of the equation, given initial conditions. Generally, estimates resulting from this method differ from that of the OLS regression since the cumulated error of the simulation is not independent from period to period. Note that by applying the values of coefficients derived through the latter method we may improve the simulation performance for the historical period but we may also disturb the economic properties of the model. For example, real or monetary disturbances may not produce the same impacts on certain endogenous variables, in terms of size and direction, that we obtained with the coefficients derived through the WEFA model simulations.

Data

The problem of availability of data consistent with the definition of variables in the theoretical model is dealt by considering the baseline solution of the WEFA world model[13] as an appropriate proxy for the assumed stationary values. By shocking various instruments in the French, German, and U.S. country-models of the WEFA world system, we obtain simulated results for the relevant variables of the models to be estimated. To conform with the notion of deviations from stationary values, all outcomes are expressed in percentage terms or differences from the baseline solution. Further, we select close proxies for those variables of the theoretical model that are not readily available in the three country-models of the WEFA system utilized for this study: real GNP(GDP) for representing output--even though output may also include production of intermediate goods; the GNP(GDP) deflator or Consumer Price Index for domestic prices; average wage or the wage rate index for the contract wage; the broader measure of money supply (M2/M3) for the demand for money--assuming, in effect, that the money market is always in equilibrium; short-term interest rates for interest rates; and nominal GNP(GDP) for nominal transactions--again, overseeing the fact that transactions incorporate value added as well as intermediate goods valuations. For the historical period, the basic data source is the WEFA Group, International Economic Services, Historical Data, January 1989.

Special attention is given to the demand-for-each-country's-bond variable, which is proxied by the respective federal government's budgetary account. The introduction of such a proxy is based on two implicit assumptions. First, that the government's budgetary account is the only available source of bond issuance. Then, a deficit in the government's budget constitutes a supply of bonds by the government to the public, provided that the government finances its deficits through bonds and not through monetization. The reverse holds true for a budgetary surplus, which, in essence, amounts to having the government paying back its bond-debt. Second, that the bond market in each country is in equilibrium. That is, the supply (demand) of the government bonds is absorbed entirely by the private sectors of both union countries. In this sense, the government's fiscal deficit/surplus is the only means in bond transactions and hence, a precise proxy of the demand for bonds.

The Estimated Models

In this section, we present the estimated models for each of the two alternative regimes. For the estimated models of four equations for each of the countries, France and Germany, which can determine endogenously an equal number of variables, we maintain the same structures but different behavioral coefficients are allowed for each exchange rate system. The list of variables differs for the exchange rate union situation, because, as stated above, under the latter regime changes of only one dollar-rate need to be determined. The estimated models replicated history (the period 1973-90) fairly closely after they had appropriately been extended to incorporate a simple dynamic structure and optimal (from minimization) coefficient estimates. These dynamic models were then used for the generation of the control solutions for the period 1991-94.

Flexible Exchange Rate Case

For the estimation referring to the flexible exchange rate regime, we choose to solve the estimated model for (the percentage changes in) domestic prices, real outputs, the dollar exchange rates and (the absolute differences in) the interest rates. Hence, the supply equations are inverted to determine prices, the demand equations in the estimated form are used to judge outputs, money demand equations are solved for interest rates and the demand for bonds equations are utilized to decide for the dollar exchange rates. The dynamic model for the flexible exchange rate period appears in Table 1.

Table 1:   Dynamic Model for the Flexible Exchange Rates Case

Domestic Prices

$p^F = 1.08 \ y^F + w^F + 0.23 \ p^F(-1) - 8.76$
$p^G = 1.03 \ y^G + w^G + 0.15 \ p^G(-1) - 7.01$

Real Output

$y^F = 0.07 \ (\ p^G + x^{FG} - p^F \ ) + 0.03 \ (\ p^u + x^F - p^F \ ) + 0.18 \ y^G + 0.03 \ y^u + 0.17 \ y^F(-1) + 2.93 + Ud^F$
$y^G = 0.04 \ (\ p^F - x^{FG} - p^G \ ) + 0.04 \ (\ p^u + x^G - p^G \ ) + 0.09 \ y^F + 0.05 \ y^u + 0.11 \ y^G(-1) \ + 1.96 + Ud^G$

Nominal Interest Rates

$r^F = -0.18 \ (\ m^F - p^F - y^F \ ) - 0.29 \ r^F(-1) + 0.13 \ y^F(-1) + 1.17 + Um^F$
$r^G = -0.23 \ (\ m^G - p^G - y^G \ ) - 0.31 \ r^G(-1) + 0.14 \ y^G(-1) - 0.98 + Um^G$

Dollar Exchange Rates

$x^F = -0.51 \ (\ H^F_t - H^F_0 \ ) - 1.07 \ (\ p^F + y^F \ ) - 0.23 \ (\ p^G + y^G \ ) + 0.57 \ r^F - 0.05 \ r^G + 0.05 \ x^G - 0.33 \ x^F(-1) + 11.93 + Uh^F$
$x^G = -0.64 \ (\ H^G_t - H^G_0 \ ) - 1.03 \ (\ p^G + y^G \ ) - 0.11 \ (\ p^F + y^F \ ) + 0.61 \ r^G - 0.06 \ r^F + 0.03 \ x^F - 0.39 \ x^G(-1) + 8.46 + Uh^G$

Notes: 1. The endogenous variables of the model appear in the LHS of the equations. F stands for France, G for Germany, and U for the United States. 2. The y denotes output; p prices; w nominal wages; x the exchange rate (e.g., $x_F = F / \$$); m money demand; r interest rates; and H demand for home bonds. 3. ud^i_t, um^i_t, uh^i_t for i = F ,G represent real demand, monetary and bond market disturbances, respectively, for each of the countries examined. 4. The subscript 0 denotes stationary values.

The coefficients calculated for the aggregate supply functions indicate that a 1 percentage point increase in real wages leads to a less than 1 percent decrease in real output for both union countries. Since the derivation of the supply equations is based on constant-returns Cobb-Douglas production functions, the coefficients estimated indicate the capital/labor shares ratio in the output of each respective country. Hence, for both France and Germany the share of capital in the production is greater than that of labor. Aggregate demand coefficients are quite small in size. Thus, a deterioration in the terms of trade with respect to either the other union country or the United States by 10 percent bring about a half of percentage point increase in the demand for the country's output. An increase in the other member country's real output, however, produces a higher response for the demand of domestic goods than would an equiproportionate increase in the U.S. real output. The impact of a U.S. real output change on the aggregate demand of Germany although quite small, is almost double than that for France.

The effect of an interest rate increase on the percentage measure of inverse velocity is negative, the correct sign, and almost of identical magnitude for both union countries. The demand for bond equations have provided us with interesting results. Interest rate changes verify the gross asset-substitutability assumption and changes in nominal transactions show a negative influence on the demand for bonds. Thus, an increase in the domestic interest rates reduces the transactions' and speculative demand for money. As a consequence, the demand for domestic bonds increases as the public is now willing to hold a higher proportion of its assets in the form of bonds. To maintain equilibrium in the financial markets, monetary authorities have to exchange domestic bonds for money and in an amount equal to the private sector's increased demand for bonds. However, an incipient exchange rate depreciation does not lead to an increase in the demand for bonds denominated in the depreciating currency.

How exchange rate changes affect the demand for bonds denominated in that currency depends on whether shocks are temporary or permanent. It is worthwhile to point out the negative sign of the lagged dependent variable in the dollar exchange rate equation of each country. This empirical finding indicates some degree of overshooting in the determination of exchange rates. It may also be interpreted to imply expectations of a future appreciation, following a period of depreciation of a currency, which in turn can cause an increase in the demand for the respective currency-denominated bonds. Note, however, that this empirical finding is also consistent with standard theoretical conclusions of the asset market approach.

Finally, extensive experimentation with various forms of the money demand function has indicated that abolition of the unitary income elasticity assumption yields general trends and changes in the interest rates that are significantly closer to the observed rates in the years 1973-78.

Exchange Rate Union Case

The structure of the model for an exchange rate union setting is modified from that existing under flexible rates. Since the dollar rates of both countries' currencies move in the same direction and proportion, and thus only one exchange rate needs to be derived, the model is now able to determine an additional variable beyond those obtained in the flexible rates case. We may solve the estimated system for either the absolute change in the demand for one union country's bond or the absolute changes incurred in the money supplies of both union countries. Changes in the demand for bonds can be interpreted as consequences of domestic policy stemming from changes in the level of public debt, while changes in the money supplies can be seen as changes in foreign exchange reserves resulting from intervention operations. We present the model determining the changes in the demand for one country's bonds in Table 2.

Table 2: Dynamic Model for the Exchange Rate Union Case

Domestic Prices

$$pF = 1.17\ y^F + w^F + 0.36\ p^F(-1) - 5.63$$

$$p^G = 1.12\ y^G + w^G + 0.19\ p^G(-1) - 4.18$$

Real Output

$$y^F = 0.13\ (p^G - p^F) + 0.03\ (p^u + x^G - p^F) + 0.12\ y^G + 0.02\ y^u + 0.24\ y^F(-1) + 2.71 + ud^F$$

$$y^G = 0.08\ (p^F - p^G) + 0.03\ (p^u + x^G - p^G) + 0.06\ y^F + 0.05\ y^u + 0.21\ y^G(-1) + 2.16 + ud^G$$

Nominal Interest Rates

$$r^F = -0.15\ (m^F - p^F - y^F) - 0.33\ r^F(-1) + 0.11\ y^F(-1) + 0.12 + um^F$$

$$r^G = -0.22\ (m^G - p^G - y^G) - 0.35\ r^G(-1) + 0.09\ y^G(-1) + 1.01 + um^G$$

Demand for Bonds of France and Dollar Exchange Rate of Germany

$$H_t^F - H_O^F = -1.14\ (p^F + y^F) - 0.72\ (p^G + y^G) + 1.63\ r^F - 0.11\ r^G - 1.21\ x^G - 0.19\ x^G(-1) + 68.31 + uh^F$$

$$x^G = x^F = -0.81\ (H_t^G - H_0^G) - 0.76\ (p^G + y^G) - 0.09\ (p^F + y^F) + 0.56\ r^G - 0.09\ r^F - 0.30\ x^G1(-1) - 0.67 + uh^G$$

and $H_t^G - H_0^G = -\dfrac{1}{X_0^{PG}}\ (H_t^F - H_0^F)$

Notes:
1. The endogenous variables of the model appear in the LHS of the equations. F stands for France, G for Germany, and U for the United States. 2. The y denotes output; p prices; w nominal wages; x the exchange rate (e.g., $x^F = F/\$$) ; m money demand; r interest rates; and H demand for home bonds. 3. The bilateral exchange rate is defined as $x^{PG} = x^F - x^G = 0$.

4. $H_t^F - H_0^F$) is the change in the demand for bonds of France. By construction, $H_0^F - H_0^F = -X_0^{FG}\ (H_t^G - H_0^G)$.

5. ud^i, um^i, uh^i for i = F, G represent real demand, monetary and bond market disturbances, respectively, for each of the countries examined. 6. The subscript 0 denotes stationary values.

In order to clarify some of the less obvious alterations in the structure of the exchange rate union simulation models, we ought to address the following point concerning the output block. The coefficient for the terms of trade with respect to the other union country variable is significantly different from that of the flexible exchange rates case for both countries. The reason is that the exchange rates involved in this term cancel out in an exchange rate union setting. This model was tested against the EMS era of 1979-90, with initial values those of 1978. The variables of the model are represented in the same manner as in the flexible period. The model was able to generate plausible simulation results for the entire simulation period.

III. Empirical Effects of Disturbances
Under Alternative Exchange Rate Regimes

The estimated models for the flexible exchange rates and exchange rate union cases are now used to determine the effects of various disturbances on specified variables. The criterion set for judging the superiority of one regime over the other is minimum variance from a control solution path for real output and the dollar exchange rate in each of the two countries. Specifically, we investigate the resulting movements of each of these variables separately under the two alternative exchange regimes. The attainment of fixed cross-exchange rates for the union currencies is to be considered of limited purpose as an adequate criterion since exchange rate stability in itself is assumed to be safeguarded by the formation of an exchange rate union. Nonetheless, it is always interesting to examine whether an exchange rate union is working in practice according to its set purpose. In regards to the second criterion, the European Community (EC) has explicitly set stability of production in the region[14] as one of the criteria for judging the efficacy of the union.

To examine the variability of these variables under the influence of each specific disturbance, we need to generate two control solutions--one for flexible rates and another for the exchange rate union case. We develop the control solutions for the period 1991-94. Each disturbance is depicted by a shock in the relevant variable of the system and its simulation effect is evaluated from the resulting size of the variance of the specified variables. In particular, variances are calculated for deviations of the simulated values from those obtained in the control solution, in the form of Mean Squared Deviations (MSD). The disturbances considered are of real, monetary, and bond market nature. Symmetric, in various types of correlation, as well as asymmetric[16] disturbances are analyzed. Real disturbances for France and Germany take the form of sustained adjustments in these countries' output variables. That is, in the case of an asymmetric positive real disturbance for France, we impose an arbitrary increase of 10 percent on its real output by specifying this change in the respective disturbance term.

Symmetric real disturbances between France and Germany, which are perfectly-positively correlated, are set up as increases in the outputs of both countries by the same (arbitrary) percentage. Simple positive correlation, however, allows for different percentage adjustments in each union country's output. Perfectly-negatively correlated disturbances are specified as diametrically opposite adjustments of the same magnitude in the two countries' real outputs. Simple negative correlation implies adjustments for France and Germany in the opposite direction and by different amounts. Finally, we implement real shocks that are of the same absolute magnitude for the two alternative regimes. Monetary disturbances are specified as certain (arbitrary) percentages of each country's money supply, while bond market disturbances are expressed as (arbitrary) changes in the countries' demand for bonds. The effects of the various disturbances, in terms of variances, on dollar exchange rates and real output of both countries, under flexible rates and in an exchange rate union setting, are presented in Table 3 and 4, respectively.

Although domestic demand disturbances in Germany generate moderate increases in the variance of its output when she joins a union, similar disturbances in France lead to decisively higher variations in the latter country's output when she abandons flexible rates in favor of an exchange rate union. However, the response of the other union country to domestic demand disturbances is almost triple variation in its output under flexible exchange rates. Therefore, the variance of the French output increases more under flexible rates when a demand disturbance raises aggregate demand in Germany only. Under flexible rates, a positive real demand disturbance in the German economy alone results in appreciation of the mark vis-á-vis the dollar and the franc that lead to an even higher output in France than that implied by the direct output-linkage between the two union countries. The channel of adjustment for France is through the net effect of its terms of trade changes with respect to the other union country and the rest of the world. In a union, however, the appreciation of the mark relative to the dollar influences French output only through that country's terms of trade with respect to the rest of the world.

Positively correlated real disturbances, that are either perfectly correlated or concentrated in France tend to produce higher variability for the output of Germany under a flexible exchange rate regime. However, Germany's output variation increases in a union when she experiences demand disturbances to a higher degree. The variance of output for France decreases under flexible exchange rates when real shocks are perfectly correlated or predominate in France, while it increases when they predominate in Germany. Thus, under such disturbances, Germany should object to forming an exchange rate union, while France should only favor it, when increases in real demand are stronger in Germany. When negatively correlated real disturbances take place, the variation in the German output is smaller under flexible exchange rates. There

Table 3: Variances of the Countries' Dollar Exchange Rates Under Flexible Rates and in an Exchange Rate Union[1]
(Sustained changes in various disturbances)

	Flexible Exchange Rates[2]	Exchange Rate Union
I. Real Aggregate Demand Disturbances		
1. Asymmetric disturbances		
a. Increase in the demand of Germany	0.85	0.32
b. Increase in the demand of France	2.17	0.19
c. Increase in the demand of the rest of the world (ROW) [3]	0.01	0.00
2. Symmetric disturbances		
a. Positively-correlated increases in the demands of Germany and France		
(1). Perfect correlation	2.69	0.43
(2). Stronger increase in Germany	8.11	1.62
(3). Stronger increase in France	11.05	1.31
b. Negatively-correlated changes in the demands of Germany and France		
(1). Perfect correlation	0.72	0.21
(2). Stronger increase in Germany [4]	0.47	0.36
(3). Stronger decrease in France	5.27	0.16
II. Monetary Disturbances		
1. Asymmetric disturbances		
a. Increase in the money supply of Germany	1.09	0.76
b. Increase in the money supply of France	2.31	0.53
2. Symmetric Ddsturbances		
a. Positively-correlated increases in the money supplies of Germany and France		
(1). Perfect correlation	3.12	0.71
(2). Stronger increase in Germany	3.40	1.92
(3). Stronger increase in France	11.72	1.09
b. Negatively-correlated changes in the money supplies of Germany and France		
(1). Perfect correlation	2.22	0.17
(2). Stronger increase in Germany	2.41	0.68
(3). Stronger decrease in France	9.07	0.09
III. Bond Market Disturbances		
1. Asymmetric disturbances		
a. Decrease in the demand for bonds of Germany	22.06	16.55
b. Decrease in the demand for bonds of France	72.27	1.32
2. Symmetric disturbances		
a. Positively-correlated decreases in the demands for bonds of Germany and France		
(1). Perfect correlation	36.35	4.90
(2). Stronger decrease in Germany	52.61	12.69
(3). Stronger decrease in France	116.82	7.15
b. Negatively-correlated changes in the demands for bonds of Germany and France		
(1). Perfect correlation	32.72	2.18
(2). Stronger decreases in Germany	39.62	7.01
(3). Stronger increases in France	94.72	0.69

Notes   1. Calculated as the average of the sum of the squared differences between the respective scenario and the baseline values

for the simulation period 1991-94. That is $V_i = \dfrac{\sum_{i=1}^{N} (S_{ij} - B_j)^2}{N} * 1000$ where V_i is the variance corresponding to scenario i, S_i and B are the scenario i and the baseline values (in levels), respectively, for the dollar exchange rate of each country and N is the number of simulation time periods--years.   2. For the flexible exchange rates case, the reported figures are averages of the variances for the dollar rates of the two currencies.   3. ROW is proxied by the United States in this analysis.

is, also, conclusive evidence that flexible rates tend to lower the variability of output for France, independently of whether these disturbances are perfectly correlated or not and of their distribution between the two countries. In this case, flexible exchange rates outperform an exchange rate union setting for both countries. Finally, since the variance of either country's output is uniformly increased in a union setting for any type of symmetric real disturbances, except for positive-correlated shocks predominating in the other union member, countries appear to be equally susceptible to real disturbances.

Monetary disturbances, in the form of money supply increases, occurring unilaterally or simultaneously in Germany and France result in an almost universal reduction of output variability for Germany when the two countries move from flexible rates to an exchange rate union. The only exception being for the case of positively correlated and more vigorous French disturbances. When asymmetric monetary disturbances originate in France or Germany alone, a higher output variability emerges for France under flexible rates. Positively-correlated disturbances that are either perfect in nature or intensified in Germany produce larger variances for the output of France in a union. However, such disturbances intensified in France result in reduced variability for its output when the two countries participate in an exchange rate union. Any type of negatively correlated disturbances between the union countries generates consistently increased variances for the output of France under a flexible exchange rate system. Hence, when monetary disturbances prevail, our simulations suggest that a strong case can be made in favor of a union for both France and Germany. A country should only prefer flexible rates over an exchange rate union on the occasion of positively correlated disturbances that have a higher intensity in the other member state. In addition, France benefits from floating rates when monetary disturbances are perfectly positively correlated.

Most bond market disturbances undertaken in this study result in favor of an exchange rate union setting, as union countries are able to pool financial shocks so as to reduce their output variability. In the presence of asymmetric disturbances, such as those arising from portfolio shifts between mark and dollar bonds, the union is found to be desirable for both France and Germany. As demand for German bonds falls, while that for French bonds remains unchanged, the variability of output in Germany increases less in a union than under flexible exchange rates. Additionally, the variance of France's output is found to be reduced in a union settings. Furthermore, if similar disturbances involve speculation between franc and dollar bonds, the output variability of both countries is reduced in an exchange rate union. Further results indicate that perfectly positively-correlated financial disturbances between France and Germany (involving only bonds) tend to produce larger variances of output for both countries under flexible rates than in a union. When positive symmetric financial disturbances take place, the country with the stronger bond market

Table 4. Variances of the Countries' Real Outputs Under Flexible Rates and In an Exchange Rate Union[1]

	Flexible Exchange Rates		Exchange Rate Union	
Sustained Changes in Various Disturbances	France	Germany	France	Germany
I. Real Aggregate Demand Disturbances				
1. Asymmetric disturbances				
a. Increase in the demand of Germany	1.16	13.29	0.41	14.26
b. Increase in the demand of France	8.14	0.19	11.33	0.11
c. Increase in the demand of the ROW [2]	0.02	0.03	0.00	0.01
2. Symmetric disturbances				
a. Pos.-corr. increases in the demands of Germany and France				
(1). Perfect correlation	19.43	13.92	23.72	12.69
(2). Stronger increase in Germany	27.64	39.47	22.16	47.65
(3). Stronger increase in France	46.55	17.21	51.32	12.22
b. Negatively-correlated change in the demands of Germany and France				
(1). Perfect correlation	7.16	6.94	14.17	9.26
(2). Stronger increase in Germany	4.01	23.07	7.92	29.32
(3). Stronger decreases in France	32.43	5.12	42.01	8.16
II. Monetary Disturbances				
1. Asymmetric disturbances				
a. Increase in the money supply of Germany	0.72	1.26	0.35	0.41
b. Increase in the money supply of France	1.14	0.12	0.08	0.07
2. Symmetric disturbances				
a. Positively-correlated increases in the money supplies of Germany and France				
(1). Perfect correlation	0.22	0.37	0.29	0.31
(2). Stronger increase in Germany	0.05	1.69	0.71	0.89
(3). Stronger increase in France	1.17	0.19	0.36	0.42
b. Negatively-correlated changes in the money supplies of Germany and France				
(1). Perfect correlation	2.17	1.12	0.04	0.05
(2). Stronger increase in Germany	5.14	5.21	0.22	0.31
(3). Stronger decrease in France	6.18	1.63	0.03	0.02
III. Bond Market Disturbances				
1. Asymmetric disturbances				
a. Decrease in the demand for bonds of Germany	11.30	25.75	5.77	6.29
b Decrease in the demand for bonds of France	47.23	3.18	1.46	2.43
2. Symmetric disturbances				
a. Pos.-corr. dec. in the demands for bonds of Germany and France				
(1). Perfect correlation	12.33	23.61	9.17	14.07
(2). Stronger decrease in Germany	2.12	39.42	22.17	28.35
(3). Stronger decrease in France	46.52	8.12	10.03	17.83
b. Neg.-corr. changes in the demands for bonds of Germany and France				
(1). Perfect correlation	44.16	29.52	6.72	8.60
(2) Stronger decreases in Germany	71.17	62.01	13.91	15.12
(3). Stronger increases in France	96.04	19.26	5.60	6.12

Notes   [1] Calculated as the average of the squared differences between the respective scenario and the baseline values for the simulation period 1991-94 . That is, $V_t = \dfrac{\sum_{j=1}^{N}(S_{ij}-B_j)^2}{N}$ where V_i is the variance corresponding to scenario i, Si and B are the scenario i and the baseline values (in levels), respectively, for real output of each country and N is the number of the simulation time periods--years. [2] ROW is proxied by the United States in this analysis variations experiences a smaller output variability, while the other union country experiences a larger output variability, in a union. Our findings show that if one country is more susceptible to financial disturbances than the other, the union may benefit the former but hurt the latter. Finally, negatively-correlated financial disturbances of any type tend to make France and Germany better off in a union.

IV. Concluding Remarks

The advantages of each exchange rate regime are evaluated in terms of short-run macroeconomic performance. The degree of variability of dollar exchange rates and real output, relative to their baseline simulation-trends, stemming from disturbances in the goods or asset markets is used as the criterion for choosing between flexible rates and an exchange rate union. That is, the ability of an exchange rate regime to operate as an automatic stabilizer for the economy is utilized as the standard for judging each exchange rate regime.

Our findings indicate that the volatility of the two countries' dollar exchange rates is reduced in a union, regardless of the type of disturbance. The diversification effect according to which the variability of the dollar exchange rates for the countries forming a union is reduced when financial disturbances are present, is clearly evident in our variance calculations. Real disturbances occurring in Germany tend to generate smaller variations in the output of Germany under floating rates. The opposite result holds true when the main thrust of real disturbances takes place in France. However, disturbances concentrating in France produce a larger variability in its output when the two countries participate in an exchange rate union. In contrast, real demand disturbances occurring primarily in Germany lead to a larger variability in the output of France under flexible rates. Thus, asymmetric real disturbances happening in Germany favor an exchange rate union, while if such disturbances appear in France flexible exchange rates are more beneficial for France. Finally, real disturbances originating in the rest of the world (United States) produce almost identical variations for the outputs of both countries under the exchange rate regimes considered.

Overall, our results show that an exchange rate union is generally preferable to flexible rates, if the average variation of the dollar exchange rates of the participating currencies is set as the criterion. In addition, if output variability for the two countries is used as the standard of judgment, our quantitative findings show that an exchange rate union is also superior to flexible rates in the presence of bond market disturbances. The latter conclusion can further be advocated by our calculations in almost all instances of monetary disturbances (except in the case of positively correlated disturbances with larger variations occurring in the other union country).

Finally, when real shocks prevail in macroeconomic activity fluctuations, evidence is offered that flexible rates seem to be the optimum exchange regime for both countries.

Michael Papaioannou, The Bank of Greece, Athens, Greece.

NOTES

1. "Pseudo exchange rate union" is referred as the arrangement in which each county retains its own central bank and has control over its national monetary policy. In contrast, a "complete exchange rate union," or "monetary union," features a single central bank and a union-wide currency. Corden, W., (1972), p. 3.

2. Marston, R., (1984) and (1985).

3. The setup is a simplistic version of the EMS, i.e., only two participants, with the rest of the world proxied by U.S.A.

4. That is, the initial levels of the corresponding variables in the two union countries are assumed to be identical, as well as the structures in each member country.

5. Stationary values are denoted by a zero subscript.

6. The aggregate supply equation is derived from a contract model of wage determination, which is based on a constant-returns Cobb-Douglas production function, inelastic labor supply and a simple indexation scheme.

7. They are derived by equating the relative prices in parentheses to zero.

8. In levels, money demand is assumed to be homogenous of degree zero with respect to wealth. Here, we present the linearized (around stationary values) version of the demand for money function.

9. Wealth enters the demand for bonds equation in levels with homogeneity of degree zero. The demand for bonds functions presented in the text are linear approximations (around stationary values) of the union countries' asset demands.

10. These assumptions mostly suffice for all b's to be positive.

11. This assertion can be easily verified by relations (I.6) and (I.7) of the text.

12. Klein, L.R., (1971), pp. 66-70.

13. The WEFA Group, World Economic Outlook, Vol. I, October 1990, PA, 19004. Alternatively, we could have employed equilibrium, trend or planned levels for the stationary values.

14. European Communities, (1973), p. 179.

15. Symmetric versus asymmetric, in the sense of taking place in both countries instead of only one.

REFERENCES

Allen, P. and P. Kenen. (1980). *Asset Markets Exchange Rates and Economic Integration.* Cambridge: Cambridge University Press.

Artis, M. and Gazioglu S. (1989). Modeling asymmetric exchange rate unions: A stylized model of the EMS. *Greek Economic Review. 11*(June), 177-202..

Bilson, J. F. O. (1985). Macroeconomic stability and flexible exchange rates. *American Economic Review. 75*(May), 62-67.

Blanchard O. and Quah, D. (1989). The dynamic effects of aggregate demand and supply disturbances. *American Economic Review.* 79(September), 655-673.

Branson, W. and Henderson, D. (1985). The specification and influence of asset markets. In R. Jones and P. Kenen (Eds.), *Handbook of International Economics.* Vol. II, Elsevier Science Publishers B.V., 749-805..

Canzoneri, M. (1982). Exchange intervention policy in a multiple country world. *Journal of International Economics.* 13(June), 267-289.

Cesarano, F. (1985). On the viability of monetary unions. *Journal of International Economics.* 19(November), 367-374.

Corden, W. (1972). Monetary integration. *Essays in International Finance*, No. 93, Princeton, N.J. Princeton University, International Finance Section.

Enders, W., and Lapan, H. (1979). Stability, random disturbances and the exchange rate regime. *Southern Economic Journal*, 46(July), 49-70.

European Communities. (1973). Treaties Establishing the European Communities; Treaties Amending these Treaties; Documents Concerning the Accession, Luxembourg.

European Communities Commission. (1979). European Economy, No. 3, July.

European Communities Commission. (1990). Exchange rate regimes in the EC: Simulations with the multimod model. *European Economy.* No. 44, (October), 303-334.

Fisher, S. (1977). Stability and exchange rate systems in a monetarist model of the balance of payments. In R. Aliber (Ed.), *The Political Economy of Monetary Reform.* (pp. 59-73). Montclair, N.J.: Allanheld, Osmun and Co.

Giovannini, A. (1988). How do fixed exchange rate regimes work: The evidence from the gold standard. Bretton Woods and the EMS. CEPR Discussion Paper. No. 282, (October).

Gray, J. (1976). Wage indexation: A macroeconomic approach. *Journal of Monetary Economics.* 2(April), 221-235.

Henderson, D. (1977). Modelling the interdependence of national money and capital markets. *American Economic Review.* 67(February), 190-199.

Klein, L. R. (1971). *An Essay on the Theory of Economic Prediction.* Chicago, Markham Publishing Company.

Marston, R. (1984). Exchange rate unions as an alternative to flexible rates: The effects of real and monetary disturbances. In J. Bilson and R. Marston (Eds.), *Exchange Rate Theory and Practice.* (pp. 407-442). Cambridge, The National Bureau of Economic Research.

_____. (1985). Financial disturbances and the effects of an exchange rate union. In J. Bhandari (Ed.), *Exchange Rate Management under Uncertainty.* (pp. 272-291). Cambridge: The MIT Press.

McKinnon, R. (1963). Optimum currency areas. *American Economic Review*, 53(September), 717-725.

Melvin, M. (1985). The choice of an exchange rate system and macroeconomic stability. *Journal of Money, Credit and Banking.* 17(November-Part 1), 467-478.

Mundell, R. (1961). A theory of optimum currency areas. *American Economic Review.* 51(September), 657-665.

Papaioannou, M. (1986). Some aspects of world financial interdependence. Ph.D. Dissertation, University of Pennsyvania.

van der Ploeg, F. (1989). Monetary interdependence under alternative exchange rate regimes: A European perspective. CEPR Discussion Paper No. 358 (November).

Tower, E., and Willett, T. (1976). The theory of optimum currency areas and exchange rate flexibility. *Special Papers in International Economics,* No. 11, Princeton, N.J., Princeton University, International Finance Section.

Vaubel, R. (1978). Real exchange-rate changes in the European community: A new approach to the determination of optimum currency areas. *Journal of International Economics.* 8(May), 319-339.

Japan's Macroeconomic Policy and
Japan-US Economic Relations During the Eighties

Chikashi Moriguchi

In my paper that was contributed to the 1980 Festschrift in honor of Professor Klein I discussed Japan's macroeconomic policies during the seventies, the present paper is in a sense a sequel to it, but not only giving a quantitative assessment of Japan's macroeconomic policy, but also I should like to deal with changed conditions of macroeconomic policy-making during the eighties which profoundly affected Japan-US economic relations. A keyword is "economic structural changes".

I. 1970's --- Macroeconomic policies in fluctuating exchange rate

Project LINK World Model started in the early seventies. L. R. Klein and A. van Peeterssen built a small LINK world model in 1971 with a simplified linkage model called "fixed trade coefficients scheme". The present author took part in this project during the Summer of 1971 and took over the job from van Peeterssen to continue with expanding the number of participating countries and improving the linkage system.

The year 1971 was a memorable one particularly for the author, an economist from Japan. Just when he was going to leave Japan, President Nixon announced his New Economic Policies, abandoning the exchangeability between the dollar and gold, letting the dollar float against major key currencies, and imposing a 10 percent import surcharge to promote currency realignment. The move by Mr. Nixon was a deliberately dramatic one together with his approach to the Chinese government, and it gave a sensation to the Japanese public ("Nikuson Shokku").

The LINK project was organized by the economic stabilization subcommittee of the Social Science Research Council. To my understanding, it was started as a U.S. initiative in order to understand and explain the international transmission mechanism of economic fluctuation under the Bretton Woods system. However, as soon as the project started we stepped into a new unexpected regime of fluctuating exchange rates.

After the world-wide recession of 1974-75, major industrial countries agreed upon holding an Economic Summit in order to coordinate macroeconomic policies among the G7. The so-called "locomotive thesis" came to attention. Towards the end of 1975 the U.S. was moving to an expansionary policy to avoid a further deterioration of labor market situation. It was a presidential election year. As the U.S. started to recover from the negative growth of two consecutive years, its import began to surge. Japanese and

German trade surplases became conspicuous by 1976.

As early as late 1976, some economists began to criticize that the Japanese monetary authority was deliberately maintaining an undervalued yen rate ("dirty float"). But intervention by the monetary authority had already been proven to be ineffective. The exchange market was simply slow in realizing that the fundamental factors were changing in favor of Japan.

However, from today's point of view, the magnitude of trade imbalance was not large; Japan's current account surplus was less than two percent of its GNP in 1977 and the U.S. deficit was smaller than one percent. At that time there seemed a prevailing view that foreign exchange rate change should be able to realign trade imbalance, and policy authorities seemed overly concerned with relatively small movements of external imbalances.

Professor Klein's testimony at Capitol Hill on the undervaluation of the yen (1977 February) was one of the series of incidences that corrected the market view.

The U.S. demanded that Japan and Germany act as locomotive countries at the 1977 London Summit. Germany was more concerned with stabilizing inflation, and could not find anything to be done for demand expansion. Japan was, perhaps, the only country that could respond positively to the request of the U.S. government. As the most senior member of G7 leaders who vividly remembered the days of world-wide depression in the 1930's, Mr. Fukuda, then Prime Minister of Japan, positively responded to the agreement of the Summit. From 1977 to 1978, Japan's expansion of public expenditure (mainly public fixed capital formation) was enormous. As the Finance Minister, Mr. Fukuda had already opened a way to finance the budget deficit by bond issuance as early as 1965. The diagram below indicates a sudden jump of government capital formation as a percentage ratio of nominal GNP. We will later touch upon the behavior of this ratio and the explosion of government debt outstandings in the eighties.

Japan had its own reason why Mr. Fukuda had to launch a massive expansion policy; a rapidly appreciating yen was dampening export industries and its deflationary impact was propagating. The J-curve effect was really remarkable and a seemingly vicious cycle of the appreciating yen and a growing trade surplus was observed. Appropriate size and timing of macroeconomic policy was a big question.

During this period, the Project LINK people at the LINK Central(the University of Pennsylvania) were engaged in simulation analysis on the effect of "locomotive scenario" or of the scenario of "massive capital transfer to non-oil-producing developing countries". A serious concern then was poor macroeconomic performance of non-oil-producing developing economies. In order to help them improve their performance, there was an agreement among G7 member countries that recycling of oil money and coordinated economic growth by industrial countries were necessary.

After energy prices had risen, the aggregate supply schedules of major industrial countries seem to have shifted downward; a simple demand management "as usual" had a higher risk of kindling inflationary expectations. However, most of the national econometric models then did not seem to be able to deal with the problem of "supply shift" and failed to grasp a trend of accumulating inflationary expectation that led to a constant upward drift of Philips-relations. Frankly speaking, the LINK model was no exception.

Demand expansion in the U.S. from 1977 to 1979 led to an accelerated inflation in the period of Carter's administration that culminated at the fuel shortage in the Summer of 1979; the U.S. inflation rate hit a two-digit range. The macroeconomic management at the time of rapidly changing relative prices lost control and the "locomotive thesis" was stranded in the desert of world-wide inflation.

Raging inflationary expectations led to the world-wide disinflation policy at the turn of the decade, led by the two new administrations in the U.K. and the U.S.. Monetarist policy armoured by the rational expectation approach found supporters among a wide range of policy makers including the people in the Finance Ministry of Japan who regretfully watched the accumulating public debt, a legacy of the locomotive thesis.

II. 1980s --- Mismatched timing in macroeconomic policies

In the first half of the eighties, individual countries seemed to be engaged in the pursuit of domestic policy targets, leaving external balances to at the working of exchange rate fluctuation. In the U.S., the Reagan administration concentrated on a disinflation policy (including jaw-boning of labor unions). In Japan "reconstruction of balanced budget" and administrative reform became a priority one public policy goal. Mr. Nakasone had been in charge of administrative reform under the previous Suzuki cabinet administration and he caught the world-wide wave of deregulation and privatization of state enterprise (a strong neoconservatism message from Mrs. Thatcher and Mr. Reagan) and rode on it through his five years in power (1982-87).

After the second oil shock the Japanese economy showed a renewed resilience; its growth rate dropped to the three percent range for the first half of the eighties under the deflationary impact of stringent fiscal policy. Government total spendings dropped from 19.6% (1978) of GNP to 16.1% by 1985. The unemployment rate rose close to the historical high of three percent, and business efforts for streamlining was intensified. Under rising engery costs, transforming the Japanese economy to a lean hi-tech oriented one with a small government became a national policy goal. Three percent growth became a national consensus view, and the social cost of industrial structural changes (potentially high unemployment, and retraining of employees etc.) was borne by private firms.

Then there came an expansion of the U.S. economy starting in 1984. Large-sized tax cuts brought out a rapid economic recovery, and U.S. imports of consumer durable goods exploded, helped by the strong dollar that was a product of tight-money/easy-fiscal policies. In addition, during the recession period the U.S. manufacturing industry had accelerated its efforts of reducing productive capacities at home and moved productive operations abroad. Japan was not an exception in booming exports to the U.S. market; Asia's newly industrialized countries and EC member countries also saw a similar export boom.

When a synchronized world-wide recession (then called Stagflation) occurred in 1974-75, there was a marked difference among G7 member countries with respect to timing to switch to a demand stimulus; the U.S., due to its domestic political situations, was the first country to go ahead with an expansionary policy that led to a sharp increase in the trade deficit. At that time, in 1984, a presidential election year, the U.S. went ahead again.

It was a repetition of a time lag, or mismatching of timing in policy coordination between the two countries. Japan, being free from a political business cycle of the U.S. style, tends to lag in switching economic policy from a stringent to an expansionary one, and to bring about a surge of trade surplus due to a preceding U.S. import expansion. It took place in 1975 and again in 1983 when incumbent U.S. Presidents switched to stimulus policies to lower unemployment. In 1987 (particularly after the stock market crash on October 19 known as Black Monday) the Republican administration, with the intention of transferring power from Mr. Reagan to Mr. Bush, was deadly in need of a continued low interest policy. In 1988 the Bank of Japan was already alerted by the warning signal from exploding land prices, but a tighter monetary policy was to be postponed until 1989 under pressure from the U.S.. The political cycle in the U.S. seems to have had an international repercussion in the age of Japan-U.S. policy coordination.

Japan's Response to the Trade Surplus: 1986-90

An accumulating balance of payments imbalance became clear by 1985. The currency realignment since the 1985 Plaza Agreement dealt a heavy blow to Japan's export industries. Nakasone proclaimed an economic policy package that tried to boost domestic demand. The pillars of his policy package were 1) expansionary fiscal/monetary policies that were aimed at depressed export industries; 2) deregulation policies that aimed at enhancing demand for imported goods and removing various institutional measures for encouraging personal saving; and 3) implementation of shorter working hours in order to create more leisure time for workers. In 1987 the government of Japan further made it clear that it would pursue a demand switching policy on the basis of economic structural adjustments and tax reform.

Actually, the expansion of public spending was not as large as it was initially planned; it turned out that the recovery was so strong that the fiscal authority actually suspended some of the planned expenditure in fiscal 1988 (see Figure 1). Under an historical low interest rate and a booming stock market which enabled a massive equity financing for corporate firms, capital costs came down as low as 2% or below. Housing investment and consumer spending expanded starting in 1987, and then business fixed investment recorded a two digit expansion for the three successive years that began in 1988.

Figure 1: Japan's Current Account Balance as Percentage Ratio of GDP

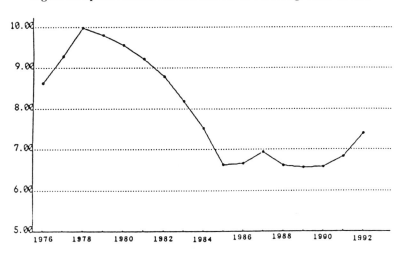

After all, the demand switching policy of Japan was successful not because of the success of economic structural adjustments, but because of booming stock and real estate markets, mainly a result of historical cheap money policies.

The Big Bubble and Real Demand

Housing investment became strong when land prices in Tokyo began to rise in 1984: first new housing starts were brought about by those who sold some land/houses in downtown Tokyo and planned to move out to the nearest suburbs. Then there came a boom of condominium construction, and land price increases spread to other major cities. Land price inflation and a stock market boom continued through 1989. Hugh capital gains took place in the following order of magnitude. In the three year period of 1986-89, the total savings of

106 trillion yen was made by total private households, while the increase in the total value of their assets (financial assets plus real estate) was 700 trillion!

The financial basis of corporate firms was strengthened for the same reasons, and supported by upwardly revised growth expectations. Business fixed investment reached nearly twenty percent of GDP. The ratio was 16% in 1985. Housing investment reached a historical high in 1973. Housing starts jumped to the level of 1.7 million units from a stagnant 1.3 million in 1985.

The public sector was no exception in enjoying capital gains from the bubble. When there began a demand switching policy, it was amidst a vicious circle of low growth, low tax revenue and even more spending cuts to balance the budget. The Ministry of Finance was quite reluctant to switch to an expansionary policy at the beginning. But what actually followed was more than a joy for MOF since the beginning of strong growth under asset inflation produced a sudden increase in tax revenue. In addition, the privatization of NTT brought in a large amount of cash through its stock sales to the public. The public sector gained most from the financial bubble!

During the period of 1986-1990 in spite of an extraordinary monetary expansion, the inflation rate did not accelerate at all. A high yen rate and freer imports have absorbed most inflationary pressures. The price of goods and services was stable, and the price of real estate which has limited flexibility of supply under the valuing taxation scheme simply absorbed speculative influxes of money.

Did that large amount of capital gains from financial and land assets positively affect consumer spending? The answer is yes, and it seems to have occurred in non-workers' households that had relatively large amounts of these assets. An aggregate consumption function was estimated using household net assets as one of explanatory variables. The results was both a positive and significant parameter. According to the estimate, 600 trillion yen capital gains should have an additional 24 trillion yen impacts on the consumption expenditure for the long run.[1] Strong consumer spending in the years 1987 and 1988 are usually associated with the capital gain effect. And various other data indicate that spending on luxury goods as well as fine art and jewelry was remarkable in this period. The following consumer spending expansion was brought about by the multiplier impacts of general overall economic expansion.

Now, after the stock market collapse at the start of 1991, almost a one hundred trillion yen capital loss must have taken place in the portfolios of households (total loss was somewhat three hundred). If the asset effect should work symmetrically, it will have a negative impact of 1.7 trillion yen. Almost one per cent of current total consumption spending will have to be eliminated.

Quantitative Analysis of Trade Sector

A simulation study was conducted on the basis of a small

macroeconometric model. In order to pick up some effects caused by structural changes in income- and price- elasticities of imports and exports, two models were built from two different sets of observations. Model 86 utilizes observations from 1976 to 1986 (all quarterly data), and Model 90 is built on the observations from 1977 to 1990.

Some of the significant differences in estimated elasticities in import and export equations are as follows:

1)   income elasticity of Japan's export became significantly smaller.

2)   price elasticity of Japan's export also came down.

3)   quantitative impact of yen rate appreciation on export volume is larger, and domestic demand in the private sector (personal consumption and business fixed investment) are affected significantly larger in Model 86 than in Model 90.

4)   price elasticity of Japan's manufactured goods imports increased significantly in the recent observation period, while income elasticity remained almost at the same level.

5)   overall impact of appreciated yen on the trade balance, BLTR@, or on current account BLCRNT% (as a percentage ratio of nominal GDP) is larger in Model 86 than in 90. As a matter of fact, in Model 90 the external (in)balance of Japan seems to be rather insensitive to yen rate appreciation.

The above-mentioned points are now almost stylized facts concerning the economic performance of Japan. Underlying factors in these changes are identified as the followings:

1)   Major categories of today's international trade are somewhat controlled by bilateral agreements in various forms. Between Japan and the U.S. steel, automobile, and machine tool industries are three major item of trade which are under the "Voluntary Export Restraint." Textile exports are under control through a multilateral scheme of MFA. For an effective export control there must be a close cooperation between the exporter's cartel and a government agency that supervises operation. That is a department or bureau of MITI. The exporter's cartel distributes a given exportable quantity among competitive exporters on the basis of the past record of exports. The number of units of automobiles is allocated to individual producers. Producers and exporters are free to choose the make and set the price of cars. Under

this scheme it is natural for exporters to shift towards exporting cars with larger value-added features. Unit prices of exports tend to rise, and hence the total value of exports from Japan also rises.

2)   Major Japanese exporters have successfully established brand images of their products and subsequently are practicing product differentiation policies. This also contributes to lowering price elasticity of demand for Japanese exports.

Table 1. Elasticities of Trade Equations			
	World Trade	Relative Price	Observations
Export 1	3.066	-2.057	1977.1 - 86.4
	(6.13)	(2.18)	
Export 2	0.9878	-1.0845	1985.1 - 92.3
	(16.6)	(3.85)	
	DOMEST.PRIV.DEMAND	Relative Price	Observations
Import 1	2.322	-0.5061	1977.1 - 86.4
	(19.5)	(1.80)	
Import 2	2.234	-0.8271	1987.1 - 92.3
	(14.1)	(2.65)	

Notes:
1) Relative price in export eq. is Japan's exp.prc/ world mfg.exp.prc.
2) Relat.prc in import eq. is mfg.import prc over domestic mfg. WPI.
3) Price elasticities are long-run elasticities estimated from Shiller's estimation method.

3)   Japan's growing overseas direct investments are accompanied by exports of capital goods and then of parts. These exports tend to form an autonomous portion of total exports that are more or less independent of income or price changes in the U.S. and South-East Asia.

4)   Consumers' access to imported manufactured goods has been improved by the government's action for opening of Japan's import market, and retail prices of imported goods are becoming more responsive to changes in exchange rates.

Table 1 shows a summary result of simulation tests of yen rate appreciation on the two models. Figures of trade balance and current accounts as percentage ratios of GDP measure discrepancies of higher yen rate cases (yen rate is fixed at 160 to the dollar) from a control case in which the rate is fixed at 180 throughout the simulation period of 1986Q1 to 1990Q4. For the export quantity, discrepancies are measured by %rate from the control case.

Contrasts between the results obtained from Model 86 and the one from Model 90 are clear. Export quantity responds to exchange rate changes by a larger magnitude in Model 86 than in Model 90. Even though imports respond to price changes more flexibly in Model 90 and negative income effects are much smaller in Model 90, the resulting changes in the balance of trade or current account balances are much larger in Model 86 than in Model 90.

A few words should be added to the 1989 Tax Reform which introduced a new 3% indirect tax. It replaced many of the old excise taxes that were levied on luxury goods, particularly on imported luxuries. The reform brought down the relative price of large-sized cars (domestic or imported) that boosted demand for durable goods of higher rank, together with large capital gains from stock and real estate markets.

Table 2. Impact of Yen Appreciation					
Model 86					
	1986	1987	1988	1989	1990
Trade Bl.	3836	-3719	-17511	-25014	-30042
Cur. Ac. as % GDP	-0.077	-0.154	-0.296	-0.366	-0.402
Export Q.	-2.66	-8.82	-17.04	-20.1	-20.59
MFG IMP.Q.	1.348	-0.67	-9.32	-13.63	-14.59
Model 90					
Trade Bl.	7430	8012	548	-3410	-4597
Cur. Ac as % GDP	-0.028	-0.014	-0.084	-0.122	-0.135
Export Q.	0.32	-2.14	-7.3	-9.86	-10.52
MFG IMP.Q.	3.242	3.638	-1.33	-4.626	-5.656

Reverse Trend Since 1990

The current account surplus showed a dramatic decline in 1990. In the service (or invisible) account, Japan's booming overseas tourism contributed to enhance its deficit. As a percentage of GDP the current account surplus came down below 1% by the end of 1990.

However, the change was short-lived. With the end of Bubble years, consumer spending became weak; consumer durable goods came to a saturation point and business fixed investment turned to decline after massive competitive behavior. The recession brought about a large cyclical downswing of import demand. In addition, the yen/dollar rate had peaked out in 1989. From 1990 to 1992 manufactured goods imports declined by nearly ten percent.

On the other hand, Japan's exports maintained a fairly stable growth rate of five percent. This was not because of the old "export drive" in recessionary times. Now Japan's exports are dominated by "voluntary restraint"; dumping practices are being closely watched. Exchange rate appreciation is more readily passed over to its dollar-denominated export price. There is no single major factor that can explain the trend. There should be at least two. First, the international division of labor among Japanese and American manufacturers (we should add Asian NIES and ASEAN countries also) are very closely and meticulously set up. The "autonomous portion" of Japan's export keeps increasing. Secondly, high growth in South-East Asian countries and reviving growth in the People's Republic of China contribute significantly to maintaining the present trend.

Thus, after a seven-year cycle that started in 1985, we have returned to a reviving situation of rising trade surpluses and a rising yen.

III. On the Trend and Cycle of Japan's Trade Surplus

Japan's balance of payments situation has shown marked characteristics since the 1970's. First, it is clearly observed that Japan's cyclical fluctuation results in a domestic economic slump that tends to produce a surge in its trade surplus with other nations. Second, there seems, nevertheless, to be a consistent upward trend of larger trade surpluses that is independent from cyclical fluctuations of the economy (see Figure 2). This trend is being discussed in relation to several factors among which the followings are worthwhile examination:

(1)   Japan's high saving ratio (or excess savings)

(2)   The so-called closedness of Japan's domestic market

(3)   A increasing trend toward managed trade that is heavily dependent upon the "Voluntary Export Restraint"(VER)

(4)   Establishment of international division of labor among major manufacturing industries of Japan and the U.S.

(1)   High Saving Thesis

The high saving ratio of Japanese households is widely known. A huge supply of personal savings goes to the banking sector that is funneled to private corporate firms for business fixed investment. When the private capital formation is highly active, it absorbs the major part of savings from households. However, excess supply has become conspicuous during the eighties from the following three factors:

i)   corporate financing became more inclined to equity-finance in domestic as well as in overseas financial markets. Particularly, equity-issuing in the overseas markets became dominant because of world-wide low interest and the general expectation for further appreciation of the yen;

ii)   prices of capital goods came down rather sharply as a result of the development of microelectronics technology. Even though corporate investment in P&E was strong, demand for funds for investment did not increase much in the eighties.

iii   the government sector became a net lender in the eighties.

"Rebalancing the public finance" was a high priority policy goal (perhaps the highest priority); the public spending was suppressed during the first half of the eighties. The General Account (current account of the central government budget) still runs deficits, but the overall account of the governments (central, local, and social security funds) now runs surpluses.

Figure 2 :Japan's Current Account Balance as percentage of Ratio of GDP

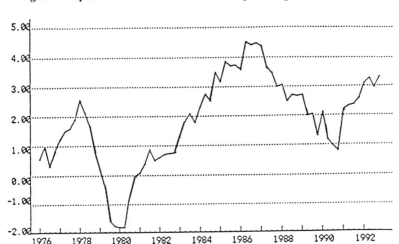

Thus, the added effects of the three above-mentioned factors are enormous. Unless the private capital formation recaptures the strength of 1987-1990, or the government sector begins to enlarge bond-financed spending (e.g. the social security fund purchases these bonds rr, American style), the domestic excess supply of savings has to be absorbed by the overseas borrowers. Actually the Japanese Government has been cooperating with the U.S. by letting some of the government agencies purchase U.S. Bonds (social security funds and public insurance bodies such as Kampo).

Many economists tend to argue that this saving-investment situation in Japan is a natural development (private households as well as social security funds are preparing for the aging of the society!) and that there is nothing wrong (or anything for Japan to be defensive about) in today's situation of fluctuating exchange rates and free capital movements. In an extreme, some monetarism-oriented economists argue that the long-run equilibrium path of the economy where the yen rate equals the PPP rate (that is about 170 yen for the dollar) produces a large net export surplus and Japan should be exporting capital to the rest of the world. In a stationary (or text-book steady-state) economy, this argument is simple and right. It also helps to explain a symmetric situation in the U.S. where the household saving rate is the lowest among OECD member countries, the government sector is running a large deficit resulting in running a huge trade deficit.

The problem is that neither Japan nor the U.S. nor any of the E.C. member countries is not in a long-run path of equilibrium path. Japan's growth rate at present is around an annual rate of one percent. Industrial production is at the bottom and the level of manufactured goods imports is eight percent below the peak level of 1990. What would have been the balance of payments situation of Japan if a high growth were maintained since 1989 and yen rate had been maintained at the rate of 125 for the dollar since the beginning of 1989? Various studies including simulation studies on macro econometric models unanimously indicate that the trade balance of Japan would have been 30 billion dollars smaller than the actual level of today. If oil price is five dollars higher than the present rate (23 dollars per barrel instead of 18), Japan's trade balance would have been 25 billion dollars less. These cyclical factors can be substantial and can explain nearly a half of today's Japanese trade surplus. The remaining half should be attributed to the high-saving thesis.

Figure 3: Relative Price of Capital Goods in Japan and the U.S.A.

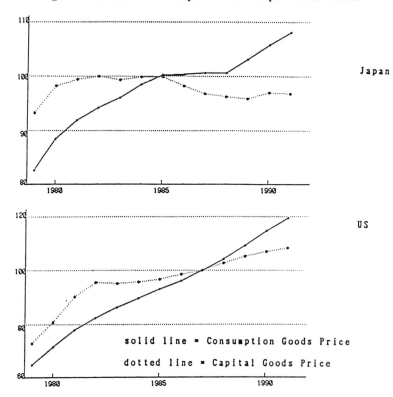

The high-saving thesis has a weak point though. Future changes in Japan's socio-economic structure that has been favorable to producers are going to be accelerated by the new administration led by Mr. Hosokawa. If some of the notorious external-internal price differentials are removed, it will certainly affect the saving behavior of households. Tax reform is half-done, and there still remains a room for fare taxation for self-employed and proprietors whose tax burden has been low and saving rate high.

2) Import Market

During the second half of the eighties, Japan's manufactured goods imports tripled in dollar values, from 41 billion dollars (in 1985) to 120 billion (1990). This cannot be explained by the appreciation of the yen or demand stimulus. A simple simulation study shows that only a 50% increase from 1985 level can be brought about by the model (20% from yen rate change and 30% from a rise of economic activity. The remaining should be attributed to the following:

Table 3: Saving-Investment Balance of Japan				
year	1980	1985	1988	1990
Household	22.2	31.4	31.42	38.71
	9.1%	9.8%	8.5%	9.1%
Corporate	-13.9	-18.1	-25.2	-45.1
	-5.7%	-5.6%	-6.8%	-10.6%
Public	-9.9	-2.6	5.59	12.62
	-4.0%	-0.8%	1.5%	3.0%
Overseas	1.6	-12.1	-11.81	-6.23
	0.7%	-3.8%	-3.2%	-1.5%

NOTE: 1) Figures in the uppercase are in trillion yen. 2) Lower case shows ratios over GNP.

i)   tax reform by which the domestic retail price of imported manufactured goods became significantly inexpensive (typical examples are imported automobiles and liquor);

ii)   changes in non-tariff barriers. Imported goods became more visible to consumers.

iii) changes in corporate policy towards utilization of more foreign products. These factors are closely related to the Government's economic structural adjustments policy.

There is an additional factor that brought in a surge of manufactured goods imports. It is a shift of production to overseas that was accelerated by a higher yen rate. Finished and intermediate products are imported from overseas production operations. The local content of electrical appliances has come down without much recognition by consumers.

In terms of estimated import equations, import price elasticity and income elasticity became larger reflecting the above-mentioned changes even though the magnitude of changes is not so large. Largest changes is observed in a shift of equations or changes in the intercept parameter.

IV. Macro Policy and Structural Issues

In 1977 when Professor Klein testified at the joint economic committee with respect to the undervalued yen, there was another "Klein Report" that appeared in a Japanese economic weekly. It was a report on the confidential paper written by Professor Klein for a group of American businessmen. Reportedly, Klein argued that the long-term labor contract (implicit or explicit) made by Japanese firms, typically called "life time employment", makes wage bills a part of fixed cost for the employers. In order to secure employment Japanese firms are inclined to stick with a long-run survival strategy in a recession period, relying on overseas markets where export prices might be set as low as a short-run marginal costs. He saw that the long-term labor contract as one of the major factors that lead to Japan's overseas "dumping" practice. As a matter of fact, it was a forerunner of Japan-US discussions on economic structural problems in the latter half of the eighties.

Figure 4: Trend of Manufactured Goods Imports (In billion U.S.$)

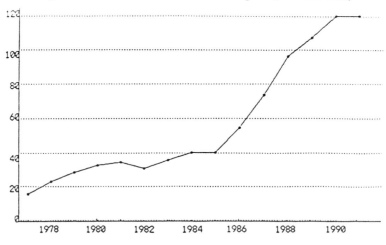

At the Tokyo Summit of 1987, Mr. Nakasone proclaimed an Emergency Policy Package to boost Japan's domestic demand. In addition to a spending package that consisted of six trillion yen spendings program, he disclosed Japan's new program of economic structural adjustments. It aimed at promoting domestic demand through various reforms of Japan's domestic institutional barriers that prevented imports growth despite a large drop in import prices. It was based upon the 1986 Maekawa Report that reviewed all sources of potential structural barriers against the improvement of standard of living of workers.

Remarkable facts of life in Japan were: 1) large consumer price differentials between home and overseas markets; 2) a large difference in working hours between Japan and Europe; 3) a long list of government regulations and interventional practices which are not stipulated by law.

International price differentials were seen not only in the prices of imported goods, but unfavorable price differentials were also seen between prices of Japan's goods sold overseas and the same goods that are sold in domestic markets.

In order to follow up with Japan's economic structural adjustments policy within a present macroeconomic framework of the two countries, Japan and the U.S. agreed to set up a negotiation on a structural policy scheme. The negotiation started in 1989 and a report was published in the following year. It is a document which identified economic structural issues in the two economies and specified policy steps for overcoming the problems. The Japanese government called this "Bilateral Negotiation on Economic Structural Issues", and the U.S. side simply named it "Structural Impediments Initiative". The implication of the latter was clear enough in stating the U.S. request to Japan to remove structural impediments. During the process of negotiation, Japan pointed out to the U.S. that the large U.S. budget deficit and low personal saving rate are the major factors of the U.S. trade deficit. Actually, it was a point made clearly by the CEA economic report of 1989.

Since then the economic structural policy has become a supplementary instrument in promoting the recovery of macroeconomic imbalances. In Japan some progress has been made in the field of door-opening of domestic markets to foreign competitors. Together with the demand switch policy, it made a significant contribution to bringing down its large trade surplus. Then what about the remaining structural issues?

Here I should like to touch only upon the working hour problem in relation to a change in a general situation surrounding Japan's labor relation.

During the rapid growth era of the 1960s, Japan's working hours came down remarkably by the widespread adoption of five-working-days-a-week at factories. Facing the shortage of a young labor supply, manufacturing employers had no choice but to accept the new practice in order to attract young workers.

During the period of slow growth period after the oil shock, this trend came to a halt. First the new practice had saturated among manufacturing firms, and secondly, firms had to reduce the number of employees and to rely more on overtime to reduce costs. Actually Japan's terms of trade worsened in the seventies and the relative income share of workers declined.

Until a strong growth trend became clear in 1987, most firms had refrained from increasing the number of workers. Company unions agreed to accept longer working hours. Total annual working hours hit 22 hundred in 1987. The long-term labor contract means that an expected future cost of additional employment is large. Unless an increase in demand is judged to be a permanent change, employers are unlikely to expand employment. Unions also agree with employers on the restraint. Workers were still willing to accept "less leisure, more income" and to avoid the situation such that expanded employment might turn out to be a future burden for a company. In this sense employees and employers formed a tacit coalition for long-term survival.

Table 4: Japan's Export to the U.S.A.

| | | JAPAN'S | EXPORT | | | |
	Chemical	Steel	Industrial Machinery	Electrical Machinery	Elecrical Parts	Ratio
1980	767	2702	3368	5135	364	7.09
1985	1407	2804	10198	14987	1024	6.83
1986	1740	2057	13618	17490	1101	6.30
1987	2081	2361	16068	17050	1496	8.77
1988	2291	2691	20230	19311	3367	17.44
1989	2524	2359	22755	20820	4326	20.78
1990	2473	2233	21258	19436	3688	18.98
1991	2762	2066	20939	19995	3880	19.40
1992	3238	1918	23408	21114	4640	21.98

	Automobiles	Parts	Ratio Units	Auto	Auto Price	USCPI	Auto Price/CPI
1980	10119	483	4.77	2469.7	4097.3	82.4	96.1
1985	19238	2475	12.87	3454.7	5568.6	107.6	100.0
1986	25889	4404	17.01	3734.8	6931.8	109.6	122.2
1987	25223	5154	20.43	3278.7	7693.0	113.6	130.8
1988	24026	4429	18.43	2879.8	8342.9	118.3	136.3
1989	23105	5088	22.02	2554.7	9044.1	124.0	140.9
1990	23096	5341	23.13	2366.7	9758.7	130.7	144.3
1991	23580	5245	22.24	2184.0	10796.7	136.2	153.2
1992	22465	5763	25.65	1890.8	11881.2	140.3	163.6

Notes: 1) Souce = Custom Office, Tokyo; 2) unit= million US $; 3) AutoUnits in thousand

From 1988 to 1990, Japan maintained a strong growth of 5 percent that brought about a change in the general perception of Japan's potential growth by industrialists. The past effort for streamlining production lines suddenly produced an acute increase in demand for labor. There came a period of massive employment expansion from 1988 to 1991 the employment of regular workers in Japan's manufacturing industry increased by six percent, while it barely increased in the preceding three year period of 1985-1988.

Total working hours showed a modest decline because the increased employment level reduced a need for overtime work. Then, production growth came to halt and the rate of decline of working hours has become clear.

Table 5: Comparison of Price Level				
	New York	London	Paris	Berlin
Food	74	60	60	59
Meat	50	50	53	52
Durables	85	134	120	129
Clothing	65	77	98	83
Energy	50	74	95	115
Transport 103	100	95	92	
Rent	56	89	90	84
Medical	121	58	62	31
Total	76	89	88	83

Notes 1) 1992 November; 2) Market rate of exchange was applied; 3) Japan is taken to be 100
4) Source: "Bukka Report 1993" by Economic Planning Agency, Tokyo

Now firms are facing an overemployment situation due to the present slump, but they are not likely to reduce employees readily, since they expect that the new trend of reduced working hours will prevail. Thus, Japan's unemployment rate remains within the two percent range despite the prolonged economic slump. This is a reason why the rate of decrease in corporate profits has been so drastic.

Figure 5: Trend of Japan's Total Working Hours

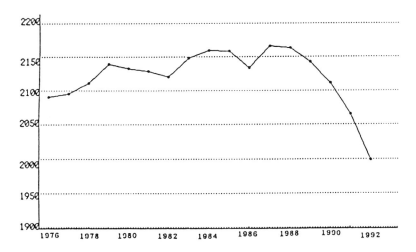

Traditionally the current situation in Japan would have led to an export growth through an aggressive "price war" in the world market which concerned Professor Klein so much in the late seventies. However, situations in the world market have changed; more than half of the total exports from Japan is said to be under control either in the form of self restraint or in the form of an international cartel. (Take a few examples such as steel, synthetic fibers, and automobiles.) We have shifted from the free trade situation to the so-called "fair trade" one where engaged players of trade game are closely watching each other. Japan's labor relations based on the long-term labor contracts seem to have started to change. It used to be run on a high growth scheme trying to take advantage of the economies of scale. With the baby-boomers reaching middle-aged, Japan's wage system faces a challenge since its steep wage profile of age is causing a serious financial burden for many firms.

Table 6: Annual Working Hours (1986)

	JAPAN	U.S.A.	U.K.	W.GERMANY	FRANCE
TOTAL HRS	2168	1924	1952	1659	1643
OVERTIME	299	172	161	83	--

Chikashi Moriguchi, Institute of Social and Economic Research, Osaka University, Ibaraki, Osaka 567, Japan.

NOTES

1. The maximum likelihood estimate of the asset effect was .0054 for the short-run (impact effect) and the adjustment speed is .11 for a quarter. This should have about a 16 trillion yen impact by the end of the first three years under the assumption that two hundred trillion capital gains were added for three years. That is to say, the level of consumption expenditure will be 8% higher than the level without any capital gains.

REFERENCES

Guidepost of Economic Structural Adjustments. (1987). *A Report of Special Subcommittee on ESA*, Tokyo: Economic Deliberation Council.

Moriguchi, C. (1988). *Changing Japanese Economy: Challenge and Response after the High-Growth Era.* (In Japanese), Tokyo: Sobunsha.

------- (1989). Driving forces of economic structural change: The case of Japan in the last decade. In W. Krelle, (Ed.), *The Future of the World Economy.* IIASA, Springer-Verlag.

------- (1989). Economic structural adjustments and macro-economic balance of Japan. *Rivista International di Scienze Economiche e Commerciali.* Milano, Italy.

------- (1991). The Japanese economy and economic structural adjustments. *The Economic Studies Quarterly. 42*, 1.

Comparative Country Performance
at Own-Prices or Common International Prices

Alan Heston, Daniel A. Nuxoll and Robert Summers

The Penn World Table (PWT) is an extremely large data set that makes possible international comparisons of output and prices of virtually all the countries of the world. Five versions (one never published) of this data set have been compiled, the most recent, Mark 5 (PWT5), in 1991.[1] These data have been used by researchers working in many different areas and not only in economics and other social sciences. In particular, a variety of studies have used the data to examine hypotheses about economic growth.[2] Because comparisons of different countries' economic performance commonly are calculated from the PWT time series, it is important to understand the differences between the growth rates of national output (i.e., gross domestic product, GDP) implicit in these data and growth rates calculated from countries' own national accounts.

I. Background

Formal, official national accounts of countries were only developed about 50 years ago, but it is now hard to conceive of running an economy or doing any kind of empirical macroeconomic research without drawing on national accounts data. For making this remarkable empirical picture of economies a reality, Colin Clark did not get a Nobel Prize, but Simon Kuznets and Richard Stone did. Our principal organized set of information about the countries of the world is now the United Nations' System of National Accounts (SNA), along with parallel data sets maintained by other international agencies such as the World Bank, the International Monetary Fund, the European Communities and the Organization for Economic Co-operation and Development.

In the SNA and parallel archives, each country's accounts are denominated in its own national currency units. Consisting of current- and constant-price time series, the accounts allow intertemporal comparisons within countries. In addition, through the use of dimensionless analytical ratios, the national accounts also permit some restricted kinds of meaningful comparisons between countries. The savings rate is an illustration of this.

But most interspatial comparisons concerned with real quantities require that the national account items of different countries be valued in a common currency. Such is the format of the Penn World Table. This revaluation, done within the United Nations International Comparison Programme (ICP), draws on appropriate currency conversion factors---that is, purchasing power parities

(PPPs)---developed from the ICP's benchmark study price surveys. In the course of the benchmark work, a set of average world prices (so-called *"international* prices") is produced.[3] When each country's outputs are valued at these international prices, aggregates now measured in "international dollars" are directly comparable across countries. This paper focuses on the most commonly used PWT variable, real GDP per capita, measured in international dollars, and specifically, on its growth rate.[4] It might be thought that being dimensionless, countries' growth rates would be directly comparable across countries without a common-currency conversion. This is an oversimplification which deserves the detailed examination below.

Empirical economists have long recognized that in time series analysis, the choice of the prices used in valuing the elements of a quantity aggregate has a great deal to do with the behavior of the aggregate over time. Particularly, Alexander Gerschenkron observed that the use of early-year prices leads to larger estimates of growth rates than the use of late-year prices. (That is, Paasche quantity indexes give lower growth rates than Laspyres.) The cross-section analogue of this is that in comparing the incomes of pairs of countries, the use of the more developed country's relative prices in valuing both countries' outputs will lead to a smaller estimated gap between the countries' incomes than if the prices of the less-developed country are used. These intertemporal and interspatial regularities have come to be known generically as the "Gerschenkron Effect."

Because of the Gerschenkron Effect, the choice of base prices in calculating growth rates is not a trivial matter. This paper examines whether using international prices results in substantially different growth rate estimates from those based on domestic prices---that is, whether growth rates derived from SNA sources differ systematically from the growth rates embedded in the Penn World Table, *and why*. The paper proceeds from here in three parts: it outlines the method used in constructing the Penn World Table first; it then examines the reasons why it might be thought that the Gerschenkron Effect would affect the PWT comparisons of income levels and growth rates; and lastly, it tests whether the Effect actually characterizes the estimates. The conclusion at the end summarizes the findings.

II. GDP Comparisons in the Penn World Table

The 138 countries of the PWT5 divide into about 70 that have participated in ICP benchmark studies and 68 that have not. The PWT5 entries for the benchmark countries were based on estimates generated in the benchmark studies. In the absence of benchmark price data, the entries for the non-benchmark countries were obtained by exploiting observed structural relationships found in the benchmark data set (See Kravis, Heston and Summers (1978b). For present purposes it is sufficient to outline only the way the benchmark country estimation is done. The growth-rate analysis is the same for both benchmark and non-benchmark countries.

A. Benchmark Studies

In an ICP benchmark study, detailed prices of identically specified final-product goods and services are collected in each of the participating countries. In addition, expenditures for each of about 150 categories of output exhausting GDP are collected.[5] By suitably combining the price information with detailed expenditure data from the countries, a set of price parities (PPs) was estimated for each country for each of about 150 categories of final goods and services covering all of GDP. With these PPs it was possible to convert the domestic currency expenditures of each of the 150 categories into international-dollar expenditures. Each country's PPs were expressed relative to United States prices. The international dollar, keyed to the United States dollar, is the numeraire currency, but it is only in a very special, nonrestrictive sense that the United States plays an asymmetric role relative to the other countries: the United States GDP when valued in international dollars is equal to the United States GDP valued in American dollars. However, the same is not so for the United States' individual components of GDP.

Only a brief explanation of the meaning of "international prices" need be given here. The international price for each of the 150 detailed categories of goods defined in the benchmark studies is a weighted average of the relative prices of that category in all of the countries of the benchmark study. Country quantity weights are used so, for instance, the prices of rice in countries with large total rice consumption have more effect on the international price of rice than countries that consume very little rice.

B. Non-Benchmark Years

The real GDP estimates of all countries were extrapolated backward and forward through time on the basis of the countries' current- and constant-price series in their national accounts. The procedures followed were the same for both benchmark and non-benchmark countries, but differed depending on the pricing concept used. (The four different real GDP per capita concepts presented in the PWT are described below.)

Base year prices In the case of the concepts geared to a base year, the benchmark estimates of a country's real Consumption (C), Investment (I), and Government (G) components were extrapolated to other years using the growth rates of those components taken from the country's component constant-price series; then these extrapolated real components were added together to get real GDP for the extrapolated years. The resulting growth rate will differ from the GDP growth rate in the national accounts because of the different price weights used. (It should be noted that the OECD follows a quite different procedure. The OECD extrapolates a country's benchmark real GDP estimate on the basis of its national accounts growth rate and then scales its real GDP components up or down to make them add up to the extrapolated GDP value. For a discussion, see United Nations (1992), pp. 62-63).

Current prices   Real GDP per capita estimates for a non-benchmark year based on that year's prices (that is, current-price estimates that are directly comparable across countries in a particular year that are based on that year's international prices rather than a base year's), are obtained by extrapolating component price parities derived from a benchmark study to the year of interest on the basis of corresponding price indexes computed from the national accounts.

A complication requiring attention is to be expected in the case of countries that have participated in more than one benchmark study.[6]   Real quantity levels of components as estimated in successive benchmark studies will not be precisely consistent with their national accounts growth rates for corresponding periods.   In the PWT4 and the PWT5, such inconsistencies were eliminated through the use of a reconciliation process based on an errors-in-variables model.[7]

C. Output Measures Based Upon Alternative Pricing Concepts

The PWT5 presents GDP per capita under four different pricing concepts. Each has its own interpretation, and is appropriate for particular applications. The growth rate discussion of Section IV, focuses on the second of the concepts described below, RGDPCH.

RGDPL, a Laspyres-type index, is GDP per capita measured in 1985 relative international prices. (Strictly speaking, while the 1985 international prices of overall C, I, and G were used in valuing the real C, I, and G quantities, the C, I, and G constant-price series of many countries were not necessarily based on the 1985 relative prices of the subaggregates of C, I, and G.)

RGDPCH is a Divisia-type index of GDP per capita based upon changing, chained international prices. It is built up from a succession of pairs of contiguous years' international prices to make cross-country comparisons in years far from 1985 suffer less from being based on international prices far from the current ones.

RGDPTT is a measure of GDP per capita that is designed to capture the effect on a country of changes in its terms of trade. Each year's value is based upon 1985 international prices for C, I, and G but current prices for exports and imports. (The domestic absorption components of RGDPTT and RGDPL are the same; however, RGDPTT's Net Foreign Balance is based on the current values of exports and imports while RGDPL's is based on the quantities. Thus even if real output has not changed, RGDPTT measures the welfare consequences of changing export and import prices.

CGDP is GDP per capita measured in current international prices. As such, CGDP's cross-country comparisons in any year are better than those based on the other measures because the international prices used are the most appropriate. However, though CGDP is useful for comparisons within a single year, any time series of CGDP reflects price changes as well as quantity changes.

III. The Gerschenkron Effect

During the course of his famous series of studies of the Soviet economy, Alexander Gerschenkron (1951) used 1939 U.S. prices to compute the value of U.S.S.R. machinery output. He calculated a 400 per cent increase in machinery output between 1927-28 and 1937, in sharp contrast with the 1500 per cent increase found in the official Soviet series which was based on 1926-27 ruble prices. As Gerschenkron investigated this vast difference in reported growth rates, he found that growth rate estimates in the United States were also sensitive to the choice of base year for prices. He found that the estimate of the change in United States machinery output between 1899 and 1939 based on 1899 prices was an increase of 450 per cent, while the estimate based on 1939 prices showed a decline of 30 per cent! His explanation for this was that if the relative price of a good fell, its output would expand more rapidly than other goods. Base prices drawn from a later period would assign lower weights to goods with rapidly growing outputs, thus producing a lower growth rate.

This point is worth some emphasis. If rapid technological progress occurs within a particular industry, the price of the industry's product is likely to fall and its output is likely to increase. Consider the impact of this on a quantity index covering the output of many industries. Choosing an early base year for prices, before the technical developments lowered the expanding industry's price, would assign a relatively large weight to each unit of the industry's product; an index using a later base year, after the expanding industry's price had declined, would assign a smaller weight to the expanding industry's growth. The index with the earlier base year would display a higher growth rate because it treats the rapidly growing industry as more significant.

It is to be expected that this effect applies also to estimates of GDP and GNP. Allan Young (1989) found that estimates of the United States annual growth rate for the six-year period 1982-88 differed by 0.3 per cent, depending on the choice of base year prices. Using 1982 prices, GNP grew by 26.2 per cent; using 1987 prices, it grew by 24.3 per cent. (Young estimated that about 90 per cent of the difference in the growth rate estimates arose from the dramatic drop in computer prices and rapid growth in computer output over the five year period.)

If prices change systematically during development, perhaps less developed countries will, in time, have the same price structure that richer countries have today. Services might be relatively cheap in low-income countries today, but as the countries develop---as capital becomes more abundant and labor becomes more scarce---the relative price of services would increase, approaching the present relative price in more developed countries.[8] This suggests that, say, Italy's price structure today foreshadows the price structures of Korea and similar countries twenty-odd years hence.

This kind of systematic price structural change has an important implication. Valuing the outputs of a less developed country using the prices of a more developed country rather than its own prices leads to a higher estimate of its relative aggregate product and to a lower estimate of its growth rate. Using the developed country's prices is like using a later base year. At least as a stylized fact, it is to be expected that both the later-year prices and the developed country prices assign relatively low weights to the same goods and services. This effect is readily observable in the income level comparisons. The ICP report of the 1975 benchmark study gives a complete set of GDP binary comparisons for 34 countries (Kravis, Heston and Summers (1982), Table 7.2, pp. 230-1). The ratio of each country's GDP per capita to that of each other country is presented, calculated in two different ways: first using the first country's prices and then using the second's. (That is, Country A is compared with Country B with both countries' outputs first valued using A's prices and then B's.) In all, 561 binary comparisons are possible (the number of combinations of 34 things two at a time). In every single case, the interspatial Gerschenkron Effect is displayed. In all 561 comparisons, the gap between the incomes of the more developed and the less developed country is smaller if the prices of the more developed country are used in valuing the outputs.

This discussion might lead one to expect that the Gerschenkron Effect would leave its mark on the Penn World Table entries in two ways. First, because the PWT constant-price series have a recent-year base (1985 in the PWT5), it should produce lower PWT growth rates than if an early base year had been chosen. The Effect would be most pronounced for rapidly growing countries where relative prices are likely to be evolving rapidly. But second, international prices are averages of all countries' national prices so they resemble the prices of a country with moderate income.[9] For poor countries with below-average incomes, national prices are like early-year prices and international prices are like late-year prices; on the other hand, for rich countries above the average, international prices are like early-year prices and national prices are like late-year prices. Therefore, it might be expected that a poor country's growth rate based on its national prices will exceed its growth rate based on international prices; and correspondingly, a rich country's growth rate based on its national prices might be smaller than its growth rate based on international prices. The proposition that flows from this discussion is that growth rates embedded in the Penn World Table might be expected to differ systematically from corresponding growth rates in the national income accounts.

The word "might" was used often in the last paragraph, in recognition of the possibility that though the arguments are sensible, the implicitly assumed empirical facts differ from reality. The level of aggregation in PWT4 and PWT5 is very high (only four components: C, I, G, and the Net Foreign Balance (NFB)) and two of the components have large non-market components. It is at least questionable whether quantities and prices are inversely related across the government sector which covers public consumption and the public construction part of investment. Together these constitute a large proportion of total GDP. (A subtlety here should be made clear. A non-negative correlation between prices and quantities in the detailed categories of government expenditure would not keep the Gerschenkron Effect from showing up in the PWT. It is negative correlation of prices and quantities *across*---not *within*---C, I, G, and NFB that the Effect depends upon for the PWT.) If C, I, and G are not sufficiently mutually substitutable within a country to make component price parities negatively correlated with component real quantities, then the inputs going into the PWT and used in rebasing national and international constant-price series may very well not meet the conditions required for a pronounced Gerschenkron Effect,[10] leaving no systematic relationship between growth rates in national and international prices. Furthermore, it would not necessarily follow that growth rates would differ using early- versus late-year (national or international) prices.

This has been a discussion of the Gerschenkron Effect implications and the pragmatic reasons why the Effect may not actually operate in the PWT. Now we look into the empirical question of what really is going on by comparing various growth rates calculated from the Penn World Table and the national accounts.

IV. PWT5 Growth Rates

This section presents an analysis of six annual average growth rates of per capita Domestic Absorption, denoted simply DA here, between 1960 and 1985. We work with the 116 market economies of the PWT5 (out of 134) for which data are available for the full 1960-85 period. The six growth rate concepts are:

1. Growth rate using national price weights of 1960: r_1
2. Growth rate using national price weights of 1985: r_2
3. Growth rate using chained national price weights: r_3
4. Growth rate using international price weights of 1960: r_4
5. Growth rate using international price weights of 1985: r_5

6. Growth rate using chained international price weights: r_6

We will compare the magnitudes of these growth rates in a number of ways.

A. Does the Difference in Choice of Base Year Matter?

The various growth rates of Domestic Absorption are presented in Table 1.[11]   Column (A) gives r_1 and Column (B) gives (r_1-r_2). The Gerschenkron Effect calls for the early-year base growth rate being greater than the later-year one, so the entries in Column (B) should be predominately positive. In fact, only 50 of the 116 differences are positive, though, as the sign test statistic given at the bottom of Table 1 indicates, this number is not significantly different from 58, the expected number in the absence of any systematic connection between the two.

Column (C) gives (r_1-r_3), the difference between the 1960-based rate and the chain index. Since the chain index uses prices drawn from years later than 1960, the Gerschenkron argument calls for the Column C entries to be predominately positive, and in fact 76 of 116 are. (This number is significantly different from 58 at the 0.1 percent level.)

Columns (D), (E), and (F), referring to growth rates based on international prices, give r_4, (r_4-r_5), and (r_4-r_6). Much the same story comes through here.   Growth rates based on early-year prices are sometimes larger and sometimes smaller than growth rates based on the later-year prices: 41 of the Column (E) entries and 29 of the Column (F) entries are positive, and both numbers are significantly below 58. The intertemporal Gerschenkron Effect is absent in these country data sets for the 1960-85 time period, and in fact it would appear that a perverse movement in prices lead to examples of an "anti-Gerschenkron Effect."

We have suggested that this might occur because C, I, and G may not be sufficiently strong substitutes to exhibit negatively correlated price and quantity changes. But the question remains: why should the Gerschenkron Effect be less pronounced in the PWT than in the national accounts? Our conjecture is that later-year price weights increase the importance of government in the calculation of growth rates more when working with international prices than with national prices. If, as is likely, real government expenditures were growing rapidly in the countries, then it would indeed be true that for a later-year base the international-price growth rate would be greater

Table 1
Growth Rates and Growth Rate Differences Calculated in Different Ways

	$A{:}r_1$	$B{:}r_1{-}r_2$	$C{:}r_1{-}r_3$	$D{:}r_4$	$E{:}r_4{-}r_5$	$F{:}r_4{-}r_6$	$G{:}r_1{-}r_4$	$H{:}r_2{-}r_5$	$I{:}r_3{-}r_6$
Country	(A)	(B)	(C)	(D)	(E)	(F)	(G)	(H)	(I)
Algeria	5.26	-0.01	-0.01	6.01	1.16	-0.04	0.75	-0.43	0.77
Angola	1.11	0.01	0.00	0.24	-0.10	-0.39	-0.88	-0.77	-0.49
Argentina	1.79	-0.13	-0.10	2.01	-0.09	0.02	0.23	0.19	0.11
Australia	3.92	-0.02	-0.03	3.82	-0.17	-0.08	-0.10	0.06	-0.05
Austria	3.40	-0.01	0.03	3.51	-0.13	0.05	0.11	0.23	0.08
Bangladesh	3.13	0.00	0.06	3.11	-0.07	-0.14	-0.02	0.06	0.18
Barbados	3.01	-0.18	-0.08	2.43	0.22	-0.06	-0.58	-0.97	-0.60
Belgium	3.07	0.03	-0.03	2.71	-0.32	-0.05	-0.36	-0.01	-0.34
Benin	2.57	0.03	0.08	2.74	-0.01	0.14	0.16	0.20	0.10
Bolivia	3.51	0.04	0.05	3.32	-0.23	-0.33	-0.19	0.08	0.19
Botswana	8.59	-0.05	0.11	8.93	-0.63	-0.32	0.34	0.92	0.77
Brazil	5.50	-0.06	-0.02	6.69	0.00	0.25	1.19	1.13	0.92
Burma (Myanmar)	4.44	-0.04	0.03	4.35	0.06	-0.01	-0.09	-0.19	-0.06
Burundi	1.53	0.15	0.20	2.43	-0.08	-0.28	0.90	1.13	1.38
Cameroon	6.64	0.01	0.36	5.87	-0.24	0.01	-0.76	-0.52	-0.42
Canada	4.15	0.04	0.01	4.06	0.13	-0.03	-0.09	-0.18	-0.06
Cape Verde	3.25	-0.01	0.06	3.50	-0.49	-0.12	0.25	0.74	0.43
Central Afrrican Repub	1.75	0.02	0.04	1.54	0.20	0.30	-0.20	-0.39	-0.46
Chad	-0.88	0.02	0.00	0.48	0.26	0.04	1.36	1.12	1.32
Chile	2.46	-0.07	0.04	2.03	-0.38	-0.09	-0.43	-0.12	-0.30
Columbia	4.76	-0.02	0.04	4.71	0.04	-0.02	-0.05	-0.11	0.01
Congo	5.80	0.02	0.15	4.97	0.61	-0.53	-0.83	-1.43	0.15
Costa Rica	4.50	-0.17	-0.09	4.50	-0.15	-0.04	-0.01	-0.02	-0.06
Cyprus	5.72	0.19	0.46	5.42	-0.04	0.15	-0.29	-0.07	0.01
Denmark	2.93	-0.09	-0.11	2.55	-0.61	-0.22	-0.38	0.13	-0.27
Dominican Republic	5.51	0.11	-0.02	4.87	-0.20	0.06	-0.65	-0.33	-0.73
Ecuador	5.38	-0.07	0.05	5.35	0.01	-0.03	-0.02	-0.10	0.06
Egypt	6.49	-0.06	0.02	7.40	0.04	-0.07	0.92	0.82	1.01
El Salvador	3.08	-0.04	0.02	3.09	-0.49	-0.23	0.01	0.46	0.27
Ethiopia	2.20	-0.13	-0.01	3.06	-0.30	-0.24	0.86	1.02	1.10
Fiji	4.29	-0.04	0.13	3.70	-0.17	-0.19	-0.58	-0.45	-0.26
Finland	3.65	-0.01	-0.04	3.46	-0.35	-0.13	-0.19	0.15	-0.10
France	3.73	-0.03	0.01	3.66	-0.18	0.09	-0.07	0.07	-0.15
Gabon	8.19	-0.08	-0.03	7.37	0.10	-0.16	-0.82	-1.01	-0.70
Gambia	4.49	-0.01	0.12	5.81	0.42	-0.15	1.32	0.89	1.59
Germany	2.99	0.00	-0.02	2.77	-0.17	-0.03	-0.21	-0.04	-0.20
Ghana	1.24	-0.25	0.27	1.34	-0.18	-0.18	0.09	0.03	0.54
Greece	5.37	0.13	0.01	5.10	-0.10	-0.03	-0.27	-0.0 4	-0.23
Guatemala	3.75	0.02	0.04	3.75	-0.27	-0.02	0.01	0.30	0.07
Guinea	1.53	0.03	0.07	1.82	-0.28	-0.24	0.29	0.60	0.60
Guinea-Bissau	3.01	-0.03	0.10	2.89	-0.25	-0.20	-0.12	0.10	0.18
Guyana	0.56	-0.08	0.01	0.71	0.12	-0.03	0.15	-0.06	0.19
Haiti	2.13	0.07	0.31	1.74	-0.07	-0.07	-0.39	-0.26	-0.01
Honduras	4.36	-0.01	-0.01	4.28	-0.33	-0.08	-0.08	0.25	-0.01
Hong Kong	9.08	0.22	0.67	8.14	0.31	-0.02	-0.95	-1.03	-0.26
Iceland	4.30	-0.14	-0.08	4.60	0.23	-0.08	0.29	-0.08	0.29
India	3.62	-0.13	-0.08	2.65	-0.07	-0.02	-0.98	-1.04	-1.04
Iran	7.00	-0.36	0.18	6.87	1.37	-0.29	-0.14	-1.87	0.34
Iraq	6.28	-0.41	0.47	5.92	0.84	0.03	-0.35	-1.59	0.09
Ireland	3.66	-0.05	-0.03	2.85	-0.30	-0.08	-0.81	-0.56	-0.75
Israel	5.77	0.10	-0.08	5.58	-0.14	-0.33	-0.19	0.05	0.06
Italy	3.88	0.10	0.19	3.92	-0.19	0.01	0.05	0.34	0.22
Ivory Coast	5.56	0.00	0.02	5.56	0.25	-0.10	0.00	-0.25	0.12
Jamaica	1.94	-0.10	0.37	2.02	-0.45	-0.35	0.09	0.44	0.80
Japan	6.47	0.08	0.33	6.16	-0.32	0.11	-0.32	0.09	-0.10
Jordan	6.69	-0.01	0.17	6.00	0.13	-0.04	-0.69	-0.82	-0.47
Kenya	3.92	-0.04	0.31	4.60	-0.16	-0.03	0.68	0.80	1.03
Korea	8.46	0.14	0.44	7.63	0.06	0.19	-0.83	-0.75	-0.58

Table 1 (cont.)

Kuwait	7.09	0.07	0.12	6.52	5.81	-0.08	-0.58	-6.32	-0.38
Lesotho	7.17	-0.12	-0.30	8.66	1.29	-0.12	1.49	0.08	1.31
Liberia	2.15	0.13	0.47	1.64	-1.16	-0.35	-0.51	0.78	0.32
Luxembourg	2.82	0.19	0.34	2.86	-0.37	-0.05	0.04	0.60	0.43
Madagascar	0.97	0.11	0.28	0.61	-0.34	-0.14	-0.36	0.09	0.05
Malawi	3.81	0.01	0.05	3.84	-0.05	0.00	0.03	0.09	0.09
Malaysia	6.99	-0.05	0.00	6.54	-0.21	-0.04	-0.45	-0.29	-0.40
Mali	2.14	-0.05	0.05	2.29	0.17	0.00	0.15	-0.07	0.21
Malta	5.38	-0.05	0.01	4.86	-0.69	0.01	-0.53	0.11	-0.53
Mauritania	2.37	-0.10	-0.05	2.92	0.89	0.27	0.55	-0.44	0.23
Mauritius	3.26	0.04	0.37	3.57	-0.19	-0.12	0.31	0.54	0.80
Mexico	5.37	-0.01	-0.01	5.22	-0.08	-0.09	-0.15	-0.08	-0.07
Morocco	6.02	-0.28	-0.08	5.66	-0.26	-0.17	-0.35	-0.37	-0.26
Mozambique	-0.01	-0.02	-0.06	0.17	-0.22	-0.20	0.18	0.38	0.32
Nepal	3.37	0.11	0.32	3.11	-0.11	-0.16	-0.27	-0.05	0.21
Netherlands	3.27	0.00	0.05	3.32	-0.20	0.08	0.06	0.26	0.03
New Zealand	2.36	-0.09	-0.03	2.42	-0.20	-0.03	0.06	0.17	0.07
Nicaragua	3.44	0.25	0.11	2.41	-1.07	-0.46	-1.03	0.29	-0.46
Niger	4.29	-0.05	-0.02	3.19	0.29	0.07	-1.10	-1.44	-1.20
Nigeria	2.88	-0.78	0.22	3.07	0.82	-0.17	0.19	-1.41	0.57
Norway	3.92	-0.04	-0.04	3.73	-0.07	-0.09	-0.19	-0.16	-0.14
Pakistan	5.73	-0.01	0.07	4.67	-0.41	-0.15	-1.06	-0.67	-0.84
Panama	5.48	0.01	-0.07	5.51	-0.25	-0.12	0.03	0.29	0.09
Papua New Guinea	3.27	0.05	0.05	3.91	0.03	0.25	0.64	0.67	0.45
Paraguay	5.63	-0.04	0.04	5.56	-0.04	0.05	-0.07	-0.06	-0.07
Peru	3.21	-0.55	0.09	3.59	0.00	-0.08	0.38	-0.17	0.55
Philippines	4.04	-0.03	0.03	4.13	-0.16	-0.07	0.09	0.21	0.18
Portugal	4.11	0.06	0.18	3.92	-0.58	-0.26	-0.19	0.44	0.24
Rwanda	5.65	0.40	0.69	4.45	0.01	-0.16	-1.20	-0.81	-0.34
Saudi Arabia	13.25	-0.93	0.18	12.07	5.75	-0.11	-1.19	-7.87	-0.90
Senegal	2.36	0.07	0.34	2.62	0.13	0.10	0.26	0.20	0.49
Singapore	7.99	0.02	-0.26	6.66	-0.33	-0.04	-1.33	-0.99	-1.55
Somalia	1.58	-0.75	0.09	3.61	0.29	0.05	2.03	1.00	2.08
South Africa	4.05	-0.05	-0.05	3.69	-0.05	-0.17	-0.37	-0.37	-0.25
Spain	4.41	-0.07	0.00	4.12	-0.33	-0.08	-0.29	-0.03	-0.21
Sri Lanka	3.89	0.03	-0.02	3.21	-0.01	0.13	-0.69	-0.65	-0.84
Sudan	2.81	-0.16	-0.07	3.36	0.66	-0.04	0.55	-0.27	0.52
Surinam	2.07	-0.24	0.07	3.47	-0.03	0.44	1.40	1.19	1.04
Swaziland	7.78	-0.14	0.05	5.21	-0.61	0.04	-2.57	-2.10	-2.55
Sweden	2.49	-0.07	-0.09	2.34	-0.52	-0.16	-0.16	0.29	-0.09
Switzerland	2.78	0.02	-0.03	2.74	0.11	-0.02	-0.04	-0.13	-0.05
Syria	7.79	-0.40	-0.07	7.71	0.19	-0.14	-0.08	-0.67	-0.01
Taiwan	7.88	0.01	0.20	7.82	-0.11	0.17	-0.06	0.05	-0.02
Tanzania	3.98	0.02	0.19	4.70	-0.94	-0.27	0.71	1.67	1.18
Thailand	6.66	0.03	0.05	5.89	-0.47	-0.28	-0.77	-0.27	-0.44
Togo	3.86	0.03	-0.10	4.46	-0.11	-0.03	0.60	0.74	0.52
Trinidad & Tobago	3.88	-0.09	0.07	4.02	0.41	-0.17	0.15	-0.35	0.39
Tunisia	6.09	-0.09	-0.07	5.51	0.21	-0.04	-0.58	-0.88	-0.61
Turkey	5.01	-0.06	0.29	4.96	-0.15	-0.01	-0.05	0.04	0.25
Uganda	1.64	0.07	-0.08	5.78	1.63	-0.08	4.14	2.57	4.14
United Kingdom	2.15	0.02	0.01	2.31	0.03	0.06	0.16	0.14	0.12
United States	3.51	0.00	0.03	3.44	0.07	0.13	-0.08	-0.15	-0.17
Uruguay	0.52	-0.19	0.04	0.27	-0.29	-0.12	-0.25	-0.16	-0.09
Venezuela	4.97	0.11	0.08	5.66	0.36	-0.03	0.69	0.44	0.79
Yugoslavia	4.69	-0.02	0.13	4.62	-0.06	0.21	-0.08	-0.03	-0.15
Zaire	2.71	0.01	0.40	2.12	-0.73	-0.22	-0.59	0.15	0.04
Zambia	2.22	0.23	0.72	0.77	-0.74	-0.40	-1.45	-0.47	-0.33
Zimbabwe	3.62	0.16	0.58	4.16	-0.73	-0.32	0.55	1.44	1.45

SUMMARY STATISTICS

Mean	4.17	0.09	-0.04	4.10	-0.07	0.04	0.06	-0.10	0.14
Standard Deviation	2.16	0.18	0.18	2.01	0.88	0.16	0.76	1.16	0.74
Median	3.88	0.04	-0.01	3.83	-0.07	-0.10	0.08	-0.05	0.03
Number Positive	---	50	76	---	41	29	47	56	62
Sign Test Probability %*	---	16.3	0.1	---	0.2	0.0	2.5	39.0	25.8

* Probability of obtaining the observed number of positive values if positive and negative numbers were equally likely.

than the national accounts growth rate. This would explain why the intertemporal Gerschenkron Effect is absent more in the international price case. An examination of the correlations between the growth rates computed for different base years also seems consistent with this conjecture.

Table 2 contains the correlations between the differences in growth rates given in Columns (B), (C), (E), and (F). Correlation coefficient estimates with an absolute value greater than 0.19 are significantly different from zero at the 5 per cent level for the present sample of 116 country observations. If 0.19 is used as a screening device---in fact, the appropriate critical value is much larger for this sort of casual fishing expedition---we see a fairly strong tendency for the difference between early- and late-year growth rates based on national prices to be associated negatively with the corresponding differences of growth rates based on international prices (Columns (B) and (E)). This is consistent with the sort of explanation offered above: when dealing with a later-year base, there is a larger government weight for the international price case than for the national price one. It is interesting that this effect is much weaker when the international chain is used, that is the correlation of (B) and (F); this provides additional support for use of the chain index for representing growth experience.

Table 2: Correlation Matrix

| | B:(r_1-r_2), | C:(r_1-r_3), | E:(r_4-r_5), and F:(r_4-r_6) | |
	(B)	(C)	(E)	(F)
(B)	1.00			
(C)	0.24	1.00		
(E)	-0.39	-0.06	1.00	
(F)	-0.06	-0.14	0.12	1.00

Further support for this conclusion is provided by the correlations of the absolute values of these differences as reported in Table 3. They are all positive. Again two of the correlations involving ABS(B) are significant but the correlation with the international price chain ABS(F) is not. The correlation between the differences between the two chain indexes and their corresponding early-year estimates (ABS(C) and ABS(F)) is positive and moderately strong. This suggests that both chain indices differ from a fixed price index whenever either does. Once more, this may be taken as support for the use of chain index in growth studies and support for the chain index from the PWT (RGDPCH) as the better representation of growth at international prices.

Still further support for this conclusion is provided in Table 4 that reports the simple correlation of the growth in the six indexes with the absolute value of the differences. The relationship between the growth rate and the size of the difference, using early-year or later-year prices, is the same whether

Table 3: Correlation Matrix: the Absolute Values of

| | B:$(r_1$-$r_2)$, | C:$(r_1$-$r_3)$, | E:$(r_4$-$r_5)$, and | F:$(r_4$-$r_6)$ |
	ABS (B)	ABS (C)	ABS (E)	ABS (F)
ABS (B)	1.00			
ABS (C)	0.28	1.00		
ABS (E)	0.38	0.09	1.00	
ABS (F)	0.05	0.21	0.08	1.00

national or international prices are used, but that relationship does not hold for the chain indexes. The correlations in the ABS (B) and ABS (E) columns show that the absolute difference between early-year and later-year growth rates is positively, and usually significantly, related to the overall magnitude of the growth rate as measured by anything but the chain index based on international prices. None of the correlations involving ABS (C) and ABS (F) are significant. The fact that growth rates associated with RGDPCH (the international price chain index in the PWT) are not correlated with the absolute differences involving the base years of any of the other growth rates again suggests why it may be regarded a preferred representation of growth across countries. However, final acceptance of this conclusion must await an examination of the differences between growth rates at national versus international prices.

Table 4: Correlations Between Growth Rates Calculated in Different Ways and Various Absolute Differences

	ABS (B)	ABS (C)	ABS (E)	ABS (F)
r_1	0.19	0.14	0.35	-0.11
r_2	0.17	0.07	0.34	-0.13
r_3	0.25	0.12	0.37	-0.12
r_4	0.21	0.09	0.35	-0.13
r_5	0.22	0.10	0.37	-0.09
r_6	0.04	0.11	-0.01	-0.07

Two basic lessons flow from this exercise: the choice of a base year matters most for rapidly growing economies, and if the choice is significant for an index using domestic prices, it is likely to be significant for an index using international prices.[12]

B. Does the Choice of International Rather Than National Prices Matter?

Columns (G), (H), and (I) present the differences between indexes using international and corresponding national prices. ((G): r_1-r_4; (H): r_2-r_5; and (I):

r_3-r_6.) In general, the absolute differences between the national and international growth rates in the three columns are greater than the differences arising from different base years or chaining. Put another way, growth rate differences arising from differences in choice of base year or use of chaining are smaller than the differences associated with the choice of international rather than national prices. The differences do not appear to be systematically positive or negative; overall, the average difference is about 0.0008, about 2 per cent of the average growth rate. (However, some large differences for individual countries will be examined below to see if they are related to country income.) Sign tests applied to the number of positive signs in the columns indicate that the 47 in (G) is significant but that the 56 in (H) and 62 in (I) are not.

Summary tables (not reproduced here) displaying the relationship between international-price and national-price growth rates for various combinations can be laid out that show an important fact: if the choice of base year matters for computing the national price growth rate, then the difference between the national and international price growth rate will be such that the choice of which should be used will also matter. (The correlation between ABS (B) and ABS (G) is 0.20, and all the correlations are positive. However, the absolute value of Columns (G), (H), and (I) are much more consistently related to growth rates. All of the correlation coefficients between the entries in these columns and the various measures of growth are positive.)

C. Does Income Level Matter?

Earlier it was observed that if the Gerschenkron Effect is evident, then the growth rates of rich countries calculated using international prices would be lower than those calculated using national prices, and the opposite would be the case for low income countries. The two-way frequency table below (Table 5) checks on whether or not the national accounts and the PWT data bear this out. The difference between the two chain indices (r_3-r_6) is displayed relative to the logarithm of per capita DA.[13]   The labels in the first column refer to ranges of the natural logarithm of DA, and the six column headings give ranges of the chain index differences. The 36 table entries are the numbers of countries out of the total of 116 with incomes falling within the first-column income range and the column-heading growth rate difference. (Example: two countries with log-DA differences between 6.5 and 7.0 had national and international chain growth rates that differed by an amount between -1.0 and -0.5 per cent.) If the Gerschenkron Effect is playing an important role here, the observations should be clustered along the principal diagonal (that is, upper-left to lower-right) cluster of cells. In fact, the cell pattern shows just the opposite pattern. The PWT's RGDPCH tends to understate richer countries' growth rates relative to national-price growth rates and vice-versa for poor countries. The two-way table's story can be summarized by the following regression result:

$$r_3\text{-}r_6 = \quad 2.057 \quad - \quad 0.279 \text{ Log(DA)}$$
$$0.794) \quad (0.102)$$

Table 5: Frequency of Differences Between International and National Chain Price Growth Rates Cross-Classified With Respect to Logarithm of Income

LOG DA	Differences Between Chain Growth Rates						
	Below -1.0	-1.0 to -0.5	-0.5 to 0.0	0.0 to 0.5	0.5 to 1.0	Above 1.0	Total
Below 6.5	1	0	2	2	1	5	11
6.5 to 7.0	2	2	6	5	5	0	20
7.0 to 7.5	0	2	2	7	2	1	14
7.5 to 8.0	2	1	6	3	2	0	14
8.0 to 8.5	3	3	9	5	1	2	23
Above 8.5	3	4	12	14	1	0	34
Total	11	12	37	36	12	8	116

The slope coefficient is significantly different from zero at the 0.3 per cent level.[14]   To judge the robustness of this result, the relationship between $r_3\text{-}r_6$ and income has been examined using a wide variety specifications variables and functional forms. For virtually all specifications, the conclusion was the same: the data show a statistically significant *inverse* relationship rather than the direct one Gerschenkron Effects would lead one to expect.

The good news is that the Gerschenkron Effect apparently does not distort the growth rates reported in the PWT. It appears that the data are so aggregate that relative price shifts do not affect the growth rates. Of course, the explanation may very well be that price structures do not evolve systematically during the course of development as assumed in the Gerschenkron Effect discussion above. At any rate, countries with smaller aggregate incomes do not have lower reported growth rates.

The bad news is that there does appear to be a significant inverse relationship between the international- and national-price chain growth rates and income. A plausible story to explain this result is as follows. In low income countries, international prices assign a higher weight (and in high income countries the opposite) to the government sector compared with national prices in the calculation of growth rates. If real growth in the government sector in low income countries is high compared with other sectors (and again the opposite in high income countries), then it would follow that international price weights would push overall growth rates of high income countries down and low income countries up. We believe this is what is happening and we plan further exploration of this possibility.

An implication of what has been shown touches on the results of empirical work with convergence models. It would appear that the PWT has some convergence built into its implied growth rates. Perhaps this is appropriate, but it certainly warrants further examination. Certainly, users of the PWT should recognize the difference between using RGDPCH in their research and using national growth rates.

V. Conclusions

This paper has compared estimates of country growth rates calculated on the basis of chaining and different base years, and based on national prices and international prices. The Gerschenkron Effect would lead one to expect to find two systematic relationships showing up in an a comparison of growth rates derived from the national income accounts and the Penn World Table (Mark 5): first, that early-year growth rates would be greater than later-year or chain rates; and that calculating growth rates using international prices rather than national prices would give higher growth rates for high income countries and lower growth rates for low income countries. While there is ample evidence that these relationships hold at a detailed level of aggregation, it is not clear conceptually whether they hold for the highly aggregate Penn World Table which operates at the level of Consumption, Investment, and Government.

In fact, a detailed empirical analysis covering 116 countries over the period 1960-85 provided no evidence of the Gerschenkron Effect being operative. Rather, the opposite was often the case. This was probably because real quantities of government and investment goods and services are not inversely related to their relative prices. A finding of some (perhaps disturbing) interest is that the difference between a country's growth rate calculated at national and international prices is negatively related to its income. This result suggests that Penn World Table growth rates will be smaller for rich countries and larger for poor ones than growth rates derived from the national accounts. Clearly, this finding deserves more investigation. To end on a positive note with a felicitous finding, the empirical analysis made a strong case for using a chain index over either an early-year or later-year base, especially when working with international prices.

Alan Heston and Robert Summers, Department of Economics, University of Pennsylvania, 3718 Locust Walk, Philadelphia, PA 19104-6297; Daniel A. Nuxoll, Department of Economics, Virginia Polytechnic Institute, Blacksburg, VA 24061-0316.

ACKNOWLEDGEMENTS

This paper is an extension of earlier work done in Heston and Summers (1989) and

Nuxoll (1992a) and (1992b). The support of the National Science Foundation for this and associated international comparison work is gratefully acknowledged.

NOTES

1. See Summers, Kravis and Heston (1980), Summers and Heston (1984), Summers and Heston (1988), and Summers and Heston (1991).

2. Recent examples include Barro (1991), Delong and Summers (1991), Mankiw, Romer and Weil (1992), and Murphy, Schleifer and Vishny (1991).

3. The benchmark studies cover about 150 categories of final expenditure that exhaust GDP. In each of the five ICP benchmark studies so far (1970, 1975, 1980, 1985, and 1990, the last still incomplete) quantity estimates are derived for each of the categories for the group of countries participating in the benchmark study.

4. In the Penn World Table, a GDP entry is the sum of estimates of the *real* values of the GDP components, Consumption (C), Investment (I), Government (G), and the Net Foreign Balance (NFB). The real values of these components are obtained for a specified base year from an ICP benchmark study, and then are extrapolated forward and backward through time on the basis of the corresponding component growth rates and price indexes from the national accounts. The Penn World Table furnishes output and price estimates for many more countries than the benchmark studies and for many more years. So far, however, estimates are only available for a few broad aggregates.

5. The studies are referred to in terms of the ICP's phases. Phase 1 was directed at working out the benchmark methodology and then applying it on a trial basis to the 1970 price data collected in 10 countries (Kravis, Kenessey, Heston and Summers (1975)). Phase 2 covered those 10 countries plus 6 more in both 1970 and 1973 (Kravis, Heston and Summers (1978a)); Phase 3 covered 34 countries in 1975 (Kravis, Heston and Summers (1982); Phase 4 covered 60 countries in 1980 (United Nations: Parts I and II (1985)); and Phase 5 covered 64 countries in 1985. The methodologies changed over the Phases, but the reported the PWT 5 estimates are based upon revised benchmark results obtained using a uniform methodology applied to revised national accounts data. A country listing of benchmark and non-benchmark countries appears in the paper which introduced the PWT 5 (Summers and Heston (1991)).

6. A table in Appendix B of Summers and Heston (1988) identifies the benchmark treatment(s), if any, of each of the 121 market economies of the PWT4.

7. This process is described in more detail in the unpublished Appendix B of Summers and Heston (1991).

8. Bela Belassa (1964) explained the Gerschenkron Effect in these terms.

9. This statement is not precisely correct. The algorithm used to calculate international prices does not simply pick the prices of a moderate income country. Even if it did, relative prices differ across countries of roughly similar income. For both reasons, international prices do not necessarily correspond to prices within any moderate income country.

10. These remarks only apply to the growth rates. The underlying conditions for the Gerschenkron Effect are built into the cross-section relationships across the countries in the PWT in any particular year. These points are more fully discussed in Heston and Summers (1989), and Nuxoll (1992a), (1992b).

11. This paper examines Domestic Absorption rather than GDP because of the way in which the PWT uses exchange rates in valuing the Net Foreign Balance. The results reported below do not change significantly if GDP is used.

12. There are several outliers in these data. If these outliers are excluded, or if a form of correlation analysis insensitive to outliers (Spearman's or Kendall's correlation coefficient) is used, the same broad conclusions follow.

13. Specifically, per capita DA in 1985 is measured in 1985 international prices. The results do not change if the income in other years is used or if different prices are used. The logarithm is used to capture the non-linearity in the relationship.

14. Neither White's nor Glesjer's test indicates the presence of heteroskedasticity. Even using heteroskedasticity-consistent standard errors, the coefficient is still significant at the 1.3 per cent level.

REFERENCES

Balassa, Bela. (1964). The purchasing-power parity doctrine: A reappraisal. *Journal of Political Economy.* 72(December), 584-596.

Barro, Robert J. (1991). Economic growth in a cross section of countries. *Quarterly Journal of Economics.* 106(May), 407-443.

DeLong, J. Bradford, and Summers, Lawrence H. (1991). Equipment investment and economic growth. *Quarterly Journal of Economics.* 106(May), 445-502.

Gerschenkron, Alexander. (1951). *A Dollar Index of Soviet Machinery Output, 1927-28 to 1937.* Santa Monica: RAND Corporation.

Heston, Alan, and Summers, Robert. (1989). An evolving international and intertemporal data system covering real outputs and prices. Paper for NBER Conference on Economic Growth.

Kravis, Irving B., Heston, Alan and Summers, Robert. (1978a) *International Comparisons of Gross Product and Purchasing Power.* Baltimore: The Johns Hopkins University Press.

_____ (1978b). Real GDP *per capita* for more than one hundred countries. *Economic Journal.* 88, 215-242.

_____ (1982). *World Product and Income.* Baltimore: The Johns Hopkins University Press.

Kravis, Irving B., Kenessey, Zoltan, Heston, Alan, and Summers, Robert. (1975). *A System of International Comparisons of Gross Product and Purchasing Power.* Baltimore: The Johns Hopkins University Press.

Mankiw, N. Gregory, Romer, David, and Weil, David N. (1992). A contribution to the empirics of economic growth. *Quarterly Journal of Economics.* 107(2), 407-438.

Murphy, Kevin M., Shleifer, Andrei and Vishny, Robert W. (1991). The allocation of talent: Implications for growth. *Quarterly Journal of Economics.* 106(May), 503-530.

Nuxoll, Daniel A. (1992a). Sufficient conditions for the Gerschenkron Effect. VPI & SU Economics Working Papers, #E92-16, June.

_____ (1992b). Differences in relative prices and international differences in growth rates. VPI & SU Economics Working Papers, #E92-17, June.

Summers, Robert, Kravis, Irving B. and Heston, Alan. (1980). International comparison of real product and its composition: 1950-1977. *Review of Income and Wealth.* 26(March), 19-66.

Summers, Robert, and Heston, Alan. (1984). Improved international comparisons of real product and its composition, 1950-1980. *Review of Income and Wealth.* 30(June), 207-262.

_____ (1988). A new set of international comparisons of real product and price level estimates for 130 countries: 1950-1985. *Review of Income and Wealth.* 34(March), 1-35.

_____ (1991). The Penn World Table (Mark 5): An expanded set of international comparisons, 1950-1988. *Quarterly Journal of Economics.* 106(May), 327-368.

United Nations and Eurostat. (1986). *World Comparisons of Purchasing Power and Real Product for 1980.* New York: United Nations.

United Nations. (1992). *Handbook of the International Comparison Programme.* New York: United Nations.

Young, Allan H. (1989). Alternative measures of real GNP. *Survey of Current Business.* 69(4), 27-34.

On the Determinants of Price Competitiveness in the World Market: A Comparative Analysis of the United States, Japan and West Germany

Soshichi Kinoshita

The last two decades have seen a drastic change in world economic environments, especially in the world oil market and the exchange rate system, which has made it inescapable for both developed and developing countries to restructure their trade and industry. As a result, one of the major issues for the policy makers and economists concerned has been to capture quantitatively the underlying mechanism working in the process of global structural changes in trade and industry, and its relationship to individual economies. It is important, therefore, to identify the determinants of international competitiveness in relation to the energy price, exchange rate fluctuation and productivity improvement. This study is an attempt to approach this goal by examining the changing price competitiveness of key export sectors in major trading countries. Here we focus our attention on the three leading world economies: the United States, Japan and Germany, and report some econometric results on the pricing behavior in the machinery industry after the first oil crisis[1].

I. Specification of the Price Equation and Data

In modeling the pricing behavior of these three countries, we take for granted that in these economies business firms as a whole behave independently as a price setter in the export market, and set their prices based on the long-run average total cost and mark-up factors. Furthermore we assume a cost-minimizing factor inputs under a Cobb-Douglas technology and constant returns to scale in production. Under these assumptions, average cost is derived explicitly from the production function and input prices by use of the following identity:

$$AC_{it} = A_o \prod (P_{jit})^{aji} (PL_{it})^{bLi} (PK_{it})^{bKi} e^{-ciT} \qquad (1)$$

where

AC_i = long-run average cost,
P_{ji} = price index of j-th intermediate input in i-th sector (equal to Pj),
PL_i = cost of labor in the manufacturing sector,
PK_i = user cost of capital in i-th sector
T = time trend

aji = elasticity of i-th output with respect to j-th input,
bLi = elasticity of i-th output with respect to labor,
bki = elasticity of i-th output with respect to capital,
ci = rate of technical progress.

Producer price and export price are given by the following two equations as:

$$P_{it} = (1 + m_i) AC_{it} \tag{2}$$

$$PE_{it} = B_0 (P_{it})^{b1i} (EXR_t)^{b2i} \tag{3}$$

where

P_i = producer price index in national currency,
PE_i = export price index in US dollar,
m_i = mark-up ratio
EXR = exchange rate in term of US dollar

In Equation (3), b_1 and b_2 are pass-through coefficients for the domestic price and the exchange rate in the foreign market. Coefficient b_2 measures the extent to which exporters adjust their export prices in national currency in response to an exchange rate change. Pass-through can be less than full if b_2 is less than unity in absolute value. Taking into account the collinear relations among individual input prices, a composite price index is constructed as a geometric mean of all intermediate input prices, with weights being derived from the base year input-output table of each country, and then used in the log-linear regression as:

$$\ln(P_{it}) = a_{0i} + a_{1i} \ln(PR_{it}) + b_{Lil} \ln(PL_{it}) + b_{Ki} \ln(PK_{it}) + c_i T_t \tag{4}$$

$$\ln(PE_{it}) = b_{0i} + b_{1i} \ln(P_{it}) + b_{2i} \ln(EXR_t) \tag{5}$$

where PR_i = a composite input price index for i-th sector, defined as

$$PR_{it} = \prod (P_{jit})^{aji} \tag{6}$$

Original price data are taken from individual national sources, and are aggregated, if necessary, to the sectors defined in this study, which roughly correspond to the two-digit International Standard Industrial Classification. Cost of labor variable is the average wage eranings for the manufacturing industry as a whole, compiled by Bureau of Labor Statistics, US Department of Labor. User cost of capital is constructed by using the following definition;

$$PK_i = pk(r + d_i) \tag{7}$$

where *pk* is price index of investment goods, *r* is interest rate and d_i is depreciation rate in *i*-th sector. Detailed description of the estimates of price indexes and other variables are given in the appendix.

II. Empirical Results for the Machinery Industries

The period used in the regression covers the years from 1970 to 1989. Preliminary tests indicated that there was a highly collinear relation among three input cost variables in the equation. To solve this difficulty, we first equate the elasticities of output with respect to inputs with the cost share of each input in the value of the output of the sector. Then the export price, adjusted for the effects of input prices is regressed to time trend and exchange rate variables as:

$$\ln(PE_{it}) - a_{il} \ln(PR_{it}) - b_{Li} \ln(PL_t) - b_{Ki} \ln(PK_{it}) = c_{oi} + c_{li}T + c_{2i} \ln(EXR_t) \tag{8}$$

where

$$a_{li} + b_{Li} + b_{Ki} = 1.0$$

Table 1: Estimated Results

COUNTRY	INDUSTRY	Period I(1970-89)		Period II(1974-89)	
		TIME TREND	EXC. RATE	TIME TREND	EXC. RATE
JAPAN	GM	-0.0354a	-1.1265	-0.0170[a]	-0.9804
		(-4.5391)	(-0.5729)	(-6.0014)	(0.2810)
	EM	-0.0471[a]	-0.8805	-0.0401[a]	-0.8247[a]
		(-14.381)	(1.3592)	(-18.264)	(3.4954)
	TE	-0.0160[a]	-0.8777	-0.0091[a]	-0.8172[a]
		(-4.1684)	(1.1491)	(-2.6868)	(2.2423)
GERMANY	GM	-0.0061[a]	-0.8962[a]	-0.0062[a]	-0.8994[a]
		(-5.7518)	(3.0804)	(-4.6071)	(2.2621)
	EM	-0.0224[a]	-0.8979	-0.0204[a]	-0.9885
		(-11.193)	(1.5697)	(-9.664)	(0.1607)
	TE	0.0051	-0.9249[b]	0.0054	-0.9506
		(0.9438)	(1.8607)	(1.4188)	(0.9598)
USA	GM	-0.0094[a]		-0.0128[a]	
		(-7.9387)		(-10.326)	
	EM	-0.0216[a]		-0.0209[a]	
		(-19.635)		(-12.950)	
	TE	-0.0031		-0.0047[b]	
		(-1.5286)		(-1.8742)	

NOTE: GM is general machinery, EM is electrical machinery and TE is transport equipment. The numbers in the parenthesis of time trend variable are t-value for testing H_0: $c_1=0$, and those of exchange rate are t-value for H_0: $c_2=-1$. a: Significant at the 5% level   b: Significant at the 10% level

The results for the three machinery insustries are shown in the table below, where a regression is run for the whole period, 1970-1989 and the period after the first oil crisis, 1974-1989. The results show some interesting features[2]. First, the time trend variable is significantly estimated with the expected sign in most cases. The exception is for the transport equipment in Germany, showing positive but insignificant value. Second, the highest technical progress is estimated for electrical machinery, followed by general machinery and transport equipment. Among them, the highest is for the Japanese electrical machinery, depicting 4.7% for the whole period from 1970 to 1989. The Japanese machinery industries have outperformed the United States and German ones by around 1 to 3.0 % in terms of technical progress. Third, the rate of technical progress in the Japanese machinery industries has been slowed down by 0.7 to 1.8% after the first oil crisis. We see no such a clear trend for the case of the United States and German machinery industries. This result suggests that Japan was more vulnerable to the oil shock than were the United States and Germany. Fourth, the absolute value of the exchange rate coefficient is significantly lower than unity in the Japanese electrical machinery and transport equipment equations for the period after the oil crisis. This implies that in these industries, around 20% of the yen exchange rate appreciation observed during this period has been absorbed domestically by the firms in the form of depressed selling price in the domestic currency unit[3]. The similar pricing behavior has worked in the German general machinery industry to a lesser extent. Within the framework of the estimated pricing model, the percentage change in dollar export price is given by

$$\Delta PE \, / \, PE_{-1} = c_1 + c_2(\Delta EXR \, / \, EXR_{-1}) + \left[a_l \, (\Delta PR \, / \, PR_{-1}) + b_l \, (\Delta PL \, / \, PL_{-1}) \right.$$
$$\left. + b_k(\Delta PK \, / \, PK_{-1}) \right] \tag{9}$$

One way to assess the relative importance of the determinants is to see how the export price movements for Japan and German are explained after the G5 agreement in the fall of 1985. The dollar depreciation in the following period was accompanied by a sharp rise in both exchange rates and export prices for both countries. Both Japan and German lost their price competitiveness over the United States, but Japan's export prices fell relative to German's ones during the same period, with the decline of 17 to 35 percent from 1985 to 1987 in relative prices. The observed differential changes in export prices for Japan and German are decomposed into individual factors ,as shown in Table 2.

As seen in Table 2, the differential price changes between Japan and German after the G5 agreement are due firstly to changes in input prices and residuals for the three machinery industries. The second important factor is the difference in pass-through ratio in both electrical machinery and transport equipment. This suggests that there must have been a large decline in relative margins on Japan's exports. The differential productivity improvement plays a small role during the 1985 to 1987 period of great exchange rate turbulence.

Table 2: Comparison of Export Price Changes Due to Individual Factors 1985-1987 %

INDUSTRY	COUNTRY	TIME TREND	EXR	INPUT PRICES	TOTAL
GM	JAPAN	-3.4	64.9	-18.0	43.5
	GERMAN	-1.2	57.4	4.5	60.7
	gap	-2.2	7.5	-22.5	-16.8
EM	JAPAN	-8.0	53.5	-15.7	29.8
	GERMAN	-4.0	63.8	5.3	65.1
	gap	-4.0	-10.3	-21.0	-35.3
TE	JAPAN	-1.8	53.0	-8.3	42.9
	GERMAN	---	63.8	3.3	67.1
	gap	-1.8	-10.8	-11.6	-24.2

NOTE. Input prices includes changes in output prices and residuals.

III. Concluding Remarks

In this study, we examined the pricing behavior of machinery industries of the United States, Japan and Germany, and showed empirically how price competitiveness depends on input costs, technical progress and exchange rate. We found that the technical progress in Japan had larger impacts on her export prices, compared to those in the Unites States and Germany. We also found that exchange rate appreciation was not fully reflected in export prices of electrical machinery and tranport equipment in Japan and general machinery in Germany.

Some qualifications should be added to the estimated results of this study. First, our price equation fails to examine the effects of scale economy on average cost changes. Second, capacity utilization and other demand factors are neglected. It is an interesting topic to shed further light on the differential impacts of technical progress in machinery industry among the three big economies studied here[4].

APPENDIX
Data Sources and Construction

The data used in this study are taken from national sources unless otherwise noted.

Japan: Export price index are from EPA's *Annual Report on National Account*. Input price indexes are from BOJ's *Price Index Annual*. Cost of Labor is measured as annual wage earnings of manufacturing industry surveyed by the Ministry of Labor. User cost of capital is constructed as $pk(r + d_i)$, where r is average contracted interested rate on loan taken from BOJ's *Economic Statistics Annual*, and d_i is computed as net retirement/gross fixed capital by using EPA's *Gross Capital Stock of Private Enterprises*. Value shares of intermediate inputs and factor inputs are estimated by using the *1980 Input-Output Table* compiled by Management and Coordination Agency.

Germany: Export price index and producer price index are taken from *Wirtschaft und Statistik*. r is average yield on corporate bond taken from *Monthly Report of the Deutsche Bundesbank*. d_i is constructed as depreciation/capital stock estimated by Ifo-Institute. Value shares of intermediate inputs and primary factors are estimated by using *the Input-Output Table 1980*, based on the European system of integrated economic account.

The Unites States: Producer price index is from *Monthly Labor Review*. Export price index in the 1970s is constructed as a weighted average of producer price index, with weights being 1980 export value share. Export price index in the 1980s is from *Monthly Labor Review*. Wage earning is BLS's weekly wage earning. r is yields on domestic corporate bonds by rating Aaa, taken from *Business Statistics*, and d_i is constructed as net retirement/gross capital stock by *using Fixed Reproducible Tangible Wealth in the United State, 1925-85*, Department of Commerce. *The 1982 Input-Output Table* is used to estimate the value shares of various input components. Exchange rate indexes for Japan and West Germany are taken from *International Financial Statistics*, International Monetary Fund.

Soshichi Kinoshita, Department of Economics, Nagoya University, 1 Furoh-cho, Chikusa-ku, Nagoya 464-01, Japan.

ACKNOWLEDGEMENTS

I am very grateful to Professor F. Gerard Adams for his comments on an earlier draft of this paper. Most of this study was done during my stay at the Department of Economics, University of Pennsylvania in the fall of 1991.

NOTES

1. There are several studies on the international competitiveness of Japan. For example, Tange (1992) compares the cost competitiveness of the Japanese and the United States industries for the 1973-85/86 period.

2. Khosla (1991) examines the pass-through of the Japanese industries for the 1975:1-1987:4 period and reports lower pass-through coefficients for the three machinery industries than this study.

3. Estimation of the equation which included a industry-specific wage earning in place of manufacturing one yields very similar parameter estimates to those of Table 1.

4. Suzuki (1992) examines the high efficiency of the Japanese electrical machinery industry by using a dynamic factor demand model with R & D externalities for the parent and subcontract firms group.

REFERENCES

Khosla, Anil. (1991). Exchange rate pass-through and export pricing - evidence from the Japanese economy. *Journal of the Japanese and International Economies*. 5(1).
Suzuki, Kazuyuki. (1992). R & D, Technology transfer and parent-subcontract firms relationships: evidence from panel data on the Japanese electrical machinry industry. *Discussion Paper*. No. 340, (April), Kyoto University.
Tange, Toshiko. (1992). International competitiveness of U.S. and Japanese manufacturing industries. In Bert G. Hickman (Ed.), *International Productivity & Competitiveness*. Oxford University Press, Oxford.

An East Asian Model of Price and Variety Competition

András Simon

This paper describes an international model system built at the International Centre for the Study of East Asian Development for the East Asian economic region. It was easy to realize at an early stage of the research that in face of the rapid growth in East Asia unparalleled in other parts of the world and its fast expansion on world markets, the "mainstream models" of demand-driven trade used in most macromodels are completely off the mark in the analysis of East Asian developments. A different specification is used therefore that has not been widely accepted as the main approach to modeling international trade, even though it is not without precedents in the economic literature. This restores the role of supply having been suppressed by demand-determined models. The usual demand-side bias of trade models has long been recognized in the literature,[1] and several models that explicitly include supply effects have been built,[2] and even more others which implicitly capture supply elements, or interpret variables or coefficients as ones reflecting supply.[3] However, these econometric models either lack the systematic discussion of the theoretical background and the functional form of the specification and consequently give the impression of being based on ad hoc ideas to achieve better fits, or give unnecessarily narrow interpretation to the supply variables (Martins 1990). To avoid a similar impression, before presenting the model itself, I present an overview of the theories in traditional models in Figure 1.

Figure 1: Market Paradigms in Empirical Trade Models

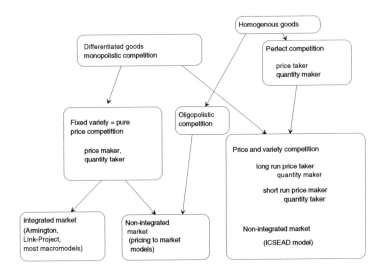

The theory of international trade has developed along the lines of general economic theory. Thinking in terms of the classical theory led to the perfect market model of international trade. With the analysis of more complex market structures and under the influence of Keynesian ideas the demand-determined model of a non-perfect market became the focus of analysis. The last two decades have seen the revival of the classical idea of supply determination, at least for long-run economic development. Empirical models of international trade seem to have been saved from this revival up to now. We shall see that recognizing the importance of non-price competition leads to a more symmetric treatment of supply and demand elements in international trade models.

I. Limitations of the Price Competitive Market Paradigm

The usual Armingtonian model of trade assumes that trade is determined by income and prices. It may be called the demand-determined model, though the term price-competitive may be more suitable to stress the contrast to models where both prices and the number of varieties of goods determine trade. The paradigm of the price competitive market is very restrictive, and it prevents us from reasonably interpreting actual macroeconomic events. The failure of the model can be best illustrated by its explanation of the expansion of Japan in the world market.

Following the logic of the model, we start from the observation, that Japanese goods were never rationed.[4] This means that independent of whether prices cleared the markets exports were determined by demand throughout the entire expansion period. Thus, gains by Japan in trade shares of the world market must be explained by two variables: income (expenditure) and relative prices. Both explanations are untenable both by common-sense and empirical evidence.

Let us see first the income variable as an explanation. According to this approach, Japanese goods are more income-elastic than other goods. We cannot rule out such a possibility *a priori*. It might happen that a detailed analysis would reveal that Japanese exports consist mainly of goods consumed by consumers of higher income. However, such findings have not been revealed. On the contrary, looking back on the early history of Japanese exports, the important products -- textiles, bicycle, steel, ships -- do not seem to be items that provide off-hand proof for such an explanation. As a matter of fact, I do not know of anyone who would support the idea that the observed high income elasticity of Japanese exports would originate from the composition of exports. On the contrary, it has been shown, that apparent income elasticities for Japanese exports are high by industries as well.[5]

There is no reasonable explanation for why Japanese brands of the same goods have generally higher income elasticities than those of other nations, notably the US. The reason why people will buy more Toyotas this year than last year probably has little to do with their rising income. The seemingly high coefficient of the income term in traditional demand equations is attributable to other factors that the demand-determined specification does not capture.[6]

If relative prices are a determining factor in Japanese expansion, prices of Japanese exports must have decreased over the past several decades. Let us compare export prices of Japan and the U.S. in Figure 2.

Figure 2: Ratio of Japanese and U.S. Export Prices

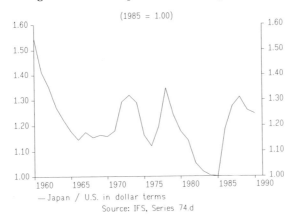

(1985 = 1.00)

— Japan / U.S. in dollar terms

Source: IFS, Series 74.d

The graph suggests, that although in the early sixties we could probably speak about an increasing Japanese price competitiveness, the experience of the last 25 years tells, that the fast expansion of Japanese exports on the world market cannot be explained by its export price dynamics.

Most econometric model-builders would probably share my doubts about the structure of their models. Many of them suppose that export expansion has something to do with domestic expansion of the supplier country. The structure of the price competitive model does not however allow them to break out of the closed triangle of income, prices, and sales volume. I will show below that this constraint on the range of explanatory variables is only a consequence of the price competitive paradigm. A more general market concept permits the supplier to increase its sales even if prices do not change.

Price and Variety Competition

The price competitive model is a unique trading post, where a given set of goods is traded. Giving up any of these two attributes may lead to a reestablishment of the role of supply. If the market consists of several isolated trading posts (micro-markets) any supplier may increase its sales without changing its prices by simply entering more and more micromarkets. Or it may come up with new and still newer goods.

Before a more precise discussion, let us pursue the Toyota example for an intuitive understanding.

People, recalling why they buy more Toyotas today than they did yesterday, even when their prices do not change, could probably list three reasons:

1) They have realized only gradually that Toyota cars are good.

The idea that demand adjusts to prices in a dynamic process is standard in econometric models and does not lead out of the price competitive paradigm.

2) Toyota has gradually expanded both its range of models and its sales network, which means that the range of offered models increased only gradually both in a geographic and a "product space" sense.

This argument leads to a purely supply-side explanation. It recognizes the fact that the market is segmented. Buyers are isolated in space by information and transaction costs. Those who live close to the new Toyota dealer will be able to test the car, get persuaded, and buy it. However, let us suppose that Toyota did not establish itself completely on the whole U.S. market. At the next dealer, who has no Toyotas yet, nobody will buy such a car or even ask for (demand) such a car. People express demand only for goods that are supplied. In this sense, no excess demand is observed, but sales are still determined by supply. We do not have to assume rationing or shortages to account for an important role of supply in total sales. For the analysis of well-functioning market economies we may assume that the market works smoothly, i.e. there are no unfilled orders in any of the micro-markets. If somebody wants to buy a Toyota, he will get it.[7] However, it is reasonable to assume that orders for Toyotas are placed only where Toyota has established its dealers. This is the essential assumption that can be used as the basis of a trade model. The market of the importer country consists of two types of segments: one, where the exporter has introduced its good, the other, where it has not entered yet. In the latter segment trade is trivially 0. Demand determines sales in any micro-market where the exporter has entered, but it is free to determine its total volume of sales by entering any number of submarkets.

3) The Toyota of the 1980s and the 1990s is not the same Toyota that was produced in the 1960s. The quality improvement has been enormous, much more, than the improvement in competing U.S. or European brands. Price statistics do not reflect quality changes adequately. Thus unchanged relative price figures may hide a widening

gap between price/performance ratios of Japanese and other cars.

Let me return to this argument after a more elaborate discussion of the consequences of case #2.

Models of Variety Choice

There are two main types of models that well represent the market structure of the above example. The Chamberlain (1933) - Spence (1976) - Dixit-Stiglitz (1977) representative consumer model and the Hotelling (1929) - Lancaster (1979) - Salop (1979) location model. The former may be called the taste for variety approach, because it relies on the fact that the utility of a representative consumer increases with the number of varieties consumed. The latter may be called the variety of tastes or ideal variety approach. Here tastes of individuals are different, and an increase in varieties increases utilities by allowing each individual to be closer to his ideal variety. Both models assume market clearing.

In the "taste for variety" approach producers offer different varieties of goods, which are non-perfect substitutes. Our example fits into the model if we consider goods offered at different locations as different varieties. The entry decision of a firm into a micro-market is equivalent to a decision to increase the number of varieties of its offered brand. Assuming that the elasticity of substitution among varieties is uniform and constant for the representative consumer, Dixit-Stiglitz arrived at a mathematically attractive simple fomula for the share of any two suppliers on a market:

$$\frac{x_{ij}}{x_{kj}} = \gamma_{ij} \frac{m_{ij}}{m_{kj}} \left(\frac{p_{ij}}{p_{kj}} \right)^{-\sigma}, \qquad \sigma > 1. \qquad (1)$$

where

x_{ij}, x_{kj} - trade flows from countries i and k to country j,
m_{ij} - number of varieties offered by country i in market j (assuming equal weights for a variety),
p_{ij} - price of the variety offered by country i in market j
σ - uniform intra-brand and inter-brand (inter-variety) elasticity of substitution.[8]

The attractiveness of this approach is in the simple analytical form of the derived demand function. Its disadvantage in our case comes from the assumption of a representative consumer making the model difficult to interpret in a case when goods are differentiated by location; the demand functions of consumers are necessarily different by location.

The "variety of tastes" model of Lancaster-Salop considers varieties to be distributed in space. The attractiveness in this approach is that it gives a unified interpretation to differentiation by both physical and locational characteristics of products. Whether we imagine that varieties are distributed in a characteristic or in a geographic space, the properties of the model do not change. The example of the dealer network fits into the locational interpretation of this model, while a diversification of Toyota car models into luxury and compact cars would need a physical interpretation of the characteristic space in the model. Whatever the nature of the characteristic space, the demand function will be the same and very similar to that derived from the taste for variety case:

$$\frac{x_{ij}}{x_{kj}} = d\,(m_{ij}\,,\ m_{kj}\,,\ p_{ij}\,,\ p_{kj}\,)\,, \tag{2}$$

where trade is affected positively by the number of varieties and negatively by the own-price of supplied goods.

As in the former case, for monopolistic competition to exist the price elasticity of demand has to be greater than 1. The expected signs of the first derivatives are similar to those of equation (1), although this model does not suggest a specific functional form.

In both cases the price elasticity of demand is > 1 to assure equilibrium on the monopolistically competitive market.

Supply Behavior

Consider a firm operating in several countries in several micro-markets or, putting it in another way, producing several varieties of a good for several markets. It has to choose the varieties and their pricing so as to maximize profits. Let us assume that this firm takes as given the variety of choices and prices of other firms. The decision problem of the profit maximizing firm will be the following:

$$\max_{[\,p_{jk}\,,\,m_j\,]} \left\{ \sum_{j=1}^{N} \sum_{k=1}^{m_j} (p_{jk} - c_{jk})\,x_{jk} \right\} \tag{3}$$

where

p_{jk} – price of kth variety offered in country j
c_{jk} – cost of producing and selling the kth variety in country j
x_{jk} – demand for the kth variety offered in country j

Subscripts by firms were omitted.

The firm has two parameters to decide upon when competing in the market: price and the number of entries into micro-markets. The price set by the firm is an increasing function, while the number of varieties offered is a decreasing function of costs. The latter becomes plausible if we recall that lower costs make more varieties produce positive profits. The number of varieties is an integer in equation (3), which formulates the problem in the spirit of the taste for the variety approach. Defining products as points in product space would allow us to consider the offered density of products as a continuous function of costs. In this case we regain the property of the classical model: on the market as a whole the supplier faces a horizontal demand curve; whatever the price, its sales are limited only by its cost function. The model differs from the classical case in that the choice of the firm has an additional degree of freedom, the freedom to decide its price in each micro-market.

Actually the new model blends the features of the clasical, price-taker and quantity-"maker", and the demand-determined, price-maker and quantity-taker, models. Looking from the aspect of the entry decision, the firm seems to be a price-taker, while in each micro-market it behaves as a price-maker. Considering the different horizons of the two decisions, we might even say, that firms are price-takers in the long run and price-makers in the short run.

Quality Changes not Reflected in the Price Index

A Toyota car of today is not the same as a Toyota car in the sixties: it is highly probable that its quality relative to its competitors has been improved much in the past decades, while its relative price did not change. In other words, the price of Toyotas in terms of a car of constant quality has fallen. According to this line of thought there is a "true" but statistically unobservable price, charged for a unit of quality. This is the price that we experienced to decrease against competitors in the case of Japanese exports in the past decades. If these prices were observed statistically, we would see, that prices are important in explaining trade, and competition is price competition not only in the short but in the long run as well.

Considering Japanese expansion as a result of quality improvement is a reasonable and probably true interpretation of events, and the conclusion conditional on the availability of statistics is correct. Stressing the role of quality in competition is a useful idea, because it enriches our knowledge on what is happening in the market, what the strong or weak points of Japanese or other competitors are. However, when setting up a quantitative model to explain exports, we pursue a pragmatic objective: to explain trade flows as functions of statistically observed variables. The idea of a constant-quality good and a price of a quality unit different from the data used in statistics does not contribute to this effort.

Cars of different vintages and therefore different qualities may be considered as different varieties. From such a point of view an improvement of quality is a case of an increase in varieties discussed before. Nevertheless, a new model of the same brand is a special variety as being a close substitute for the old model. This drives the old model rapidly out of the market, making it difficult for statistics to observe the marginal rate of substitution between the old and the new model. As a result, existing practical methods may not be able to distinguish between the marginal rates of substitution in cases of faster and those in cases of slower quality changes. However regretful we feel that statistics do not record (relative) quality improvement as an increase in the value of a good, we have to adjust to the facts of life and settle without explaining unmeasured variables. Instead it is more useful to consider how the existence of quality changes effect trade flows as measured by statistics. From this aspect we can say that the general model of price and variety competition does include elements that may allow for the quality change phenomenon, while the purely price competitive model cannot capture this element.

The price competitive model has the same deficiencies as before, there is no way of interpreting unrecorded quality changes in the framework of the model. Prices do not reflect them, and relating their effects to income would make no sense. In contrast, in the variety and price competitive model, the decision on a new -- better -- quality may be treated analogously to a decision about a new variety. Both depend on the same cost-revenue comparison of the supplier. The fact that a new model is a close substitute may make it reasonable to interpret the introduction of a new model as an "incremental variety": the improvement over an old model may be considered as an independent variety, which the firm may introduce or not. Its decision will be based on the costs of improving the model and the expected additional revenue from the change. The same elements of supply decision appear, and consequently the same form of equation applies as in the general case.

II. Empirical Model Specification

Entries

The number of entries and exits of a firm are determined by costs. In macroeconomic applications this variable is not directly observable. Even in the case when suitable aggregate cost data are available the explanation of productivity changes is usually out of the range of econometric models. My suggestion for a quantification of cost effects is to use sales capacity as an indicator of cost efficiency. The choice seems to be appropriate because, under profit-maximizing conditions, the decision on an increase in capacity is a direct consequence of cost efficiency. More specifically, we may assume, that the firm makes a decision on its total capacity on the basis of comparing expected costs and revenues. This decision is not explicit in macroeconomic models, capacity is derived usually from inputs of capital, labor and an exogenous technical progress, and I do not have any better idea either. However, we may explicitly model the decision of the supplier on how to allocate its production capacity by destination of sales. Entry is a condition of sales on a micro-market. Thus, the number of entered micro-markets shows directly the intention of the supplier about the distribution of its output capacity. As markets are non-integrated, prices differ. So may differ local sales and distribution costs. The entry decision depends on these local conditions. In a country, where the price level increases, profit-maximizing firms will enter more micro-markets.

Let us assume that $\bar{p}$ is the average (world) market price that determines the capacity-decisions of firms. The planned share of country j in total sales originating from country i then depends on the ratio of the expected price level in country j to the average price:

$$\frac{\overline{x_{ij}}^{\,s}}{\overline{x_{i.}}^{\,s}} = s\left(\frac{\overline{p}_{.j}}{\overline{p}_{..}}, e^{\lambda_{ij}t} \right), \qquad (i = 1, \ldots, N, \quad j = 1, 2, \ldots, N) \quad (4)$$

where

$s\,(\ )$ - behavioral function of the supplier firm (negative in the relative price term)

$\overline{x_{ij}}^{\,s}$ - planned export sales of country i to country j

$\overline{x_{i.}}^{\,s}$ - planned output capacity of country i (exogeneous)

$\overline{p}.j, \overline{p}..$ - expected prices in market i and in the world respectively, in a common currency unit

t - time.

The decision may be influenced by changes in the "distance" of market i from market j : risks, transaction costs, trade barriers. These effects are captured by the trend factor. A λ rate of growth may be considered as an indicator of the rate of integration. It is probably reasonable to assume that the rate of international integration is a parameter depending on the policy and institutions of the importing country. This means, that if the country does not discriminate among exporters, $\lambda_{ij} = \lambda_{kj}$ for $k \neq j$. There is, of course, no reason to assume, that this rate is equal to the rate of internal integration, λ_{ij} .

Let us assume that in the long run competition is purely non-price competition: firms expand their shares only by entering new markets or improving quality at the same price. This means, that in a given micro-market firms do not plan to change their share. Accordingly, the share in a market of the micro-markets entered by a firm or a country is equal to the share of planned volumes.

$$m_{ij} = m_{.j} \frac{\overline{x_{ij}}^{\,s}}{\overline{x_i}^{\,s}} , \qquad (i = 1, 2 \ldots, N , \quad j = 1, 2, \ldots, N) \tag{5}$$

where

m_{ij} - number of entries in market j by supplier i

$\overline{x_i}^{\,s}$ - total expected sales in market j

$m_{.j}$ - total number of micro-markets (varieties) in market

Demand and Sales.

The demand function of a differentiated market was formulated in equations (1) and (2). For econometric estimations the functional form of the Dixit-Stiglitz model seems to be most convenient. If we had direct observations of the number of entries, equations (1), (5) and a price equation would give the model of sales. In the absence of data, however, the variables of the number of entries have to be substituted. By substitution from (5) and restricting the function of the supplier to be additive in logarithms, we arrive at the function of the real share of country i to country k in imports of country j :

$$\frac{x_{ij}}{x_{kj}} = \alpha \, \frac{m_{ij}}{m_{kj}} \left(\frac{p_{ij}}{p_{kj}} \right)^{-\phi} = \alpha \, \frac{\overline{x_i}s}{\overline{x_k}s} \left(\frac{p_{.i}}{p_{.k}} \right)^{-\phi} \left(\frac{p_{ij}}{p_{kj}} \right)^{-\sigma} e^{(\lambda_{ij}\,\lambda_{kj})^{\,t}} t \tag{6}$$

where

α is a constant, and ϕ is the price elasticity of supply.

($\phi_{ij} = \phi$ is assumed for the sake of simplicity.)

It means, that the rate of shares of two supplier countries depends on the ratio of the numbers of their entries as well as on demand. The former depends on their relative output capacities and the expected relative domestic prices in the two countries.

Expected Prices.

Expected relative prices of markets depend on productivity and exchange rate trends. We assume: (1) expectations are backward-looking, and (2) expectations are formed on the basis of domestic prices and exchange rates, with identical pass-through coefficients.

Goods and markets are heterogenous, especially in a macroeconomic model. This means that exportables and domestic goods differ in the rate of productivity and price growth. In most countries productivity is growing faster in the exports sector than in the economy as a whole, which means that the relative price of exportables is decreasing with time. Unfortunately, price statistics do not distinguish domestic prices of exportables and non-tradeables. When using domestic prices instead of prices of exportables, we had to take this difference in their price trends into account. The simplest way was to include a time trend by exporting countries into the equations of relative expected domestic prices. Results of the estimations will suggest that this method might not have been fully satisfactory.

$$\log \frac{\overline{p_{i.}}}{p_{.j}} = \beta_{0ij} + \beta_{1j} \log \left(\frac{p_i^d / r_i}{p_j^d / r_j} \right)_{-1} + \beta_{2j} t \quad (j = 1, 2, \ldots, N, \ i \neq j) \quad (7)$$

where

$$p_{i.} \ , \ p_{.j} \quad - \quad \text{expected future average domestic price converted into common currency unit}$$
$$p_i^d \ , \ p_j^d \quad - \quad \text{local-currency based domestic price}$$
$$r_i \ , \ r_j \quad - \quad \text{exchange rate (local/dollar)}$$

We used a simple one-year lag structure to capture formation of expectations. As expected prices cannot be observed directly, the equation was not estimated, but only substituted into the trade volume equation.

Trade Volumes.

Equation (6) does not automatically satisfy the adding-up restriction. To ensure adding-up, the simplest procedure is used in this model: to express the share of the domestic supplier in the market as a residual. This is the same method used in demand-determined trade models.

Our assumption means that the flow x_{ij} is determined from the identity:

$$x_{jj} = x_j - \sum_{i \neq j, \ i = 1}^{N} x_{ij}. \quad (8)$$

while for $i \neq j$ we substitute (7) into (6) in a way, that the denominator in the relative share formula is always the importing country. This means that the denominator on the left-hand side is domestic sales, and domestic price is used as "export" price to the domestic market.

$$\log \frac{\overline{x_{ij}}}{x_{jj}} = \delta_{0ij} + \log \frac{x_{i.}^s}{x_{j.}^s} + \delta_{1j} \log \left(\frac{p_i^d / r_i}{p_j^d / r_j} \right)_{-1} + \delta_{2j} \log \frac{p_{ij}}{p_j^d / r_j} + (\delta_{3j} + \delta_{4i}) \, t$$

$$(\ i = 1, 2, \ldots, N, \quad j = 1, 2, \ldots, j - 1, j + 1, \ldots, N \ ), \qquad (9)$$

where

p_{ij} , p^d_j - export price (in dollar), domestic price (in local currency).

$\overline{x_{i.}}^{s}$ -- total output capacity (planned sales) of country i.

Planned sales are again unobservable directly. By definition they are equal to planned capacity. Assuming that planned and actual capacity data are highly correlated, we used actual capacity data as indicators of total planned sales. This means that $\overline{x_{i.}}^{s}$, $\overline{x_{j.}}^{s}$ are output capacities of country i and j respectively.

This flow-by-flow modeling approach has the advantage that we do not have to cope with the difficulties of two-level budgeting. Having solved the problem of the total budget constraint, we do not have to add up trade flows to an imposed total import variable.

This convenience, of course, does not make up for the many deficiencies of the model in failing to meet most of the requirements of demand theory: cross-price responses, etc.

Export Prices

Naturally, the purely cost-based pricing models of the integrated market paradigm are highly inadequate in this approach, which assumes the existence of *separated* markets as the main scene of non-price competition. In the specification of pricing behavior I use a "pricing to the market" specification following from recent models of international market structures. It means, that for each country N different prices are determined, one for the domestic market, and N-1 for exports.

Pricing behavior is modeled asymmetrically. Domestic price equations are specified in the country models individually, explaining prices by cost factors. These prices serve in the trade model as anchors of the price level, as prices of trade flows depend only on other prices. Each trade-flow price depends on the domestic prices prevailing in the corresponding exporter and importer country. Prices of both the exporter and the importer country affect trade prices through a distributed lag function. These prices are defined in a common currency, and price changes are not broken down into cost or exchange rate changes, which means that domestic cost effects have the same pass-through properties as exchange rate changes. Values of the distributed lag coefficients are constrained to ensure long-run terms of trade neutrality of devaluations. If weights of export prices are relatively more front-heavy than those of import prices, the terms of trade deteriorate in the short-run.

$$\log p_{ij} = \alpha_{2ij} + \gamma_{1j} \log p_i^d / r_i + \gamma_{2j} \log p_j^d / r_j$$
$$+ \gamma_{3j} \log (p_i^d / r_i)_{-1} + \gamma_{4j} \log (p_j^d / r_j)_{-1}$$
$$+ \gamma_{5j} \log (p_i^d / r_i)_{-2} + \gamma_{6j} \log (p_j^d / r_j)_{-2}$$
$$+ \gamma_{7j} \log (p_{ij})_{-1} + \gamma_{8j} t$$

(10)

$$\gamma_{1j} + \gamma_{3j} + \gamma_{5j} = \gamma_{2j} + \gamma_{4j} + \gamma_{6j}$$

where

p_{ij} - dollar-based export price
p^d_i, p^d_j - domestic price
r_i, r_j - exchange rate (local/dollar).

The coefficients are restricted to be the same across exporters in a market.

Let me discuss the virtues of this specification in comparison with more traditional models.

In most of the Keynesian models the rigidity of export prices is usually assumed to be in terms of the domestic currency, which means that devaluation creates a terms of trade deterioration. The price discrepancy between export prices and import prices persists for ever, unless the country models are specified in a way that a price adjustment mechanism is incorporated. After the price adjustment, if the specification followed international trade theory, the monetary approach of the balance of payments would become the only relevant explanation to the trade balance. However, in most of the country models the monetary blocks are not sophisticated enough to meet this theoretical requirement and the terms of trade remains the only factor that is changed by a devaluation. This means that the terms of trade effect is a necessity in these models because there is no other mechanism that could bring about more imports and less exports; as trade is determined purely by prices and incomes, and the income effect does not work out lacking a suitable monetary mechanism in the models, the only way to decrease imports is to increase their relative price in the importing country.

This will not be true any more if supply decisions are incorporated into the model. A switch of demand needs a relative price change *within* a country, but a switch of supply occurs if prices differ *across* countries. The trade balance will change if price *levels* differ, whatever the terms of trade may be, because arbitrage on part of the *producers* (traders) takes place. The idea is 400 years old in theory, it is a wonder how it could escape so many empirical models. It is a wonder, even though it has its explanation: the idea of the market of price-makers has entrenched itself in the minds of model builders in its purely price-competitive form, replacing the classical model not only by enriching it, but throwing out its useful elements as well.

If such a price level effect is incorporated into the model, there is no need for the artificial assumption of a steady terms of trade *effect*. It is more realistic to assume the terms of trade to be steady, which means terms of trade neutrality in the long run.

In this paper I do not meet the challenge to estimate how long this long run means, I rather force upon the equation an assumption, that prices adjust towards terms of trade neutrality after a three-years period. Terms of trade neutrality is not equivalent to the law of one price, which means equalization of price levels as well. The assumed 3-year adjustment of the terms of trade may be faster than the adjustment of the price levels. More importantly, this adjustment does not have to involve monetary sectors of country models, which are usually not suitable to handle international price level adjustments anyway.

To sum up briefly the features of the trade model, the terms of trade effect works in the short run, but it fades away within 3 years. The price level effect may work indefinitely or as long as the structure of the country models permits them to persist. It means, that the trade model itself is not able to ensure PPP. This problem has to be addressed in the country models.

Other Variables

Each current-price flow (the diagonal x_{ij} not included) is calculated as a product of volumes and prices:

$$x\$_{ij} = x_{ij}\, p_{ij} \quad \text{for } i \ne j \tag{11}$$

Total exports in constant and current dollars:

$$x_{.i} = \sum_{i \ne j,\, j = 1}^{N} x_{ij} \tag{12}$$

$$x\$_{.i} = \sum_{i \ne j,\, j = 1}^{N} x\$_{ij} \tag{13}$$

Total imports in constant and current dollars:

$$x_{j.} = \sum_{i \ne j,\, i = 1}^{N} x_{ij} \tag{14}$$

$$x\$_{j.} = \sum_{i \neq j,\, i=1}^{N} x\$_{jj} \tag{15}$$

Balance of trade in manufactures:

$$b\$_j = x\$_i - x\$_j \tag{16}$$

Export and import price aggregates:

$$p_{i.} = \frac{x\$_i}{x_i} \tag{17}$$

$$p_{j.} = \frac{X\$_{.j}}{X_{.j}} \tag{18}$$

III. The Model System

East Asia is remarkable in the world economy for its high rates of economic growth and the expansion of its share in world trade. It is an area whose performance seems to make it particularly appropriate for testing the role of supply in trade theories to explain shares in world trade, rather than demand. In the following I describe a model built along these lines. The arguments for the specification are presented in the first part of this paper. Here I shall explain only some technical details of the model structure, data, model estimation techniques, as well as results of the estimation and some simulation exercises.

The model system consists of the trade model itself and a series of country models linked together through trade. The trade model is *limited to trade flows of manufactures among countries within the area.* The system comprises an area of 11 countries.

For seven of the countries we installed into the system full-fledged econometric models, built either at ICSEAD or at national research centers. The following country-models were used:

USA-Japan: ICSEAD model (Inada-Wescott)
Korea: a yearly version of the KDI quarterly model
Taiwan: model of Academica Sinica (Lo et al.)
China: ICSEAD model (Simon)
Philippines: ADB model based on PIDS model
Thailand: ADB model based on TDRI model

For Hongkong, Singapore, Indonesia, and Malaysia inputs into the trade model were treated as exogenous.

Simulation of the system proceeds sequentially, country by country and by trade. There are three types of variables:

1. variables internal for the country models.
2. variables internal for the trade model
3. linkage variables.

When simulating a country model, only variables internal to the country model as well as the linkage variables are calculated. When simulating the trade model, only variables internal to the trade model as well as linkage variables are calculated. Linkage variables are either

a) outputs (endogenous) of the country models and inputs (exogenous) to the trade model:

- output capacity
- domestic demand
- domestic prices, or

b) outputs (endogenous) of the trade model and inputs (exogenous) to the country models:

- export prices
- import prices
- export quantities
- import quantities

The trade model determines prices and quantities for each trade flow, but passes only aggregates of exports and imports to the country models. These variables contribute to determine domestic demand, prices and capacities, which again feed back to the trade model. Trade with other areas or of other goods are not included in the trade model, although these variables may be explained within the country models independently.

This paper does not discuss the structure of the country models.

Data Constraints and Data Creation

The statistical requirements of estimating the described model are very high, flow-by-flow trade and price data are needed. The Department of Development Research and Policy Analysis at the U.N. Secretariat has compiled a series of trade matrices and unit value indexes. This database was used in the model as the basis of trade data. However, the statistical principles followed in setting up the U.N. trade matrix were specifically designed to make the data suitable for the use by one class of models, the demand-determined, full price-pass-though paradigm. (It has been serving this purpose, representing the fundamental data source for the Project-Link Model.) According to this paradigm, export prices differ only by exporting countries; import prices are weighted averages of export prices. Because this principle has been accepted by the compilers of the data, unit value indexes are calculated from the raw data only for exporters, and import deflators are derived from these by averaging. Therefore, flow-by-flow price indexes are defined to be statistically identical across importers. Needless to say, it does not make much sense to estimate equation (10) on the basis of this data.

A correct estimation of the price equation system needs flow-by-flow price data. Compiling such a database from the raw data of 5 digit SITC unit value indexes would have exceeded our resources, so the following short-cut method has been chosen.

Import deflators for total manufactures in the 11 countries considered were readily available from national sources. In deriving the import deflator for the rest of the world the $\sum_{j=1}^{N} x_{.j} \equiv \sum_{i=1}^{N} x_{.i}$ identity was used. By deflating import values from the UN database by these deflators we arrived at import volumes by countries. However, these import volumes were not consistent with the matrix of trade volumes derived by the U.N.. An iteration by the RAS algorithm was used to create compatibility. Certainly this procedure created flow-by-flow data, but it did not create additional flow-by-flow statistical information. Thus the empirical basis of the model is much weaker than the number of "observations" would suggest.

Estimation Results

We decided to use the cross-section information of country-observations in estimation, assuming that the parameters of price responses are the same in each flow of trade and that trends are different. The reason for using panel data was not only our desire to gain degrees of freedom in the estimation. We would overstretch the limits of our statistical information if we tried to derive flow-by-flow parameters from a database that does not in fact contain original flow-by-flow information.

Eleven countries were included into our model and the time period covered by the data is 1972-1987. Originally we had preferred to build one panel from the 11x11x16 observations. Because of computing limitations however, we had to split the data. We grouped trade flows by importing countries and created two groups. The first group included U.S.A., Japan, Korea, Taiwan, and Singapore. In this group we tried to cover the most industrialized or major countries. The second group comprised China, the Philippines, Hong-Kong, Thailand, Malaysia, and Indonesia. In addition, we had to cut both panels into halves by observations. Thus, we estimated coefficients for the period 1972-1979 and 1980-1987.

For the estimation we used the iterating version of the estimated generalized least squares method (EGLS), which means a weighted least squares with empirically determined heteroscedasticity correction for each cross-section.

Table 1 shows the estimation results of equation (10). Coefficients are significant. The lag structure of the price coefficients gives relatively more weight to the prices of the exporter country in the early periods, indicating an early deterioration in the terms of trade, which levels off gradually in an adjustment process.

Table 1 Pass-through Elasticities in Price Equations

	Price in exporter country			Price in importer country			lagged endogenous	R^2	*DW*
	(t)	(t-1)	(t-2)	(t)	(t-1)	(t-2)			
Group 1,	0.58	0.04	-0.45	0.21	-0.13	0.09	0.37	0.979	1.99
period 1	(10)	(0.4)	(8.8)	(5.2)	(2.5)	(2.5)	(7.1)		
Group 1,	0.75	-0.3	-0.34	-0.06	0.15	0.01	0.62	0.958	2.55
period 2	(18)	(4.4)	(7.6)	(2.4)	(5.3)	(0.3)	(12.3)		
Group 2,	1.06	-0.57	-0.15	0.28	-0.01	0.07	0.3	0.979	2.19
period 1	(16)	(5.6)	(2.3)	(4.9)	(0.2)	(1.0)	(5.9)		
Group 2,	0.55	-0.5	-0.04	0.01	-0.01	-0.00	0.12	0.917	1.84
period 2	(21)	(9.5)	(1.0)	(0.8)	(0.3)	(0.01)	(2.1)		

Table 2 shows some of the estimation results of the trade equations. The price elasticities seem to be rather low. As it will be seen at the simulations, this leads to a situation where the Marshall-Lerner conditions need five years to be manifested after a devaluation of the Japanese Yen. Even though there are many reasons to substantiate a long adjustment period, the estimated coefficients of prices may have a downward bias. The problem might be, that the numerator and the denominator of the price rate does not refer to the same basket of goods as the denominator comprises non-traded goods. As prices of non-traded goods are relatively more independent from those of exports than prices of tradables, this fact may exaggerate the fluctuations in relative prices and distort the measured price elasticity downward.

For the second group of countries, a supply response to domestic price changes could not be detected. This might mean that, in the small markets comprised by this group, market power of supplying exporters is strong enough so that they do not have to adjust to the local price level. In this case the statistically observed price differences reflect only differences in baskets of goods.

Table 2: Price Elasticities of Demand and Supply			
Import price / domestic price (demand response)	Relative domestic price (supply response)	R^2	DW
Group 1, period 1	-0.60	-0.44	0.999	1.60
	(6.1)	(6.2)		
Group 1, period 2	-0.23	-0.22	0.999	2.31
	(2.8)	(6.6)		
Group 2, period 1	-0.96	--	0.999	1.64
	(8.0)			
Group 2, period 2	-1.05	--	0.999	1.50
	(11.5)			

IV. A Comparative Simulation

Simulation exercises of a trade model are suitable for calculating international multiplier effects of country shocks. A common question for example is, how does a change in the United States economy affect the East Asian countries? How much truth is there in the saying that "a sneeze in the US makes East Asia catch cold"?

The traditional demand-determined models would predict a high sensitivity of East Asia to a U.S. disturbance. If we assume that exports of a typical country to the U.S. have an income elasticity of 2 or 3, a 1 percent recession on the other side of the Pacific Ocean will create a 2 or 3 percent setback in exports, which may have an effect on output in the range of 0.5 percent. Our model does not suggest such a dependency of these countries on the U.S. economy. It may be interesting to see how much the predictions of the two approaches differ. To make such a comparison on a common basis, we have set up a model, using the same country models, but with an alternative, demand-determined, trade model specification.

The Demand Determined Alternative

This model is in fact not a typical Keynesian trade model. To save the effort of maintaining an alternative trade-prices databank created along the lines of the full pass-through assumption, we kept the trade equations of the "pricing to the market" approach. Only equation (9) was replaced by a traditional demand equation:

$$x_{ij} = \varepsilon_{0ij} + \varepsilon_{1j} \log x_{jj} + \varepsilon_{2i} \log \left(\frac{p_{ij}}{p_j^d / r_j} \right) \qquad (9b)$$

Even this half-way move towards the Keynesian model changes the behavior of our model significantly.

Estimation results of the demand equations are summarized in two tables.   Table 3 contains price elasticities and statistics for the estimated equations, while Table 4 gives the income elasticities. Price elasticities were forced to be equal across country groups, while income elasticities may differ by countries.

Table 3: Price Elasticities in the Demand Model			
	Price Elasticities ($\epsilon 2$)	R^2	DW
Group 1, period 1	-0.77 (9.5)	0.999	1.46
Group 1, period 2	-0.92 (11.7)	0.999	2.52
Group 2, period 1	-0.84 (8.4)	0.999	1.59
Group 2, period 2	-1.23 (15.1)	0.999	1.62

To compare the behavior of the two models, we ran the same scenarios with both the demand-determined and supply-determined model.[9]

Table 4: Income Elasticities in the Demand Model			
	Period 1		Period 2	
Countries	elasticities	t-ratios	elasticities	t-ratios
USA	4.3	(19)	4.2	(28)
Japan	19	(6.4)	1.5	(6.3)
RKorea	1.4	(12)	1.3	(9.6)
Taiwan	0.8	(6.8)	1.0	(2.6)
Philippines	3.9	(18)	1.9	(2.4)
Thailand	0.7	(8.9)	1.3	(10)
Hongkong	1.2	(16)	1.6	(14)
Indonesia	0.2	(1.0)	2.2	(7.6)
Malaysia	1.0	(10)	1.4	(12)
Singapore	1.4	(8)	0.5	(2.8)
China	2.3	(4.6)	2.2	(7.6)

Demand Shock

First we simulated the effects of a demand shock on East Asia. The following question was answered: how would the economy change if U.S. propensity to consume decreased by 1 percent in a given year? As a sample of the results, computed values of some variables of the U.S. and Japanese economies are given here. Results for other East Asian countries are very similar. For each variable the results are given in two lines. The first line contains the values derived from the model of ICSEAD, while the second line shows computed values from the demand-determined model.

Table 5: Model Validation (Simulation period : 1972 - 1987)							
Supply Model				**Demand Model**			
exports							
Variables	RMSE	AAE	RMSPE	AAPE	RMSE	AAE	RMSPE	AAPE
US	1244.8	11010.98	.97	.81	1854.16	1498.41	1.45	1.18
Japan	5387.99	4017.43	3.76	3.06	5197.26	4063.72	4.14	3.38
Korea	608.19	493.81	4.54	3.76	1351.45	931.15	10.78	7.37
Taiwan	754.71	619.25	4.62	3.74	1733.84	1090.45	6.27	5.01
Singapore	740.98	585.14	8.88	7.61	830.98	533.22	6.82	5.01
Philippines	261.31	155.45	14.51	9.14	357.84	276.55	22.71	15.56
Thailand	281.13	183.65	7.01	5.76	340.19	204.56	7.35	5.86
Hongkong	1868.68	1206.14	7.11	5.58	1990.07	1467.30	9.43	7.64
Indonesia	482.83	307.67	15.38	13.02	358.54	240.18	12.78	10.58
Malaysia	179.66	158.88	4.91	4.17	305.43	239.53	6.80	5.70
China	726.35	535.52	5.40	4.71	1560.76	824.69	7.22	5.55

	Supply Model				**Demand Model**			
export prices								
Variables	RMSE	AAE	RMSPE	AAPE	RMSE	AAE	RMSPE	AAPE
US	.00	.00	.42	.36	.00	.00	.42	.36
Japan	.03	.02	2.74	2.19	.03	.02	2.72	2.17
Korea	.01	.01	2.11	1.49	.02	.01	2.19	1.52
Taiwan	.02	.02	2.63	2.02	.02	.02	2.63	2.01
Singapore	.04	.03	4.79	3.56	.04	.03	4.90	3.65
Philippi	.04	.03	5.17	4.08	.04	.03	5.28	4.11
Thailand	.02	.02	3.01	2.27	.02	.02	3.02	2.27
Hongkong	.02	.01	2.45	1.73	.02	.01	2.34	1.71
Indonesia	.05	.04	7.57	6.37	.05	.04	7.38	6.26
Malaysia	.02	.02	3.20	2.81	.02	.02	3.17	2.78
China	.03	.02	3.76	3.03	.03	.02	3.68	3.01

The two models imply a very different sensitivity between the Japanese economy and the U.S. economy. A one-year shock in consumption in the United States has a multiplicative effect, which is to a large extent passed through to Japan in the demand-determined case. In the first year the shock creates a 1.3 percent increase in U.S. G.D.P. and a 0.22 increase in the G.D.P. of Japan. This seems to be an unreasonably high rate, as Japanese exports do not represent more than about 1.2-2.0 percent of U.S. G.D.P. and 3-5 percent of Japanese G.D.P.. In the Supply-model there is less pass-through. The multiplier effect is higher within the country and lower in Japan. In the first-year the multiplier of G.D.P. is 0.06 percent. This difference in the multipliers arises from differences in import elasticities. In the demand model any demand shock has an exaggerated effect on the international market by the assumed high import elasticities.

V. Simulation Results

The distortion of the demand model is obvious without running any simulations if we consider the case of a recession in the U.S.. In case of a drop in demand the demand model predicts a larger than average drop in exports for the fast growing East-Asian counties. In fact, experience shows that they are more resilient against recessions. Analysts thinking in terms of the demand model have to speak about an asymmetry between contraction and expansion in the market. In fact, demand does not behave asymmetrically, it is only omitted supply that makes the demand model behaving asymmetrically, as it adds to sales both in peaks and troughs of the business cycle.

Figure 3: Simulation Statistics Root Mean Square Percentage Error (Simulation Period: 1972 - 1987)

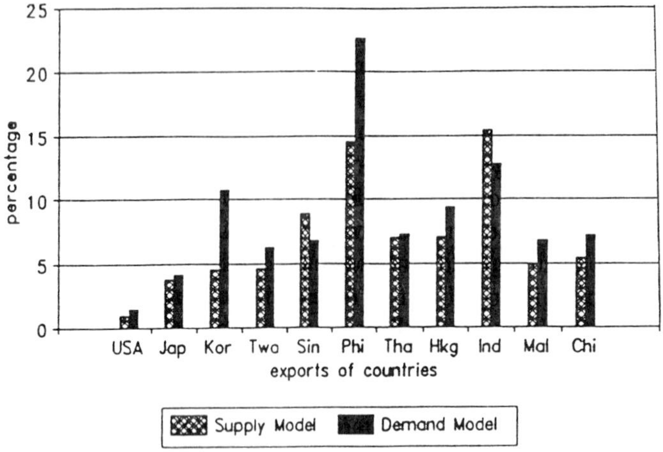

RMSE = root-mean-square-error
AAE = average-absolute-error
RMSPE = root-mean-square-percentage-error
AAPE = average-absolute-percentage-error

Exchange Rate Shock.

The exchange rate shock simulation calculates the effects of a 10 percent devaluation of the yen against all other currencies in the region

It is striking how long it takes for the trade balance to improve by the prediction of the supply model. The result is in contrast to the "optimistic" prediction of the demand model, where there is no lag at all and the volume effect is pervasive from the first year. In the supply model in the first 4 years the terms of trade deterioration outweighs the effects of trade volume changes and it is only in the 5th year that the trade balance starts to improve in value terms. Even though some bias of the estimates of price elasticities may exaggerate this length of adjustment, as was mentioned above, the events of the last 6 years in Japanese-U.S. trade seem to confirm this result. Only recently has the impact of the dramatic drop in the value of the dollar against the yen starting in 1985 been showing up in the bilateral trade balance. No wonder, if we take into account that the forces behind the switch are on the side of supply. Firms had to reevaluate their development plans, redirect market strategies, establish new factories by direct foreign investment to replace exports, etc. On the surface of the statistics these events may appear to be a decrease in income elasticities, as it was noted by several analyses in Japan.[10] In fact, it is a supply effect, a change in export strategies of Japanese firms.

Table 6: Multiplier Effects of a 1 Percent Shock in U.S. Propensity to Consume in the 1st year						
U.S.A		year 1	year 2	year 3	year4	year 5
Import	Supply model	0.51	0.05	0.06	0.07	0.05
	Demand model	1.89	0.65	0.16	0.18	0.11
Export	Supply model	0.10	0.06	-0.03	-0.02	0.02
	Demand model	0.25	0.06	-0.06	-0.04	0.03
Trade Balance	Supply model	-1.04	0.09	0.04	-0.07	.011
(bill. dollars)	Demand model	-3.08	-1.22	-0.33	-0.46	0.36
G.D.P.	Supply model	1.45	0.40	0.07	0.09	0.05
	Demand model	1.30	0.30	0.03	0.04	0.00
Domestic Price Level	Supply model	-1.44	0.18	0.24	0.25	0.29
	Demand model	-1.40	0.17	0.20	0.20	0.21
Terms of Trade	Supply model	-0.21	0.04	0.13	0.08	0.05
	Demand model	-0.16	0.02	0.10	0.07	0.04
Japan		year 1	year 2	year 3	year 4	year 5
Import	Supply model	0.10	0.12	-0.01	-0.00	0.01
	Demand model	0.46	0.17	-0.02	0.02	0.05
Export	Supply model	0.33	0.04	0.04	0.04	0.03
	Demand model	1.21	0.49	0.11	0.11	0.06
Trade Balance	Supply model	0.63	-0.06	-0.00	0.05	0.06
(bill. dollar)	Demand model	1.87	0.76	0.20	0.26	0.19
G.D.P.	Supply model	0.06	0.02	0.02	0.02	0.03
	Demand model	0.22	0.14	0.08	0.07	0.08
Domestic Price Level	Supply model	-0.03	0.02	0.04	0.04	0.04
	Demand model	0.01	0.07	0.10	0.10	0.10
Terms of Trade	Supply model	0.42	-0.04	-0.20	-0.12	0.07
	Demand model	0.45	0.04	-0.14	-0.07	0.03

It is interesting to find that both models used in our simulations fit statistical observations well, but they give very different predictions. This fact warns us to be very careful when quantifying shock effects in the economy; past experience alone does not give any decisive answer. The correct answer will depend on the economic theory applied in our calculations..

Table 7: Multiplier Effects of a 10 percent Devaluation of the Yen

U.S.A		year 1	year 2	year 3	year 4	year 5
Import	Supply model	0.16	0.46	0.40	0.33	0.26
	Demand model	0.50	0.68	0.50	0.24	0.02
Export	Supply model	-0.06	-0.15	-0.14	-0.14	0.13
	Demand model	-0.32	-0.27	-0.23	-0.26	0.23
Trade Balance	Supply model	1.44	1.20	0.69	0.43	0.22
(bill. dollar)	Demand model	0.40	0.54	0.48	0.44	0.86
G.D.P.	Supply model	-0.09	-0.07	-0.01	-0.02	0.03
	Demand model	-0.16	-0.13	-0.05	-0.03	0.00
Domestic Price Level	Supply model	-0.28	-0.39	-0.37	-0.36	0.35
	Demand model	-0.24	-0.38	-0.41	-0.40	0.40
Terms of Trade	Supply model	1.04	1.12	0.76	0.53	0.35
	Demand model	1.00	1.12	0.81	0.51	0.35

Japan		year 1	year 2	year 3	year 4	year 5
Import	Supply model	-0.42	-1.17	-1.09	-1.07	0.95
	Demand model	-2.13	-2.01	-1.72	-1.78	1.55
Export	Supply model	0.87	1.62	1.33	1.12	0.99
	Demand model	1.74	2.34	1.73	1.24	0.86
Trade Balance	Supply model	-1.70	-1.20	-0.53	-0.15	0.24
(bill. dollar)	Demand model	0.16	0.17	0.24	0.40	0.24
G.D.P.	Supply model	0.48	0.12	0.10	0.09	0.26
	Demand model	0.71	0.35	0.31	0.24	0.38
Domestic Price Level	Supply model	4.27	4.25	4.55	4.65	4.80
	Demand model	4.27	4.30	4.64	4.75	4.91
Terms of Trade	Supply model	-2.14	-2.25	-1.33	-0.93	0.69
	Demand model	-1.91	-2.12	-1.28	-0.88	0.69

András Simon, Department of International Economics, Budapest University of Economics, H -1093, Budapest, Fóvám tér, Hungary; International Centre for the Study of East Asian Development Kitakyushu, Japan

ACKNOWLEDGEMENTS

I am indebted to F.G. Adams, M. Fujita, L. R. Klein, B. Gangnes, L. Lau, R.F. Wescott for helpful comments on earlier drafts of this paper. The computational work was done by Jin-Myon Lee (International Centre for the Study of East Asian Development, Japan, and Korea Development Institute, Seoul)

NOTES

1. See, for example, Orcutt (1950), Harberger (1953), Lawrence (1979).

2. See, for example, Moriguchi (1973), Helkie-Hooper (1988), Hooper-Mann (1989), Martins (1990).

3. Numerous models include trend variables in trade equations to capture supply effects, Adams-Shishido (1988), Hickman-Lau (1974), Krugman-Baldwin (1987) among others. Junz-Rhomberg (1973) impute the long lags of price responses in world export markets to slowness in

response of supply.

4. There have been some anecdotic cases of rationing, like Honda cars recently, but they are not more frequent than in any other segments of the market.

5. The unsatisfactory coverage of my readings does not allow me to refer to others than Adams-Shishido (1989) as an example.

6. Analyses that I know refrain from interpreting the expansion of the share of Japanese exports as a result of an income effect. However, they do not give alternative explanations. Adams-Shishido (1989) -- apparently unsatisfied with the income-elasticity explanation -- includes a time trend into the equation explaining shares of Japan in U.S. imports.

7. In a non-walrasian economy in any micro-market there may exist either excess offers over actual sales or unfilled orders (requests). The ratio of micro-markets in excess demand to those in excess supply is an indicator of market frictions. Hansen (1970), Muellbauer (1977), McCafferty (1977) elaborated this idea. Simon (1989) applied it to analyze shortage economies.

8. Bismut-Martins (1987), Martins (1990) formulate a model applying a multilevel CES utility function worked out by Sato (1967) to determine demand. This allows them to distinguish between horizontal (here intra-brand) and national (here inter-brand) substitutability.

9. I use the term supply-determined for the sake of simplicity, knowing, that in fact the model equally has demand and supply determinants. "Price and variety competitive model" would be a more precise term. I do not use it partly because it is too long, partly because it does not refer directly to the important role played by supply in this model.

10. See for example Nomura Medium-term Economic Outlook for Japan and the World (1989) or Industrial Bank of Japan (1990).

REFERENCES

Adams, F. G., Shishido, S. et al. (1988) *Structure of Trade and Industry in the U.S.-Japan Economy.* NIRA Research Output. NRS-85-1.

Almon, C. S. and Nyhus, D. (1977) The INFORUM International System of Input-Output Models and Bilateral Trade Flows. *IIASA 5th Global Modeling Conference,* Laxenburg.

Armington, P. S.(1969) *A Theory of Demand for Products Distinguished by Place of Production.* IMF Staff Papers, May, XVI, No 1.

Bismut, C. and Martins, O. J. (1987) Compétitivité-prix, parts de marché et differenciation des produits. In *Actes du Colloque Aix-en-Provence,* 1987, Commerce international en concurrence imparfaite, eds. D. Laussel et C. Montet, *Economica* (1989), Paris.

Bosworth, B. P. (1988) Comment on Krugman and Baldwin, "The Persistence of the U.S. Trade Deficit," in *Brookings Papers on Economic Activity, 1,* 44-47.

Chamberlin, E. H. (1933) The Theory of Monopolistic Competition. *Rand Journal of Economics, 16,* 473-486.

Dixit, A. K. and Stiglitz, J. (1977) Monopolistic Competition and Optimum Product Diversity. *American Economic Review, 67,* 297-308.

Dornbusch, R. (1987) Exchange Rates and Prices. *American Economic Review, 77,* 1, March.

Feenstra, Robert C. (1987) *Symmetric Pass-through of Tariffs and Exchange Rates under Imperfect Competition: an Empirical Test.* NBER Working Paper No. 2453, December.

Fisher, E. (1989) A Model of Exchange Rate Pass-Through. *Journal of International Economics, 26,* No. 1/2, February.

Gabszewicz, J., Shaked, A., Sutton, J. and Thisse, J. F. (1981) International Trade in Differentiated Products. *International Economic Review, 22,* 527-535.

Gangnes, B. (1989) *Industry Structure and Exchange Rate Pass-Through: An Empirical Analysis.*

University of Pennsylvania. Working Paper.

Goldstein, M. and Kahn, M. S. (1985) Income and Price Effects in Foreign Trade, in R. W. Jones and P. B. Kenen, eds., *Handbook of International Economics*, Vol II. Elsevier Science Publishers.

Harberger, A. C. (1953) *A Structural Approach to the Problem of Import Demand.* American Economic Review Papers and Proceedings: 1480160.

Hanazaki, M. (1989) *Industrial and Trade Structures and the International Competitiveness of Asia's NIEs.* Japan Development Bank Report.

Helkie, W. H. and Hooper, P.(1988) The U.S. External Deficit in the 1980s: An Empirical Analysis, in R. C. Bryant, G. Holtham, P. Hooper, eds, *External Deficits and the Dollar: The Pit and the Pendulum*, Washington, The Brookings Institution.

Helpman, E. and Krugman, P. (1985) *Market Structure and Foreign Trade.* MIT Press, Cambridge, U.S.A.

Hickman, B. and Lau, L. J. (1974) Elasticities of Substitution and Export Demnads in a World Trade Model. *European Economic Review, 4,* 347-380.

Hooper, P. (1987) Comment on Krugman and Baldwin, "The Persistence of the U.S. Trade Deficit," in *Brookings Papers on Economic Activity, 1,* 47-51

Hooper, P. and Mann, C. L.(1989) The U.S. External Deficit: Its Causes and Persistence, in A. Burger, ed., *The U.S. Trade Deficit: Causes, Consequences and Cures.* Boston, Kluwer Academic Publishers.

Hooper, P. and Mann, C. L.(1989) *Exchange Rate Pass Through in the 1980s: The Case of U.S. Imports of Manufactures.* The Brookings Institute, Washington D.C.

Hotelling, H.(1929) Stability in Competition. *Economic Journal, 39,* 41-57.

Industrial Bank of Japan(1991) Forecast of the Change in Japanese Trade Structure and Trade Surplus, *Trends of Economy and Industry* 1991/8, (In Japanese).

Junz, H. B. - Rhomberg, R. R.(1973) Price Competitiveness in Export Trade Among Industrial Countries. *American Economic Review.*, Papers and Proceedings, 412-418.

Kitamura, Y. (1989) Restoring External Balances. Chapter 7. in *Nomura Medium-term Economic Outlook for Japan and the World*, 1989. Nomura Research Institute.

Krugman, P. (1987) "Pricing to Market When the Exchange Rate Changes." Chapter 3 in S. W. Arndt and J. D. Richardson, eds, *Real-Financial Linkages Among Open Economies.* Cambridge: MIT Press.

Krugman, P.- Baldwin, R.(1987) The Persistence of the U.S. Trade Deficit. *Brookings Papers on Economic Activity, 1,* 1-44.

Krugman, P. (1989) Differences in Income Elasticities and Trends in Real Exchange Rates. *European Economic Review, 33,* May: 1031-1046.

Lancaster, K. (1979) *Variety, Equity, and Efficiency.* New York: Columbia University Press.

Lawrence, R. Z. (1979) Toward a Better Understanding of Trade Balance Trends: The Cost-Price Puzzle. *Brookings Papers on Economic Activity.* pp. 191-212.

Mann, C. L. (1986) "Prices, Profit Margins, and Exchange Rates." *Federal Reserve Bulletin*, June.

Marston, Richard C. (1988) "Pricing to Market in Japanese Manufactures." Mimeo, University of Pennsylvania, September.

Martins, J. O. (1990) Comportement à l'exportation avec différenciation des produits. *Revue D'Économie Politique, 100,* 3, 416-438.

McCafferty, S. (1977) Excess Demand, Search and Price Dynamics. *American Economic Review.* 67: 228-235.

Moriguchi, C. (1973) Forecasting and Simulation Analysis of the World Economy. *American Economic Review, 63/2.*

Muellbauer, J (1978) Macrotheory vs. Macroeconomics: the Treatment of Disequilibrium in Macromodels. Discussion Paper No. 59. Birckbeck College.

Nomura Medium-term Economic Outlook for Japan and the World (1989), Nomura Research Institute.

Ohno, K. (1988) Export Pricing Behavior of Manufacturing: a U.S.-Japan Comparison. Working paper WP/88/78, International Monetary Fund, August.

Orcutt, G. H. (1950) Measurement of Price Elasticities in International Trade. *Review of Economics and Statistics,* 117-132.

Petersen, C., Pedersen, K. D., Riordan, E. J., Lynn, R. A. and Bradley, T. (1991) BANK-GEM: A World Bank Global Economic Model. World Bank, IECAP. Draft.

Salop, S. (1979) Monopolistic Competition with Outside Goods. *Bell Journal of Economics, 10,* 141-156.

Sato, K. (1967) A Two-level Constant Elasticity of Substitution Production Function. *Review of Economic Studies, 34,* August, 201-218.

Simon, A. (1989) A Search Model of Shortages. *Acta Oeconomica, 41,* 137-156.

Spence, M. E. (1976) Product Selection, Fixed Costs, and Monopolistic Competition. *Review of Economic Studies, 43,* 217-236.

Modeling the Household Economy

Duncan S. Ironmonger

The household economy can be defined as the system that produces and allocates tradeable goods and services by using the unpaid labor and capital of households. Recent surveys of time use indicate that in most developed countries the hours of work absorbed in the household economy are of the same order of magnitude as those absorbed in the market economy. This essay comments on the current stage of modeling the household economy and its linkages with the market economy and outlines the framework for an economywide model of household production and household leisure which can be linked through household input-output tables to a Klein-type dynamic macro model of market industries.

Lawrence Klein has played a major role on a world-wide scale in the development of national econometric models. This econometric revolution of the last 40 years has had a profound impact on our understanding of how market economies actually work. Millions of dollars are spent each year by business and government in developing, maintaining and using these models to make forecasts of how national economic systems will evolve and to explore the outcomes of actual or prospective changes in policy.

Broadly speaking, the economy-wide models of our market systems are of two main types (1) dynamic macro models - which are associated with the work of Lawrence Klein, and (2) input-output models - which are associated with the work of Wassily Leontief. The Leontief-type input-output models derive from the very detailed estimates of the flows of labor, energy and materials between the tens, if not hundreds, of industries of the market economy. The models are usually based on annual data and are often concerned with understanding micro-economic issues and long-term structural change. The Klein-type dynamic macro models are more concerned with understanding the short-term dynamics of business cycles than with long-term structural change, and although less detailed in terms of industry structure, are often based on quarterly data.

Both types of models could usefully be expanded to cover the *household economy*. The household economy can be defined as *the system that produces and allocates tradeable goods and services by using the unpaid labor and capital of households*. The most obvious point is that both Klein and Leontief models depict the activities of households as being concerned mainly with the *consumption* of commodities provided by the market. Time-use research shows that this is a very incomplete view of the role of households in economic activity.

Households are very large users of labor time and capital resources in *production.* If we examine the facts, in terms of labor and capital, and hence in terms of both inputs and outputs, the household economy is approximately the same order of magnitude as the market economy (Gershuny 1988, Ironmonger 1989, Chadeau, 1992). In other words, the household economy transforms intermediate commodities provided by the market economy into final items of consumption through the use of its own unpaid labor and its own capital goods.

The second major point is that the depiction of the consumption of *commodities* (goods and services) by households as the driving force for economic activity, ignores the necessity of *leisure time* in which consumption of commodities can take place. In other words, the arguments of the utility functions which drive economic models should include measures of consumption *time* as well as the quantities of commodities consumed. These quantities would include not only those provided by the market but also those produced in the home.

I. Modeling the Household Economy

Aggregate models of economic behaviour often ignore two features of the micro-economic structure; first, the most numerous and ultimately influential decision-making units in the economy are not governments, firms, or multinational corporations, but households which have different methods of organisation, management and decision making. Second, while it is traditional to consider the decisions of households as being chiefly about consumption expenditure and labor supply, they are better considered in a way with more far-reaching implications for structural and policy analysis.

Household decisions are about the allocation of time between market production, household production and leisure. Households allocate money between the purchase of leisure commodities and the purchase of intermediate goods for household production. Households also make decisions about the acquisition or disposition of real and financial assets.

Modeling of household time-allocation decisions goes beyond understanding the simple work-leisure trade-off. It confers other advantages. It provides knowledge of the detailed interactions within households between production and leisure activities. It also provides a framework for the analysis of the derived demands for commodities implicit in household production as well as in the use of leisure time. Most importantly, at the macro-economic level, it permits the investigation of the cyclical relationships between market and household production. One hypothesis is that the two spheres of production vary in a counter-cyclical pattern (Ironmonger, 1989) and a verification of this hypothesis is worth undertaking.

The principal official measures of economic activity, employment and gross national product, omit household work. These omissions obscure about one-third of GNP and about half of productive work (Hawrylyshyn, 1976; Ironmonger 1987; 1989). It can be argued that these distortions have seriously misled policy making at both the national and international levels (Waring, 1988). Until now, the lack of household data has prevented the development of economy-wide models based on household production. Changes in our statistical practices, which would remove the present cloak on reality, would also facilitate household model building. In turn, the use of these models for policy deliberations, would justify the collection and processing of household time-use data.

Data on labor-time inputs to household production have also been used to produce estimates of the imputed value of household production in Australian GDP (Ironmonger, 1987; Australian Bureau of Statistics, 1990). For estimates in other countries see for example Hawrylyshyn (1976), Murphy (1978), Kendrick (1979) and Peskin (1982). A notable advance in this area is the work at Northwestern University to produce extended national accounts (Eisner,1978, 1991). These integrate the productive activities of the household economy with those of the market and provide the appropriate social accounting framework for household modeling.

By providing a more realistic place to the household economy and to the role of leisure, the enlarged economic models should supply a better understanding of what is actually happening in the market economy. The new household-based economy-wide models should provide better understanding on issues of work and leisure, the gender inequalities that persist in both paid and unpaid work, and the apparent inability of our economic systems, with high and increasing levels of unemployment, to provide equitable access to paid work.

The point is frequently made that the increase in women's participation in paid work leads to overstatement in measured economic activity, because the reduction in household production is not counted. There is the further point that the increase in the proportion of young people in education in early adult years leads to understatement of measured economic activity, because, although teachers' time is counted, the time students spend is not.

A third point is that over the booms and slumps of business cycles, the increases and decreases in market output and income are offset by opposite changes in household production and unpaid income.

One hypothesis that can be explored at both the micro and macro levels is the question of the effects of the diffusion of new commodities produced by the market for the household. Clearly, product innovation in the household economy and in household leisure activities has been, and will continue to be, a major source of economic growth for both the market and the household. Moreover, many of the new commodities will have significant impacts on the use of time as they are adopted. New economic models, which include a more realistic portrayal of household activities in relation to both production and leisure, should help unravel these effects on the uses of time. They should also provide an improved understanding of the environmental resource impacts of household activities and how these can be minimized.

II. A Framework for Modeling Household Activities

The theoretical antecedents of the framework are essentially derived from the "new household economics" of Becker (1965, 1981), Ironmonger (1961, 1972), Lancaster (1966, 1971) and Muth (1966). Following this theory, the household economy is regarded as a productive sector with household activities modelled as a series of industries.

The distinctive elements of the new approach to household economics are that commodities are used in household production processes designed to satisfy wants, the commodities produced possess characteristics, or want-satisfying qualities, which define a production and consumption technology for households. Given there are changes in this technology through the introduction of new commodities, households are engaged in the process of finding out about changes in the technology matrix of coefficients connecting characteristics to wants. Households alter household production and expenditures as they find out about changes in this matrix (Ironmonger, 1983: 51). A useful review of the development of this theory is by Nerlove (1974).

The activities approach from the theory of the new household economics readily combines with the input-output approach of Leontief (1941) to establish a series of household input-output tables as the framework for modeling household activities.

A. Household Input-Output Tables

Household input-output tables combine data on commodity inputs derived from conventional surveys of household expenditure with time inputs derived from data from surveys of household time use (Ironmonger 1987).

The new framework of household input-output tables has made use of the internationally standardized methodology for the collection of household time use data which was developed for the multi-national project reported in Szalai (1972). In Australia the first survey to use this methodology was conducted in Melbourne and Albury/Wodonga in 1974 (Cities Commission, 1975). The Australian official statistical organisation completed a further survey in Sydney in 1987 (Australian Bureau of Statistics, 1988) and conducted the first Australia-wide survey of household time use in 1992.

Some initial steps to develop the input-output framework for building a dynamic macro model of household industries have been made with Australian data on the uses of time and money. The first input-output table of household productive activities of Australian households was published in a research discussion paper in 1987 (Ironmonger and Sonius). The detailed table, for the year 1975-76, has been republished in *Households Work* (1989) with additional tables in Ironmonger (1989) and Ironmonger and Richardson (1991).

The relative importance of the major groups of inputs into each type of household productive activity is shown in Table 1 which is a condensed version of the 1975-76 household input-output table. The input-output coefficients showing the proportions the inputs are to each dollar's worth of output are shown in Table 2. Household work time has an overall coefficient of 70.0 per cent, materials 20.7 per cent, capital equipment and housing 8.0 per cent and energy 1.3 per cent.

In recent years there have been significant demographic shifts involving changes in fertility, the ageing of the population and the composition of households. Australia is experiencing an especially large increase in the number of households without children and a dramatic increase in the number of single person households (Ironmonger & Lloyd-Smith, 1992). Counting children as those under the age of 15 years, the number of households without children has doubled in the last 20 years from under 2 million to almost 4 million, whilst the number of households with children has remained static a less than 2 million. One in five of all households are now single-person households; and half of these are women over the age of 50.

Table 1: Household Productive Activities, Total Inputs and Outputs, Australia 1975-76 ($ million)

INPUT -- OUTPUT --	Time	Materials	Energy	Equipment	Space	TOTAL
Prepared meals	7922	8219	144	861	279	15425
Housework	8140	271	32	525	288	9256
Repairs, maintenance	1825	483	7	222	61	2599
Other domestic work	1743	362	8	49	73	2609
Education, voluntary work	2088	392	8	49	73	2609
Child care	4717	1208	187	445	143	6700
Shopping	4969	745	224	453	-	6390
Gardening	1496	46	-	133	49	1725
TOTAL	32900	9726	609	2799	954	46986

Source: Ironmonger and Sonius (1989: Table 2.3)

Modeling the implications of these shifts requires household models to use a typology of households. Distinctions need to be made between households of different types, for example according to the numbers and ages of members, of the allocations of time and money. This allows model-builders to go beyond the customary portrayal of all decisions of the household sector being made by a singular "representative consumer" in the familiar Marshallian sense.

Accordingly, the input-output table for all households has been disaggregated into two tables, one for households with children and the other for those without children. In later versions of this work it is intended to extend this typology along life-cycle lines and to consider the decisions and processes that lead to transitions between the various stages of household formation and re-formation.

Table 2:   Household Productive Activities, Input-Output Coefficients, Australia 1975-76 (Per cent)

INPUT --	Time	Materials	Energy	Equipment	Space	TOTAL
OUTPUT --						
Prepared meals	51.3	40.3	0.9	5.6	1.8	100.0
Housework	88.0	2.9	0.3	5.7	3.1	100.0
Repairs, maintenance	70.2	18.6	0.3	8.5	2.3	100.0
Other domestic work	76.3	15.9	0.3	4.9	2.7	100.0
Education and						
voluntary work	80.0	15.0	0.3	1.9	2.8	100.0
Child care	70.4	18.0	2.8	6.6	2.1	100.0
Shopping	77.8	11.7	3.5	7.1	-	100.0
Gardening	86.7	2.7	-	7.7	2.8	100.0
TOTAL	70.0	20.7	1.3	6.0	2.0	100.0

Source: Ironmonger and Sonius (1989: Table 2.4)

B. Extended Household Input-Output Tables

The tables have been extended to cover the complete range of household activities, not only household production, but also leisure activities, sleeping and personal care, and market work. Table 3 shows the classification of household activities used in preparing the extended household input-output tables.

The extended household input-output tables for Australia for the two types of households, with children (C) and without children (NC), and shown in Tables 4 and 5.

These household input-output tables provide the framework for developing a model of Australia based on household activities (Ironmonger, Sawyer, Lloyd-Smith, Donath and Tran van Hoa, 1990). The model aims to explain the leisure and household production activities of two types of Australian households - those with children (which are falling in numbers both proportionally and absolutely) and those without children (which have almost doubled in 25 years) and to provide forecasts of the main variables which describe the aggregate activities of these two types of household.

This household sector model is designed to be linked to a time-series model of the market production sector.

Table 3: Household Activities

Leisure Activities		(l_k) k = 1,9
	1	Meals at Home
	2	Entertainment
	3	Friends
	4	Active Leisure
	5	Television
	6	Radio & Records
	7	Reading
	8	Other Passive Leisure
	9	Travel for Leisure
Household Production Activities (h_j) j = 1,8		
	1	Meal Preparation
	2	Cleaning & Laundry
	3	Repairs & Maintenance
	4	Other Domestic Work
	5	Education & Community Work
	6	Child Care (incl travel)
	7	Shopping (incl travel)
	8	Gardening
Market Production Activities		
	1	Paid Work Time
	2	Paid Work Breaks
	3	Paid Work Travel
Sleeping and Personal Care		
	1	Sleeping
	2	Personal Care at Home

C. Variables in the Model

The definitions, dimensions and units of the variables in the model are listed in Table 6.

D. Household Decisions

In building this model it is intended to start from the theoretical position that household time and money allocation decisions are made on the basis of an intertemporal, separable utility function. Using the terminology of the variables specified in Table 6, the decision of each household can be considered to be based on the maximization of an overall utility function of the form

$$\text{Maximize} \quad U \,[\, \mathbf{l}, \, \mathbf{h}, \, M, \, \mathbf{a} \,] \tag{1}$$

where the arguments of the function are a vector of leisure times $\mathbf{l}$ and a vector of asset holdings $\mathbf{a}$ (with positive utilities) and a vector of times in household work $\mathbf{h}$ and the total time in market work M (with negative utilities). In the present specification of the model **l** is of order 9x1, and **h** is of order 8x1.

The maximization of the overall utility function of a household can be considered as taking place in at least three hierarchical decision stages. The first stage covers long-term decisions about occupation and employment in paid work and asset portfolio selection; the second concerns the short-term allocation of the income that flows from the paid work and the day-to-day allocation of time to household production. A third stage allocates the remaining time and resources between alternate leisure activities.

Table 4: Households With Children, All Household Activities, Total Inputs and Outputs, Australia 1975-76

INPUT	Materials	Energy	Equip-ment	Housing	Total	Time	Time
	$ m	$ m	$m	$m	$m	Hours m	$m
Leisure Activities							
Meals at Home	2	35	221	199	458	2,246	
Entertainment	1,930	10	61	55	2,056	623	
Friends	224	18	110	99	451	1,115	
Active Leisure	883	-	599	71	1,553	801	
Television	345	519	255	822	2,873		
Radio & Records	0	2	115	12	129	135	
Reading	110	262	57	330	647		
Other Passive Leisure	299	12	72	65	448	734	
Travel for Leisure	720	270	489	-	1,478	691	
Household Production Activities							
Meal Preparation	3,892	63	455	154	4,564	1,732	4,889
Cleaning & Laundry	190	14	278	159	640	1,793	5,042
Repairs & Maintenance	286	3	134	40	463	449	1,341
Other Domestic Work	199	3	52	30	284	338	979
Educn & Commty Work	248	5	81	56	390	630	1,828
Child Care (incl travel)	816	125	333	113	1,387	1,473	4,186
Shopping (incl travel)	398	135	258	-	791	1,049	3,016
Gardening	49	-	73	23	145	255	753
Market Production Activities							
Paid Work Time	-	-	428	-	428	6,066	17,947
Paid Work Breaks	-	-	69	-	69	973	2,886
Paid Work Travel	707	312	567	-	1,586	801	2,366
Sleeping and Personal Care							
Personal Care at Home	507	92	113	102	815	1,152	
Sleeping	15	-	404	1,279	1,698	14,400	
ALL ACTIVITIES	11,369	1,155	5,692	2,769	20,985	40,970	55,979

Source: Households Research Unit, The University of Melbourne

This would imply a first-stage utility function with arguments being the disutilities flowing from a vector of hours of paid work (**m**), the positive utility of the money income from work, (M\$ = ω i′**m**, where ω is the market wage rate and **i**′ is the transpose of the unit vector) and the vector of asset stocks, **a**. The first stage utility function is

$$\text{Maximize } U(\textbf{ m}, \text{M\$}, \textbf{ a}) \tag{1a}$$

The stage two maximization is

$$\text{Maximize } U [\textbf{ u}, \textbf{ h}] \tag{1b}$$

where **u** is the vector of the leisure commodities produced in the household, which provide positive utilities, and the vector **h** of household work time, involving disutilities.

Table 5: Households Without Children, All Household Activities, Total Inputs and Outputs, Australia 1975-76

	Materials	Energy	Equip-ment	Housing	Total	Time	Time
	$ m	$ m	$m	$m	$m	Hours m	$m
Leisure Activities							
Meals at Home	2	25	159	164	351	2,262	
Entertainment	1,606	9	57	58	1,730	807	
Friends	204	19	117	121	461	1,669	
Active Leisure	757	-	378	83	1,218	1,147	
Television	2	33	505	213	753	2,942	
Radio & Records	0	4	156	29	189	399	
Reading	1	15	255	95	366	1,317	
Other Passive Leisure	328	8	51	52	438	718	
Travel for Leisure	797	257	477	-	1,531	895	
Household Production Activities							
Meal Preparation	2,386	93	416	119	3,015	1,640	4,650
Cleaning & Laundry	132	17	237	112	498	1,542	4,345
Repairs & Maintenance	198	5	116	33	352	459	1,338
Other Domestic Work	164	5	63	30	262	411	1,181
Educn & Commty Work	53	10	101	63	227	864	2,557
Child Care (incl travel)	17	64	18	6	105	105	302
Shopping (incl travel)	378	101	228	-	707	1,176	3,385
Gardening	48	-	60	22	130	310	917
Market Production Activities							
Paid Work Time	-	-	199	-	199	4,121	12,118
Paid Work Breaks	-	-	35	-	35	730	2,162
Paid Work Travel	452	174	322	-	948	605	1,779
Sleeping and Personal Care							
Personal Care at Home	650	71	93	96	910	1,325	
Sleeping	11	-	318	1,037	1,366	14,344	
ALL ACTIVITIES	8,177	910	4,360	2,332	15,791	39,788	34,734

Source: Households Research Unit, The University of Melbourne

The stage three maximization of utility is

Maximize   U [l, s]   (1c)

where **l** is the vector of time in different leisure activities, which provide positive utilities, and the vector **s** of the commodity-intensities of leisure activities, also with positive utilities.

Thus, the utility functions provide the framework for an allocation of time and money in several spheres - leisure activities, household production and the interchange with the market through work, commodity and asset transactions. A further benefit of this approach is that the input-output structures provide the levels of additional flows of intermediate commodities between household activities entailed in the levels of final activity chosen by households.

As there are two types of household in the present model, the utility functions assume two forms U^c , for households with children, and U^{nc}, for households without children, although in the preliminary estimates reported in this essay, the U^c and U^{nc} functions are of the same form.

The maximization decision in equation (1) is constrained by the total household time (T) available for either productive or leisure activities and by a measure of total wealth.

The constraint of total household time (T) is clear. It is

$$i'l   +   i'h   +   M   =   T$$ (2)

where i' is the transpose of the unit vector used to effect a summation of the elements of a vector. Thus, $i'l = L$, total leisure time and $i'h = H$, total household work time.

The total wealth constraint, and how it operates, is not so clear. It could be stated in the form of *fully extended* income minus *fully extended* consumption equals the change in total assets, dA. *Fully extended* means not only putting a value on household work time but also putting a value on leisure time. If the market work wage rate, ω, is used for these valuations the following form for the overall wealth constraint follows

$$\omega L   +   w H + w M   +   r'a   -   p'l   = dA$$ (3)

where **p** is a vector of order 9x1 of the time prices of leisure activities and **r** is a vector of order nx1 of assets prices.

The time price p_k of leisure activity k will be defined by $p_k = s_k + \omega$ where s_k is the intensity, in \$ per hour, of the commodity use component of leisure activity k, and r_i is the *initial* price of asset i for a holding A_i, so that the model is most properly regarded as a one-period model. In an intertemporal setting, the constraint expressed by equation (3) would be regarded as an *initial wealth constraint*.

Table 6: Variables in A Model of Australia Based on Household Activities

Variable	Definition		Order	Units
i	the unit vector	$i' = [1,1,1,1,1,\ldots\ldots,1]$		
l	leisure times vector		9x1	hrs./yr.
l_k	leisure time in k-th activity		1x1	hrs./yr.
L	total leisure time	$L = i'l$	1x1	hrs./yr.
h	household work times vector		8x1	hrs./yr.
h_j	household work time in j-th activity		1x1	hrs./yr.
H	total household work time	$H = i'h$	1x1	hrs./yr.
m	market work times vector		ix1	hrs./yr.
m_i	market work time in i-th industry		1x1	hrs./yr.
M	total market work time	$M = i'm$	1x1	hrs./yr.
T	total time for work or leisure	$T = L + H + M$	1x1	hrs./yr.
ω	market time wage rate		1x1	$/hour
H$	total cost of household work	$H\$ = \omega H$	1x1	$/year
M$	total market work income	$M\$ = \omega M$	1x1	$/year
a	assets stocks vector		nx1	$
r	assets prices vector		nx1	$/$/year
A$	total income from stocks of assets	$A\$ = r'a$	1x1	$/year
A$/A	average rate of return on assets	$A\$/A = r'a / i'a$	1x1	$/$/year
Y$	total income from wages and assets	$Y\$ = M\$ + A\$$	1x1	$/year
E$	extended total income	$E\$ = Y\$ + H\$$	1x1	$/year
C$	total household (consumer) expenditure		1x1	$/year
n	intermediate purchases for household production		8x1	$/year
N$	total intermediate purchases	$N\$ = i'n$	1x1	$/year
v	leisure commodities purchases vector		9x1	$/year
V$	total leisure commodities purchases	$V\$ = i'v$	1x1	$/year
	$N\$ + V\$ = C\$$			
o	all household commodities produced vector		9x1	$/year
o_j	household output of j-th activity		1x1	$/year
O$	total value of household output	$O\$ = i'o$	1x1	$/year
	$O\$ = H\$ + N\$$			
u	leisure commodities produced in household		9x1	$/year
u_k	production of k-th leisure commodity		1x1	$/year
U$	total value of leisure commodities produced	$U\$ = i'u$	1x1	$/year
w	leisure commodities used vector	$w = u + v$	9x1	$/year
w_k	commodities used in k-th leisure activity	$w_k = u_k + v_k$	1x1	$/year
W$	total value of leisure commodities used	$W\$ = i'w$	1x1	$/year
	$W\$ = U\$ + V\$$			
s	commodity intensity of leisure vector		9x1	$/hour
s_k	commodity intensity of k-th leisure activity	$s_k = w_k / l_k$	1x1	$/hour
p	leisure prices vector		9x1	$/hour
p_k	price of k-th leisure activity	$p_k = s_k + \omega$	1x1	$/hour

Useful Constants		
Hours per week	168	(7x24)
Hours per year of 365 days	8,760	(365x24)
Hours per year of 52 weeks	8,736	(52x168)
Minutes per day	1,440	(24x60)
Minutes per week	10,080	(7x1,440)

However, with the formalisation of the decision process as taking place in three hierarchical stages, the wealth constraint can also be considered as having separable components, one appropriate to each decision stage.

There is a `money' income constraint in the sense of the constraint on purchases in the market being limited by income from market time and from the market return on assets, "earned" and "property" income in the customary sense. This could be written:

$$M\$ \; + \; A\$ \quad = \quad Y\$ \tag{3a}$$

where $M\$ = \omega \, M$, and $A\$ = \mathbf{r'a}$.

The labor time input component $(\omega \, H \; = \; H\$)$ of the cost of household production, which is also included in constraint (3) above, is only part of the constraint on household production. This constraint is probably best expressed in the form that the total cost of the inputs to household production equals the total value of the outputs of household production ($O\$$). That is

$$H\$ \; + \; N\$ = O\$ \tag{3b}$$

where $N\$$ is the total value of purchases of intermediate commodities from the market ($N\$ = \mathbf{i'n}$), and $H\$$ is the total cost of household labor input, as before.

The total cost of the purchases of intermediate goods from the market, $N\$$, are a part of the customary aggregate, "consumer expenditure", $C\$$. Hence, $C\$ - N\$ = V\$$, the total value of commodities used 'directly' in leisure activities. These direct leisure commodities and the leisure commodities produced in households together provide the commodity-intensity of leisure activities, s.

If all outputs of household production were used in leisure activities, then the total value of leisure commodities $S\$ = V\$ + O\$$. The ratio of $S\$$ to total leisure time L is the aggregate *commodity-intensity of leisure* expressed in dollars per hour and the total value of leisure time, measured at   the opportunity cost of the market wage rate is $L\$ = \omega \, \mathbf{l}$.

Equation (3) also includes the value of leisure time, $L\$$, as a component of both fully extended income and as a component of the consumption item **p'l**. Thus (3) can be expressed as

$$L\$ \; + \; H\$ \; + \; M\$ \; + \; A\$ \; - \; (\, O\$ \; + \; V\$ \; + \; L\$ \,) \quad = \quad dA \tag{3c}$$

This formulation of the overall constraint on household decisions is somewhat unrealistic. Market time and household time produce commodities (goods and services) which could be exchanged, either in the market or in the household. Hence it is sensible and realistic to include both $M\$$ and $H\$$ in the income term and O\$ and V\$ in the consumption term. However leisure time is not exchanged. Hence, including L\$ in both the income and consumption components is no improvement to our formulation of the overall wealth constraint.

Consequently, a more realistic statement of the wealth constraint is in terms of *extended income*, E\$ = Y\$ + H\$, is

$$E\$ - s'l = dA \tag{3d}$$

No matter how the household decision-making processes are formalized in terms of utility maximization subject to constraints, the outcome is that these processes determine:

(1) the supply of market work, M

(2) the supply of household work h_j to the j-th productive activity and hence,

(3) the residual amount of total leisure time, L;

(4) the demand for leisure time l_k in the k-th leisure activity;

(5) the production by households of leisure commodities u_k for the k-th leisure activity;

(6) the purchases of intermediate commodities n_j for the j-th productive activity;

(7) the purchases of leisure commodities v_k for the k-th leisure activity; and

(8) changes in, and levels of, the disposition of real and financial asset portfolios of households.

The present essay reports on estimates of the second and fourth of these sets of outcomes, the use of labor in various household production activities and the disposition of leisure time between leisure activities.

E. Interaction With the Market

In the complete model, for a given household (or household type), T will be regarded as exogenous, but the variables L, ω, Y are derived as a result of interaction with the market sector of the economy. Specifically, total leisure time, L, is determined as the residual from T - M - $\dot{i}h$ where M is total market time and will need to be determined from the interaction between the supply of market time from households and the demand for that time from market production sector of the economy.

Statistics on market times are available year by year for the Australian economy. Consequently, the model can be used to determine total household work times, total leisure times, disaggregated leisure times and disaggregated household work times.

Large (n >1000) cross-sectional data sets for two periods 1974 and 1987 (n= 1500 and n=3300) have been relied upon to estimate the parameters of household work and leisure share equations.

F. The Complete Model

The model is intended to be configured to operate in three stages, each stage comprising a section of the complete model. Stage I covers a macro-model of the market economy, Stage II is a model of household production activities and Stage III is a model of leisure activities.

The macro market model of Stage I would be expected to have all of the main features of a Klein-type macro model. This model determines, amongst other variables, the market output for a set of industries and the market work time absorbed in those industries. The market work is distributed across the two types of households and both sexes according to cross-section information on the use of time. A market model with four industries would have 20 main variables of interest - 4 market outputs and 16 market work times.

Table 7 shows the numbers of variables of interest in each of the three stages of a model with two types of individuals and two types of households.

The household production model of Stage II determines the allocation of work time to household productive activities, the purchases from the market of intermediate inputs for household production and the output of household commodities. This stage of the model contains 64 variables of interest.

The leisure activities model of Stage III determines the use of leisure time, the use of leisure commodities produced in the household, and the purchases of leisure commodities from the market. This stage of the model contains 72 variables of interest.

G. Household Demography

The model starts with a standard demographic approach comprising the usual fertility, mortality and immigration assumptions of the population projections issued by the Australian Bureau of Statistics. These give annual forecasts of the (five-year) age distributions for both males and females.

A new feature of the model is to project the number of households by using data on the distribution of the population into households according to the bivariate size distribution of households of the number of adults per household (A) and the number of children per household (C). Children are those less than 15 years of age. This procedure gives forecasts of the numbers of households and the household populations for each of 37 household types (Ironmonger and Lloyd-Smith, 1992). For analysis, household are aggregated into two types - those with children (C > 0) and those without children (C = 0).

The methodology used in making these projections of households and household populations departs from the traditional method which use headship ratios. In many households the concept of headship is no longer valid since there is shared responsibility for the management of the household. This change has led Census authorities in many countries to substitute the concept "reference person" for "head". As a consequence, reported headship/reference person ratios are less stable and therefore less valid as a basis for forecasting the numbers of households.

III. Estimating the Model

The initial estimates of the model parameters have been derived from a data set on the allocation of time from 24-hour time diaries for men and women of different ages and on different days of the week. Consequently separate equations for men and women are estimated. The data are for adults aged 15 or more years. The estimates are for the input of time into household production activities and for the shares of time on leisure activities.

Variables	Persons		Households		Activities		Total
Stage I. Market Economy							
Market Output ($)					4	=	4
Market Work Time (Hours)	2	*	2	*	4	=	16
Stage II. Household Production							
Household Work Time (Hours)	2	*	2	*	8	=	32
Purchase of Intermediate Commodities ($)			2	*	8	=	16
Output from Household Production ($)			2	*	8	=	16
Stage III. Leisure Activities							
Household Leisure Time (Hours)	2	*	2	*	9	=	36
Leisure Commodities from Household Production ($)			2	*	9	=	18
Purchase of Leisure Commodities from Market ($)			2	*	9	=	18

Table 7: Variables of Interest

A. Household Production Activities

These activities absorb more labor time (unpaid) than all (paid) work time in the market (business and government). The analysis is in terms of the two types of households - those with and those without children.

Household production activities (the household economy) are in competitive relationships with market activities over a range of industries - particularly in meals, in child care, in cleaning & laundry and in repairs & maintenance. Households have a large investment in capital and do not pay taxes on their own labor or production, hence, despite each household's small scale, they have some advantages in competing with the larger scale units of the market.

The model has eight household production activities which transform purchased inputs of materials and energy, using (unpaid) household time to produce outputs for input either into leisure activities or as additions to and maintenance of human and physical capital.

In households with children, women on average spend about 50 hours per week in household production activities; men about 23 hours. In households without children the averages are 33 hours per week for women and 22 hours for men. More details of these hours according to the eight activities identified in our model are shown in Table 8.

Table 8: Mean Hours of Household Work Time, Australia, 1987 (Hours per Week)

	Households With Children		Households Without Children	
	Men	Women	Men	Women
Meal Preparation	2.75	10.90	3.31	8.85
Cleaning & Laundry	1.58	9.98	1.59	8.40
Repairs & Maintenance	2.25	0.36	1.95	0.22
Other Domestic	1.52	1.78	1.60	1.59
Education & Community	6.47	7.25	6.65	6.50
Child Care (incl. travel)	3.70	12.49	0.39	1.22
Shopping (incl. travel)	3.32	6.19	3.98	5.19
Gardening	1.10	0.62	2.09	1.01
Total Household Work	22.68	49.57	21.55	32.99

It is clear, both from the data and from considerations of our theory, that the prior decisions about participation in the labor market, particularly if that participation decision is for full-time work, and the prior decisions about children and about household cohabitation which determine the numbers of adults and children in each household, set major constraints on the allocation of time to household production activities.

Accordingly, estimates of the effects of these constraints on the use of household time have been made by including dummy variables in the equations for employment status (FT), the number of adults per household (NA) and the number of children (NC).

The estimated functions for times spent in household work are of the following form

$$h_j = \alpha_{0j} + \alpha_{1j} \, FT + \alpha_{2j} \, NA + \alpha_{3j} \, NC + \alpha_{4j} \log \omega$$
$$+ \alpha_{5j} \log Y + \alpha_{6j} \, AGE + \alpha_{7j} \, DAY \tag{4}$$

with dummy variables, AGE and DAY, included to capture the effects of age and day of the week.

Table 9 shows estimates of the regression estimates for three factors of interest, labor market status (FT), the wage rate (log ω) and income (log Y).

Labor market status clearly has a highly significant effect on the total time in household productive activities. In particular, the education and community work category is increased on average by an hour per day if the individual is not in full-time paid-work status. This effect is for all categories - men and women and whether in a household with children or without children.

Women who are not in full time attachment to the labor market also have increased participation in the major household industries of meal preparation, of cleaning and laundry, and, for those in households with children, of child care. The additional labor input in these three industries for women not in full time paid work amounting to over 10 hours per week for those in households with children and 6.4 hours per week for those in households without children.

For men not in full-time paid work, there is a small, but significant, increase in the time spent in meal preparation. Shopping time also increases for both men and women not in full-time paid work, significantly for women in households with children and for men in households without children.

The effects on hours of household work of the level of money income Y\$, and the wage rate, ω, are mainly to be seen in households with children. There, apart from the mixed education and community work category, the signs of the effects are generally in the direction expected, negative for ω and positive for Y\$.

B. Leisure Activities

This set of activities includes eating and all social and entertainment activities both active and passive whether at home or away from home.

The model covers the leisure time of both adult males and adult females in terms of the hours spent in nine types of leisure activities and the total inputs of commodities (materials, energy, equipment and housing) into each of these activities whether from household production or from market production.

The analysis is for the two types of households, those without children (C = 0) and those with children (C > 0)

The dependant variable, g_k (the share of total leisure time (L) devoted to the k-th activity), is estimated as a function of the logarithm of total leisure time (log L) with the remaining variables being essentially the same as in the equation for household work. Thus we have

$$g_k = \beta_{0k} + \beta_{1k} \log L + \beta_{2k} FT + \beta_{3k} NA + \beta_{4k} NC + \beta_{5k} \log \omega$$
$$+ \beta_{6k} \log Y + \beta_{7k} AGE + \beta_{8k} DAY \tag{5}$$

where $g_k = l_k / L$ is the share of leisure time in the k-the activity over total leisure time, and , as before, L is total leisure time, ω is the wage rate, and Y is total income.

This formulation was adopted because separate identification of the intensity of use components i_k was not possible. They would instead be estimated and incorporated in the constant term a_k. The share equations (5) are estimated for each of the nine leisure activities; one equation can be deleted and the well-known invariance condition applies. These equations determine shares of total leisure time, but total leisure time itself must be determined as residual from both the market time and household work time models.

Table 9: Household Work Functions Regression estimates of coefficients
(t values in parentheses, significant values bold font)

	Households With Children		Households Without Children	
	Men	Women	Men	Women
1. FT (Labor market status: not in full time paid work = 1)				
Shopping (incl. travel)	.050 (0.3)	**.379 (3.1)**	**.470 (3.6)**	.127 (1.1)
Meal Preparation	**.257 (2.3)**	**.484 (4.1)**	**.206 (3.4)**	.448 (4.3)
Cleaning & Laundry	-.102 (1.0)	**.584 (4.1)**	.112 (1.5)	**.380 (2.8)**
Repairs & Maintenance	-.088 (0.5)	**.101 (2.4)**	-.068 (0.6)	-.011 (0.5)
Other Domestic	-.145 (1.8)	.105 (1.8)	.066 (0.9)	**.110 (2.1)**
Child Care (incl. travel)	-.113 (0.8)	**.380 (2.0)**	-.010 (0.2)	.090 (1.0)
Gardening	.015 (0.2)	.032 (0.9)	.166 (1.6)	.020 (0.4)
Education & Community	**.994 (2.9)**	**.879 (4.1)**	**.888 (3.7)**	**1.205 (5.6)**
Sum (hours/day)	.868	2.944	1.830	2.369
2. Log ω (Wage Rate)				
Shopping (incl. travel)	**-.130 (2.1)**	-.052 (1.1)	-.059 (1.2)	-.151 (0.9)
Meal Preparation	-.053 (1.3)	**-.157 (3.5)**	.017 (0.5)	-.061 (1.3)
Cleaning & Laundry	**-.112 (3.1)**	**-.122 (2.3)**	.052 (1.9)	-.027 (0.4)
Repairs & Maintenance	**-.216 (3.4)**	**.031 (2.1)**	.017 (0.4)	.002 (0.2)
Other Domestic	.016 (0.5)	-.009 (0.4)	-.031 (1.1)	.031 (1.3)
Child Care (incl. travel)	-.096 (1.9)	**-.427 (6.0)**	.009 (0.5)	-.062 (1.5)
Gardening	-.009 (0.3)	.000 (0.0)	**-.076 (2.0)**	-.002 (0.1)
Education & Community	.190 (1.5)	**.175 (2.2)**	**.210 (2.3)**	-.064 (0.7)
3. Log Y (Income)				
Shopping (incl. travel)	.036 (1.2)	.011 (0.7)	**.060 (2.5)**	.005 (0.3)
Meal Preparation	**.046 (2.4)**	-.018 (1.2)	.018 (1.2)	-.018 (1.3)
Cleaning & Laundry	**.047 (2.7)**	-.009 (0.5)	.002 (0.2)	-.026 (1.5)
Repairs & Maintenance	**.071 (2.3)**	.002 (0.4)	.012 (0.8)	-.003 (1.3)
Other Domestic	.073 (0.5)	.007 (1.0)	.019 (1.4)	.004 (0.5)
Child Care (incl. travel)	.006 (0.2)	.033 (1.4)	-.003 (0.3)	-.010 (0.9)
Gardening	-.013 (0.8)	.000 (0.1)	**-.076 (2.0)**	-.002 (0.1)
Education & Community	**-.336 (5.6)**	-.040 (1.5)	**-.598 (13.6)**	**-.053 (2.0)**

Estimates of the parameters α_{0j}, α_{1j}, α_{2j}, α_{3j}, α_{4j}, α_{5j}, α_{6j} α_{7j} of equation (4) and β_{0j}, β_{1j}, β_{2j}, β_{3j}, β_{4j}, β_{5j}, β_{6j} of equation (5) , are derived conditional on values of the variables L, ω, Y, FT, NA, NC and of course for men and women of a specific age on a given day of the week.

Estimates of the leisure share functions are set out in Table 11 which shows the coefficients for three factors of interest, total leisure time (L), the wage rate (ω) and income (Y).

The coefficients of the total leisure time are generally highly significant. As expected, meal time at home is a decreasing share. Television time has a surprisingly constant share of total time. Television time share is also unaffected by other factors, the only significant influences seem to be the day of the week. Entertainment, friends, active leisure and leisure travel time all have significant positive coefficients as total leisure time increases.

Table 10: Household Leisure Time: Mean Shares of Total Leisure Time, Australia, 1987, Per cent

	Households With Children		Households Without Children	
	Men	Women	Men	Women
	%	%	%	%
Meals at home	26.4	26.0	22.6	23.7
Entertainment	1.2	1.5	1.0	1.4
Friends	9.7	12.4	12.8	11.3
Active Leisure	6.8	6.3	8.3	8.6
Travel for Leisure	4.6	4.3	5.1	4.5
Television	34.4	28.9	30.8	29.9
Radio & Records	1.6	1.0	2.7	2.2
Reading	5.2	4.8	6.3	6.2
Other Passive Leisure	10.1	14.9	10.4	12.2
Total	100.0	100.0	100.0	100.0
Total Leisure Time				
Hours per week	32.52	33.95	33.22	42.00

The opportunity cost and income variables (ω and Y$) have little impact on the underlying share equations. For both men and women in households with children, the only exceptions being the highly significant negative effect for women of the wage rate on the share of leisure time spent on meals at home and for men some small, but significant positive income effects on time with friends or in leisure travel balanced by a negative income effect on television viewing.

For men in households without children there are also some exceptions to our preliminary general conclusion that wage rates and income have little significant effect on the underlying linear logarithmic leisure share functions of our model.

IV. Linking the Model to the Market Economy

The eventual model will be a model of "total" economic activities in the sense of the extended national accounts which include the household economy as a productive sector (Eisner, 1978). Thus, productive activities resulting from unpaid labor in households are included in the model as eight additional industries producing output in competition with the usual industries of the market economy.

Table 11: Leisure Share Functions Regression estimates of coefficients
(t values in parentheses; significant values in bold font)

	Households With Children		Households Without CHildren	
	Men	Women	Men	Women
1. Log L (Total Leisure Time)				
Meals at home	**-0.195 (16.0)**	**-0.184 (15.2)**	**-0.174 (16.4)**	**-0.181 (18.1)**
Entertainment	**0.011 (2.7)**	**0.012 (2.4)**	**0.010 (3.0)**	**0.021 (4.7)**
Friends	**0.073 (5.9)**	**0.090 (6.9)**	**0.063 (5.1)**	**0.070 (6.2)**
Active Leisure	**0.054 (5.1)**	**0.042 (4.5)**	**0.081 (8.3)**	**0.047 (5.0)**
Travel for Leisure	**0.034 (5.6)**	**0.042 (7.8)**	**0.036 (7.0)**	**0.042 (8.7)**
Television	0.022 (1.2)	**0.034 (2.0)**	-0.006 (0.4)	-0.027 (0.2)
Radio & Records	**-0.011 (2.1)**	-0.002 (0.6)	0.002 (0.5)	0.002 (0.2)
Reading	0.013 (1.7)	0.009 (1.4)	0.013 (1.8)	**0.016 (2.4)**
Other Passive Leisure	-0.001 (0.1)	**-0.043 (3.7)**	**-0.037 (3.6)**	-0.012 (1.3)
2. Log ω (Wage Rate)				
Meals at home	-0.011 (0.9)	**-0.022 (3.2)**	-0.013 (1.6)	-0.009 (1.1)
Entertainment	0.003 (0.1)	0.001 (0.1)	**0.006 (2.1)**	0.003 (0.8)
Friends	-0.002 (0.2)	0.014 (1.8)	-0.005 (0.5)	0.013 (1.5)
Active Leisure	-0.001 (0.1)	-0.002 (0.4)	**0.019 (2.5)**	**-0.011 (2.5)**
Travel for Leisure	-0.005 (1.0)	0.001 (0.4)	**0.011 (2.8)**	0.004 (1.2)
Television	0.015 (0.4)	0.014 (1.4)	**-0.045 (3.6)**	-0.008 (0.7)
Radio & Records	-0.000 (0.0)	0.001 (0.4)	-0.003 (0.8)	-0.001 (0.2)
Reading	0.011 (1.6)	0.003 (0.9)	**0.013 (2.5)**	**0.012 (2.5)**
Other Passive Leisure	-0.008 (0.8)	-0.009 (1.3)	**0.017 (2.1)**	-0.004 (0.5)
3. Log Y (Income)				
Meals at home	0.001 (0.3)	0.002 (1.0)	0.002 (0.5)	-0.002 (0.9)
Entertainment	0.003 (0.1)	0.001 (0.2)	**0.006 (2.1)**	0.003 (0.8)
Friends	**0.011 (2.1)**	0.001 (0.0)	**0.012 (2.6)**	0.002 (0.9)
Active Leisure	0.000 (0.1)	0.002 (0.9)	**-0.010 (2.7)**	-0.001 (0.6)
Travel for Leisure	**0.007 (2.7)**	-0.001 (1.0)	0.001 (0.4)	0.002 (1.8)
Television	**-0.021 (2.6)**	-0.005 (1.5)	-0.001 (0.1)	0.001 (0.3)
Radio & Records	0.001 (0.4)	0.000 (0.2)	-0.003 (1.4)	0.001 (0.9)
Reading	0.001 (0.2)	-0.001 (1.0)	0.001 (0.2)	-0.001 (0.4)
Other Passive Leisure	0.003 (0.5)	0.003 (1.4)	-0.006 (1.8)	**-0.004 (2.0)**

The model also accounts for leisure time which is jointly determined within the total time constraint on leisure, household work and market work.

Changes in technology continue to influence the rates of change of productivity in both household and market sectors. Within both, the development of information based activities with less environmental resources impact needs to be studied. The model will attempt to capture some of these effects.

Duncan Ironmonger, Department of Economics, The University of Melbourne, Victoria 3052, Australia.

ACKNOWLEDGEMENTS

The author acknowledges the helpful contributions of Michael Bittman, Susan Donath, Vic Jennings, Bill Lloyd-Smith, Jim Perkins, Amelia Preece, Kim Sawyer, Tran Van Hoa and Ross Williams.

REFERENCES

Australian Bureau of Statistics. (1988). *Information Paper: Time Use Pilot Survey, Sydney, May-June 1987* (Catalogue No. 4111.1) Sydney: Australian Bureau of Statistics.

Australian Bureau of Statistics. (1990). *Measuring Unpaid Household Work: Issues and Experimental Estimates,* (Catalogue No. 5236.0) Canberra: Australian Bureau of Statistics.

Becker, G. (1965). A theory of the allocation of time. *Economic Journal. 75*, 493-517.

Becker, G. (1981). *A Treatise on the Family.* Cambridge: Harvard University Press.

Chadeau, A. (1992). What is households' non-market production worth? *OECD Economic Studies. 18*, 85-103.

Cities Commission. (1975). *Australians' Use of Time, Albury-Wodonga and Melbourne 1974: A Preliminary Report.* Canberra: Cities Commission.

Eisner, R. (1978). Total incomes in the United States, 1959 and 1969. *Review of Income and Wealth. 24*, 41-70.

Eisner, R. (1991). *The Total Incomes System of Accounts.* Chicago: University of Chicago Press.

Gershuny, J. (1988). Time, technology and the informal economy. In R. E. Pahl (Ed.). *On Work: Historical, Comparative & Theoretical Approaches.* (pp. 579-597). Oxford: Basil Blackwell.

Hawrylyshyn, O. (1976). The value of household services: A survey of the empirical estimates. *Review of Income and Wealth. 22*, 101-131.

Ironmonger, D. S. (1961). *New Commodities and Quality Changes in the Theory and Measurement of Consumer Behaviour.* Ph.D Dissertation, Cambridge University.

Ironmonger, D. S. (1972). *New Commodities and Consumer Behaviour.* Cambridge: Cambridge University Press.

Ironmonger, D.S. (1983). A conceptual framework for modelling the role of technological change. In S. Macdonald, D. Lamberton & T Mandeville *The Trouble With Technology: Explorations in the Process of Technological Change.* (pp. 50-55). London: Frances Pinter.

Ironmonger, D.S. (1987). Research on the household economy. Research Discussion Paper Number 1, Melbourne: Centre for Applied Research on the Future, The University of Melbourne.

Ironmonger, D.S. (1989). Australian households: A $90 billion industry. Research Discussion Paper Number 10, Melbourne: Centre for Applied Research on the Future, The University of Melbourne.

Ironmonger, D.S. (Ed.). (1989). *Households Work: Productive Activities, Women and Income in the Household Economy.* Sydney: Allen & Unwin.

Ironmonger, D.S. and Lloyd-Smith, C. W. (1992). Projections of households and household populations by household size propensities. *Journal of the Australian Population Association.* 9(2): 153-171.

Ironmonger, D. S and Richardson, E. (1991). Leisure: An input-output approach. Research Discussion Paper Number 16, Melbourne: Centre for Applied Research on the Future, The University of Melbourne.

Ironmonger D, Sawyer K, Lloyd-Smith W., Donath S., and Tran, Van Hoa. (1990). A model of Australia based on household activities. Paper presented to the Workshop on Asia-Pacific Modelling at the Second Convention of the East Asian Economic Association, Bandung, Indonesia, 26 August 1990.

Ironmonger, D.S. and Sonius, E. (1987). Household productive activities. Research Discussion Paper Number 2, Melbourne: Centre for Applied Research on the Future, The University of Melbourne.

Ironmonger, D.S. and Sonius, E. (1989). Household productive activities. In D. S. Ironmonger, (Ed.). *Households Work: Productive Activities, Women and Income in the Household Economy.* (pp. 18-32). Sydney: Allen & Unwin.

Kendrick, J. W. (1979). Expanding imputed values in the national accounts. *Review of Income and Wealth.* 25, 349-363.

Lancaster, K. (1966). A new approach to consumer theory. *Journal of Political Economy.* 74, 132-157.

Lancaster, K. (1971). *Consumer Demand: A New Approach.* New York: Columbia University Press.

Leontief, W. W. (1941). *The Structure of the American Economy , 1919-1939.* Oxford: Oxford University Press.

Murphy, M. (1978). The value of nonmarket household production: Opportunity cost versus market cost estimates. *Review of Income and Wealth.* 24, 243-255.

Muth, R. F. (1966). Household production and consumer demand functions. *Econometrica.* 34(July), 699-708.

Nerlove, M. (1974). Review of Ironmonger (1972) and Lancaster (1971). *Journal of Political Economy.* 82, 1084-1090.

Peskin. J. (1982). Measuring household production for GNP. *Family Economics Review.* 2:

Szalai, A. (Ed.) (1972). *The Use of Time: Daily Activities of Urban and Suburban Populations in Twelve Countries.* The Hague: Mouton

Waring, M. (1988). *Counting for Nothing: What Men Value and What Women are Worth* Wellington: Allen & Unwin.

Econometrics and Environmental Policy Analysis

Richard F. Kosobud

As an observer who believes that the question of the optimal environmental policy on a specific issue is at bottom an empirical matter to be based on estimates of the benefits of pollution reduction balanced against associated abatement costs, I am invariably startled by the vehemence and qualitative character of the debate over the appropriate environmental action. It is not simply an issue of vested interests. It is as if one group sees expansive economic activity ever degrading finite stocks of clean air, water, and land, which must stop, compared with another group that sees abatement costs far in excess of any benefits of prevention or clean-up of pollution. Perhaps the lack of convincing empirical work is not the whole story of the controversy, but it is some significant part of it, I will argue. This essay gives my views on the problem, and what might be done to alleviate it.

I. Origins

I can trace the origin of my interest in the application of econometric techniques to environmental issues to Larry Klein's lectures in the period 1959-1960, his first years at Penn. I was absorbed by his careful, lucid efforts to open the econometric toolbox to his students, and motivated by the feeling he conveyed that what was inside could, by proper application, yield quantitative information of value in reducing the uncertainty and debate surrounding many policy issues.

Reading a recent review by Klein of the Cowles Commission days at the University of Chicago, I find themes written that are clear reminders of those influential lectures. Sophisticated estimation methods are not to be given undue emphasis at the expense of "... data base, economic analyses (institutional as well as theoretical), political insight, and attention to the steady flow of information" (Klein, 1990, p 114). The quote could serve as a credo for econometric advances into the area of environmental policy analysis, as I will argue.

II. The Mostly Empty Basket of Econometric Knowledge

Missing from much of environmental research are one or more features characteristic of econometric methodology; lacking is the search for structural relationships and their stochastic implications, the identification and statistical testing of parameters, the continuing effort to extend and gauge the reliability of data, the drawing out of the theoretical and institutional consequences, and the continuing checking and tests of existing knowledge.

Environmental research covers a wide range of studies: at the micro level there have been clinical and epidemiological studies of health effects, risk assessments, modeling of enterprise and industry abatement cost functions, analysis of types of market failures, and indirect estimates of consumer demand for environmental quality. At the macro level there have been efforts to model the consequences for aggregate variables of policies designed to deal with problems such as urban smog, acid rain, ozone depletion, and global warming.

My argument is that econometric techniques have been used but sporadically in these areas. My focus is on the macro-modeling aspects with which I am most familiar and where the gaps seem most evident. I will use examples from studies of global warming to illustrate the main points.

Several general observations based on my own survey of the extent of econometric research set the stage. Given the growing public debate, and interest, in environmental matters, and given the possible large economic impact of many of the policy issues, an observer could expect a wide range of research efforts underway to adapt or extend econometric tools to environmental issues, and to develop appropriate sample survey methods to measure consumer views on non-market goods and services.

The number of such studies seems small, however, both in estimating the benefits of avoiding environmental problems, or restoring quality levels, and in estimating the costs of achieving these benefits. These benefit-cost results ought to be, in the opinion of the mainstream of environmental economics, of central importance in policy decisions.[1]

In checking the last five years of *Econometrica* (from January 1988 to September 1992), I find less than 2 percent of all articles that deal explicitly with environmental topics, or relate methodology to environmental problems. It is true that articles develop methods that could be of use in environmental analysis, but where applications are discussed, they are in such areas as inventory cycles, efficiency wages, the interest rate curve, and various market failures, but not those failures giving rise to environmental concerns. There are specialized refereed journals that do present environmental research, but the premier methodological journal remains the main source of development of the discipline's tools.

Survey research might also be expected to play an important role in this research. Conceptually, the benefits of action to reduce the decline in the quality of particular resources ought to be measured by the consumer's willingness to pay for that action. Probing that willingness seems well suited to sample surveys, but the efforts to date are meager.

Similarly, the abatement costs of actions designed to reduce a decline in quality can be measured by the income required to maintain the consumer at the same level of welfare that was enjoyed prior to the action. Again, designing survey instruments would seem an attractive way to obtain this information. Asking the individual by means of specially designed survey techniques would seem to offer an attractive source of information on both benefits and costs. Yet, the number of sample surveys that concentrate on these questions is also limited.[2]

My own check of the Archives of the Inter-University Consortium for Political and Social Research (1991-1992) reveals that less than one percent of the 2,521 entries on surveys listed appears to deal significantly with environmental issues. My survey likely missed those studies with only an occasional question, but I included studies that dealt with topics like cancer on the chance that environmental problems caused by increased ultraviolet sun rays were included (the ozone depletion problem).

Alternatives to survey investigation do exist; for example, estimating willingness to pay by hedonic methods. But this approach appears to be limited to available data on land and housing prices (distance from pollution) or similar information largely generated for purposes other than environmental analysis.

The lack of solid sample survey results means that alternative quantitative information must be used. For the cost side, various kinds of expenditures on pollution control have been used. For the benefit side, impact studies have been used based on values assigned to sickness, agricultural output, and the like. Or, in some studies to be discussed, highly aggregated impact or damage functions have been specified on a speculative basis to provide a clue about the benefits of reducing these impacts. The methods frequently used to obtain these estimates or fit these functions all too often fail to accord with econometric practice.

III. The Difference It Makes

The *Economist* (August 8, 1992, p. 14) has argued that environmental policy has moved from a period of undercontrol of pollution to a period of the opposite (without, evidently, passing through the intervening stage), a statement guaranteed to generate a vehement qualitative argument.

One of the areas frequently cited as being overcontrolled is urban smog. A recent benefit-cost assessment and review of existing studies led the authors to conclude that there was substantial agreement that the costs of reducing tropospheric or ground-level ozone levels were in excess of benefits. The U. S. Clean Air Act Amendments (CAAA) of 1990, however, called for more stringent controls on the precursors of this pollutant. Why have the empirical studies, if they were right, carried so little weight in the legislative process?

The authors' comment was revealing: "It was impossible to provide anything approaching statistical confidence intervals for . . . (the costs of improvements)." And they further called for more appropriately tailored research action. "It is important to find ways . . . to ascertain individual's willingness to pay for any such improvements" (Krupnick and Portney, 1990, p. 252). The quality of the research results being suspect surely must constitute one cause for their neglect by legislators, and others.

The U. S. Clean Air Act of 1970 was explicit in denying the relevance of cost studies in determining standards when questions of health were involved. The CAAA of 1990 while containing an echo of that prohibition did provide for trading in sulfur dioxide emission permits in the hope of controlling acid rain in a more cost-effective manner. Both environmental and industrial groups supported this move. The growing share of environmental protection expenditures in GDP, now estimated around two percent, may be building support for econometric benefit-cost analysis and cost-effective approaches to environmental control.

The major government supported research effort into acid rain that was to provide the scientific basis for legislation, the National Acid Precipitation Assessment Program, yielded a number of scientific assessments of the sources and impacts of the pollutants, but little in the way of policy analysis of various courses of action. The CAAA of 1990 was drawn up largely independent of that study. A naive approach to policy analysis, I hasten to add, could do more harm than good threatening as it does to usurp the political decision; but a benefit-cost appraisal of policy options based on sound econometric research interjected into the heat of debate surely ought to be of value.

A dramatic example of the lack of credible econometric knowledge is provided by the recent Montreal Protocol agreement to begin the phasing out of certain stratospheric ozone depleting chemicals. The chief U. S. negotiator after writing that governments may have to act about the environment amid much scientific uncertainty, responsibly balancing the risks and costs of acting and not acting, concludes, "Unfortunately, the current tools of economic analysis are inadequate aids to this task, and can even be deceptive indicators; they are in urgent need of reform" (Benedick, 1991, p. 205). To all appearances, a remarkable international agreement was reached to achieve, and to continue to monitor, stabilization of the stratospheric ozone concentration without having the benefit of econometric advice one way or the other.

Many other areas in economics have thrown up problems not dramatically different from those being listed. However, environmental issues being, in most cases, market failures require government intervention in some form to correct them even in the best functioning decentralized economy. The argument for this intervention would seem to provide a strong motivation and justification for the application of econometric techniques as has happened, for example, in the case of inflation control, but has not happened in environmental control.

IV. The Difficulties

Since econometric work requires a theoretical foundation, the first thought to cross the mind is that environmental or resource economics lacks this base. But from Pigou's classical formulation of the appropriate tax response to an externality, there have been efforts to introduce intertemporal considerations, uncertainty, and asymmetries in information that have led one observer to write, "During the decade of the 1960s, and even more in the 1970s, what we today call environmental economics, and more generally resource economics, was developed and codified. ... the analytical elements are all available for use" (Dasgupta, 1990, p 52).

Do the analytical elements available for use need new econometric tools, or adaptation of old ones for their policy implementation? This question raises interesting issues. I have already mentioned that survey research instruments seem well suited to discovering evidence of individual environmental preferences and welfare; but, discovering the "willingness to pay" or the welfare costs of policy are likely to required more sustained research into survey methods. The time horizons are exceptionally long in some instances (global warming), the uncertainties exceptionally wide (in most), and the implications exceptionally complex (urban smog and others). Answers to questions are likely to vary enormously depending on how they are posed, and their accuracy requires sophisticated checks. The challenge to create and adapt tools seems clear.

One builder of very long-run models of economic growth and global warming has written that the econometric techniques are not available for the estimation of the parameters of his model. "In the place of a formal statistical procedure, we have simply chosen parameters so that the values taken by the model in the first three periods are tolerably close to the actual data" (Nordhaus, 1992, p 15). The author notes that in principle it would be better to form a likelihood function over the observations and to estimate the uncertain parameters by means of an iterative procedure, but such a procedure seems presently unavailable for small optimization models.

If it is not for lack of theory, and perhaps only partly for lack of new or adapted tools, then it must be for lack of data that so little persuasive evidence is available. Survey research generates its own data, so the absence of studies there remains a problem of effort, interest, and funding. Environmental time series analysis requires something akin to the National Income and Product Accounts data so vital to business cycle, economic growth and macro-policy studies. And that something is not yet available.

Environmental problems may be viewed as failures to establish appropriate accounting prices that could serve as a guide for policy considerations. The discussion is a topic not for deep ecology, but for deep social accounting; that is, what explains the absence of time series data on environmental matters in a social accounting framework?

Many of the principles for extending the standard social accounting scheme to renewable or environmental resources have been worked out (Statistical Office of the United Nations, 1992). The implementation, however, has been very slow. Valuation of stocks of air, water, and land would seem to present formidable problems, although many of them seem no more intricate than those encountered and more or less resolved in the valuation of private capital and investment.

In an extended system, net domestic product could be relabeled "sustainable income." That is, net national product would be that measure of income (welfare) after making allowance for deterioration in the quality of air, and other resources similar to the treatment of the economic value of private capital. There is considerable resistance, however, to broadening the accounts in this way. Some social statisticians appear to pale at the thought of developing appropriate measures. Users of standard data may fear dilution of the measures they want; the late Arthur Okun is reported to have argued strongly against the extension on the grounds that vital data necessary to understanding inflation would be contaminated (reported by R. Solow in Moss, 1973, p. 102).

There is a related argument that is best kept separate from this controversy, and that argument is over the elimination of environmentally "defensive expenditures" from final product. For example, expenditures for nasal steroid spray to minimize an asthma attack during a smog alert could logically be deleted from national income.. Trying to distinguish final from intermediate output does open Pandora's box, and it may be the better part of valor to postpone that particular debate.

To return to the matter of developing time series data, the problem probably has to do with the incentives to gather information on quality changes in the "public commons." No property rights having been established, the emissions of pollutants in important instances has gone unrecorded. However, the trend toward greater interest in environmental matters, likely to continue in my view, should increase demands for more comprehensive information presented in an integrated social accounting manner.

The valuation of unit changes in the stock of a renewable resource such as the atmosphere is admittedly no simple task. Physical measures of the quality of the stock, in many cases, seem to be a straightforward matter of instrument recording; for example, the concentrations of atmospheric carbon dioxide and the chlorofluorocarbons are monitored with reasonable accuracy. How value unit changes in these measures? One approach is to estimate the present discounted value of future impacts of present emissions, which means, in the case of carbon dioxide, estimating the complex changes induced into the climate system and then the subsequent impacts of temperature elevation, among other changing climate features, on health, agricultural productivity, hydrological systems, energy industries, and the like.

If that does not present problems enough there remains the issue of the discount rate. For long-run changes it is probably the social rate of time preference that is wanted. One approach is to use a weighted combination of the intertemporal rate on private consumption foregone and the different intertemporal rate on private investment foregone (the shadow price of capital). Cline (1992) has argued that discount rates in use tend to be too high reflecting either rates of return on capital before taxes or consumer borrowing rates. These are not back-of-the-envelope calculations for the faint hearted, but they could be the estimates, with confidence intervals, of solid econometric work.

The other blade of the econometric scissors is the estimation of the marginal abatement cost of reducing a unit change in concentrations. I have heard some grumbles from colleagues that public policy is overcontrolling ground-level ozone and acid rain, as mentioned previously, which means that, presently, marginal abatement costs are greater than marginal social benefits. My own current position is agnostic; that is, I feel it is time to call in the econometricians.

Once defensible estimates of the impacts of market externalities become more available they can flow into the empty lines of an extended framework of social accounts reflecting environmental features of the economy. Some time will be required to build up the number of observations necessary for sophisticated study.[3] The results of research based on these data could be especially important to developing economies.

Another area that can be cited as a cause for lack of progress in environmental research is lack of funding, private and public--notably federal funding. There may be more than a modicum of truth in this. Research support in the area of global warming, as an illustration, has grown rapidly but mainly in the hard sciences rather than in the policy sciences. Funding of impact studies or more broadly benefit-cost analyses has been the poor stepchild.

I will show, after discussing opportunities for econometric studies of global warming in more detail, that the value of additional uncertainty-reducing information about economic impacts is relatively high. That is, the return per dollar of research expenditure, while high in all areas, is higher in econometric studies of the benefits and costs of action on global warming than in studies of other aspects. This may be true in many other environmental areas. Self-serving as this may read, my estimates are based on an explicit methodology open to review and critique.

V. The Case of Global Warming

Induced climate change provides a good example for a test of the main hypothesis of this essay--that environmental research has suffered from insufficient econometric attention. It possesses many of the characteristics of environmental problems in the extreme, and can, therefore, shed more light on the difficulties delaying the application of econometric techniques.

Global warming is a very long-run problem in several senses. The climate change induced by the build-up of atmospheric trace gases, due mainly to economic activity, lags many years behind the increase in gas concentrations. If policy action to reduce fossil fuel consumption is decided upon, it will be years before direct time series data will become available to test the impacts.

There are enormous uncertainties surrounding all aspects of the problem, from scientific modeling of gas cycles and induced climate change, to the economic and other impacts of this change, to policy costs of action. The stakes could be very high. Impacts of global warming, many scientists believe, may be significant: on agriculture, on hydrological systems (salt water level increases and fresh water decreases), on storm frequency and hence damages, on climate sensitive non-agricultural industries, and on ecological systems (Intergovernmental Panel on Climate Change, Working Paper II, 1990).

The costs of abatement could also be significant. If sharp cut-backs in fossil fuel consumption, the main source of CO_2 emissions, are decided upon the costs of reduced use and substitutes have been estimated to be very large (Energy Modeling Forum, 1990).   Given the uncertainties, decision makers face the uncomfortable and unavoidable choice of one of two risks: the mistake of acting now and regretting it later, or of not acting now and regretting it later. The dilemma cannot be avoided.

A policy controversy has broken out between those advocating insurance-type actions, and those advocating a wait and see posture-the classic environmental debate. Several models familiar to econometricians would seem applicable among them being Bayesian learning approaches.

These controversies surfaced in another form at the Rio meetings of the United National Conference on the Environment and Development in 1992 tending to divide developed (who have emitted much) and developing countries (who have emitted little CO_2). In the face of the many uncertainties, such divisiveness is not surprising. As trace gases circulate quickly in the global atmosphere after emission, and as one country by unilateral action can do little to alleviate matters, an effective policy calls for international negotiation, and agreement.   Regional climate changes are likely to differ creating country and regional winners or losers. These attributes of the problem pose difficult questions for research, but also pose challenges that could stimulate the econometric imagination.

An Overview of the Econometric Tasks

Diagram 1 provides a convenient framework for describing major technical, behavioral, institutional, and definitional relationships characterizing global warming, and for highlighting the econometric challenges confronting researchers.

The first stage, reading from left to right, involves the long-run projection of trace gas emissions due to economic activity. An economic growth model is required with energy-environment sectors sufficiently disaggregated to give detail on fossil fuel resources (with their associated CO_2 coefficients), and on alternatives to fossil fuels. A number of models, some of them adapted from earlier research into energy and economic growth issues, have been utilized; others have been specially designed for climate change.

Heavy demands are made upon model builders. For most purposes, the models must be global in scope yet be capable of being disaggregated by country or region. They must be specified in a manner appropriate to making long-run projections frequently 100 years or more into the future. The appropriate models must have sufficient detail to reveal activities of the energy sector, to describe the capital accumulation and consumption processes, and to have linkages to climate models. The drive has been toward general equilibrium modeling for obvious theoretical reasons, but the issues involved in the choice of model seem far from being resolved.

The second stage, or column, of Diagram 1 indicates how emissions become inputs into trace gas (CO_2) cycle submodels that specify how gas concentrations build-up. Essentially this is a matter of keeping track of sources of emissions, chemical transformations in the atmosphere, and sinks, For example, about half of CO_2 emissions remain in the atmosphere for some 150 to 200 years.

The third column indicates how the thickening greenhouse blanket affects climate. What is required at this stage is a link to a general circulation model (GCM) of the atmospheric climate system. Even econometricians may be awed by the ambitions of climatologists to capture the complex physical and heat processes of the planet's climate in a system of nonlinear differential equations that comprise the heart of the GCMs. Research on these models has been very active; they could be one of the hopes of the future in improved policy analysis of the greenhouse effect (IPCC, 1992, Section B). Difficulties have been encountered, as pointed out by researchers in the area, in modeling cloud cover changes and ocean feedbacks, among other aspects, so that greenhouse effect theory, quite well developed, has not been matched by empirical implementation of these GCMs. Regional climate change may be more variable than planetary averages; unfortunately regional modeling has proved even more difficult. There is no question that complex interrelationships and feedbacks make development of models of sufficient reliability for policy analysis a slow process. Climatologists appear to be making a more concerted attack on their environmental problems than econometricians, however.

At some point in the future an integration of a fully developed GCM

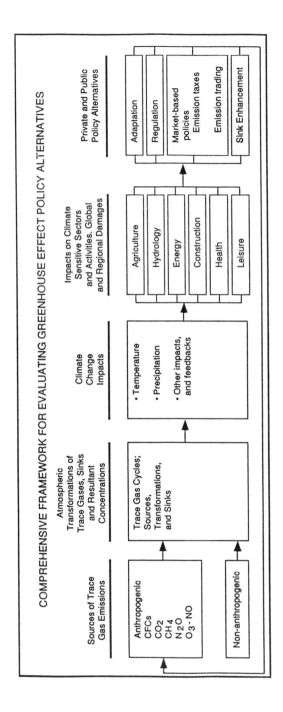

DIAGRAM 1

with fully developed gas cycle and emission-economic models will await the ambitious econometrician, and her climatologist colleague, aiming for a complete specification. As a compromise, for purposes of current policy analysis, scaled down climate models have been used in which induced temperature change has been selected as the dependent variable. Temperature change is treated as a sufficient statistic to represent the entire vector of climate variables that will be perturbed by trace gas concentration increases.

The fourth column provides for the development of impact (damage) relationships by which climate change is translated into effects on health, agriculture, water systems, and the like. Climate variables have long been studied for their influence in agriculture, although it is surprising the extent to which new problems are raised by induced climate change. Questions about the speed and magnitude of climate changes and consequent impacts on agriculture are not yet well understood (National Academy of Sciences, 1983).

Study of other types of impacts remain mostly unexplored. In an effort to get a feeling for damages, several researchers have added a simple aggregate damage function to other equations of their economic-climate models so that induced temperature change translates into GDP losses. This permits policy experimentation, but is no replacement for systematic micro-studies of impacts, studies that will test the econometrician's skills in detecting the possibly subtle influence of climate as distinct from other variables affecting human activity (Cline, 1992; Nordhaus, 1992).

The last column of Diagram 1 represents the modeling effort required to introduce policy constraints into the model, and estimate their consequences. Two policy options attracting attention currently are the imposition of carbon taxes to curtail fossil fuel use, and the issuance of marketable CO_2 emission permits in the volume that will bring emissions to the desired level. Both represent departures from past U. S. environmental policy choices, and both raise numerous questions associated with a new policy tool.

The history of environmental legislation suggests strongly that less cost-effective command-and-control policy options have been more frequently adopted than pollution taxes or tradeable permits. The near future is unlikely to see a dramatic switch to the new market-based approaches, although sulfur dioxide trading (to control acid rain), and nitrogen oxide trading (to control a precursor of urban smog) are being more actively explored (Kosobud, 1994). How to estimate the combined effects of a pattern of command-and-control regulation covered with a new layer of market-based methods in order to gauge their cost-effectiveness is unexplored econometric territory.

The abatement costs estimated from simulations of particular tax rates or permit price paths may be compared to the associated reductions in impact effects or damages. Studies of abatement costs have been more plentiful, but researchers continue to entertain dramatically different views on the costs associated with reductions of greenhouse gas emissions. For example, one group of studies finds significant energy inefficiencies in the economy that permit reduction of emissions at low cost--simply remove the inefficiencies. Another group finds inefficiencies much less prevalent, hence abatement costs are significantly higher (National Academy of Sciences, 1993, Nordhaus, 1990.).

The framework of Diagram 1 is, therefore, filled with econometric tasks: the specifying of structural relationships; the compounding of errors from emissions to estimated impacts; the estimating of parameters by adapting econometric methods to varied time series and cross-section data; the testing of estimated models; extending and improving data; and the endless revising of hypotheses.

It also bristles with problems: data base gaps, intertemporal issues that stretch the imagination, untested policy tools, international cooperation or conflict potentials, and a steady flow of information altering the uncertainties year by year. I am sure that the area must hold out both attractions and repulsions to careful econometric workers.

A Simple Illustration

A very simple model that will bring out these issues more explicitly is set out in this section. Assume that the stream of CO_2 emissions from the first stage specification of Diagram 1 results in a steady growth of concentrations as follows:

$$M_t = M_o \, e^{\delta t}, \tag{E1}$$

where M_t is the stock at time t, M_o is the initial, preindustrial stock, and δ is the growth rate of fossil fuel consumption generated by economic activity.

The radiative consequence of increased concentrations for sunlight falling on the earth's surface is approximately given by

$$R = \alpha \, \ln \, (M_t/M_o), \tag{E2}$$

where R is measured in watts per square meter, and α is a parameter that converts concentrations to radiative impact. The implication for climate change as represented by surface temperature is given by

$$\Delta T = R \, \beta \, \Gamma, \tag{E3}$$

where ΔT is equilibrium change in degrees Celsius, β is a coefficient converting watts per square meter into degrees Celsius, and Γ is a feedback parameter.

Γ is the object of much research and surrounded by great uncertainty as it embodies a number of responses to climate change. Among the most significant are thought to be cloud cover which may have a net cooling effect, ocean absorption of heat and consequences for the release of CO_2, and albedo change implications (snow and ice reflect back sunlight).

The IPCC has used three estimates for Γ in their scenario analyses, a high of 3.4, a low of 1.1, and a central value of 1.9. Inserting these numbers into (E3), and a radiative forcing value of 4.4 watts per square meter when CO_2 concentrations double, and a β value of .3 degrees Celsius per watt per square meter, yields the familiar estimates of an equilibrium global warming ranging from 1.5 to 4.5 with a central estimate of 2.5 degrees Celsius (Cline, 1992, p. 22). While the concentration increases are expected to occur, if nothing is done about the matter, within the next century, and by mid-century if the warming potential of the non-CO_2 trace gases is taken into account, the realized temperature increase will be less than estimated because of various lags of realized behind equilibrium change.

For our purposes Γ captures the uncertainty range (the other estimates are known to a greater degree of reliability) presently surrounding climate modeling. Substitute (E1) into (E2) and the result into (E3) to obtain

$$\Delta T = \alpha \ \beta \ \Gamma \ \delta \ t \qquad \text{(E4)}$$

which conveys, simplified as the message is, the idea that induced climate change is an interrelated process of economic activity, carbon cycle, and climate system events, all stochastic in nature and imperfectly understood.

The simple example may be completed by inserting (E4) into an impact or damage equation, usually non-linear, that relates temperature change inversely to GDP. Finally, an abatement cost function may be specified relating the degree of policy constraint of emissions, fossil fuels, to GDP. When done in the context of a long-run optimizing model there results an "optimal" carbon tax or marketable CO_2 emission permit price path that equates marginal social damage to marginal social abatement cost (Kosobud, 1992).

The Mostly Full Basket of Econometric Problems

For a start, consider the problem of testing the energy-environment models used for projecting emissions. The most frequently cited test is that of prediction. There are various ways of constructing these tests none of which seem adaptable for models that must project 100 years into the future. Most researchers, consequently, take care not to label what they do as predictions, rather calling them projections or scenarios.

Attempts have been made to backcast, but the lack of data, already mentioned, hampers this effort. Other efforts have been made to start-up a model prior to the present date to see how close projections come to current observations. It is hard to find successful efforts in either attempt.

An alternative is to vary parameters and exogenous variables over some sensible range and project a set of outcomes. In one approach, the "sensible" range of values can be estimated from expert opinion. These estimates can be treated as a distribution from which values can be selected for a probabilistic scenario analysis (National Academy of Sciences, 1983).

Another point of contention in the field of global warming studies, that may sound familiar to econometricians, is the issue of small versus large models: which are best? Are the small in size long-run models of a few equations worth serious attention? Partly, one may guess, it depends on the task at hand. Conclusions based on projections going out a hundred years or so raise questions of the range of error, the transparency of elements driving the models, and the importance of sensitivity analyses of changes in parameters or exogenous variables.

The Energy Modeling Forum (1992) gave a number of leading model builders the assignment of running standardized base case assumptions with the intent of comparing the time paths of state variables. Then, a set of policy scenarios constraining the models were specified with the same objective. The smaller optimization type models--linear and nonlinear programming models, for example--could be altered in clear-cut fashion and the consequences investigated. A large block recursive model with market clearing sectors provided projections that were less transparent. And a large general equilibrium model was not extended for the full horizon as projections, apparently, moved outside plausibility limits.

The construction of the general equilibrium model (for the U. S.) on neoclassical principles of production, consumption, and growth theory and estimated with post-war data using standard econometric techniques (Jorgenson et al, 1992) ought to have run away with all the prizes, but the internal workings of the elaborate edifice were not at all transparent, compared with the simpler models, and it was difficult to ascertain the sensitivity of results to parameter changes.

This model was fitted to data from the 1947 to 1985 period. However, that period may not resemble the long-run future in significant respects. One observer has expressed concern that new environmental legislation will make electricity costs higher, nuclear energy less important, gas both more cheap and available, and renewable energy resources an attractive option. That is, structural changes will make prior econometric estimates based on the past inappropriate, (Joskow, 1992). What may be required is respecification of models by inserting engineering and other information that reflects judgements about the future. Can this be done while maintaining econometric values?

Questions for the Econometrician

I propose to present some current results produced by a model constructed primarily on non-econometric principles in order to ask two general questions: what credibility can be given to the results, and what econometric methods are available to strengthen that credibility?

The model is a large intertemporal global linear programming framework that incorporates all the features of Diagram 1. It has been extended and updated from work by Nordhaus (1973) and is now being maintained by Argonne National Laboratory (Kosobud 1992).

Demand-side relationships have been partially estimated by econometric methods (for example, price and income elasticities of demand for four final energy commodities). Supply-side relationships are primarily engineering-type estimates of a wide range of energy technologies and expert estimates of costs and resource stocks. The objective function is global consumer energy welfare (surplus). Solutions are sensitive to relative input prices and generate intertemporal equilibrium paths of prices and quantities of final and intermediate energy commodities that maximize consumer surplus at least energy cost.

Three experiments are performed. The first scenario is a base case (no explicit carbon policy) run with parameter and exogenous variable settings that assume low to moderate income growth close to scenario IS92a of the IPCC which in turn incorporates many World Bank assumptions (IPCC, 1992, Table A3.1, p 77).

The second scenario, a policy scenario, embodies a frequently proposed effort to stabilize 1990 CO_2 emissions in the near term, say by the year 2000. The third scenario embodies a policy of setting a long-run climate stabilization goal comparable to the long-run climate consequences of the second scenario. What this means requires clarification. The second scenario while stabilizing emissions, a flow variable, does not stabilize concentrations, a stock variable. These increase steadily over the long-run future as will be shown reaching a certain level at the year 2100 and then continuing upward.

The third scenario takes the level of concentrations reached by the second scenario in 2100 and sets this as a *concentration* stabilization goal to be reached by 2100, but not necessarily before that date. The program is allowed to choose a least energy cost and emissions path to achieve that long-run goal.

The two policy scenarios achieve by 2100 the same concentration or temperature elevation control goal, but one does it by making near-term cuts in emissions and the other permits the program to choose a least cost path of emissions.

Three indicators of the resultant scenarios are presented in Figure 1. The first panel (a) depicts the concentration time paths, the second (b) depicts emissions, and the third (c) depicts shadow prices interpreted here as tax rates or tradeable emission permit prices. Not reported for the scenarios are the different time paths of energy resource use and mix changes and energy commodity prices available from this large model.

Note that both policy scenarios reduce concentrations one-third by 2100 from base case, and hence reduce temperature elevation by a like proportion. But policy A, the near-term reduction in emissions, permits concentrations, and temperature, to continue to increase beyond 2100. Policy A' stabilizes concentrations and temperature at 2100, bringing about a stabilization of anthropogenic climate change. Policy A' is in this sense a more restrictive effort than policy A. For ease of visual reading a line of no change in concentrations is presented.

The time path of emissions for the three scenarios in panel (b) tells the story of fossil energy resource use compared with the non-polluting backstops (nuclear and solar). The scenario requiring near-term cuts imposes on the energy industry short-term major adjustments from baseline. The scenario aiming for a long-term goal permits emissions to remain close to baseline before decreasing about the year 2030, and then decreasing below the other policy path at about the year 2065. More gradual adjustments are permitted the energy industry.

Panel (c) of Figure 1 gives a closer look at economic implications of the policy scenarios. Imposing short-run constraints on emissions can be made to yield, in the programming model as pointed out, a shadow price on fossil fuels. This shadow value can be interpreted as the tax rate that could be levied in a decentralized market system to achieve the desired reduction in emissions, or as the emissions permit price if permits were issued each year for use and not to be banked. The rate or price is calculated as a tax per ton of CO_2 by carbon weight.

Figure 1 shows that this tax rate or price in scenario A rises to over $100 per ton in the year 2000 then declines during the midcentury as new technologies come on stream only to rise again as substitutes must be sought for fossil fuel used in transportation (the model has both hydrogen and electricity driven transportation systems, but they are expensive). A tax of about $100 per ton roughly doubles the 1993 per barrel price of crude oil of standard grade.

In scenario A', the constraint in the model is now placed on the 2100 CO_2 concentration reached in A (about a fifty percent increase over the 600 billion tons in the atmosphere in 1980). A shadow price results which may be converted into a time path of tax rates or permit prices as before. These values in contrast with the other scenario start low, about $5.00 per ton, and rise as the rate of interest (5% in the scenarios) until fossil fuels are replaced by the backstops, around 2080. The path of A' stays below A until about 2050

Figure 1: Model Results: Stabilize by 2000 at 1990 Levels

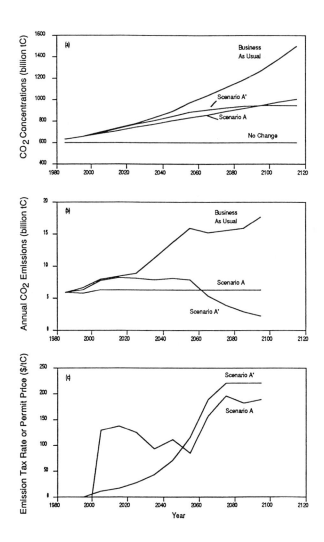

requiring less extensive adjustments of the energy industry. The path then rises above A as the fossil fuel industry is reduced to that level that stabilizes climate change.

An interesting way to think of policy option A' is to pretend that the nations of the world agreed to a marketable emissions permit system in which a large block of permits were allocated (somehow) among all parties in just the appropriate volume that would allow for the fifty percent increase in concentrations and no more. Put another way, an "evaporative" permit would have to be designed that would be canceled when a unit of fossil fuel, containing a given amount of CO_2, were extracted. Issuing these permits in just the right number creates a stock of valuable but exhaustible entitlements to emit. A further pretense will be that efficient markets will price this stock so that the permit price will rise as the long-run rate of interest (the shadow price of capital).

The implications for welfare of the two policies may be calculated from the model's objective function. A discounted saving of close to one trillion 1980 dollars is realized in A' compared with A. A trillion dollars is only a small fraction of total consumer surplus over 100 years, of course, but as the late Senator Dirksen of Illinois might say today, a trillion here and a trillion there and you are soon talking big bucks.

The saving is achieved largely through intertemporal cost-effectiveness. The low initial permit price, or tax rate, rising over time enables emitters to make current decisions and future plans that are more optimal than in the alternate scenario, especially since the model has built into it putty-clay capital stock features and constraints on the growth of new technologies that mimic realistic market situations. The permit markets are assumed to be seeking competitive equilibrium.

However, Hurwitz (1992) in commenting on these results observed that when negative commodities are adjoined to conventional ones, as in this case, nonconvex production possibility sets result and something less than the perfectly competitive equilibrium is achieved. Prices that vary as the volume of permits changes may correct this problem, but more government intervention into the private sectors than the simple issuance of a block of permits may be required.

How much reliance can be placed on such results? Long-run projections of this character do not constitute evidence to stand up in the econometric court. No range of error estimates are available. The deep parameters within the model, elasticities of demand and production coefficients, and the exogenous information, resource stocks and costs, have no confidence limits around them or estimates of their range.

One course of action is to attempt a probabilistic scenario analysis by estimating a distribution over parameters and exogenous variables, by choosing values from that distribution, and by generating a set of scenarios. Expert opinion can be drawn upon to get ranges of values. How to incorporate possible technology developments of the future about which only a clue or two is presently available? How to incorporate learning behavior into a policy model? The reader may add more questions to the list.

VI. Conclusions

The research agenda that has been laid out for econometric work in the area of global warming, and in other environmental areas, could be refined. It is a challenging one.

The value of additional information garnered by such research in these highly uncertain areas ought to be large. Take the linear programming model as way to prepare a crude estimate. Select critical uncertain parameters at each phase of Diagram 1; for example, the economic growth rates or air borne CO_2 retention rates in the first two columns of the diagram, the feedback parameter for the climate change phase, the damage parameter for the impact relationship, and the consequence for GDP of imposing tax rates at the last policy stage. For each parameter select a "correct" and an "incorrect" value allowing for a substantial difference.

The "correct" values inserted in the model yield a "correct" path of variables with an associated value of the objective function. One by one the "incorrect" values may be inserted and the model run out for ten years or so before the correct readings are discovered by research. Then, the correct value can be inserted yielding a path partially correct and partially incorrect. The difference in objective function value between the totally correct and partially incorrect scenarios provides a measure of the value of reducing uncertainty about important parameters or exogenous variables.

Nordhaus (1991) has prepared these estimates with a smaller model that reveal, first, a large return to research in the global warming area and, second, a markedly varying rate of return for research in the different phases of this area. Table 1 provides a breakdown.

Table 1: Percent of Total Welfare Gain from Improved Information

Phase of the Model	Percent of Total Gain
Carbon cycle sector	4
Climate sector	33
Economic impact sector	47
Policy cost sector	16
	100

Increased econometric research into impacts of global warming seems highly desirable. I cannot resist adding that federal research funds have been concentrated in the first two areas of Table 1, the "hard science" areas, That is, while the rewards for research appear to justify increasing expenditures on all types of global warming studies, the distribution of that support leaves room for improvement.

Richard F. Kosobud, Department of Economics, University of Illinois at Chicago, Chicago, Illinois 60670.

NOTES

1. There are exceptions to this statement. Baumol and Oates in their influential book (1988) abandon the two-bladed version of benefit-cost analysis as a determinant of policy recommendations substituting environmental targets determined by the political process. What remains for econometric analysis is the estimation of cost-effective ways to achieve these targets.

2. A recent excellent account of the present state of knowledge about contingent valuation studies and the distance yet to go in producing sturdy results may be found in Cummings et al. 1986. The problems confronting research in this area present challenges to econometricians; problems such as determining the accuracy of answers and overcoming answer biases of various kinds.

3. Time series data on the physical aspects of the quality of environmental resources are being developed more rapidly than valuation data. For example, data on atmospheric trace gas concentrations extending back over 100 years or more are now available (Carbon Dioxide Information Analysis Center, 1990).

REFERENCES

Baumol, W. J., and Oates, W. E. (1988). *The Theory of Environmental Policy.* New York, NY: Cambridge University Press.

Benedick, R. E.. (1991). *Ozone Diplomacy.* Cambridge, MA: Harvard University Press.

Carbon Dioxide Information Analysis Center. (1990). *Trends '90, A Compendium of Data on Climate Change.* Oak Ridge, TN: Oak Ridge National Laboratory.

Cline, W. R. (1992). *Economics of Global Warming.* Washington, D.C.: Institute for International Economics.

Cummings, R. G., Brookshire, D. S. and Schulze, W. D. (1986). *Valuing Environmental Goods: An Assessment of the Contingent Valuation Method.* Totawa, NJ: Rowman and Allanheld.

Dasgupta, P. (1990). The environment as a commodity. *Oxford Review of Economic Policy.* 6(1), 51-67.

_____ (1992). Costs of Environmentalism. *The Economist.* August 8, 1992, 12-14.

Energy Modeling Forum. (1991). *Interim Report.* Palo Alto, CA: Stanford University.

Hurwitz, L. (1992). Problems of Institutional Analysis. In Midwest Consortium for International Security Studies, *Global Climate Change: Social and Issues.* (pp. 95-100). Chicago, IL.

Intergovernmental Panel on Climate Change. (1992). *Climate Change 1992.* The Supplemental Report to the IPCC Scientific Assessment, New York, NY: World Meteorological Organization and United Nations Environment Program.

Intergovernmental Panel on Climate Change. (1990). *Potential Impacts of Climate Change*, Report Prepared for IPCC Working Group II. New York, NY: World Meteorological Organization and United Nations Environment Program.

Jorgenson, D. W., Slesnick, D. T. and Wilcoxen, P. J. (1992). Carbon taxes and economic welfare. *Brookings Papers on Economic Activity: Microeconomics*. 393-431.

Joskow, P. L. (1992). Comment on Jorgenson, Slesnick, and Wilcoxen. *Brookings Papers on Economic Activity: Microeconomics*. pp. 432-443.

Klein, Lawrence R. (1991). Econometric contributions of the Cowles Commission, 1944-1947: A Retrospective Review. *Quarterly Review*. Banca Nazionale Del Lavoro, No. 177, June, 107-117.

Kosobud, R. F. (1992). The problem of benefit-cost analysis. In Midwest Consortium for International Security Studies, *Global Climate Change: Social and Economic Issues*. Chicago, IL, 5-25.

_____ (Ed.). (Forthcoming). *Market-Based Approaches to Environmental Policy*, Chicago, IL: The University of Illinois Press.

Krupnick, A. J., and P. R. Portney, P. R. (1992). Controlling urban air pollution: A benefit-cost assessment. *Science*. 252(April), 522-528.

National Academy of Sciences. (1983). *Changing Climate*. Washington, D. C.: National Academy of Sciences Press.

National Academy of Sciences. (1991). *Policy Implications of Greenhouse Warming*. Report of the Mitigation Panel, Washington, D. C.: National Academy Press.

Nordhaus, W. D. (1973). *The Efficient Use of Energy Resources*. New Haven, CT: Yale University Press.

Nordhaus, W. D. (1990). To slow or not to slow: The economics of the Greenhouse Effect. *The Economic Journal*. 101(6), 920-37.

Nordhaus, W. D. (1992a). Uncertainties and research priorities in global environmental issues. Hearings October 8 & 10, 1991 before the Subcommittee on Science of the Committee on Science, Space, and Technology, U. S. House of Representatives, Washington, D. C.: U. S. Government Printing Office.

Nordhaus, W. D. (1992b). Rolling the 'DICE': An optimal transition path for controlling greenhouse gases. New Haven, CT: Yale University, mimeographed.

Solow, R. M. (1973). Comment. In M. Moss, (Ed.), *The Measurement of Economic and Social Performance. Studies in Income and Wealth. 38*. New York, NY: Columbia University Press.

Statistical Office of The United Nations, SNA Draft. (1992). *Handbook on Integrated Environmental and Economic Accounting*, New York, NY: United Nations.

Modeling Inventory Investment and Profits in A Macro-econometric Model

Joel Popkin

One of the first economists cited by Bodkin, Klein and Marwah (1992) in their history of macroeconometric model-building is Michael Kalecki. His work in the 1930s, apart from contributing important insights into macro theory, was the first to demonstrate that macro models of economic behavior have dynamic properties that flow from the specific economic relationships they contain. Much has been done to reflect and analyze the dynamic properties of macro-economic behavior. But more remains to be done, if model accuracy is to be improved. This paper is intended to focus research attention toward two variables where the payoff in terms of improved accuracy is likely to be high. Strategies for the improvement of the specification of these variables are then provided. The two variables are corporate profits and inventory investment.

These variables are critical in determining cyclical turning points. In the case of inventories, it is well-established empirically that they are an important source of business cycle movements (Klein and Popkin, 1961). But, neither inventory nor profit change is predicted well. The strategy to pursue is to tap the considerable information these variables themselves contain about their future direction. To do so should yield substantial improvements in the accuracy with which short-term quarterly movements in GDP are forecast, and in so doing, would improve cyclical turning-point predictability.

The considerable information these variables contain is their composition. It is the thesis of this paper that the level of these variables and changes in them are not independent of their composition. If the composition is not in equilibrium, the variables are almost certain to change. The reason such improvements have been slow in coming is that the variables needed to capture current-period inventory and profits disequilibria, and those that occur with a lag, are not contained in the set of variables specified in most macroeconometric models. That is because such variables are not usually among those that comprise the components of final demand. Thus the disequilibrium in consumption may be adequately captured by lagged values of consumption and income. But not all inventories are of finished goods ready for sale to final purchasers, nor are all profits earned only by the industries that sell directly to final demand. The determination of these two variables is much more complex structurally and requires the incorporation of variables behind the final demand level, i.e., intermediate sector variables. This paper demonstrates the likely existence of such disequilibria and its effect on final demand variables. Along the way, it points out approaches to modelling these variables.

I. Inventories

Inventory investment and the stock measure to which it relates are presented in three ways in the national income and product accounts (NIPAs). One is by broad industry of ownership--farm, nonfarm business, autos, etc. A second is by the three industrial sectors that hold most of the stocks: retail trade, wholesale trade and manufacturing. The third disaggregation is by stage of fabrication--materials and supplies, work-in-process and finished goods; it is presented for the manufacturing sector only. The emphasis in all these disaggregations is on the sector holding the inventories and, in the case of inventories by stage of fabrication, by the extent to which holders of the stocks have processed them.

All three types of disaggregations ignore the importance of measuring stocks of like items wherever they are held. Refrigerators are found in the inventories of the manufacturers that produce them and the wholesalers and retailers who distribute them. To explain the behavior of refrigerator inventories (and production), it is necessary, for the purposes at hand, to define the variable to include all refrigerators, regardless of which business sector holds them. It is also likely that the production of refrigerators, both for stock and sales, will be determined in part by the stock of refrigerators in the pipeline (Childs, 1967). It will also be determined by the stock of refrigerators already on order.[1]

For manufacturing as a whole, the finished goods inventories in some industries may contain some of the same items that are in the materials and supplies inventories of another. Steel sheet is an example. It is a finished goods inventory of the steel industry and a materials and supplies inventory of metal fabricating industries.

While there is a need to group inventories of like items regardless of where they are held, the data series are as yet unavailable. Their structure can be conceptualized, however, by the following taxonomy:

1) Define finished goods inventories as those that will never be touched again by the manufacturing process;

2) Define raw materials inventories as those that have never been touched by that process; and,

3) Characterize the remainder as inventories of partly processed goods.

The taxonomy can be implemented approximately by classifying 4-digit industries as to whether they produce largely finished goods, are the first processors of raw materials or fall somewhere in between.[2] When this is done, finished goods inventories can be measured as the sum of wholesale and retail trade inventories and the finished goods inventories of finished goods producers. Inventories of unprocessed goods can be approximated by the materials and supplies inventories in those industries that are the first processors of raw materials.[3] All other inventories--from the finished goods stocks of primary processors through the materials and supplies holdings of finished goods producers--are stocks of semi-processed goods.

A detailed study based on data for 1967 suggests that one-half of all inventories are of finished goods, one-third are of semi-processed goods and the remaining one-sixth are of crude materials that are awaiting processing by the manufacturing sector.[4]

There is no reason to expect each of these three stocks to react with the same lag structure to changes in final demand. Research by Popkin (1978) indicates that each inventory stock adjusts with its unique timing and amplitude. As a result, it is unproductive to ignore the stage-of-process structure of inventories in the inventory investment component of final demand.

Until the necessary data become available, a reduced form inventory investment equation can be derived from final demand variables. This can be accomplished in the following way using, for simplification, that version of Metzler's (1941) pioneering inventory-accelerator model in which all inventory change is assumed to be involuntary:

$$Y_t = C_t^A + I_t + E_t \qquad (1)$$

where

Y_t = income;
C_t^A = anticipated consumption;
I_t = inventory investment; and,
E_t = autonomous spending.

$C_t^A = C_{t-1}$ where C_t is actual consumption $\qquad (2)$
$C_t = \beta Y_t$, where β is the marginal propensity to consume. $\qquad (3)$
$I_t = IF_t + IP_t, \qquad (4)$

where

IF = investment in finished goods inventories, and
IP = investment in semifinished good inventories.

$IF_t = C_{t-1} - C^A_{t-1}, \qquad (5)$
$XF_t = \lambda C^A_t + \lambda (C_{t-1} - C^A_{t-1}) \qquad (6)$

where XF is production of finished goods and λ is the inverse of the ratio of the value of a unit of finished goods at the final demand level to the value at the finished manufacturers level.

$$IP_t = \gamma (XF_t - \lambda C_t) \qquad (7)$$

where γ is the inverse of the ratio of the value of a unit of finished goods output to a unit of semiprocessed materials input.

These six equations reduce to:

$$Y_t = \beta Y_{t-1} + \beta \Delta Y_{t-1} - \gamma \lambda \beta \Delta^2 Y_t + E_t \qquad (A)$$

This contrasts with the following equation if there is only one kind of inventory--IF:

$$Y_t = \beta Y_{t-1} + \beta \Delta Y_{t-1} + E_t \qquad (B)$$

The difference is $\gamma \lambda \beta \Delta^2 Y_t$, a second difference term with a negative sign. Assuming γ, λ and β are less than one as they should be, system A will converge, like system B. But beyond that, the similarity ends. In system A, Y fluctuates more than in system B (See Figure 1). And system A lags system B at turning points. The lag lengthens with successive cycles. Reflecting both differences, system A converges more slowly than system B.

FIGURE 1

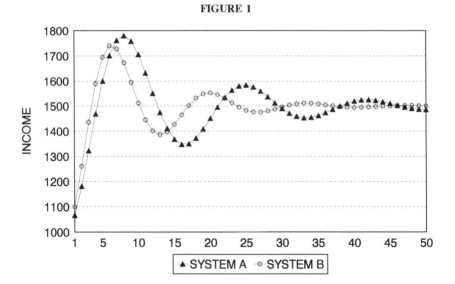

Figure 2 depicts how the two systems generate inventory investment. System A's two vertical sectors compound inventory shortfall when income is initially shocked upward. Inventory excess develops later in "A" than in "B," and takes longer to work off. Mirroring the income movements in Figure 1, inventory investment in system A both fluctuates more than in system B and lags more.

The Metzler involuntary inventory investment model determines inventories as the sum of production for current consumption and to make up previous involuntary changes in the stock. That production is plotted in Figure 3 for finished and semiprocessed goods separately. With respect to make-up of involuntary changes, finished stocks respond several time periods sooner than unfinished stocks. Thus, both positive and negative involuntary accumulation may be taking place at any point in time. The picture would become even more complicated if voluntary accumulation were introduced.

The still simple model of system A needs augmentation to explore the properties of inventory behavior in a more realistic setting. This model is primarily an inventory accelerator model in which desired inventory stock always equals its initial value. (That is why investment in both kinds of inventory goes to zero in Figure 3.) The effect of relaxing this assumption should be evaluated. But there are other augmentations that as yet have not been explored. One is to incorporate the aforementioned concept of ownership (inventory on hand and on order) introduced by Mack (1967). Another is to explore further the concept of inventories measured on a wherever-held basis, beyond the analysis possible with system A. The impact of both the "wherever-held" and "ownership" concepts should be applied separately both to materials and supplies and to finished goods inventories in each major producing sector.

FIGURE 2

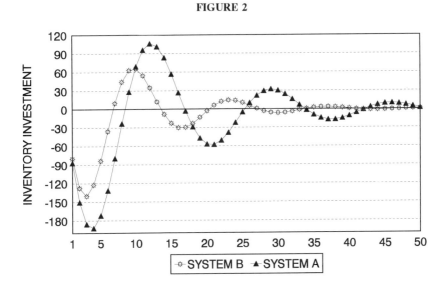

But these permutations do not consider other aspects of inventory theory such as production smoothing (Blinder and Maccini, 1991) or of joint determination of finished inventories along with production, order-fill rates and prices (Hay, 1970). Clearly, much more work needs to be done at two levels. The first is to specify and estimate detailed simultaneous inventory models. The second is to use the information obtained at the detailed level as a guide to the selection of a smaller set of variables (perhaps including those such as XF in system A) that can be introduced efficiently into the typical macroeconometric model.

FIGURE 3: Production of Inventories for Consumption

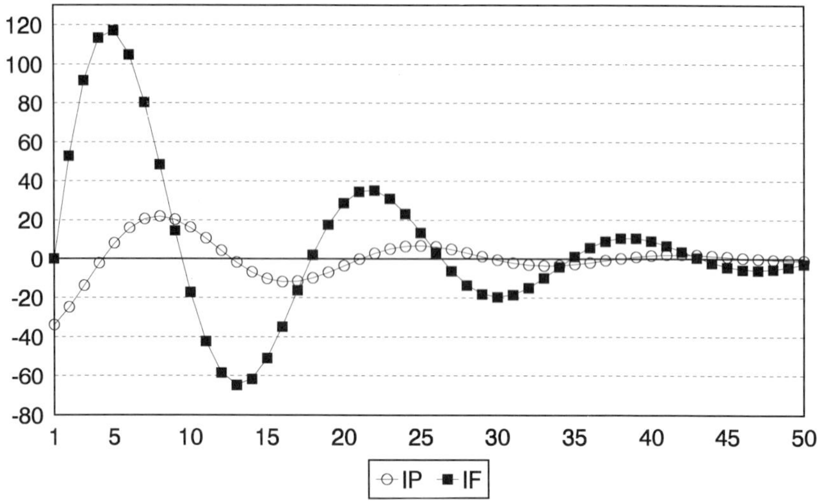

II. Profits

The problem with corporate profits is that it is an aggregation over many industries in contrast to most of the other components of the National Income and Product Accounts (NIPA) which are aggregations over all categories of *final demanders*. Profits are earned selling to both intermediate business and to final demand. In any period, profits in an industry may be an amalgam of long-run and short-run equilibrium components and of disequilibria. The short-run equilibrium component is normally the result of variable cost changes. Short-run disequilibria reflect the failure of markets to clear, usually as a result of price, but sometimes quantity, disequilibria. Both types of short-run influences, but particularly disequilibria, mean that buyers in one industry of the output of another may be paying a recently increased price which they have not yet passed through. As a result profits are below equilibrium in the buying industry and above it in the selling industry. The determination of aggregate profits in macro models does not take account of the extent to which each quarterly profits magnitude may reflect tensions (disequilibria) which will be resolved by the redistribution of profits among industries, or those tensions which will result in increases or decreases in the level of aggregate profits.

If inventories can be in disequilibria across stages of process, prices are likely to be as well. Prices serve to distribute profits across stage-of-process industries, so the profits variable in a traditional Keynesian model is likely to reflect disequilibria across stage-of-process industries. For example, in the early stages of expansion, prices of semiprocessed materials may rise but the increase may not yet be reflected in finished goods prices. Only when they are, will profits rise. That increase could be explained if primary product prices and those at other stages of process were in the model and fed into the aggregate profit variables.[5]

The substantial variability of profits at different stages-of-process can be seen in Figure 4. It contains the ratio of prices of outputs relative to prices of materials inputs for industries at various stages of processing. There are four stages. The first is the ratio of prices charged final demand purchasers by distributors relative to those they pay for the same goods when they buy them from domestic producers. The second is of prices charged by those producers relative to the prices they pay for semimanufactured parts and other materials used in production. The third ratio is of prices of semimanufactures relative to prices of primary manufactures, and finally, there is a ratio of primary goods prices to those of crude materials--e.g., steel ingot prices relative to those of iron ore, limestone, steel scrap and coal.

FIGURE 4: Ratio Net Output to Net Material Input Prices

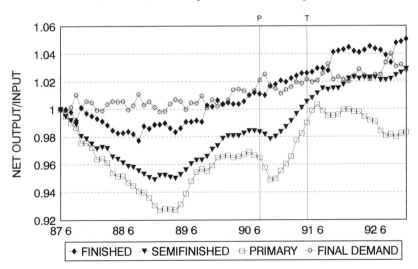

The price ratios in Figure 4 show that the greatest variability is among those at the semimanufactures and primary levels. They narrowed during the latest recession (July 1990 to April 1991). Similar ratios for retailers and finished goods manufactures went through the recession and the lengthy period of slow growth that preceded and followed it, without showing any signs of falling. Thus, the pressure on unit profits during the recession was greatest at the early stages of manufacture.

The ratio for semifinished and primary goods also dipped in the slow growth period preceding the recession. These ratios respond to both the rate of change of demand and to changes in that rate. That is why early stage prices are relatively reliable leading indicators of economic growth.   Another characteristic of early-stage ratios is that they may rise toward the end of a recession. Thus they began to rise half way through the recession the 1990-91. That these ratios respond to both first and second order changes is not unlike the finding for inventories by stage of processing depicted in system A. That system, unlike the simple Metzler system, has a second order term as well. Both findings suggest that, in the absence of specific intermediate sector variables, higher order terms can be used to represent developments behind final demand.

A number of models currently contain early stage-of-process prices, such as those for crude oil and farm products. They are typically exogenous. But more is known about down-stream price behavior, so it should be possible to link such exogenous crude materials prices through several stages of process, to the vector of final demand prices. Some of the same linkages could also be broadened to included various kinds of inventories. The resulting endogenous variables could then be used to improve the structural relationships determining aggregate profits and inventory investment.

Joel Popkin, Joel Popkin & Company, 1101 Vermont Avenue, N.W. # 201, Washington, D.C. 20005.

ACKNOWLEDGEMENTS

The author wishes to thank Dr. Jack L. Rutner for his work in simulating the inventory investment models presented in this paper.

NOTES

1. See Mack, 1967.
2. This approximation becomes more exact for an economy, the greater the triangularity properties of its I-O table. Popkin (1978) has shown the rows and columns of the U.S. I-O table for 1987 can be rearranged so that less than 5 percent of transaction are below the main diagonal.
3. Following the convention of Abramowitz (1950), inventories of goods-in-process may be split evenly between materials and supplies and finished goods inventories.
4. Inferential calculations from data presented by Blinder and Maccini (1991) suggest

there has been little change in these proportions.

5. The reader should not infer that adopting such a specification would require restricting the determination of profits as other than a residual. That is because the price variables reflect only changes in unit profits, which still must be multiplied by (variable) quantities of sales to yield total profits.

REFERENCES

Abramowitz, Moses. (1950). *Inventories and Business Cycles*. New York: National Bureau of Economic Research.

Blinder, Alan S. and Maccini, Louis J. (1991). Taking stock: A critical assessment of recent research on inventories. *The Journal of Economic Perspectives*. 5(1), 73-96.

Bodkin, Ronald G., Klein, Lawrence R., and Marwah, Kanta. (1992). *A History of Macroeconometric Model-Building*. Vermont: Edward Elgar Publishing Company, 12, fn. 42 and p. 26.

Childs, Gerald L. (1967). *Unfilled Orders and Inventories: A Structural Analysis*, Amsterdam: North-Holland.

Hay, George A. (1970). Production, price, and inventory theory. *American Economic Review*. September, 531-545.

Mack, Ruth. (1967). *Information, Expectations, and Inventory Fluctuation*. National Bureau of Economic Research.

Metzler, Lloyd A. (1941). The nature and stability of inventory cycles. *Review of Economic Statistics*. 23(August), 113-129.

Popkin, Joel. (1978). The integration of a system of price and quantity statistics with data on related variables. *The Review of Income and Wealth*, 24(1), 25-39.

_____ (1984). The business cycle at various states of process. *Journal of Business & Economic Statistics*. 2(3), 215-223.

_____, and Klein, Lawrence R. (1961). An econometric analysis of the post-war relationship between inventory fluctuations and changes in aggregate economic activity. In Joint Economic Committee, Congress of the United States. *Inventory Fluctuations and Economic Stability*. Part III. _____: University of Pennsylvania.

Actual and "Normal" Inventories of Finished Goods: Qualitative and Quantitative Evidence from the Italian Manufacturing Sector

Paolo Sestito and Ignazio Visco

It is widely claimed that Italian firms gained in flexibility in the eighties, thanks to technological advance (shared with the other industrialized countries) and a reduction of the labor market rigidities that had characterized the previous decade. The increase in flexibility should have led to less reliance upon inventories. A further stimulus in this direction is also likely to have come from the high real interest rates that prevailed in the eighties.

This paper considers the above claim, examining the variability over time of industrial production and sales and the evolution of actual and intended inventories of finished goods. To this end use is made of a qualitative indicator derived from monthly surveys, that serves to reflect the divergence between actual and "normal" inventories of finished goods.[1] A basic problem concerns the measurement of the actual inventory level, which is not directly observed in Italy and cannot be approximated by simply cumulating the difference between industrial production and sales. The index of sales shows a spurious upward trend that has to be removed to take account, as we argue, of the reduction of the degree of vertical integration that has occurred in the Italian industry.

The paper is organized as follows. Section 1 presents a number of measures of the variability of production and sales. Section 2 considers the behaviour of the inventory indicator. Its relationship with the actual change of inventories is used to obtain a correction of the sales index in Section 3; this allows an estimate to be made of the actual level of finished goods inventories, whose accumulation is examined using the production-smoothing model as a reference. Section 4 concludes.

I. The Variability of Industrial Production and Sales

As is well known, comparison of the variability of production and sales may shed light on the process governing inventories of finished goods. Convex production costs (and the costs associated with changing the level of production) tend to reduce its variability vis-à-vis that of sales, while the risk of stockouts and the existence of fixed costs in production work in the other direction. The evidence provided by such a comparison is not entirely conclusive because one would need to identify the properties of the shocks impinging upon both demand and cost functions. Even in a production-smoothing model (based on convex production costs), where the role of inventories is to reduce the variability of production, the latter may exceed the variability of sales because of shocks to the cost function (intertemporal substitution induces firms to produce more in favourable times) or correlation over time of demand shocks (innovations in the demand process lead firms to revise expectations for the near future).[2]

Blinder and Maccini (1991) in their thorough survey of evidence and models of inventory behaviour conclude that the traditional production-smoothing model may have received too much attention in the literature. On the one hand, the bulk of inventories (and inventory changes) is not made up of finished goods in manufacturing, where the model may work best. On the other hand, even in the case of inventories of finished goods, several facts seem to work against the production-smoothing model: production seems to be more variable than sales in most industries; sales and inventory changes are often not negatively correlated; and the estimated speed of adjustment to the desired level of inventories turns out often to be implausibly low.[3] As Blinder and Maccini argue, this evidence is quite impressive, even if not conclusive. Notwithstanding their arguments, however, the standard production-smoothing model continues to provide a useful reference point in the literature, especially when finished manufacturing goods are considered and when data availability limits the possibility of testing other specifications. Furthermore, as it will be shown, our preliminary evidence for Italy is not so devastating for this model.

Tables 1 and 2 present evidence on the variability of indexes of industrial production and sales in Italy, separately for the seventies (1973-1980) and the eighties (1981-1991), considering both 19 subsectors and manufacturing industry as a whole.[4] In particular, the following regressions have been estimated

$$log(X_t) = \sum \alpha_s d_s + \beta t + \mu_t \qquad (1)$$

where X_t is the monthly index of production or sales, d_s are seasonal dummies, t is a time trend, μ_t is a zero-mean residual and α_s and β are parameters. Simple variances of estimated residuals and seasonals have been computed. The contemporaneous correlation between the production and sales residuals has also been computed.

The comparison between the two indexes produces the usual picture that production is not unambiguously less variable than sales, contrary to what a simple production-smoothing model would predict. Moreover, innovations in production and sales (as measured by the residuals of the estimated equations) are positively correlated in almost every case.[5] In particular, the evidence for the 19 subsectors hints that no general pattern prevails, pointing to possible aggregation problems when analyzing the total. This is a caveat to bear in mind when looking at the estimates presented in the next sections, where use is made of aggregates.

Table 1: Production and Sales: Variances and Correlation by Sector (1973-1980)(1)

Sectors	Production		Sales		Correlation (2)
	Residuals	Seasonals	Residuals	Seasonals	
1	0.3	0.3	1.2	0.5	0.18
2	0.7	1.5	0.8	3.4	0.53
3	0.5	1.8	0.9	1.8	0.80
4	0.5	2.3	1.0	2.6	0.76
5	3.2	2.5	0.9	5.1	0.54
6	3.2	2.6	2.0	5.0	0.30
7	0.9	8.3	0.9	7.1	0.75
8	2.8	11.3	2.2	3.8	0.64
9	1.0	5.5	3.4	6.1	0.22
10	0.4	3.4	0.3	1.3	0.70
11	0.1	0.8	1.8	0.1	-0.34
12	1.4	2.5	1.2	2.3	0.65
13	0.8	2.4	2.1	2.7	0.21
14	1.0	7.4	0.9	6.6	0.86
15	0.9	10.2	0.9	6.7	0.77
16	1.5	7.3	1.2	5.4	0.76
17	1.0	2.3	1.4	3.4	0.75
18	1.1	11.0	1.8	7.8	0.81
19	2.5	11.2	6.5	5.1	0.53
Total	0.5	2.9	0.5	3.0	0.91

Notes: (1) The residual and seasonal variances are the variances (multiplied by 100) of estimated μ_t and α_s from the regressions $\log(X_t) = \Sigma \alpha_s \text{seasonals} + \beta \text{trend} + \mu_t$, where X_t = production or sales. (2) Correlation between the residuals of production and sales.

Comparing the two periods, there is no clear evidence of higher variability of production residuals in the eighties, but production was generally more variable in its seasonal component. This greater variance of the seasonals during the eighties could be linked with a gain in flexibility by industrial firms. Indeed, there is anecdotal evidence that firms were able to concentrate their cuts in production when a recessionary period was ahead, by lengthening vacations and taking advantage of the *Cassa Integrazione Guadagni*, a public wage supplementation fund that allows firms to put workers on a temporary layoff scheme. The reduction in the variability of sales and the lower cyclical turbulence in the eighties may also have played a role in reducing the variance of production residuals. The previous decade was characterized by more pronounced cyclical episodes,[6] probably leading to more erratic behaviour by both production and sales (Figures 1 and 2).

Besides examining the detrended level of production, the short-term variability of production was also considered, after having accounted not only for a log-linear trend but also for cyclical movements. The hypothesis that, for a given level of cyclical turbulence, production was more flexible during the last decade was examined, looking at the variance of the "irregular" component of production, as measured by the difference between the seasonally adjusted index of production and an estimate of its "trend-cycle", obtained from a 9-term Henderson average produced by the X11 seasonal adjustment procedure (Table 3). For the aggregate index, the erratic component of production is actually more variable in the eighties than in the seventies. This result does not extend, however, to all the subsectors considered.

Figure 1: Industrial Production (1973 = 1; seasonally adjusted)

Figure 2: Industrial Sales (1973 = 1; seasonally adjusted)

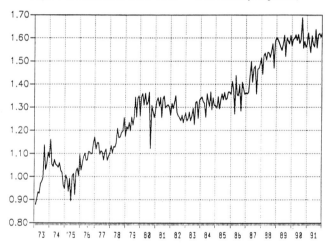

II. The Deviation of Actual from "Normal" Inventory Levels

The measures of variability reported in Tables 1 and 2 were not invalidated by the difference in the trends of production and sales, which clearly suggests the presence of a spurious component. The ratio between the production and sales indexes falls from 1 in 1973 to .86 in 1991 (see Figure 3).

Table 2: Production and Sales: Variances and Correlation by Sector (1981-1991) (1)

Sectors	Production		Sales		Correlation (2)
	Residuals	Seasonals	Residuals	Seasonals	
1	0.5	0.1	0.9	0.7	0.30
2	0.5	4.0	0.6	6.3	0.55
3	0.4	3.3	0.7	3.3	0.92
4	0.2	4.6	0.3	4.4	0.57
5	2.0	3.7	0.4	7.1	0.13
6	2.1	3.6	1.0	7.4	0.61
7	0.6	16.1	0.4	9.4	0.54
8	3.4	32.2	1.9	14.6	0.58
9	1.3	7.6	3.1	8.1	0.10
10	0.3	3.7	0.4	1.2	0.57
11	0.3	0.5	0.2	0.2	0.17
12	0.7	2.9	0.7	2.4	0.27
13	0.8	5.7	1.3	2.4	-0.11
14	0.4	12.5	0.4	9.5	0.53
15	0.4	14.1	0.6	10.2	0.55
16	1.0	13.6	0.8	12.8	0.83
17	0.3	3.2	0.3	3.0	0.45
18	0.5	15.6	0.4	12.9	0.70
19	3.2	14.3	1.2	12.2	0.24
Total	0.2	5.9	0.2	3.9	0.90

Notes: (1) The residual and seasonal variances are the variances (multiplied by 100) of estimated μ_t and α_t from the regressions $\log(X_t) = \Sigma \alpha_s \text{seasonals} + \beta \text{trend} + \mu_t$, where X_t = production or sales. (2) Correlation between the residuals of production and sales.

Table 3: Variance of the Irregular Component of Production

Sectors	1973-1980		1981-1991		Ratios
	Variance of the irregular component	Share of total variance	Variance of the irregular component	Share of total variance	Seventies Eighties
1	4.0	10.9	4.2	5.1	0.95
2	3.6	5.4	3.1	5.2	1.19
3	3.2	4.9	1.7	2.6	1.92
4	2.9	3.3	2.0	1.8	1.47
5	5.4	2.7	12.2	1.8	0.44
6	5.6	2.9	12.7	1.8	0.44
7	4.8	8.5	3.5	1.9	1.37
8	32.1	21.0	9.2	4.0	3.49
9	1.6	1.4	7.3	1.3	0.22
10	3.6	6.6	3.9	5.9	0.93
11	4.2	7.8	4.8	6.5	0.87
12	6.1	10.6	8.8	2.4	0.69
13	6.5	20.3	2.0	1.9	3.19
14	6.2	7.5	4.2	20.2	1.47
15	8.7	15.9	2.7	3.3	3.24
16	4.9	2.5	7.4	6.6	0.66
17	5.0	6.7	6.1	2.6	0.81
18	9.2	12.7	2.7	2.2	3.46
19	35.4	7.6	11.2	2.6	3.16
Total	124.8	57.6	271.6	67.2	0.46

Note: The irregular component is the difference between the seasonally adjusted index and the trend-cycle, obtained from a 9-term Henderson average produced

For any plausible values of the starting inventory level and the production-to-sales ratio in the first year, this divergence would produce negative inventories, obviously a nonsensical result. One possible interpretation is that the sales index was influenced by the restructuring process and by the reduction of the degree of vertical integration of the industrial sector that occurred mainly in the late seventies and early eighties. The sales index refers in fact to all the products sold by a firm (including those not directly manufactured), while the production index is a quantity index measuring specific items. [7]

In the absence of a direct measure of stocks in Italy, a correction for the spurious trend in sales has to be made before an estimate of actual inventories can be obtained by cumulating the difference between production and sales. Our research strategy is to exploit the qualitative information on the difference between actual (unknown) and "normal" inventories of finished products in manufacturing, surveyed monthly since 1962 by the Istituto nazionale per lo studio della congiuntura (ISCO), in order to derive an indirect estimate of inventories, given the relationship that must link the time series of industrial production, sales and inventories.

In the ISCO surveys firms are asked about their actual (end-of-period) vis-à-vis "normal" levels of finished goods inventories (taking seasonal factors into account); answers are coded, as percentages, into four groups: "above", "equal" and "below normal", and "no inventories" at all. Assuming a distribution for the continuous (unknown) quantitative responses, a series σ_t can be constructed (following Conti and Visco, 1984, in using the Theil, 1952, and Carlson and Parkin, 1975, procedure) such that:

$$I_t^* - I_t = - c\sigma_t I_t \qquad (2)$$

where I_t is the actual (end-of-period) level of inventories, I_t^* is the "normal" level and c is a positive constant signalling a threshold value below which there is no perceived divergence between normal and actual inventories (i.e. an answer "equal to normal" is given).

More specifically, given the percentages, A and B, of those answering above and below normal, one can write $A = Pr(I \geq I^* + \delta)$, $B = Pr(I \leq I^* - \delta)$. Assuming that $\delta = cI$, where c defines the boundaries of the normality interval as percentages $\pm cI$, one can obtain estimates of $(I^* - I)/I$ for a given form of the distribution of the answers. Under the assumption of a normal distribution, the series σ can be constructed as a function of the abscissa values of the standardized normal Z_1 and Z_2 (corresponding, respectively, to the areas $1 - A$ and B) such that $(I - I^*)/I = c\sigma$. In particular, $\sigma = -(Z_1 + Z_2)/(Z_1 - Z_2)$. Assuming instead a uniform distribution, one obtains $\sigma = BAL/(1 - A - B)$, where $BAL = A - B$ is the "balance" statistic usually computed in these qualitative surveys. [8] Series of σ have been constructed under both hypotheses;

they are very similar and those based on the normal distribution are the ones actually used in what follows. In the computations, the "no inventories" answers have been added to those "below normal". However, ignoring these answers, the pattern of σ over time does not change much. It should also be observed that the simplifying assumption of equal and constant limits for the "equal to normal" class has been made for all firms and all time-periods. This assumption should be among the first to be tested in future work. In fact, measurement errors may be induced by neglecting the heterogeneity (and possible asymmetry) of the scaling factor (and more generally of the "normal" level of inventories) across firms.

Figure 3: Ratio of Production to Sales Index (1973 = 1; seasonally adjusted)

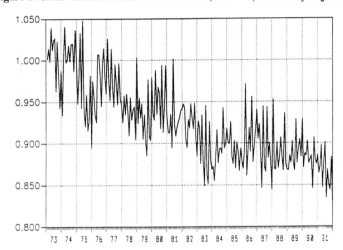

As Schlitzer (1993) has shown, the deviation between actual and "normal" inventories of finished products coming from the ISCO surveys appears to be strongly correlated to industrial output fluctuations for which it also acts as a leading indicator. Figure 4 shows the behaviour over time of the measure of σ_n, based on the assumption of a normal distribution. Going back to the issue of flexibility in the eighties, it is immediately evident that deviations of actual inventories from the level considered to be normal became smaller, in absolute size, during the last decade.

A visual inspection of the ISCO variable shows that the normal level firms have in mind is likely to be neither a long-run target (say the steady-state level) nor a very short-run objective. In both cases this variable would have been much more erratic. In the former case because all the shocks to actual inventories should have been translated, one for one, in that variable. In the latter case because firms would have attempted to promptly eliminate any divergence from the "normal" level, with the result that these differences would again be quite erratic and less persistent.

Figure 4: Deviations of Finished Goods Inventories From Normal (*)

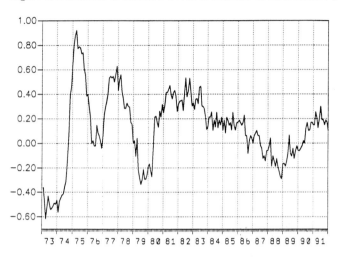

(*) The survey answers have been transformed on the assumption of a normal distribution.

In searching for the determinants of the discrepancy between actual and normal inventories, the first problem therefore arises from the ambiguity in the concept of the "normal" level of inventories: is it more similar to a long-run or to a short-run target? A direct examination of the normal level is not possible because it cannot be estimated without knowing both the actual level of inventories and the scaling factor c. Accordingly, rather than examining the determinants of I_t^* (conditional on some assumption for c and an estimate for I_t),[9] the determinants of the discrepancy between normal and actual inventories have been considered, with particular reference to the effects of short-term demand conditions and innovations in real interest rates. Basically, σ_t has been regressed on the deviation of the *ex-ante* real interest rate (r_t) from its lagged six-term moving average (mr_{t-1}) and other ISCO variables reflecting the expected change of demand in the coming 3-4 months (T_t) and the current level of demand vis-à-vis its "normal" level (L_t).[10] The possibility of breaks in the equation was also investigated, leading to the following estimate:[11]

$$\sigma_t = -.028 + (.885 - .224d_t)\,\sigma_{t-1} + .005(r_t - mr_{t-1})$$
$$\quad\;\;(3.43)\;\;(29.80)\;\;(3.67)\qquad\qquad (2.10)$$

$$\quad -.313d_tL_t - .425(1 - d_t)T_t \qquad\qquad\qquad (3)$$
$$\quad\;\;(5.52)\qquad (5.30)$$

adj. R^2 = .933, SER = .0742, DW = 2.15, h = 1.25,
MLM(1) = 1.83, MLM(12) = .14, MLM(1-12) = .85,
ARCH(1) = 1.41, ARCH(12) = 1.76, ARCH(1-12) = 7.15

where d_t is a dummy variable equal to 0 prior to 1978 and to 1 afterwards.

Two interesting results should be noted. First, a significant and strong impact of real interest rates on the intended change in inventories has been obtained. Second, the time break that has been uncovered is consistent with the hypothesis of greater flexibility in the production process: in the more recent period, the autoregressive component of σ_t becomes less important, while the important demand variable appears to be the discrepancy between the current and the normal level of orders rather than the expected change of demand in the near future.[12] This is in line with the interpretation that the flexibility gained in production management allowed firms to reduce the accumulation of inventories with which to meet future movements of demand.

III. Actual inventory accumulation

To examine whether the production-smoothing model works as a reasonable approximation of actual aggregate inventory behaviour in Italian manufacturing, an estimate of actual inventories needs to be obtained. As mentioned in the previous section, a correction for the spurious trend in the sales index needs to be made before an indirect measure can be derived of the actual change in inventories of finished products.

To recover the level of actual inventories, we thus start from the identity:

$$I_t = I_{t-1} + QP_t - QS_t = I_0 + \sum_{i=1}^{t} (QP_i - QS_i) \qquad (4)$$

where QP_t and QS_t are production and sales in month t. In (4) we have that $QP_t = aP_t$ and $QS_t = bS_t$, where P and S are the quantity indexes of production and sales and a and b are their values in the base year (1973). We then assume that the proper sales index, S_t, can be expressed as $S_t^* G_t$ where S_t^* is the unadjusted index and G_t is a correction factor assumed to be a simple function of time.

Whatever the determinants and the precise definition of "normal" inventories are, we can further assume that the investment in inventories for period t is somehow related to $(I_{t-1}^* - I_{t-1})$. This implies that whenever σ_{t-1} is (approximately) zero, production and (adjusted) sales tend to be approximately equal to each other. The evolution over time of the correction factor G_t can then be identified by looking at the production-to-sales ratio for those months when σ_{t-1} was approximately equal to zero. In doing so different benchmarks have been used, selecting those months with σ_{t-1} included in intervals from ±.05 to ±.20. The picture does not change much across the different cases (Figure 5). We ended up by using the one with most observations, even if it should be observed that in the last period almost all observations fall in the range.[13]

Considering only the months with absolute values of σ_{t-1} below .20, the ratio between the two indexes in the 1973-1976 period is on average roughly consistent with the production-to-sales ratio observed for the average of 1973 (a/b = 1.018, i.e. the ratio of the actual values in the base year) as reported in Conti and Visco (1984). This result corroborates our procedure, which provides an average value of .889 for the ratio of the two indexes during the 1984-1991 period. A few problems arise for the intermediate period: for the years from 1977 to 1980 there seems to be intermediate values for the production-to-sales ratio; for the years from 1981 to 1983 we basically have no observation with σ_{t-1} approximately equal to zero. We therefore decided to assume that G_t followed a linear trend in the years 1977-1983 going from 1 to .905 (obtained from .889 by using the known value of a/b). We ended up with the following values for G_t :

$$G_t = 1 \qquad\qquad\qquad\qquad \text{for the 1973-1976 period}$$
$$G_t = 1 - .001116(t - 48) \qquad \text{for the 1977-1983 period}$$
$$G_t = .905 \qquad\qquad\qquad\qquad \text{for the 1984-1991 period}$$

where t is a linear trend equal to 1 in January 1973.

Using then 1.018 for the production-to-sales ratio in 1973 and 1.607 as the starting value for inventories in units of 1973 average sales (also taken from Conti and Visco) the actual level of inventories (also expressed in units of 1973 average sales) can be obtained as:

$$I_t = 1.607 + 1.018 \sum_{i=1}^{t} P_i - \sum_{i=1}^{t} S_i^* G_i \qquad\qquad (5)$$

Figure 6 shows that the level of actual inventories of finished goods resulting from (5) presents marked cyclical fluctuations as well as a gently declining long-run trend.

To examine whether production-smoothing reasonably approximates actual inventory behaviour, a very simple specification has been considered, consistent with the way the estimate of I_t was obtained:

$$\Delta I_t = \lambda(I_{t-1}^* - I_{t-1}) + x_t + \epsilon_t \qquad\qquad (6)$$

where x_t summarizes all the elements not included in the level of inventories that firms considered to be "normal" at the end of month $t - 1$ (the buffer role of inventories, news leading to revisions in the planning of inventory accumulation, expectations about cost and demand shocks not already taken into account by I_{t-1})[14] and ϵ_t is a stochastic error.

Figure 5: Ratio of Production to Sales Indexes in the Months With Normal Inventories Approximately Equal to Actual Inventories (1973 = 1)

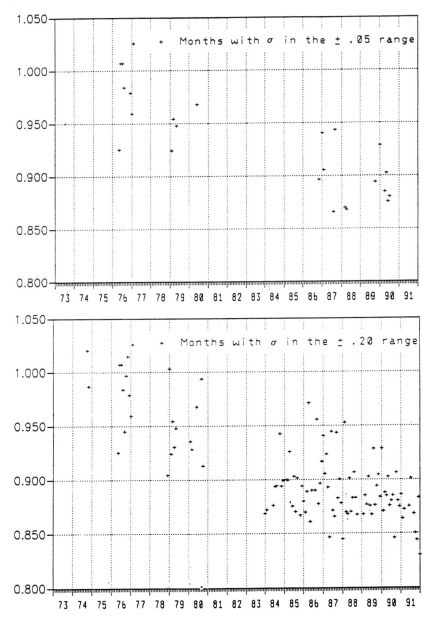

Recalling that $-\sigma_t I_t = (I_t^* - I_t) / c$, equation (6) can then be estimated.[15] Neglecting the x_t component, an estimate of $\lambda c = .01$ is obtained; the fit of the equation is, however, rather low, residuals are highly serially correlated and the t-value of the estimated coefficient is about 1.5. Roughly the same estimate, with a t-value equal to 1.95,[16] is obtained when x_t is simply specified as $\beta \Delta (T_t S_t)$, which proxies the acceleration of expected sales for the near future.

Figure 6: Actual Inventories (*)

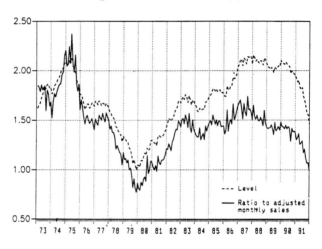

(*) Expressed in units of 1973 average monthly sales.

The preferred specification, which completely eliminates the serial correlation present in the previous estimates, is the following:

$$\Delta I_t = .035 + .013 \ (I_{t-1}^* - I_{t-1}) / c + .073 \ \Delta(T_t S_t)$$
$$\quad \ (2.23) \quad (3.96) \qquad\qquad\qquad\qquad (2.50)$$

$$\quad - .705 \ S_t + .143 \ S_{t-1} + .034 \ S_{t-2} - .109 \ S_{t-3} \qquad (7)$$
$$\quad \ (16.66) \quad (3.45) \qquad (0.78) \qquad (2.26)$$

$$\quad +.216 \ P_{t-1} + .125 \ P_{t-2} + .270 \ P_{t-3}$$
$$\quad \ (3.17) \qquad (1.83) \qquad (4.35)$$

adj. $R^2 = .668$,   SER $= .023$,   DW $= 2.02$,
MLM(1) $= .79$,   MLM(12) $= .31$,   MLM(1-12) $= 1.10$,
ARCH(1) $= .78$,   ARCH(12) $= 1.37$,   ARCH(1-12) $= 14.03$

where x_t includes current and past values of sales and past values of production, catching both the buffer role of inventories and the presence of cost shocks not taken into account by I_{t-1}.[17]

The significant effect of the past discrepancy between actual and normal inventories provides a remarkable result, given the use of aggregate data.[18] As a caveat one has to keep in mind that the final specification used for x_t is not fully consistent with the way the measure of I_t was obtained because the presence of x_t was neglected at that stage. Moreover, from (7) we only get an estimate of λc, so that it is difficult to evaluate the actual performance of the production-smoothing model unless some outside guess on the unknown scaling factor c is made or some further identifying restriction is introduced. From the estimate of λc in (7) we can nonetheless conclude that, for reasonable values of c, the implied values of λ suggest that a sufficiently rapid adjustment to the desired target is not incompatible with the data. For $c = .025$, for example, one would obtain $\lambda = .52$, so that about eighty percent of the gap between desired ("normal") and actual inventories would be closed in just two months.

An independent estimate of λ could be obtained by putting more structure into our model.[19] Denoting by P^*_{t-1} the level of production planned at the end of month $t - 1$ in the absence of adjustment costs and assuming that the actual production process might be described as a simple geometric adjustment to its desired level, we can write:

$$\Delta P_t = \lambda(P^*_{t-1} - P_{t-1}) + x_t + \eta_t, \qquad 0 \le \lambda \le 1 \qquad (8)$$

where x_t again summarizes all the elements not included in the desired level of production and η_t is a stochastic error. A simple specification for P^* could be the following:

$$P^*_{t-1} = \gamma S_t + (1 - \gamma)\hat{S}_{t+1} + I^*_{t-1} - I_{t-1} \qquad (9)$$

where the level of production planned in month $t - 1$ for month t is intended to close the gap between the desired and actual levels of inventories and to meet anticipated demand. Given obvious production lags, the latter is expressed in (9) as a weighted average of the sales that are actually going to take place in month t (presumably known and basically decided by firms at the end of $t - 1$) and those expected by firms for the immediate future (say month $t + 1$). Equations (8) and (9), together with the identity $\Delta I_t = P_t - S_t$, lead to:

$$\Delta I_t = \beta_1(I^*_{t-1} - I_{t-1}) + \beta_2 \hat{S}_{t+1} + \beta_3 S_t + \beta_4 P_{t-1} + x_t + \eta_t \qquad (10)$$

where $\beta_1 = \lambda$, $\beta_2 = \lambda(1-\gamma)$, $\beta_3 = \lambda\gamma - 1$, $\beta_4 = 1 - \lambda$, so that $\beta_2 + \beta_3 + \beta_4 = 0$. For a particular specification of $\hat{S}_{t+1}$ in terms of past values of S_t, equation (7) could be parameterized to produce proper estimates of β_2, β_3 and β_4, with x_t being essentially a function of ΔP_{t-1} and ΔP_{t-2}. In this case an estimate of λ could be obtained by simply summing the coefficients of the production terms present in the equation, ending up with an estimate near .4, which is not a low speed of adjustment at a monthly level. Alternatively, we have considered the possibility of proxying expected demand as:

$$\hat{S}_{t+1} = S_t (1 + \hat{\gamma}_0 + \hat{\gamma}_1 dst + \hat{\gamma}_2 ds_{t-1} + \hat{\gamma}_3 ds_{t-2})   \quad (11)$$

where $ds_t = \Delta S_t / S_{t-1}$ and the $\hat{\gamma}$'s have been obtained from the third-order autoregression (over the 1973.6 - 1991.12 sample period):

$$ds_t = .006 - .733 ds_{t-1} - .365 ds_{t-2} + .130 ds_{t-3}$$
$$ (2.52) \ (10.99) \quad\quad (4.61) \quad\quad\quad (1.95) \quad\quad\quad (12)$$

adj. $R^2 = .450$,   SER = .033,   DW = 1.99,   h = 1.10.

Using (11) the following estimate of equation (10) has been obtained:

$$\Delta I_t = .040 + .014(I^*_{t-1} - I_{t-1}) / c + .077\Delta(T_t S_t)$$
$$ (2.53) \ (4.33) \quad\quad\quad\quad\quad\quad\quad (2.64)$$

$$+ .219\hat{S}_{t+1} - .802 S_t + .554 P_{t-1}   \quad (13)$$
$$ (3.23) \quad\quad (17.82) \ \ (9.83)$$

$$- .329\Delta P_{t-1} - .230\Delta P_{t-2}$$
$$(5.29) \quad\quad\quad (3.97)$$

adj. $R^2 = .662$,   SER = .023,   DW = 2.01,
MLM(1) = .49,   MLM(12) = .17,   MLM(1-12) = 1.04,
ARCH(1) = .62,   ARCH(12) = .05,   ARCH(1-12) = 15.85.

In the first place it should be noted that the fit, general characteristics and parameter estimates of equation (13) are very similar to those of equation (7). Two points are worth emphasizing. On the one hand, the constraint on the coefficients of $\hat{S}_{t+1}$, S_t and P_{t-1} given by equation (10) is substantially satisfied, with the estimates of $\beta_2 + \beta_3 + \beta_4$ equal to .04,[20] and a significant buffer stock role revealed. On the other hand, the estimate of the adjustment coefficient turns out to be about .45, which implies that the production (and inventory) gap appears to be closed at a sufficiently high speed, seventy percent of the adjustment taking place in two months. From this estimate, and that of the coefficient of $(I^*_{t-1} - I_{t-1}) / c$, we also obtain an extremely reasonable value for c, slightly above 3 percent. Finally, it should be observed that, through the inventory disequilibrium variable, the gains in flexibility discussed in the previous section (as well as the changes in real interest rates) obviously also affect the actual inventory accumulation process.

IV. Conclusions

In this paper the evolution and determinants of the stock of finished goods inventories held by Italian manufacturing firms have been examined. The only direct information available on inventories allows a qualitative indicator, that reflects the divergence between actual and "normal" levels, to be derived from monthly surveys. This indicator probably reflects the intended change in the stock of inventories and appears to be significantly affected by changes in real interest rates and demand. Consistently with the increased flexibility of the production process, it turns out that during the last decade the intended change in inventories of finished goods was geared more to the current level of demand than to expectations of change in the near future. The greater flexibility obtained in production management allowed firms to reduce the need to accumulate inventories in advance to meet expected demand.

An indirect estimate of the actual stock of inventories was obtained by cumulating over time the difference between the production and (adjusted) sales indexes. As the latter shows a spurious upward trend, presumably linked to the restructuring process and the reduction of the degree of vertical integration that has occurred in Italian industry, an adjustment was made that also took account of the qualitative information coming from the indicator on the divergence between actual and "normal" inventory levels. As a ratio to current sales, the stock of inventories of finished goods that results from this procedure shows both marked cyclical fluctuations and a gentle downward trend. Estimating the process of actual inventory accumulation, the traditional production-smoothing/buffer-stock model seems to remain a suitable reference point, at least in the case of the stock of finished products; the disequilibrium signalled by firms has a significant effect on their actual investment in inventories and the speed at which the disequilibrium is closed appears to be sufficiently high.

The evidence presented in this paper is certainly preliminary. Even if we believe that the aim of showing the usefulness of survey information in the analysis of inventory accumulation was essentially achieved, a number of steps and assumptions should probably be further analyzed and tested. With only a limited amount of aggregate information, a wide spectrum of issues was put under scrutiny. As the various caveats we advanced should have indicated, it would be especially important to allow for the heterogeneity and asymmetry in the way firms signal the presence of disequilibria in their inventory stock; moreover, as regards the construction of the actual level of inventories, better ways should be found to deal with the spurious trend that plagues the currently available index of sales.

In particular, the use of disaggregated data would probably reduce the measurement problems that showed up in the estimation of the actual level of inventories, since it would allow the identification of industries where the spurious trend in the sales index was either less important or could be corrected by means of available external information. Perhaps more importantly, it could attenuate the impact of some of the simplifying assumptions that have been adopted to transform the qualitative evidence on the discrepancy between actual and "normal" inventory levels into a quantitative time series. Furthermore, it might also make it possible to separate firms producing for stock from firms producing to order, making it possible to compare the performance of the production-smoothing model directly with that of other models of inventory investment.

In any case, to conclude, the evidence presented in this paper supports the idea that, at least in the aggregate, production-smoothing and buffer-stocks are still important factors in the determination of finished goods inventory accumulation in Italy. An important role for real interest rates has also been identified, and some support has been found for the claim that increased flexibility in production management in the eighties allowed Italian firms to rely less on the accumulation of inventories of finished products to meet short-run changes in demand.

Data Appendix

P: index of production in manufacturing industry; computed monthly by Istat in quantity terms since 1953, adjusted by the Bank of Italy for differences across months in the number of working days and for seasonality, 1973=1 (see Bodo, and Pellegrini, 1993).

S': unadjusted sales index in manufacturing industry; computed monthly by Istat in nominal terms since 1973, deflated and seasonally adjusted by the Bank of Italy, 1973=1.

T: expected change of demand in the next 3-4 months; weighted balance of plus and minus answers in the ISCO monthly surveys; seasonally adjusted by the Bank of Italy.

L: level of current demand and orders vis-à-vis a "normal" level; weighted balance of answers above and below normal in the ISCO monthly surveys; seasonally adjusted by the Bank of Italy.

I: actual level of (end-of-period) inventories of finished goods, obtained by cumulating the difference between production and (adjusted) sales as described in the text and expressed in units of monthly average 1973 sales.

r: *ex-ante* real rate of interest, computed as $(i - \pi)/(1 + \pi)$, where the monthly loan (prime) rate, i, has been computed joining different series elaborated by the Bank of Italy, and π is the rate of change of wholesale prices expected at the end of month t for the next six months by the industrial businessmen participating in the *Mondo Economico* forum (see Visco, 1984, for the construction of this series from the survey answers); the π series is originally semi-annual and has been interpolated using the qualitative answers to another monthly ISCO question on price changes as indicator.

σ: percentage deviation of actual and normal inventories (divided by an unknown scaling factor) obtained by assuming a normal distribution of the answers to the monthly ISCO surveys.

Paolo Sestito and Ignazio Visco, Banca d'Italia, Via Nazionale 91, 00184, Roma, Italy.

NOTES

1. The importance of survey methods and data in the analysis of economic fluctuations has often been emphasized by Lawrence Klein. For an early contibution, see Klein (1954).

2. Exact bounds for the comparison between sales and production variability have been derived by West (1986).

3. However, the low speed of adjustment may well depend on aggregation bias (see Seitz, 1993) or measurement errors in the construction of inventory data (see Fair, 1989).

4. See the Data Appendix for details about the time series utilized in this paper. The 19 subsectors are: 1) Fuel and power products; 2) Ferrous and non-ferrous ores and metals; 3) Non-metallic mineral products; 4) Chemical products; 5) Metal products (excluding transport equipment), agricultural and industrial machinery; 6) Office and data-processing machines, precision and optical instruments; 7) Electrical equipment; 8) Motor vehicles and engines; 9) Other transport equipment; 10) Meats and other food products (excluding beverages and milk products); 11) Milk and dairy products; 12) Beverages; 13) Tobacco products; 14) Textiles and clothing; 15) Leather, leather and skin goods, footwear; 16) Timber, wooden products and furniture; 17) Paper and printing products; 18) Rubber and plastic products; 19) Other manufacturing products.

5. These results are confirmed when seasonally adjusted data are used directly.

6. The reference period for cyclical upturns can be found in Schlitzer (1993).

7. Iacoboni and Sestito (1987) have presented evidence of a downward trend for the ratio of sales of own-manufactured goods to total sales, the latter being measured by the sales index currently available.

8. For further methodological details, see Visco (1984), especially pp. 65-74. It should be observed that in the case of a uniform distribution the balance statistic would be a reasonable indicator provided that the "equal to normal" class was roughly constant over time: this appears not to be a bad approximation in the present case.

9. This alternative route was followed by Conti and Visco (1984). For the period they examined, however, the estimate of I_t was likely less subject to measurement errors. Moreover, the heterogeneity across firms neglected in the construction of σ may have been less important in the seventies than in the eighties. In the next section a measure of the actual level of inventories of finished goods is in any case obtained and some preliminary evidence of the determinants of actual inventory accumulation presented.

10. See the Data Appendix for further details also on these time series.

11. Allowing for the lags present in the equation, the sample period goes from July 1973 to December 1991. *t*-statistics are reported in absolute value in parentheses. Statistical inference would not be much affected by the use of the White procedure to estimate the standard errors. The other reported statistics are: the adjusted R^2 and the standard error of the residuals (SER); the Durbin-Watson (DW) and the Durbin (h) tests for autocorrelation of first order in the residuals; the modified Lagrange multipliers (MLM) and Engle (ARCH) statistics appropriate to test for autocorrelation and autoregressive conditional heteroschedasticity of different order in the residuals.

12. The restriction implied in our treatment of the time breaks is not rejected, with an $F(4,212) = 1.42$.

13. Observe that for plausible values of c (say $c = .05$) $σ = ± .20$ implies $(I - I^*)/I = ± .01$.

14. This reflects the possibility that I^* might account for only part of all the relevant short term elements.

15. The sample period goes again from July 1973 to December 1991.

16. 1.98 using the White procedure, which takes account of potential heteroschedasticity.

17. In (7) as well as in what follows the production index P has been multiplied by the production-to-sales ratio in the base year, $a/b = 1.018$.

18. For the bias arising from the use of aggregate data in the estimate of the production-smoothing model, see Seitz (1993), who also relies on survey data of the kind considered in this paper.

19. For a similar derivation, see Fair (1989).

20. The restriction, however, is rejected at the 5 percent level, with a t-value equal to 2.33, pointing to the possible presence of some slight misspecification, as is almost bound to occur in the construction of the expected demand variable.

REFERENCES

Blinder, A.S. and Maccini, L.J. (1991). The resurgence of inventory research: What have we learned? *Journal of Economic Surveys*. 5(4), 291-328.

Bodo, G. and Pellegrini, G. (1993). L'indice di produzione industriale in base 1985: ricostruzione storica e depurazione stagionale. Banca d'Italia. *Supplemento al Bollettino Statistico,* nuova serie. 3(5).

Carlson, J.A. and Parkin, M. (1975). Inflation expectations. *Economica*. 42(166), 123-138.

Conti, V. and Visco, I. (1984). The determinants of "normal" inventories of finished goods in the Italian manufacturing sector. In A. Chikàn (Ed.), *New Results in Inventory Research*. Amsterdam: Elsevier, 73-89.

Fair, R.C. (1989). The production-smoothing model is alive and well. *Journal of Monetary Economics*. 24(3), 353-70.

Klein, L.R. (Ed.). (1954). *Contributions of Survey Methods to Economics*. New York: Columbia University Press.

Iacoboni, R. and Sestito, P. (1987). Produzione e fatturato nell'industria italiana (1973-1986). Banca d'Italia. *Bollettino Statistico*. 42(3-4), 389-434.

Schlitzer, G. (1993). Nuovi strumenti per la valutazione e la previsione del ciclo economico in Italia. Banca d'Italia. *Temi di Discussione*. 200.

Seitz, H. (1993). Still more on the speed of adjustment in inventory models: A lesson inaggregation. *Empirical Economics*. 18(1), 103-127.

Theil, H. (1952). On the time shape of economic microvariables and the Munich business test. Revue de l'Institut de Statistique. 20(2-3), 105-120.

Visco, I. (1984). *Price Expectations in Rising Inflation*. Amsterdam: North Holland.

West, K.D. (1986). A variance bounds test of the linear quadratic inventory model. *Journal of Political Economy*. 94(2), 374-401.

The Flow-of-Funds Equations of Japanese Banks

Mitsuo Saito, Kazuo Ogawa, and Ichiro Tokutsu

The purpose of this paper is to present an empirical flow-of-funds model of the banking sector of Japan. This model is estimated on the basis of the panel data of 1975-82 and annual time series data of 1953-86. The model has three features. Firstly, it is the banking sector of a flow-of-funds model of the Brainard-Tobin type (1968). It consists of a system of portfolio equations of banks in which holdings of each financial asset and liability are explained by available funds and a spectrum of real rates of return. Combined with equations of other sectors, it can serve as a systematic and exhaustive description of flow of funds in the economy as a whole.[1]

Secondly, the model is estimated by pooling panel and time series data. In estimating this type of portfolio equations, one of the most serious problems is multicollinearity.[2] The pooling method is a useful way of avoiding the possible presence of multicollinearity in available data.

Thirdly, the model is constructed by taking into account the structural changes which the Japanese banking sector has experienced during the past three decades.

It is found as a result of our estimation that, in contrast to the results of most of the other empirical studies, the discount rate is a significant variable in many of the demand equations for assets and liabilities of private banks. This evidence suggests that the discount rate is an effective policy tool for the Bank of Japan.

This paper is organized as follows. Section I presents the skeleton of the model. The basic concepts in the theoretical model and the empirical data corresponding to them are explained in relation to the "Financial Institutions" sector of the System of National Accounts. Sections II and III present the estimated results of the model by the panel data and time series data, respectively. Finally, Section IV is devoted to the summary of the results and the discussion of the remaining problems.

I. Flow-of-Funds Accounts and Banks' Behavior

The banking sector in this paper implies the private financial institution as a whole, including all banks (city and regional), financial institutions for small business, financial institutions for agriculture, forestry and fishery, insurance institutions, trust companies, and security companies. Excluded are the Bank of Japan and public financial institutions, such as postal savings, trust fund bureau and government financial institutions.

Table 1 lists financial assets and liabilities in the balance sheet of the Financial Institutions of the System of National Accounts (SNA). Since the Financial Institutions of the SNA includes the Bank of Japan and public financial institutions, this list has loans by the Bank of Japan (BOJ) and loans by government on the credit side, while it has transfers from general government and currency on the debit side. For the same reason the amount of loans by the Bank of Japan is equal to that of loans from the Bank of Japan, and the amount of call loans and bills bought is equal to the call money and bills sold.

Table 1: Closing Balance-Sheet Items of Financial Institutions in the System of National Accounts

Assets	Liabilities
Net fixed assets	
Land	
	Currency
Demand deposits(including deposits at the Bank of Japan)	Demand deposits
	Other deposits[1]
	Life insurance policies
Short-term securities	
Long-term securities	Long-term bonds
Corporate bonds	
Loans by the Bank of Japan	Loans from the Bank of Japan
Call loans and bills bought	Call money and bills sold
Loans by private financial institutions	Loans by government
Loans by government	Transfers from general government
	Trade credit and advances
Other financial assets	Other liabilities
	Net worth and Corporate shares[2]
Total assets	Total liabilities

Note: 1) including trusts; 2) Shares on the asset side are expressed in terms of market value while those on the liability side are in terms of face value.

Since our definition of the banking sector is restricted to private financial institutions, the balance-sheet of the sector is compiled from the balance-sheet of the SNA as follows. The real fixed assets is presented in terms of gross concept rather than of net concept. Ordinarily the behavioral function of banks is related to financial assets and liabilities but not to real assets. In this study, however, it is assumed that the office buildings and equipment of banks are expanded as a result of bank's portfolio selection. Currency on the debit side, which is a liability item of the Bank of Japan, and currency held by public financial institutions on the asset side are excluded from the original SNA balance-sheets to obtain currency held by private financial institutions. Demand deposits on the asset side are divided into two parts : required reserves and excess reserves.

Much work on the theory of banks' behavior has been written over the past two decades, e.g. Santomero (1984). In many of them, banks' behavior is described as choosing an optimal portfolio among assets and liabilities so as to maximize an objective function (profits or the total value of assets) given the total of available funds (the budget restraint).[3] Under the assumption of a competitive market, such behavior is represented, as shown in Equations (1) and (2) below, by a set of portfolio equations of the Brainard-Tobin type. Holdings of each asset and liability are a function of available funds and the rates of return of all the assets and liabilities. For the purpose of estimation of such portfolio equations, the SNA items in the balance sheet of Table 1 is reduced to those of Table 2.

Table 2: Closing Balance-Sheet Items of Banks in this Study

Assets		Liabilities	
A1:	Gross fixed assets(including land)	L1(=-A8) :	Loans from the Bank of Japan and bills rediscounted
A2:	Currency	L2(=-A9) :	Call money and bills sold
A3:	Demand deposits (Deposits in the Bank of Japan and in other financial institutions)	L3(=-A10) :	Other liabilites - other financial assets
A4:	Bonds	DE :	Bank deposits, trusts and life insurance policies
A5:	Shares		
A6:	Call loans and bills bought		
A7:	Loans by banks		
		W :	Net worth and bonds issued by banks
	Total assets		Total liabilities

Available funds of banks consist of deposits, bonds issued by banks, and net worth. Concerning deposits, we assume that banks accept as given and beyond their control the quantity of time and demand deposits. For time deposits one may argue that a bank sets the yield it is willing to pay and thereby adjusts the quantity of deposits. In Japan, however, the interest rate of bank deposits has been institutionally fixed at a relatively low level for a long period and the change of the rate has been infrequent, the magnitude of change being small.[4] Therefore, we adopt the above assumption. The balances of trusts and life insurance policies are similarly assumed as given for banks.

The bonds issued by banks and net worth (*W*) are available funds for the long-term. They are also assumed as given for banks.

Since part of deposits must be kept as required reserves where the required ratio is set by the Bank of Japan, deposits in the available funds (*DE*), or deposits on the liability side, are total deposits less required reserves. Accordingly, demand deposits (*A3*) on the asset side are total demand deposits less required reserves, or so-called excess reserves.

Finally, it is assumed that banks are subject to no money illusion: the quantity of every financial item is deflated by the GNP deflator.

II. The Estimated Results: Panel Data Analysis

Static Equations

The behavior of banks, as described in the preceding section, may be represented by a set of portfolio equations of the Brainard-Tobin type as follows.

$$Ai = h_{1i}^* DE + h_{2i}^* W + \sum_k g_{ki}^* Rk + h_{Oi}^* + u_i^*, \quad \text{for i} = 1,2,...,10, \quad (1)$$

$$W + DE = \sum_i Ai, \quad (2)$$

where

Ai = end of period holdings of asset i; items, i=1, 2,..., 7 are true assets, and items, i=8, 9, 10 are "negative assets" or minus the value of liabilities. (See Table 2 for the list giving what each Ai represents);

DE = deposits, demand and time;

W = net worth and bonds issued by banks;

Rk = real rate of return of Ak (asset for k = 1, 2,..., 7, or liability for k = 8, 9, 10);

u_i^* = disturbances; and

$h_{1i}^*, h_{2i}^*, h_{Oi}^*, g_{ki}^*$ = parameters.

(Required reserves are excluded from both $A3$ and DE.)

In the tables and discussion, $-A8$, $-A9$, and $-A10$ are often listed as $L1$, $L2$, and $L3$, where L stands for liability.

The total of available funds of a bank is the sum of W and DE. In view of the difference in the effect on asset and liability holdings between W (net worth and bonds issued by banks) and DE (deposits), W and DE are treated as different independent variables in bank's portfolio equations. Therefore, the budget restraint is defined as shown in Equation (2). The restraint is satisfied when the parameters and the disturbances are subject to the following relations.

$$\sum_{i=1}^{10} h_{1i}^* = 1, \quad \sum_{i=1}^{10} h_{2i}^* = 1, \quad \sum_{i=1}^{10} h_{Oi}^* = 0, \quad \sum_{i=1}^{10} g_{ki}^* = 0 \quad \text{(for all k's), and} \quad \sum_{i=1}^{10} u_i^* = 0. \quad (3)$$

Optimal asset demand equations which satisfy the two convenient properties of the adding-up constraints and linearity in expected asset returns are assumed here. These can be rigorously justified in the framework of mean-variance analysis.[5] Friedman and Roley (1979) derived optimal asset demand equations with such properties under the assumption that agents maximize expected utility characterized by constant relative risk aversion with asset returns jointly distributed as normal.

Dynamic Equations

The system of equations as shown above is the representation of banks' behavior in equilibrium. Many empirical papers on the Japanese bank loan market have recently been published, e.g. Hamada-Iwata (1980), Kamae (1980), Furukawa (1985), Tsutsui (1982), Ito-Ueda (1981). Most of them focused particularly on whether observed interest rate and loan transaction are in equilibrium or in disequilibrium. They generally adopted the procedure of selecting observations on the dual decision hypothesis by Fair-Jaffee (1972). Instead of such an approach, let us describe the actual movement of the stocks of assets and liabilities by the following set of dynamic equations based on the stock adjustment principle

$$Ai_t - Ai_{t-1} = \sum_{j=1}^{10} a_{ij} (Aj_t^* - Aj_{t-1}), \quad \text{for } i = 1,2, \ldots, 10, \quad (4)$$

where Ai_t^* = end of period desired (or equilibrium) holdings of asset i.

Ai_t^* may be defined as the level of Ai_t, which is obtained by substituting *DE, W,* and *Rk* in period t into Equation (1). That is to say,

$$Ai_t^* = h_{1i}^* DE_t + h_{2i}^* W_t + \sum_k g_{ki}^* Rk_t + h_{0i}^* + u_{it}^*, \quad \text{for } i = 1,2,\ldots,10. \quad (5)$$

The budget restraint of the dynamic equations is defined as

$$W_t + DE_t = \sum_{i=1}^{10} Ai_t^* = \sum_{i=1}^{10} Ai_t. \quad (6)$$

Therefore, in addiiton to Equation (3), the parameters in Equation (4) must satifsy the following relations:

$$\sum_{i=1}^{10} a_{ij} = 1, \qquad \text{for } j = 1, 2, \ldots, 10. \quad (7)$$

The Method of Estimation and the Data Used

For estimation purposes, substitute Equation (5) into Equation (4) to get:

$$Ai_t = \sum_{j=1}^{10} b_{ij} \, Aj_{t-1} + h_{1\,i} \, DE_t + h_{2i} W_t + \sum_k g_{ki} \, Rk_t + h_{0\,i} + u_{it}, \text{ for } i = 1,2,\ldots,10 \quad (8)$$

where $b_{ij} = \delta_{ij} - a_{ij}$, ($\delta_{ij}$ is a Kronecker delta).

$$h_{ij} = \sum_{j=1}^{10} a_{ij} \, h_{1j}^{*}, \; h_{2\,i} = \sum_{j=1}^{10} a_{ij} \, h_{2j}^{*}, \; g_{ki} = \sum_{j=1}^{10} a_{ij} \, g_{kj}^{*} \, (\text{ for all } k's) \; h_{0\,i} = \sum_{j=1}^{10} a_{ij} \, h_{0j}^{*},$$

$$\text{and } u_{it} = \sum_{j=1}^{10} a_{ij} \, u_{jt}^{*}$$

The budget restraint in this system of equations implies the following relation in the parameters and the disturbances:

$$\sum_{i=1}^{10} b_{ij} = 0, \; \sum_{i=1}^{10} h_{1i} = 1, \; \sum_{i=1}^{10} h_{2i} = 1, \; \sum_{i=1}^{10} g_{ki} = 0, \, (\text{for all } k's), \; \sum_{i=1}^{10} h_{0i} = 0,$$

$$\text{and } \sum_{i=1}^{10} u_{it} = 0, \, (\text{for all } t's). \quad (9)$$

Equation (8) consists of a large system of equations, particularly when the number of assets and liabilities totals ten.   Therefore, we have tried to collect as many data as possible to obtain a stable and reasonable magnitude for estimates.[6]   More specifically, in order to avoid the possible presence of multicollinearity, we adopt the method of pooling cross-section and time series data.[7]   At the first stage of estimation of Equation (8), the parameters, h_{1i}, h_{2i}, h_{0i} and b_{ij}, are estimated from bank panel data.   For a particular year, all banks face the same market rate of return for each financial asset or liability.

Therefore, the part $\sum_k g_{ki} \, Rk_t$ is assumed to be common for all banks.   In addition, since this part moves from year to year, it may be represented by a year dummy variable and its coefficient:

$$\sum_{t'=2} f_{it'} \, Z_{t',t} = \sum_k g_{ki} \, Rk_t \, ,$$

where $Z_{t't} = 1$ for $t' = t$, $= 0$ otherwise.

The panel data used are the closing balance-sheets of 156 private financial institutions for the period 1975-82.[8]   Each bank has its own characteristics ; we take this into account by replacing the constant term with a term using a dummy variable for each bank. Thus, the constant term, h_{0i}, in Equation (8) is replaced by :

$$\sum_{r'=2}^{156} h_{0\,ir'} \, ZB_{r'r} \, ,$$

where $ZB_{r'r} = 1$, for $r' = r$, $= 0$ otherwise.

It has frequently been argued in Japan that the role and behavior of city banks are significantly different from those of regional banks. City banks are much larger in the scale of management than regional banks. Regional banks accept more deposits in proportion to the size of management, while city banks give more bank loans. Thus, in the call market, city banks have more demand for call money (liabilities) while regional banks provide more call loans (assets). This feature of Japanese banks may be allowed for by a dummy variable for the coefficients of DE and W as follows :

$$h_{1\,i} = (\,h'_{1\,i} + h''_{1\,i}\,ZC_r\,), \qquad\qquad h_{2\,i} = (\,h'_{2\,i} + h''_{2\,i}\,ZC_r\,),$$

where

$\qquad ZC_r = 1$, for r = 1, . . . 13, i.e., 13 city banks, and

$\qquad ZC_r = 0$, for r = 14, . . . 156, i.e., regional banks and other financial
$\qquad\qquad\qquad\qquad\qquad\qquad$ institutions.

Finally, we allow for the possible presence of heteroscedasticity by dividing both sides of Equation (8) by DE.[9] All these considerations will lead to the following equations to be estimated:

$$(Ai\,/\,DE)_{t,r} = \sum_{j=1}^{10} b_{ij}\,(Aj_{t-1,r}\,/\,DE_{t,r} + (\,h'_{1\,i} + h''_{1\,i}\,ZC_r\,)$$

$$+\ (\,h'_{2\,i} + h''_{2\,i}\,ZC_r\,)\,(W\,/\,DE_{t,r}) + \sum_{r'=2}^{156} h_{0\,ir'}\,(ZB_{r'r}\,/\,DE_{t,r}) \qquad\qquad (10)$$

$$+\ \sum_{t'=2}^{7} f_{t\,i'}\,(\,Z_{t't}\,/\,DE_{t,r}\,) + v_{it,r}, \qquad \text{for } i = 1, 2, \ldots , 10.$$

Equations (10) are estimated by the ordinary least squares method.[10] As is well known, the ordinary least squares method is equivalent to the SUR (Seemingly Unrelated Regressions) method for the system of Equations (10), since all the equations share common regressors.[11]

Estimated Results

The estimated results are shown in Tables 3 and 4.[12] Since the data used satisfy the adding-up constraints, the least squares rule gives a set of estimated equations satisfying the adding-up restraints.

The effect of net worth on assets and liabilities are shown in Columns (1) to (3) of Table 3 ; Columns (1), (2) and (3) present the coefficients of regional banks, the dummy variable giving the difference in regional and city behavior, and city banks respectively. It is seen that the coefficients of assets of both regional and city banks are all positive except for two cases, while those of liabilities are all negative, except one case. This implies that, as the scale of banks expands, the level of assets Ai and liabilities Lj are increased, except for a few cases.

The effect of deposits on assets and liabilities are similar as shown in Columns (4) to (6) of the same table. The figures of Columns (4) and (6) reveal that the coefficients of assets of both regional and city banks are all positive with no exception, while those of two liabilities $L1$ and $L2$ are all negative with one exception. The coefficients of liability $L3$, other liabilities, are positive. This item includes holdings of foreign assets. The positive sign of the coefficient for this item implies that, as bank deposits are increased, banks will increase holdings of assets and decrease holdings of liabilities.

The matrix of coefficients of dynamic adjustments are shown in Table 4. The coefficients on the diagonal are positive, as expected. A smaller value of the diagonal element implies a quicker adjustment to the desired level of the relevant asset. The diagonal coefficients for $A1$ (gross fixed assets), $A4$ (bonds), $A5$ (shares), $L1$ (loans from the Bank of Japan) and $L2$ (call money) are relatively large ; they range between 0.6466 and 0.7091. Those for $A2$ (currency), $A3$ (demand deposits), $A6$ (call loans), and $L3$ (other liabilities) are relatively small, ranging between 0.1791 and 0.4701. That for $A7$(loans by banks), 0.4979, falls between. By and large, the above results are quite reasonable, since liquid assets and liabilities tend to exhibit quicker adjustment.

The adjustment coefficients of call loans and call money are not symmetric ; the coefficient of call loans, 0.1791, is the smallest among all thediagonal coefficients, while that of call money, 0.6565, takes a larger value.

The dynamic property of the system of banks' portfolio equations is stable ; A^t becomes nearly a zero matrix when t = 9 : all the elements turn out to be less than 0.1.

From the adjustment coefficients, the long-run (or equilibrium) effects of net worth and deposits on asset and liability holdings may be calculated as:

$$(I - A)^{-1} \, h_1 \quad \text{and} \quad (I - A)^{-1} \, h_2 \, ,$$

respectively, where A is the matrix of the adjustment coefficients a_{ij} and I is an identity matrix ; h_1 and h_2 are the vectors of the coefficients h_{1i} and h_{2i} respectively.

The calculated values of the above vectors are shown in Table 5A. Column (1) of the table presents the weighted average of the net worth coefficients of regional and city banks, where the weights are the proportions of net worth held by each type of banks.[13] The long-run coefficients $(I - A)^{-1}$ h_1 in Column (2) are calculated by using h_1 of Column (1).

Similarly, Column (3) presents the weighted average of the deposits coefficients of regional and city banks. Here, the weights are the proportions of deposits held by each type of banks. Column (4) shows the calculated $(I - A)^{-1} h_2$.[14]

A comparison between the short-run and long-run effects of net worth on asset holdings reveals that the long-run coefficient of bank loans($A7$) is 166 % larger than the short-run one. The long-run coefficients of demand deposits ($A3$), shares ($A5$) and call loans ($A6$) are about twice as large as their short-run coefficients. A negative value of the long-run coefficient of currency($A2$) is -

Table 3: The Results of Panel Data Analysis (1): The Effects of Net Worth and Deposits

		the effect of net worth (W)				the effect of deposits (DE)		
		(1)	(2)	(3)	(4)	(5)	(6)	(7)
		regional banks	dummy	city banks	regional banks	dummy	city banks	
		h'_{1i}	h''_{1i}	$h'_{1i} + h''_{1i}$	h'_{2i}	h''_{2i}	$h'_{2i} + h''_{2i}$	R^2 /Se
$A1$:	Gross fixed assets	0.006 (0.002)	0.014 (0.013)	0.020	0.008 (0.001)	-0.004 (0.001)	0.005	0.946 0.001
$A2$:	Currency	0.051 (0.013)	0.044 (0.097)	0.095	0.102 (0.008)	-0.019 (0.011)	0.083	0.851 0.010
$A3$:	Demand deposits	0.276 (0.021)	-0.463 (0.161)	-0.187	0.143 (0.013)	0.188 (0.019)	0.331	0.880 0.017
$A4$:	Bonds	0.060 (0.023)	0.354 (0.173)	0.414	0.171 (0.014)	-0.121 (0.020)	0.050	0.971 0.018
$A5$:	Shares	0.131 (0.006)	0.128 (0.048)	0.258	0.024 (0.004)	0.015 (0.006)	0.039	0.990 0.005
$A6$:	Call loans and bills bought	0.107 (0.017)	-0.126 (0.129)	-0.019	0.075 (0.011)	0.056 (0.015)	0.131	0.706 0.014
$A7$:	Loans by banks	0.391 (0.024)	0.278 (0.179)	0.670	0.435 (0.015)	-0.100 (0.021)	0.335	0.997 0.019
$A8$ (-$L1$):	Loans from the BOJ and bills rediscounted	-0.041 (0.009)	-0.132 (0.068)	-0.174	0.012 (0.006)	-0.020 (0.008)	-0.008	0.882 0.007
$A9$ (-$L2$):	Call money and bills sold	-0.070 (0.011)	0.061 (0.081)	-0.009	-0.014 (0.007)	-0.011 (0.009)	-0.026	0.929 0.009
$A10$ (-$L3$):	Other liabilities	0.088 (0.013)	-0.158 (0.097)	-0.070	0.043 (0.008)	0.016 (0.011)	0.058	0.934 0.010
	Total	1.000	0.000	1.000	1.000	0.000	1.000	

The figures within the brackets are the standard errors of the coefficients; $\overline{R^2}$ is the coefficient of determination adjusted for degrees of freedom; and *Se* is the standard error of the equation.

0.245. This, at least, seems too low. This is not necessarily unreasonable, however, since currency and excess reserves (demand deposits at the BOJ($A3$)) may serve a common purpose. The sum of the long-run effects of net worth

Table 4: The Results of Panel Data Analysis (2): The Adjustment Coefficients, b_{ij}

		(1) $A1_i$	(2) $A2_i$	(3) $A3_i$	(4) $A4_i$	(5) $A5_i$	(6) $A6_i$	(7) $A7_i$	(8) $-L1_i$	(9) $-L2_i$	(10) $-L3_i$
A1:	Gross fixed assets	0.6543 (0.023)	-0.0047 (0.004)	-0.0059 (0.003)	-0.0034 (0.002)	-0.0305 (0.007)	-0.0060 (0.003)	-0.0042 (0.001)	0.0123 (0.005)	-0.0051 (0.006)	0.0010 (0.004)
A2:	Currency	0.5040 (0.180)	0.3333 (0.030)	-0.0817 (0.020)	-0.0820 (0.013)	-0.1472 (0.052)	-0.0057 (0.026)	-0.0798 (0.011)	-0.0563 (0.035)	0.0123 (0.047)	0.0747 (0.032)
A3:	Demand deposits	0.0158 (0.300)	0.1868 (0.050)	0.4701 (0.034)	-0.1772 (0.022)	0.2445 (0.087)	0.0093 (0.043)	-0.1797 (0.018)	-0.2149 (0.059)	-0.4553 (0.078)	0.1086 (0.054)
A4:	Bonds	-0.6032 (0.321)	-0.0060 (0.054)	-0.0869 (0.036)	0.6466 (0.024)	-0.3033 (0.093)	0.0294 (0.046)	-0.1118 (0.019)	-0.1571 (0.063)	-0.2314 (0.083)	0.0175 (0.058)
A5:	Shares	-0.0308 (0.090)	-0.0200 (0.015)	-0.0666 (0.010)	-0.0014 (0.007)	0.6872 (0.026)	0.0073 (0.013)	-0.0224 (0.005)	0.0363 (0.018)	0.0565 (0.023)	-0.0784 (0.016)
A6:	Call loans and bills bought	-0.0775 (0.241)	-0.1688 (0.040)	-0.0012 (0.027)	-0.0563 (0.018)	0.0609 (0.070)	0.1791 (0.035)	-0.0395 (0.014)	-0.0008 (0.047)	-0.1213 (0.062)	0.0413 (0.043)
A7:	Loans by banks	-0.5370 (0.334)	-0.2349 (0.056)	-0.1377 (0.037)	-0.3087 (0.025)	-0.1572 (0.097)	-0.1812 (0.048)	0.4979 (0.020)	-0.3368 (0.065)	-0.1343 (0.087)	-0.4536 (0.060)
A8 (-L1):	Loans from the BOJ and bills rediscounted	-0.1739 (0.126)	-0.0403 (0.021)	0.0334 (0.014)	-0.0095 (0.009)	0.0356 (0.036)	0.0026 (0.018)	-0.0026 (0.007)	0.7091 (0.025)	0.0232 (0.033)	0.0282 (0.023)
A9 (-L2):	Call money and bills sold	0.2255 (0.151)	0.0259 (0.025)	-0.0148 (0.017)	0.0161 (0.011)	-0.3102 (0.044)	-0.0456 (0.022)	0.0240 (0.009)	0.0461 (0.030)	0.6565 (0.039)	0.0348 (0.027)
A10 (-L3):	Other liabilities	0.0228 (0.181)	-0.0713 (0.030)	-0.1088 (0.020)	-0.0242 (0.014)	-0.0797 (0.053)	0.0107 (0.026)	-0.0818 (0.011)	-0.0379 (0.036)	0.1989 (0.047)	0.2261 (0.033)
	total	0.0000	0.0000	0.0000	0.0000	0.0000	0.0000	0.0000	0.0000	0.0000	0.0000

The figures within the brackets are the standard errors of the coefficients.

on both assets is practically zero, as might be expected.[15]

It is interesting to note that the effect of an increase in deposits (*DE*) on bank loans (*A7*) is greater by 73 % in the long-run than in the short-run. For all other assets (*A1* to *A6*) the long-run impact of deposits is less. Specifically, initial holdings of bonds and shares will be halved in the long-run. This is consistent with the following practice of Japanese banks: holding a large amount of government bonds, corporate bonds, or corporate shares is almost compulsory at the time of new issue; later banks sell these assets to invest in bank loans.[16]

Table 5B shows the long-run and short-run effects of net worth and deposits on asset and liability holdings in terms of elasticity. It is also seen that the long-run elasticities of bank loans with respect to net worth and deposits are much higher than those in the short-run.

III. The Estimated Results: Time-Series Analysis

The Method of Estimation and the Data Used

As the second stage of the pooling method, we estimate the remaining parameters of Equation (8) from annual time series data for 1953 to 1986. We divide the both sides of Equation (8), except the terms Rk_t, by the total of available funds F_t, or the sum of W_t and DE_t. Then, using the estimates from panel data, we may have the following equations :

$$Ai_t / F_t - \sum_{j=1}^{10} \hat{b}_{ij} (Aj_{t-1} / F_t) - (\hat{h}_{1i} - \hat{h}_{2i}) (DE/F)_t - \hat{h}_{2i}$$

$$= \sum_k g_{ki} Rk_t + h_{0i} (1/F_t) + d_i + w_{it}, \quad \text{for } i = 1, 2, \ldots, 10, \quad (11)$$

where $F_t = W_t + DE_t$,

$\hat{h}_{1i}$, $\hat{h}_{2i}$ and $\hat{b}_{ij}$: the estimates respectively for h_{1i}, h_{2i} and b_{ij}, based on panel data,

d_i : a constant term added to make the estimates more stable, and

w_{it} : disturbances.

The dependent variable of the regression, or the left-hand side of Equation (11), is calculated from the observed data and the estimates. Thus, the time series regression will yield the estimates for g_{ki} and h_{0i}. Equation (11) derived from Equation (8) should have no constant term. We introduce, however, a constant term d_i in estimating the equation, in order to allow for discrepancy between the concepts for panel data and for time series ones and to make estimates more stable.

Division of Equation (8) by F_t is done to minimize heteroscedasticity.

When most of other variables are ratios (less than unity in absolute value), Rk_t rather than Rk_t / F_t may be appropriate for this regression in the light of the order of magnitude of the variables.

		original estimates				modified estimates			
		effect of net worth (W)		effect of deposits (DE)		effect of net worth (W)		effect of deposits (DE)	
		(1)	(2)	(3)	(4)	(5)	(6)	(7)	(8)
		short-run	long-run	short-run	long-run	short-run	long-run	short-run	long-run
A1:	Gross fixed assets	0.012	-0.026	0.007	0.005	0.013	-0.028	0.008	0.005
A2:	Currency	0.069	-0.245	0.094	0.019	0.072	-0.251	0.106	0.037
A3:	Demand deposits	0.088	0.229	0.225	0.200	0.048	0.165	0.121	0.004
A4:	Bonds	0.204	0.265	0.119	0.072	0.213	0.292	0.135	0.125
A5:	Shares	0.183	0.404	0.031	0.015	0.191	0.433	0.035	0.051
A6:	Call loans	0.056	0.131	0.099	0.083	0.059	0.136	0.112	0.093
A7:	Loans by banks	0.505	1.344	0.392	0.677	0.527	1.411	0.444	0.788
A8 (-L1):	Loans from the BOJ	-0.095	-0.291	0.004	0.016	-0.099	-0.312	0.004	-0.006
A9 (-L2):	Call money	-0.045	-0.521	-0.019	-0.036	-0.047	-0.550	-0.022	-0.059
A10 (-L3):	Other liabilities	0.024	-0.289	0.050	-0.050	0.025	-0.295	0.056	-0.037
	total	1.000	1.000	1.000	1.000	1.000	1.000	1.000	1.000

Table 5A: The Effects of Net Worth and Deposits: Short-run and Long-run

The rates of return used are real rates, i. e. nominal rates of return minus the rate of change of the GNP deflator. The nominal rates of return are the bond yields ($R4$), the rate of return on shares ($R5$), call rate ($R6$), the interest rate on bank loans ($R7$), the discount rate ($R8$) and the interest rate of the U. S. Treasury Bills (3-month) plus the rate of change of the exchange rate ($R10$).[17] The last variable is used for other liabilities because foreign assets occupy a large share in this item.[18, 19] As for currency ($A2$) and demand deposits ($A3$) the rate of return is irrelevant. The rate of return on gross fixed assets is neglected since it is likely that the inclusion of too many rates of return may lead to multicollinearity.

Four dummy variables are added as a part of the constant term in Equation (11) for the period of 1954-55, 1962-65, 1975-85, and 1986, respectively. The dummy for 1954-55 characterizes the early stage of development of financial institutions in Japan. The dummy for 1962-65 corresponds to the temporary closing of bond markets except for telephone and telegraph bond during this period. The dummy for 1975-85 is intended to account for the effect of the large volume of bonds issued by the government on the portfolio behavior of banks. Finally, the dummy for 1986 is introduced to take into account the shock of the 1985 Plaza accord: the sudden, big appreciation of the yen exchange rate and its strong aftermath in 1986.

Since time series data is the aggregate for all the private financial institutions, the average of the estimates for regional and city banks may be appropriate for $\hat{h}_{1i}$ and $\hat{h}_{2i}$. These are in Columns (1) and (3) of Table 5A, as described above. Demand deposits (*A3*) in the time series equations, however, include only deposits in the Bank of Japan; deposits among private banks are netted out. In the panel data *A3* includes deposits at both the Bank of Japan and private banks. Therefore, we make a modification of the estimates of $\hat{h}_{1i}$ and $\hat{h}_{2i}$ given above. The value of $\hat{h}_{13}$ (or $\hat{h}_{23}$) is decreased by 46 %, and this amount of decrease in $\hat{h}_{13}$ (or $\hat{h}_{23}$) is distributed among coefficients of other assets and liabilities $\hat{h}_{1i}$'s (or $\hat{h}_{2i}$'s)(i ≠ 3) as an addition to $\hat{h}_{1i}$ (or $\hat{h}_{2i}$).[20] These modified estimates are given in columns (5) through (8) of Table 5A and the corresponding elasticities in columns (5) through (8) to Table 5B.

Table 5B:	The Effects of Net Worth and Deposits(in terms of elasticity): Short-run and Long-run							
		original estimates				modified estimates			
		effect of net worth (W)		effect of deposits (DE)		effect of net worth (W)		effect of deposits (DE)	
		(1)	(2)	(3)	(4)	(5)	(6)	(7)	(8)
		short-run	long-run	short-run	long-run	short-run	long-run	short-run	long-run
A1:	Gross fixed assets	0.077	-0.168	0.364	0.273	0.081	-0.177	0.412	0.278
A2:	Currency	0.146	-0.518	1.655	0.331	0.152	-0.530	1.876	0.647
A3:	Demand deposits	0.147	0.383	3.134	2.792	0.079	0.275	1.692	0.054
A4:	Bonds	0.122	0.159	0.594	0.360	0.128	0.175	0.673	0.624
A5:	Shares	0.344	0.759	0.484	0.237	0.359	0.814	0.548	0.801
A6:	Call loans	0.188	0.437	2.537	2.303	0.196	0.456	3.131	2.590
A7:	Loans by banks	0.083	0.222	0.540	0.933	0.087	0.233	0.612	1.087
A8 (-L1):	Loans from the BOJ	-0.698	-2.136	0.220	0.986	-0.729	-2.290	0.251	-0.380
A9 (-L2):	Call money	-0.103	-1.191	-0.365	-0.695	-0.108	-1.258	-0.414	-1.130
A10 (-L3):	Other liabilities	0.171	-2.072	2.966	-2.984	0.179	-2.117	3.361	-2.224

Estimated Results

The estimated results of Equation (11) by the SUR are presented in Table 6. It is preferable to include the six rates of return in every equation to obtain all the own and cross effects of them and to make the equations subject to the adding-up restraints. The results of such estimation, however, yield multicollinearity, making some of the cross effects unreasonable in sign and magnitude. Concerning the signs of coefficients, g_{ki}'s, it is expected, in principle, that the own effect is positive and the cross effects are negative, implying the mutual substitutability among all the assets and liabilities.[21] Therefore, we exclude the rates of return which show incorrect sign in the preliminary estimation, although we leave a positive sign for a few cross effects.[22]   It is to be noted, however, that the SUR method of estimation guarantees that the adding up restraints are satisfied.

Table 6 shows that the own effect of the rate of return is statistically significant in all of seven assets and liabilities.[23,24] On the other hand, the cross effects are smaller and less significant in many cases. It is seen from the table that statistically significant cross effects are:

1) bank loan rate($R7$) for currency ($A2$) and borrowing from the BOJ ($L1$),
2) call rate ($R6$) for borrowing from the BOJ ($L1$),
3) the discount rate ($R8$) for bonds($A4$), call loans ($A6$), and call money($L2$),
4) the rate of return on shares ($R5$) for gross fixed assets ($A1$), bonds ($A4$), and bank loans($A7$), and
5) the bond yields($R4$) for bank loans ($A7$).

This finding is consistent with the institutional evidence of the Japanese banks.

The bank loan rate ($R7$) has positive cross effects on gross fixed asset ($A1$). This is interpreted as follows : when the bank loan rate rises, the demand for branches as well as the supply of loans by banks increases, which in turn leads to an increase in the demand for real assets, such as land and buildings. The rate of return on foreign assets($R10$) also exerts positive effects on call loans ($A6$), which includes dollar calls.

The constant dummy variables for the period of 1975-85, which measures the impact of recent structural change in the financial markets on the banks' behavior, are statistically significant in most of the equations. Most striking are the shift of banks' portfolio from bank loans ($A7$) to bond holdings ($A4$) and to investment on foreign assets($L3$). This reflects a large volume of bonds issued by the government since 1975 and a rapid pace of liberalization of international capital transactions in the financial scene since 1980.

The own and cross effects of nominal rates of return are presented in terms of elasticity in Table 7. The values in parentheses are long-run elasticities calculated on the basis of inverse of matrix of $(I - A)$, above. The magnitude of elasticities seems reasonable for most of estimates. All the own elasticities are positive, and most of the cross elasticities are negative, suggesting that substitutability is dominant among assets and liabilities of private banks. The elasticities for borrowing from the BOJ are very large ; however, they are not unreasonable if one takes into account the large variance in the movement of borrowing from the BOJ.[25]

Table 6: The Results of Time Series Analysis: The Effects of Rates of Return

		(1) g_{4t} R4	(2) g_{5t} R5	(3) g_{6t} R6	(4) g_{7t} R7	(5) g_{8t} R8	(6) g_{10t} R10	(7) h_{2t} I/F	(8) d_t const.	(9) Z54,55	(10) Z62-65	(11) Z75-85	(12) Z86	(13) Se/DW
A1:	Gross fixed assets		-0.0073 (.0021)		0.0265 (.0143)			64.4 (57.4)	0.0205 (.0010)	-0.0015 (.0026)	0.0004 (.0014)	-0.0053 (.0013)	0.0101 (.0024)	0.0023 1.9952
A2:	Currency		-0.0045 (.0028)		-0.1136 (.0201)			294.7 (80.3)	-0.0367 (.0014)	-0.0047 (.0036)	0.0064 (.0020)	0.0073 (.0018)	-0.0028 (.0034)	0.0033 1.4066
A3:	Demand deposits (Deposits in the BOJ)				-0.0056 (.1392)	-0.1099 (.1500)	-0.0056 (.0131)	296.3 (193.3)	0.0052 (.0039)	0.0065 (.0089)	-0.0186 (.0042)	0.0065 (.0039)	-0.0334 (.0084)	0.0073 1.0404
A4:	Bonds	0.4279 (.1025)	-0.0135 (.0058)			-0.5260 (.1099)	-0.0122 (.0155)	241.2 (167.2)	0.0078 (.0044)	-0.0019 (.0075)	-0.0042 (.0040)	0.0264 (.0035)	-0.0185 (.0079)	0.0063 1.1959
A5:	Sharess		0.0676 (.0138)				-0.0020 (.0289)	489.4 (339.9)	-0.0359 (.0063)	-0.0183 (.0163)	-0.0006 (.0085)	0.0013 (.0071)	0.0274 (.0174)	0.0141 2.1304
A6:	Call loans and bills bought			0.2120 (.0398)		-0.1365 (.0455)	0.0053 (.0101)	-78.9 (128.0)	-0.0486 (.0022)	-0.0012 (.0055)	0.0011 (.0028)	0.0040 (.0026)	0.0143 (.0059)	0.0045 1.8070
A7:	Loans by banks	-0.3942 (.1646)	-0.0376 (.0137)		0.9761 (.2198)	-0.3841 (.2723)		1443.5 (369.8)	0.0096 (.0088)	-0.0684 (.0175)	-0.0043 (.0088)	-0.0454 (.0076)	-0.0468 (.0151)	0.0147 1.7227
A8: (-L1)	Loans from the BOJ and bills rediscounted	-0.0337 (.1264)	-0.0046 (.0069)	-0.3935 (.0540)	-0.8834 (.2192)	1.3998 (.2526)	-0.0249 (.0170)	-1311.6 (178.7)	0.0594 (.0055)	0.0740 (.0093)	0.0005 (.0041)	-0.0017 (.0036)	0.0021 (.0080)	0.0064 2.4380
A9: (-L2)	Call money and bills sold			0.1815 (.0590)		-0.2047 (.0707)		-256.7 (208.6)	0.0159 (.0035)	0.0041 (.0090)	0.0104 (.0045)	-0.0020 (.0042)	0.0043 (.0082)	0.0076 1.4217
A10: (-L3)	Other liabilities					-0.0385 (.0585)	0.0394 (.0210)	-1182.2 (243.3)	0.0028 (.0041)	0.0113 (.0106)	0.0089 (.0053)	0.0089 (.0050)	0.0432 (.0116)	0.0093 1.3028

See footnote for Table 3.

Table 7: The Effects of Nominal Rates of Return in Terms of Elasticity

		nominal rates of return					
		R4	R5	R6	R7	R8	R10
A1:	Gross fixed assets	0	-0.0330	0	0.0463	0	0
		(-0.0066)	(-0.1579)	(-0.0954)	(-0.0675)	(0.2480)	(-0.0007)
A2:	Currency	0	-0.1013	0	-1.0018	0	0
		(-0.0997)	(-2.5023)	(0.3197)	(-2.7484)	(1.2088)	(0.0767)
A3:	Demand deposits	0	0	0	-0.0946	-1.4834	-0.0444
		(3.9030)	(8.6274)	(-2.5452)	(-15.954)	(7.5391)	(0.0769)
A4:	Bonds	0.2208	-0.0162	0	0	-0.1938	-0.0026
		(0.9743)	(-0.0768)	(0.1763)	(0.2058)	(-1.2544)	(0.0047)
A5:	Shares	0	0.1194	0	0	0	-0.0006
		(0.0251)	(0.3077)	(-0.0724)	(-0.2772)	(0.2919)	(-0.0094)
A6:	Call loan and bills bought	0	0	0.5537	0	-0.2783	0.0063
		(-0.0290)	(0.4489)	(0.3690)	(-0.6346)	(0.4954)	(0.0000)
A7:	Loans by banks	-0.0481	-0.0106	0	0.1074	-0.0335	0
		(-0.2415)	(0.0096)	(0.0508)	(0.4871)	(-0.1904)	(-0.0011)
A8(-L1):	Loans from the BOJ	-0.2249	-0.0717	-2.4060	-5.3249	6.6809	-0.0694
	and bills rediscounted	(-0.8461)	(0.6330)	(-8.0599)	(-19.084)	(23.206)	(-0.2311)
A9(-L2):	Call money and bills sold	0	0	0.3881	0	-1.1570	0
		(-0.2267)	(-1.1420)	(0.9943)	(0.6071)	(-0.4781)	(0.0212)
A10(-L3):	Other liabilities	0	0	0	0	-0.6881	0.0587
		(0.1788)	(-0.7728)	(0.4994)	(-0.1788)	(-0.2788)	(0.0923)

Values in parentheses are long-run elasticities.

Some Implications of the Estimated Results

Using a set of portfolio equations for banks estimated above, it is possible to evaluate the impact of exogenous change in the supply of assets and liabilities on the spectrum of interest rates. These evaluations, however, are made under the assumption that an autonomous change in the financial market is adjusted within the banking sector alone. This exercise serves to reveal the characteristics of the working of our financial model.

Our system of estimated equations, i.e. estimated version of Equation (11), may be written in the matrix form :

$$A = F (R, X), \qquad (12)$$

where $A = (Ai)$, $R = (Rk)$, $X = (DE, W)$.

By vector B, let us denote the sum of asset(or liability) holdings by the other sectors and an exogenous part of the demand for end of period asset(or liability) holdings by the banking sector.[26] Suppose that B and X are given exogenously; that is, any change in the financial market is adjusted within the banking sector alone.

Then, in equilibrium, the following relation holds :

$$F (R^*, X) + B = 0 , \tag{13}$$

where R^* is a set of equilibrium rates of return.[27] Assuming that X is kept unchanged, differentiating Equation (13) will yield to:

$$dGdR^* + dB = 0, \tag{14}$$

where $dB = (dBi)$,

$$dG = \left(\frac{\partial F_i}{\partial Rk} \right) ,$$

$$dR^* = (dRk^*),$$

and $\partial F_i / \partial Rk$ is evaluated at the equilibrium points. Solving (14) gives :

$$dR^* = - dG^{-1} dB . \tag{15}$$

Equation(15) shows how far the equilibrium point shifts by a given change in B. Therefore, the i-j element of dG^{-1} indicates the change in equilibrium rate of return of asset i induced by a unit change in the supply of asset j.

Table 8: The Change in Rates of Return Induced By a Change in the Supply of the BOJ Loans and Government Bonds
(unit: percent)

		(1) increase of 0.740 trillion yen in the BOJ loans: $dB8 = 0.740$ trillion*	(2) increase of 1.0 trillion yen in government bonds: $dB4 =- 1.0$ trillion*
R4 :	Bond yields	-1.26	2.18
R5 :	The rate of return on shares	-0.03	0.00
R6 :	Call rate	-0.85	1.05
R7 :	Bank loan rate	-0.90	1.37
R8 :	The discount rate	-1.00	1.23
R10:	The rate of return on foreign assets	-0.98	1.21

* $dB8$ and $dB4$ represent the exogenous changes in the supply of financial instruments, which correspond to $R8$ and $R4$ respectively.

The matrix *dG* is constructed from the marginal coefficients on asset returns in the time series regression. It is a 6 X 6 matrix, where the effects on gross fixed assets(*A1*) and loans by banks(*A7*) are combined in one item, and the call loans (*A6*) and call money (*L2*) are netted out.[28] The inverse matrix of *dG* is non-negative. In the *dG* matrix, the demand for shares(*A5*) is decomposable from the other assets and liabilities, so that the rate of return on shares is not affected by exogenous change in the supply of other assets or liabilities.

The calculated results of *dG*[-1] shows that an increase of supply of loans by the BOJ by 740 billion yen will lead to a change in the rates of return as shown Column (1) of Table 8.[29] This implies that a decrease of one percent in discount rate will be followed by decreases of 0.90, 1.26, 0.85 and 0.98 percent in bank loan rate, bond yields, the call rate, and the rate of return on foreign assets, respectively.[30]

On the other hand, an increase of one trillion yen in the supply of (government) bonds will raise the interest structure as given in Column (2) of Table 8. The increases are 1.37, 2.18, 1.05, 1.23, and 1.21 percent in the bank loan rate, bond yields, call rate, and the discount rate, and rate of return on foreign assets, respectively. These unidirectional changes in the interest rates seem reasonable in relative magnitudes.

Suppose that the market adjustment is made in the whole financial market, or that similar portfolio equations in the other sectors are combined with those of the private financial sector. Then the responses of the rates of return to a policy variable will be less than the estimates above.

Comparison with Other Studies

Next, we turn to a comparison of our estimated results with those of other empirical studies. The estimates by other researchers are tabulated in Table 9, together with our corresponding estimates. The table presents short-run and long-run elasticities (the estimates for long-run elasticities are given within brackets). The results for excess reserves are shown in Rows (1) to (3). Hamada-Iwata (1980) estimated demand equations for excess reserves for city banks and for regional banks. Furukawa's estimates (1985) are for total reserves, currency plus deposits in both the BOJ and other banks. One estimate is for city banks, and the other for mutual loan and savings banks. Our estimates for the elasticities of currency and total reserves with respect to deposits (DE) are 11.46 and 2.04 for the short-run and the long-run, respectively. Our short-run elasticity is much larger than those in other studies, while the change to the long-run results in a greatly reduced estimate. This finding may be reasonable, since a temporary increase of cash and reserves will be eventually reallocated among more profitable assets.

As for the bank loans equation, Tsutsui's estimates (1982) are available. His estimate of the semi-elasticity of bank loans with respect to bank loan rate,

2.00, falls between our short-run estimate, 1.38, and long-run estimate, 6.27. Our long-run estimate for the elasticity with respect to deposits, 0.78, is close to Tsutsui's, 0.97.

		Table 9: A Comparison of Our Estimates with Other's				
			(1) deposits	(2) bank loan rate	(3) call rate	(4) discount rate
excess reserves	(1)	Hamada-Iwata	0.19[2]	0.51[3]		
			(0.67)	(0.93)		
	(2)	Furukawa[1]	0.54[2]	1.07[4]		
	(3)	Ours	11.46			
			(2.04)			
bank loans	(4)	Tsutsui	0.97		2.00[5]	
	(5)	Ours	0.44		1.38[5]	
			(0.78)		(6.27)	
call money	(6)	Hamada-Iwata			0.15[2,6]	
					(1.51)	
	(7)	Ours			0.18[6]	
					(0.46)	
call loans	(8)	Hamada-Iwata			0.06[3,5]	
					(0.11)	
	(9)	Ours			0.21[6]	
					(0.14)	
borrowing from BOJ	(10)	Furukawa			-0.23	0.32
					-1.28)	(1.75)
	(11)	Ours			2.41	6.68
					-8.06)	(23.21)

Notes: All the values are in terms of elasticities. The values in parentheses are long-run elasticities.
1. currency and total reserves; 2. for city banks; 3. for regional banks; 4. for mutual loan and savings banks
5. semi-elasticity; 6. marginal coefficient

Hamada-Iwata (1980) estimated a call money equation for city banks and a call loans equation for regional banks using monthly data. Our estimates are comparable with Hamada-Iwata's in order of magnitude. It should be noted that the call rate has a different effect on the demand for call loans and for call money. The demand for call money is more responsive to the call rate than that for call loans. This is more striking in the long-run. In this respect our finding is consistent with Hamada-Iwata's.

Furukawa's study (1985) is one of the few that estimates the demand equation for borrowing from the BOJ. Two interest variables, call rate and discount rate, are included in the equation. Both exert significant effects on the demand for borrowing from the BOJ. Our estimates for the elasticities of borrowing from the BOJ with respect to both the call rate and discount rate greatly exceed Furukawa's counterparts both in the short-run and in the long-run. However, there is a similarity between the two in that the own interest rate elasticity is larger than the cross interest rate elasticity.

Finally, emphasis should be laid upon the special feature of our model: the discount rate is a significant variable in many of the demand equations for assets and liabilities, specifically for bonds($A4$), shares($A6$), borrowing from the BOJ($L1$), and call money($L2$). The upshot is that the discount rate is an effective policy tool for the BOJ. Most of the empirical studies of the private banks have failed to find a significant role for the discount rate in the behavioral equations.[31]   This result has been frequently cited as indirect evidence for perpetual excess demand for loans from the BOJ, and hence for BOJ loans being rationed among private banks. According to this view, a change of the discount rate has little effect on the portfolio behavior of banks, or private banks are more responsive to credit availability from the BOJ. Our result contrasts with this traditional view.

IV. Concluding Remarks

Let us summarize the results.

(i)   The effect of net worth and available fund on banks' portfolio behavior is successfully estimated from panel data of 1975 to 1982.   An increase in both net worth and available funds leads to an increase of holdings for most assets and liabilities, both in the short-run and in the long-run.

(ii)   The estimation of the effect of the rate of return on banks' portfolio behavior from time series data enables us to obtain estimates of own and cross effects. These are of reasonable magnitude and turn out to be dominated by gross substitutability among assets and liabilities.

(iii)   The system of equations subject to the adding-up constraint is successfully estimated by the application of the SUR method to both the panel and time series data.

(iv)   The impact of institutional changes in the Japanese financial market during the past three decades are quantitatively measured by use of dummy variables in the banks' behavioral equations.

(v)   It is found that the discount rate has a considerable effect on many of the banks' portfolio equations, indicating that the discount rate is an effective policy tool of the BOJ.

(vi)   Although special care is taken to avoid multicollinearity, it is likely that some of the estimates are still subject to it.   When more data, both panel and time series, are available, the estimates can be refined.

Appendix

This appendix describes the data sources and computational procedures used to construct the variables of this study. For panel data analysis, the primary source of data is the NEEDS data file. This file consists of annual financial statements of 156 private banks from the fiscal year of 1975 through 1985. Covered are the five major types of banks in Japan.-- i.e. city banks, regional banks, mutual loan and savings banks, trust companies, and long-term credit banks. Our sample in panel data analysis utilizes observations up to 1982, which is consistent with the sample period in the time series analysis. The time series data are drawn from the Closing Balance-sheet of Financial Institutions from 1953 through 1986 in the System of National Accounts in Japan (Annual Reports on National Accounts, Government of Japan). Mutual correspondence of variables among our study, the System of National Accounts, and NEEDS data file is available from the authors upon request. The interest rates corresponding to assets and liabilities are listed in Table A1

Table A1: Definitions of Rates of Return

R4:	Market yield on telephone & telegraph coupon bonds in Tokyo Stock Exchange, *Economic Statistics Annual* (The Bank of Japan).
R5:	Market rate of return on common stocks on Tokyo Stock Exchange First Section, *Rate of Return on Common Stocks* (Japan Securities Reseach Institute).
R6:	Call rate(unconditional), *Economic Statistics Annual* (The Bank of Japan)
R7:	Average contracted interest rate on loans of all banks, *Economic Statistics Annual* (The Bank of Japan)
R8:	Official discounts rate, *Economic Statistics Annual* (The Bank of Japan)
R10:	$(1+RF)(ER/ER_1)$, where RF is U.S. treasury bill rate(three months) and ER is the Yen-Dollar exchange rate at home currency, *The Economic Report of the President*(The United States Government Printing Office) and *Economic Statistics Annual* (The Bank of Japan).
P:	GNE implicit deflator(1980=1.00), *Annual Report on National Accounts*(The Economic Planning Agency, Government of Japan).

All the interest rates are in real term, i.e. net of the rate of change in P.

Mistuo Saito, Faculty of Economics, Tezukayama University, Tezukayama, NARA 631, JAPAN, Kazuo Ogawa, Faculty of Economics, Kobe University, Rokko, Nada, KOBE 657, JAPAN, and Ichiro Tokutsu, School of Business Administration, Kobe University, Rokko, Nada, KOBE 657, JAPAN.

ACKNOWLEDGEMENTS

The authors are grateful to Yuzo Honda and Peter Mayer for helpful comments on earlier drafts. However, the responsibility for any remaining errors is ours.

NOTES

1. Klein (1983, Chapter1) emphasized the importance of such a flow-of-funds model, presenting its illustrative framework.
2. See Smith (1975).
3. For example, Pyle (1971), Brainard-Tobin (1968), and Tobin (1969). For Japanese banks, see Suzuki (1966), Horiuchi (1980), Hamada-Iwata (1980), Honda (1984), and Furukawa (1985).
4. The use of certificates of deposits (CD), whose interest rate is negotiable, has recently been extended. Its share in total deposits, however, is still small (3.5 percent in 1985 for all banks).

5. It is assumed, here, that banks form their expectations for future asset returns in a static fashion. Thus, the observed current real rates of return may be explanatory variables in the portfolio equations.

6. It is very hard to obtain the estimates with correct signs and reasonable magnitudes for all the parameters in such a large system of equations when the estimation is executed from the time series data alone. For example, see Backus, Brainard, Smith, and Tobin (1980) for such an attempt. Also, it seems to us that the empirical testing of various assumptions of the consumer demand system by Deaton and Muellbauer (1980) is made on the basis of an insufficient number of observations.

7. The method of pooling cross-section and time series data was developed by Tobin(1950) and was extensively applied to the estimation of consumer demand functions of individual commodities by Stone, et al. (1954). The procedure was also adopted by one of the authors for estimating the U.S. household flow-of-funds equations, Saito (1977).

8. Our sample of 156 institutions covers the greater part of this category.

9. This amounts to assuming that the variance of u_{it} for bank r in Equation (8) is proportional to the square of DE_t for bank r, or the variance of $u_{it}/DE_{t,r}(=v_{it,r}$ in Equation (10)) is constant for any t and r. See Johnston (1984), chapter 8, Goldberger (1964), chapter 5.

10. In estimating Equation (10) there are two types of stochastic models applicable to the panel data: the fixed effects model and the random effects one, e.g. Hsiao (1986). The basic assumption of the former is that the characteristics of each individual(bank) may be best explained by the difference in constant term, or the use of a dummy variable for each individual, and stochastic factors are represented by an ordinary error term with zero mean and a common variance. On the other hand, the assumption of the latter is that the stochastic term is decomposed into two parts: the individual effect, whose variance is common to all individuals, and the remainder effect.

We adopt the former rather than the latter on the basis of the following preliminary calculations. First, the maximum likelihood estimation of the random effects model, in line with Balestra and Nerlove (1966), was applied to our data. The results indicated that the estimate for the ratio of the variance of the individual effect to that of the total effect was negative; this is unreasonable. Next, the Best Asymptotically Normal (BAN) estimation in Balestra and Nerlove (1966) was also tried by using all the exogenous variables with one-period lag as instrumental variables. The results revealed that the period of convergence to the equilibrium state calculated by the estimated b_{ij} (see on p.10 below) was more than three times as long (28 years) as that of the fixed effects model, implying an unreasonably slow adjustment. In contrast, the estimated results based on the fixed effects model are plausible in view of economic theory and empirical findings, as described in the text. Therefore, we concluded that the fixed effects model is more suitable to our panel data than the random effects model.

11. See Zellner (1962).

12. For the detailed description on the sources and computational procedures of the data used in estimation, see the Appendix.

13. I. e., the weighted average of the figures of Columns (1) and (3) in Table 3.

14. I. e., the weighted average of the figures of Columns (4) and (6) in Table 3.

15. A negative long-run value of gross fixed assets, though its absolute value is small, may be unreasonable. It is likely, however, that the land and other tangible assets held by banks are disposed of in the long-run.

16. See also the discussion on modified estimates, below.

17. The exchange rate is the rate of yen to the dollar.

18. The rate of return on foreign assets is introduced after 1971, when the fixed exchange system was virtually abandoned.

19. It has been argued in the literature that the more relevant concept of bank loan rate is what is called "effective" loan rate, which takes account of the fact that a portion of the loans made by the banks is required to be kept in the deposits of borrowers. Instead of the bank loan rate ($R7$),

we also tried the effective loan rate for estimation. The results turned out basically unchanged.

20. The ratio of deposits at the Bank of Japan in total deposits in 1980 was 0.54 in all banks. The addition to $\acute{\kappa}_{1i}$(or $\acute{\kappa}_{2i}$)($i = 3$) is made in proportion to original $\acute{\kappa}_{1i}$(or $\acute{\kappa}_{2i}$)($i = 3$).

21. The magnitude of the coefficients might depend on the risk structure among assets and liabilities. This is expressed in terms of the elements of the variance-covariance matrix of asset returns. For further discussion, see Friedman and Roley (1979). Ogawa (1984) computed the degree of substitutability among assets in the portfolio of Japanese non-financial corporations based on this idea.

22. In effect about a half of the cross effects have been suppressed to zero. The exclusion of some of the cross effects is subject to arbitrariness, but it is stressed that the adoption of the pooling method has greatly increased the stability of the estimates over those from a simple time series regression. Smith (1975) stated that in the postsample forcasts *a priori* constrained model outperformed the unconstrained model.

23. The standard errors for g_{ki}, h_{0i}, and d_i in Table 6 are calculated under the assumption that b_{ij}, h_{1i}, and h_{2i} in Equation (11) are the exact extraneous information, although they are the estimates obtained from the panel data and are subject to sampling errors. In principle, the standard errors in Table 6 should be computed by taking into account the standard errors of the panel data regression. See Goldberger (1964, chap.5, sec.6), or Theil and Goldberger (1961). It is to be noted, however, that the calculation of the standard errors of SUR, combined with the sampling errors of the panel data estimates, is formidable and almost impossible to execute in the large system such as ours. Instead, we made the following supplementary computaion. Equation (11) with the same specification as in Table 6 were estimated by a single equation method under the two alternative assumptions: (1) that the extraneous coefficients are exact, and (2) that they are subject to the sampling errors of the panel data estimates. The ratios of the coefficient standard errors of the latter to the former turned out to be less than 1.14 in 27 out of 29 coefficients of the real rates of return, while the remaining two are 1.36 and 1.42, implying that the difference in the standard errors between the two methods was not large. This may be due to the fact that, by and large, the coefficient standard errors of the panel data were relatively small. These results may give us a general idea of the size of the correction of the standard errors in Table 6, when the estimation of the SUR combined with the sampling errors by the panel data were to be executed.

If the standard errors in Table 6 are corrected by the ratios calculated above, any conclusion derived from the following significance tests does not change at all.

24. The call rate has a significant effect both on call loan (*A6*) and call money(*L2*).

25. For example in the period from 1971 to 1972, borrowing from the BOJ increased by 212 percent while the discount rate was lowered by one percentage point.

26. The other sectors include the households, non-financial business, and the public sectors.

27. In the following numerical analysis equilibrium implies a short-run equlibrium. Thus, lagged asset or liability holdings, $A_{j,t-1}$, are included in vector **B**.

28. The six financial assets (or liabilities) in this calculation correspond to the six rates of return, which are shown in the first column of Table 8. The merging of the two assets(*A1* and *A7*) in constructing the matrix dG is justified when the demand for fixed assets changes nearly proportional to the supply of loans.

29. The figure of 740 billion yen is chosen so as to lead to one percentage point change in the discount rate. In Japan the discount rate is decided by the BOJ. In view of this fact suppose that the discount rate (*R8*) is an exogenous variable and the BOJ loans supply (*B8*) an endogenous one in the above model. Then the 740 billion yen is the required increase of BOJ loans to keep the market in equilibrium for one percent fall in the discount rate. In actual practice a change in the discount rate is accompanied by an accommodating change in the BOJ loans supply. If actual accommodation from a one percent discount rate cut is substantially short of the 740 billion yen, credit rationing may come into effect.

30. Many econometric models have the "term-structure" equations which relate a key interest

rate (e.g. the discount rate) to the interest rates of other financial instruments (e.g. the bank loan rate, the call rate, etc). From the angle of our financial model, they may be viewed as the equations which describe the tendency of the unidirectional changes in the interest rates, as presented here, or as the reduced form equations of the financial model of our type, in which the key rate is treated as exogenous. In the MPS model the key rate (the short-term U.S. TB rate) alone is determined by the supply and demand relation, while other rates are given by the term-structure equations. See Ando and Modigliani (1975) and Tobin (1975).

31. For example, in the studies presented above the banks' portfolio equations do not include the discount rate variable, except for the Furukawa's one (1985). The difference between the estimates of the effect of the discount rate in other studies and in ours may be largely due to our adoption of pooling methods of estimation. This enables us to use more information and to by-pass some of the collinearity problems.

REFERENCES

Ando, A. and Modigliani, F.(1975). Some reflections on describing structures of financial sectors. in G. Fromm and L. R. Klein (Ed.), *The Brookings Model: Perspective and Recent Development*, (pp. 524-563). Amsterdam: North-Holland.

Backus, D., Brainard, W. C., Smith, G., and Tobin, J. (1980). A model of U.S. financial and nonfinancial economic behavior. *Journal of Money, Credit, and Banking. 12*, 259-93.

Balestra, P. and Nerlove, M. (1966). Pooling cross-section and time series data in the estimation of a dynamic model: The demand for natural gas. *Econometrica. 34*, 585-612.

Brainard, W. C. and Tobin, J. (1968). Pitfalls in financial model building. *American Economic Review, Papers and Proceedings. 58*, 99-122.

Deaton, A. S. and Muellbauer, J. (1980). An almost ideal demand system. *American Economic Review. 70*, 312-26.

Fair, R. C., and Jaffee, D. M. (1972). Method of estimation for markets in disequilibrium. *Econometrica. 40*, 497-514.

Friedman, B. M. and Roley, V. V. (1979). A note on the derivation of linear homogeneous asset demand functions. *NBER*, mimeographed.

Furukawa, A. (1985). *Gendai Nippon no Kin'yu Bunseki (Financial Analysis of Japan)*. Tokyo: Toyokeizai Shimposya.

Goldberger, A. S. (1964). *Econometric Theory*. New York: John Wiley.

Hamada, K. and Iwata, K. (1980). *Kin'yu Seisaku to Ginko Kodo (Monetary Policy and Banks' Behavior)*. Tokyo: Toyokeizai Shimposya.

Honda, Y. (1984). The Japanese banking firms. *The Economic Studies Quarterly. 35*, 159-80.

Horiuchi, A. (1980). *Nippon no Kin'yu Seisaku (Monetary Policy in Japan)*. Toyokeizai Shimpo-sya.

Hsiao, C. (1986). *Analysis of panel data*. Cambridge: Cambridge University Press.

Ito, T., and K. Ueda (1981). Test of the equilibrium hypothesis in disequilibrium econometrics : An international comparison of credit rationing. *International Economic Review. 22*, 691-708.

Johnston, J. (1984), *Econometric Methods*. 3rd edition. New York: McGraw-Hill.

Kamae, H. (1980). Nippon no Kashidashi Shijo no Fukinko no Keisoku -- Kaizen Sareta Data o Mochiite (The Measurement of Disequilibrium in the Japanese Loan Market --Using Revised Data). *Keizai Kenkyu. 31*, 81-87.

Klein, L. R. (1983). *Lectures in Econometrics*. Amsterdam: North-Holland.

Ogawa, K. (1984). Kigyo no Shisansentaku Kodo: Jitsubutsu Shisan to Kin'yu Shisan no Sentaku o Megutte (A Portfolio Selection Behavior of a Firm: Selection among Real Assets and Financial Assets). *Kokumin Keizai Zasshi. 150 *, 75-95.

Pyle, D. H. (1971). On the theory of financial intermediation. *Journal of Finance. 26*, 737-47.

Saito, M. (1977). Household flow-of-funds equations: Specification and estimation. *Journal of Money,*

Credit, and Banking. 9, 1-20.

Santomero, A. M. (1984). Modeling the banking firm: A survey. *Journal of Money, Credit, and Banking. 16*, 576-602 and 697-712.

Smith, G. (1975). Discussion. In G. Fromm and L. R. Klein (Ed.), *The Brookings Model: Perspective and Recent Development.* (pp. 568-572). Amsterdam: North-Holland.

Stone, R., et al. (1954). *The Measurement of Consumers' Expenditure and Behavior in the United Kingdom, 1920-38*, Vol.I, Cambridge: Cambridge University Press.

Suzuki, Y. (1966). *Kin'yu Seisaku no Koka --- Ginko Kodo no Riron to Seisaku (The Effect of Financial Policy).* Tokyo: Toyokeizai Shimposya.

Theil, H., and A. S. Goldberger (1961). On pure and mixed statistical estimation in economics. *International Economic Review. 2*, 65-78.

Tobin, J. (1950). A statistical demand function for food in the U.S.A.. *Journal of the Royal Statistical Society, 113*, Series A, General, Part 2, 113-141.

Tobin, J. (1969). A general equilibrium approach to monetary theory. *Journal of Money, Credit, and Banking, 1*, 15-29.

Tobin, J. (1975). Discussion. In G. Fromm and L. R. Klein (Ed.), *The Brookings Model: Perspective and Recent Development*, (pp. 565-67). Amsterdam: North-Holland.

Tsutsui, Y. (1982). Wagakuni Ginko Kashidashi Shijo no Fukinko Bunseki (The Disequilibrium Analysis of Bank Loan Market in Japan). *The Economic Studies Quarterly, 33*, 33-54.

Zellner, A. (1962). An efficient method of estimating seemingly unrelated regressions and tests for aggregation bias. *Journal of American Statistical Association, 57*, 348-68.

Modeling Transition from the Centrally Planned Economies to the Market Economies: A Case Study of Poland

Władysław Welfe

I. The Transition Path

The transition from a centrally planned to a developed market economy seems to be a long lasting process. Let us consider the Polish example. The 1980s were characterized by a series of attempts to implement economic reforms. Although they did not reach their general targets - they brought departure from the command type planning and central allocation of resources on behalf of direct links between the firms (prevailing state-owned) and other economic agents. They broadened the area of freely negotiated wages and then prices and, at last, increased significance of economic policy financial instruments.

In some countries, like Hungary, it has brought about a step by step elimination of the shortage economy. In others, like Poland, the price reforms, because of the social pressures towards full wage indexation - ended up with higher rates of inflation and persistent shortages. This called for reversing the policy measures, i.e. introducing nearly complete freedom of price formation and tight control over wage increase associated with deflationary fiscal and monetary policy (high interest rates). These unpopular measures could be taken up only by a new, democratic government, which came to power in autumn 1989, gaining the confidence of majority of the population.

The above measures were introduced in 1990 under the stabilization program. They resulted, after bringing hyper-inflation to its momentum, in a decline of the rate of inflation (down from 60%-80% to still high 5% on a monthly basis). The crucial result was, however, the transition from a shortage economy towards a demand determined economy. Thus, the economic equilibria, first of all, the market equilibria, were regained within a short period of time not exceeding two-three quarters. After the introduction of a free domestic market for foreign currencies, the official exchange rate was raised to the free market level and rationing abolished, the opening of the economy brought about a growing competition from imports. It helped, to a large extent, to undermine the monopolistic or oligopolistic position of prevailing state-owned firms. Hence, the new market adjustment mechanisms began to work due to rising sensitivity of firms to price signals. What is even more important, the demand constraints became the decisive factor in producers' adjustments. On the macrolevel, the constraints in the final and intermediate demand imposed by the restrictive financial policy and collapse of intra-CMEA trade were followed rising by a deep decline in domestic activities and rising unemployment.[1]

The behaviour of state-owned firms tending to maximize the wage-bills rather than profits, and showing inertia in responding to market signals - had important implications, however. At the macro policy level - there were fears to allow an increase in the final incomes. The following demand increase might not have been followed by an output increase (and relevant increase in the low rate of capacity utilization), but might be superseded by the crowding out effects, i.e. additional inflation. The remedy was sought at the microlevel. This is in the deregulation of the public sector (its commercialization, and privatization) as well as in the development of private sector. The privatization of small firms (trade, services) followed within a short time. However, the deregulation of the large, industrial state-owned firms - along with developing joint ventures - may last several years. It is not clear yet which empirical model of market economy will be approached - liberal, or social market economy (as experienced by Germany). For all these reasons, the market mechanisms of adjustments may differ from those observed in developed market economies.[2]

The financial and telecommunication infrastructure needs serious development and the fiscal institutions deep restructuring (because of the introduction of personal income tax, value added tax, etc.). The financial flows and price adjustments play an increasing role, and what is striking - the frequency of changes and the government responses are rising, calling for quarterly and monthly observations and forecasts (in contrast to the previous annual central planning systems).[3]

II. The Major Tendencies in
Modeling the Economies in Transition

The econonometric model builder thus faces a difficult situation, especially when dealing with annual models. On the one hand, the structures of quasi-supply determined macromodels are adequate for the past (W. Welfe, 1991), and for the future - the demand-oriented models seem appropriate. On the other hand - switching from one economic regime to another may follow "over-night" (like in Poland) or take several years (like in Hungary) and is not irreversible. Hence, we advocate the view that the adequate tool for economic analysis and medium-term forecasting for a period of transition are the complete models (W. Welfe, 1983). They generate both final and intermediate demand as well as capacity output and commodity supplies (the same for production factors), assuming the possibilities to switch between economic regimes. They define the relevant adjustment mechanisms specific for the above economic regimes.

For the demand determined systems, except for quantitative adjustments (for instance observing that supply follows demand), the typical market mechanisms need to be introduced. Hence, the prices will follow not only the cost-push, but their changes will also reflect the demand pressures. On the other hand, the relative prices and their expectations will increasingly affect the current and future decisions of the producers and investors. Simultaneously, the increasing role of the private sector should be acknowledged. Then the implications of the theory of production could be introduced into the model equations, especially production functions (L. R. Klein, 1983).

Description of the financial flows must be extended, following the development of financial institutions, including money and capital markets. Applications of the portfolio theory have to be tested. The changes in the fiscal system must be acknowledged, emphasizing the interdependence between state budget incomes and expenditures, and the activity levels of production sector.

The short-term macro-economic policy analysis and forecasting rests even more on the financial sector development. Hence, the quarterly models' structure must accentuate, first of all, the financial flows. However, one should not forget their links with the real flows representing macroaggregates, despite the fact that the official statistics do not yet generate the quarterly information about their behavior yet.

The model builder faces also serious data problems. First, many of the existing time series reflect realizations of different regimes (in some periods of effective supply, in others effective demand). This difficulty might be partly solved by using extraneous information and the disequilibrium approach. In the past we made use of suitable disequilibria indicators.[4] The classic switching regression and newly developed simulation based estimation methods[5] may help in the future to estimate small-size disequilibrium submodels. We have in mind small blocks composed of equations which enter the latent (unobservable) variables as excess demand or excess supply (price, wage and inventory equations) being recently represented by proxies such as the rate of capacity utilization or rate of unemployment.

Secondly, some of the time series appear new (for instance on interest rates changes) not necessarily based on systematic reports but on censuses or sample surveys (extended to cover firms of households anticipations); some are discontinued (for instance, on wholesale commodity supplies). In this case, the new information typically covers enough quarters to fit simple equations of the quarterly model (except for lags), but, of course, is by no means sufficient to adjust the parameter estimates of the annual models (some of them covering also the 1960s or the 1970s). Hence, the model builder has to look for extraneous information. Some prefer to assume geographical analogies, i.e. to take into account the estimates for neighboring market economies (for instance from Austria for Slovakia). Others try to extract the information from local data of higher frequency, either by making use from estimates obtained from quarterly models or by means of adding additional annual observations and testing regression coefficients for stability. The typical calibration procedures are not frequently used because of the fear that underlying assumptions as to the price generation process are not fulfilled and because of the limited number of years of a market system functioning.

The use of quarterly and monthly data calls for broader application of the techniques developed in the last decade for estimation and testing dynamic models (Cuthbertson et al., 1992). The use of procedures aimed at testing the time series for stationarity and their relevant transformations allowing for cointegration of regressed variables must be advocated. On the other hand, the error correction models may help to separate the short-term adjustments from the long-term relationships. Further, there is a need to introduce expectations (including rational expectations), especially in equations explaining price and wage formation, investment decisions.

To meet these requirements is not an easy task, as typically there are only short time-series at our disposal. Several heuristic approaches are available, among them the procedures aimed at reducing the number of explanatory variables by means of defining composite variables being the (linear) combinations of initial explanatory variables. The "weights" might either stem from extraneous information (in some cases from I-O tables) or be established using the search procedures.

The above major topics of the macromodelling activities are fairly general. It seems that they apply to a majority of former CPE's in transition. However, they will be discussed in some more detail, taking as the point of reference the macroeconometric models of the Polish economy. This is because of the author's personal involvement in their construction and use. The annual large model W-5 (more than 1000 equations) built in the early 1980s was a complete model and was systematically used in the 1980s (its supply version) and is used in the 1990s (its demand version).[6] It has undergone serious reconstruction recently, following the general guide lines formulated above. Its final version (W-10) will allow to link variables representing the former Material Product System (MPS) with the new System of National Accounts (SNA) macroaggregates (data being available from 1980). It will also separate public and private sector across major subdivisions of the model. The sample covers data from 1965-1992 and is systematically extended. Hence, the forecasts and policy simulations will have to rely heavily on coefficient adjustments based on the updated information.

The quarterly model WK was built in 1991.[7] Its structure takes into account the non observable macrovariables from the SNA (GDP, its components and use) which had to be estimated by the authors themselves.[8] Its first, experimental version was based on 1989.1-1991.2 sample and included 83 equations. It is systematically updated and revised. As the specific questions dealing with the quarterly data and quarterly adjustments have been already discussed in detail in a separate paper[9] we shall concentrate on the more general specification problems characteristic for both types of models.

III. The Production Potential

For the economies in transition the major medium- and long-term problem is the reconstruction of the production potential associated with improvement in managerial skills in order to meet the high technological and ecological requirements for the production process and quality of products to be competitive in the world. To model the additions to the production potential - it is necessary to generate the additions to the fixed capital. They depend on current and previous (lagged) investment outlays. The crucial question, how the investment outlays are generated, will be discussed below in the next section.

In order to evaluate the productive capacities we need a production function linking fixed capital and potential output. If the sample covers periods characterized by shortages of production factors (other than fixed capital), then the list of variables needs to include not only labor input but also energy and material inputs (at the macro-level, they can be reduced to imports of intermediate commodities), determining the rate of capacity utilization (W. Welfe, 1989 and 1991). We have:

$$X = f \ (K, \ NK, \ T) \ . \ u \ (NH/NK, \ MQ/MK) \tag{1}$$

where: X = value added, K = fixed capital, NK = employment ensuring full utilization of fixed capital, T = time for technical progress, $u(.)$ = rate of capacity utilization, NH = labor input, MQ = material input from imports, MK = material use of imported goods ensuring full utilization of fixed capital (error terms omitted).

The estimation of parameters of (1) may follow different patterns. We used a transformation of (1), relating the observed output directly to the observed factor inputs:

$$X = f \ (K.u_k, \ NH, \ T) \tag{1a}$$

where u_k = rate of fixed capital utilization (e.g. represented by a ratio of shifts worked to maximum number of shifts). The rates of utilization of the factor inputs depend on shortage indicators, including $IM = m \ (MQ/MK)$ = indicator of shortages in material inputs. (For details see W. Welfe, 1989 and 1992c).

Assuming the availability of labor (i.e. unemployment) and of intermediate commodities, we can ignore in simulation excercises, the second term of (1) and thus obtain the capacity output:

$$XK = f \ (K, \ NK, \ T). \tag{2}$$

In order to determine the number of potential jobs NK we need an equation relating it to the fixed capital K. The parameters can be estimated from a function, where labor input NH depends on the fixed capital adjusted for its use: $K.u_k$. Assuming $u_k = 1$ we obtain NK. It can be compared both with the available labor force NS to check whether full employment is theoretically feasible and with the effective demand for the labor force ND. Should there be constraints in the labor force availability - then we shall get an estimate of excess demand for labor.

If, however, the major role is played by the final demand XD constraint, then the output tends to be equal to demand: X = XD. Hence, the rate of capacity utilization is given by

$$UX = XD/XK\text{-}1 = X/XK\text{-}1 \tag{3}$$

where the capacity output XK is obtained either using (2), if the estimates were based on a sample covering shortage economy or using (3) if UX is directly observable. Then in this case the difference NK - ND = NK - N will show the underutilization of available jobs. It can be accompanied by unemployment if the labor demand ND is lower than labor supply NS.

The labor supply (NS) depends on demographic factors adjusted for changes in variables representing social and economic pressures affecting households. The expectations with respect to availability of jobs must be introduced as an additional factor.

Labor demand (ND) can be derived from the production function (1), where production is equal expected value added (X). In supply determined systems it will reflect the major factor constraint. In demand determined systems it will approach the demand for production (defined in the next paragraph).

The rate of unemployment UN will be determined from the following identity: UN = ND/NS - 1. This solution leads in simulation excercises to low accuracy of estimates of the rate of unemployment and variables depending on it. Hence, one may try to explain the rate of unemployment directly, relating it first of all to the rate of capacity utilization and calculate then the labor supply indirectly from the above identity, defining UN.

The measures of utilization of capacities and unemployment are important characteristics of potential disequilibria in commodity and labor markets, affecting the price and wage formation, respectively.

The annual model W-5 of the Polish economy provides, in principle, the possibilities of computation of the above potential measures. They show that during the recession of 1990-91 the rates of utilization of industrial capacities declined to 30%-50% in manufacturing industries and the number of potential jobs exceeded the unemployment levels (more than 2.5 million by the end of 1992). This result may be questioned to some extent, as in the last years (including the recession period), the scrappings rate in industry was extremely low (2% reflecting pessimistic expectations of state-owned firms with respect to the modernization of their equipment). Hence, not all components of the existing fixed capital can be considered a potential source of production.

Generally for the quarterly models there exists the possibility to construct approximate production functions, using (deflated) industrial sales, labor and fixed capital inputs. However, at the moment, reliable information on hours worked is not available and the sample is too small to reconstruct the time series of fixed capital from accumulated data on investments.

IV. The Commodity Supplies and the Final and Intermediate Demand

The supplies of domestically produced commodities in a shortage economy are typically determined allowing for the major production factor constraint as in (1). Value added has to be transformed next into gross output or sales:

$$Q = X + QM \tag{4}$$

where Q = gross output, QM = material use.

In the market economies, it is assumed that the output (supplies) follow the total demand, which may be expressed in its simplest way by using the following balance identity:

$$QS = QD = Q = QM + H + E - M \tag{5}$$

where QS = gross output supply, QD = gross output demand, H = final domestic demand, E = exports, M = imports. This identity holds for particular industries and branches as well.

The above assumption has to be modified, however. First, the producers are oriented on expected (and not actual) demand for their products (QD^E), and second - especially in the transition period, their response to a demand increase is lagged and/or, instead of offering additional supplies, they tend to require higher prices. This behavior may be assymetric i.e. a decline in demand may induce an immediate decline of output. Hence, we might have:

$$QS = \begin{cases} \lambda_1 \, QD^E + (1 - \lambda_1) \, Q_{-1} & \text{if} \quad QD^E - Q_{-1} > 0 \\ \lambda_2 \, QD^E + (1 - \lambda_2) \, Q_{-1} & \text{if} \quad QD^E - Q_{-1} < 0 \end{cases} \tag{6}$$

where $0 < \lambda_1 < 1, \quad 0 < \lambda_2 < 1$.

Now, the supplies need not exactly be equal to effective demand $QS \neq QD = Q$ and this difference will affect the inventory changes.

Note that the price equations must be sensitive to the above distortions between demand and supply (which will be partly reflected in the changes in the rate of capacity utilization). In this case, the above impact of temporary disequilibria will be transmitted to the final and intermediate demand, provided the demand equations will incorporate the price changes as explanatory variables.

The blocks of equations explaining the domestic final demand (H) cover the demand of households (C), government institutions (G), investors (J), and inventory increase (DR). In general we have

$$H = C + G + J + DR. \tag{7}$$

The consumer demand functions follow the more or less demanding theory of consumers' behavior. The demand financed from personal disposable income is separated from the requirements for public goods. In annual models covering the periods with prevailing commodity shortages, leading to systematic deviations of effective demand from notional demand in particular markets, the well known disequilibrium approach must be used.[10] On the other hand, for periods characterized by market equilibria, expectations, especially of the rate of inflation must be introduced. It can be more easily done within the quarterly models and the estimation results transferred to the annual model.

In general, we have for consumer demand function (CD):

$$CD = c((YP - TYP)/PY, PC/PY, PC^E) \tag{8}$$

where YP = nominal personal income, TYP = personal income tax, PC = deflator of consumption, PY = personal income deflator, superscript E means expectations; and for (real) expenditures used in estimation process:

$$C = CD - C(I) \tag{9}$$

where I - suitable disequilibrium indicator.

The government final consumption treated frequently in the past as the exogenous policy instrument needs to be explained mostly in terms of government budget real current expenditures (except for subsidies, wages). In general, we have

$$G = f((BCP - SUP - BFP)/PBC) \tag{10}$$

where BCP = budget current expenditures, SUP = net subsidies, BFP = wage bill, BPC = deflator.

The investment demand functions need much reconstruction and extensions. In the past the demand of investors was, in principle, related to the expected increase in capacities, plus replacements. The capital-output ratio and lags in investment completion played the decisive role (in the shortage economies the disequilibria in the investment goods market had to be observed too).[11] When approaching market systems - the constraints in the availability of investment funds must be introduced into the equations explaining investment outlays. These are the banking loans, firms' own funds, government subsidies. Further step will be introduction of the ratio of expected profits to expected costs, including the long-term interest rate. The latter variable was used in the equation explaining investment demand in the quarterly model WK.

The most general forms of investment equations would be

$$J = \begin{cases} j \ (DX, \ JFP \ / \ PJ, \ UX) \\ j \ (DX, \ UX, \ PR^{E} / RD) \end{cases} \tag{11}$$

where J = investment outlays, JFP = f(AFP, BJP, DBZJP) and AFP = profits, BJP = government investment expenditures, BZJP = long-term claim of banks; letter D before the symbol indicates first difference, PJ = investment deflator, PR = rate of profits, RD = long-term interest-rate, superscript E means expectation.

The inventory increase must in principle follow the general economic activities. However, because of inaccuracies of expectations and lag in market adjustments it will also reflect the impact of temporary disequilibria. Hence, in general we can write

$$DR = d \ (\ DQ, \ QD - QS, \ R_{-1} \ ). \tag{12}$$

Assuming, that the total inventory increase will be obtained as a residual from the balance identity (5) using (7), we can split equation (12) defining, on the one hand, the demand for inventory increase (DRD):

$$DRD = d \ (\ DQ, \ R_{-1}) \tag{13}$$

and on the other hand the residual component reflecting the changes in the market pressures.

In an open economy, the model is extended to cover the foreign trade. Typically exports follow the foreign demand. This is partly true also for the periods when shortages prevailed as receipts from exports were regarded as the major source of financing imports needs which never had the chance to be met. Hence, in W-5 (and earlier) models the demand functions were defined. They need to-day some extentions in order to determine the impact of likely market adjustments (including the quality changes). Hence, we have

$$ED = e \ (\ EW, \ PE\$ \ / \ PEW \ ) \tag{14}$$

where EW = world trade or another proxy for foreign potential demand, PE\$ = PE/WZ\$, where PE = export price in zloty, WZ\$ = exchange rate zloty vis a vis US \$, and PEW = world prices in US \$.

The export supply functions might be useful also in a market economy, in order to determine the (equlibrium) export prices. The supply function is given by

$$ES = e \ (\ QS, \ PE \ / \ PQ \ ) \tag{15}$$

where PQ is domestic producers price and the changes in the ratio PE/PQ show the changes in exports profitability (notice, that despite the changes in the exchange rate also tax exemptions will affect the profitability changes).

Assuming ED = ES = E we can solve the system (14) and (15) for exports prices PE.

Imports were in the past mainly complementary. Hence, in the imports equations the major role was played by the level of those activities which were decisive - of consumption with respect to the imports of consumers goods, of investment with respect to the imports of investment goods and of GDP (or NMP) with respect to the imports of intermediate commodities. These potential values of imports were adjusted taking into account total available funds for imports (exports plus exogenous foreign financing).[12]

The opening of the economies in transition to the world competition brought about enormous inflow of imported commodities, partly because of their higher quality or broader variety of assortments, but to a large extent due the stabilization of exchange rates under high rates of domestic inflation.

This calls for careful specification of the impact of changes in imports prices versus domestic prices (allowing, wherever possible, for quality changes). Hence, the typical specification of the import function for the period of transition will be as follows:

$$M = m \ (\ ACT, \ PM \ / \ PQ, \ QU \ ) \tag{16}$$

where ACT = relevant activity level (Q, X, J, C), PM = PM\$.WZ\$ ($1 + v_1 + v_2$), where PM\$ = imports price in US \$, v_1 , v_2 = rates of custom duties and turnover tax, respectively ($0 < v_1 < 1$, $0 < v_2 < 1$), and QU = indicator of quality changes.

In this way a decrease of relative import prices will bring an imports increase and by virtue of identity (5) it will negatively affect the domestic output.

The balance of trade is now becoming residual. The government can affect its value indirectly - either by adjusting the exchange rate or by changing the customs duties, taxes or subsidies.

The mechanisms governing the real flows of the economic system depend on the prevailing economic regime. For the *supply determined system* we could distinguish the following major feedbacks linking, first of all, the supply side (see Figure 3.1). There were: - long-term growth mecahnism, which we called *supply type accelerator* (W.Welfe, 1985): an increase in supplies of investment goods yields after respective time lags (needed for completion of investment projects) an increase in fixed capital formation and the output and supplies of investment goods (this mechanism is frequently assumed in the theory of growth); - short-term *foreign trade multiplier*: an increase of exports brings an increase in the imports of material inputs and hence, an increase in output and exports; it may be combined with supply accelerator if additional imports cover investment goods; - short-term *supply consumption multiplier*: growing shortages in consumer goods market bring decline in labor supply, hence decline of supplies of commodities but also decline of incomes and demand (Barro-Grossman hypothesis - see Charemza-Gronicki (1985); this hypothesis has been positively verified only partially (W. Welfe, 1991).

Assuming that the system is *demand determined*, we can easily identify the familiar feedbacks. However, for an economy of transition to the market economy we have to allow, first of all, for the low sensivity of the state-owned producers to the market signals, their assymetric responeses to demand changes etc. Hence, these feedbacks have to be modified in an appropriate way (see Figure 3.2): - the well known *consumption multiplier*: an autonomous increase in household incomes will result in an increase in consumers demand. This income increase mya have various sources: a relaxation of wage control in the production sector, an increase of wage bill in the public sector (increase in wages and/or employment), pensions and social aid. It will be followed by an output increase, unless it is absorbed by increased imports and reduced due to likely price increase. The output increase will affect the employment increase, thus the compensation of employees, their incomes and consumers demand. An autonomous increase in other final demand components (investments, exports) will lead - under the above mentioned constraints - to an output increase and initiate a multiplier response; *the extended multiplier* such that an increase in goverment budget spending increases its final demand, which will be transformed into domestic output increase under the same constraints as above and bring about an increase in the tax base and government budget receipts and - given budget deficit predetermined - its spending. Note, that the consumers' income increase lead to an increase in the tax base not only directly but also indirectly due to the consumption multiplier; - the *accelerator* - an expected increase in capacity output co-determines the demand increase for investment goods and for total final demand; if, once again, the import shares do not increase and the crowding out effects will not become stronger - additional increase in output will follow, which may positively affect the expectations with respect to the capacity growth as well as firms profits.

FIGURE 3.1
Complete Supply Determined Models - Real Flows

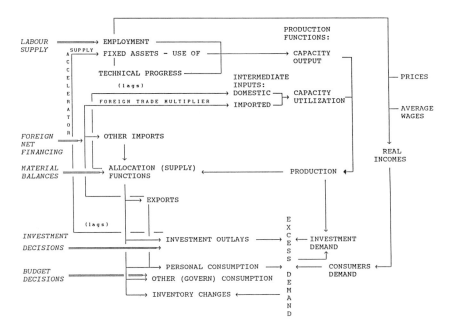

V. The Financial Flows

The financial flows can be analyzed by groups of economic units (agents) - households, firms (business), government, foreign firms or by looking into alternative markets (money, capital etc.). The first approach seems more promising, at least, in the first stage of transformation towards market economies.

The households incomes and expenditures were in the past a subject of many modelling exercises. The major change in the specification of wage equations is that it has to allow for the impact of unemployment (following Philips curve approach) instead of demand pressures in the labor markets (shortages of labor)[13]. As these impacts seem assymetric, much testing will be indispensible and also one can make use of the results obtained from quarterly model and evidence provided by other authors. For the productive sector we shall make use of the following equation:

$$ZP = f (\overset{.}{PY}, \overset{.}{X} / N, 1 / \overset{.}{UN})$$ (17)

where ($\cdot$) is the rate of growth, ZP = average wages, X/N = labor productivity, and UN = rate of unemployment.

FIGURE 3.2
Complete Demand Determined Models - Real Flows

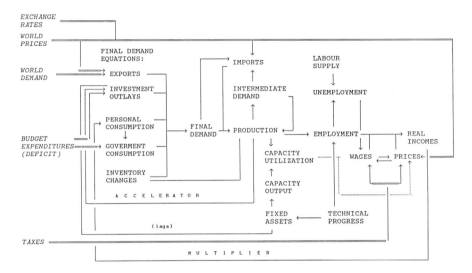

The wage policy frequently imposes constraints on the rates of wage increases. Under low rates of inflation, the wages increase is linked to the productivity growth, otherwise to the rate of inflation (introducing specific indexation schemes). Fiscal control is introduced on the assumption that firms exceeding the prescribed norm of wage increase will have to pay from their profits highly progressive tax. Hence, we have to modify (17) assuming that:

$$\overset{.}{ZP} = f (\lambda \overset{.}{ZP}{}^{*}),$$ (17a)

where $\overset{.}{ZP}{}^{*}$ = e $\overset{.}{PY}$ is the norm i.e. tax free rate of increase in wages, e = the assumed government indexation coefficient (elasticity), 0 < e < 1, and λ is an adjustment coefficient. This coefficient represents the average tendency to adjust the wage increase above (or below) the prescribed level, because of the productivity changes, general pressures on the labor market and specific pressures in a particular industry:

$$\lambda = g \ (\overset{.}{X} / N, \ 1 / \overset{.}{UN}, \ ZP_{-1}).\tag{17b}$$

The final rate of wage increase will, of course, depend on the frequency of indexation and likely lags between the current and base period used to establish the rate of inflation.[14]

The data problems may arise as the wages in the fast developing private sector are notoriously underestimated. Similar data problems are known with respect to personal incomes of private enterpreneurs. Their personal income (this applies also to farmers) is assumed to depend on either total sales (output) or value added. The other components of personal income include transfers from the government budget (mainly pensions related to wages) and incomes from property (including interest rate payments related to savings deposits).

The expenditures of the households are obtained by adding up their consumption (and investment outlays) multiplied by respective deflators, personal income tax and other taxes. Total saving is thus obtained as a residual, unless a simultaneous expenditure equation system is used to determine all expenditures, including saving.

The coverage of financial flows of firms can be hardly complete. It mainly shows their total receipts (QP) being obtained by multiplying the volume of sales by respective deflators and, in more detail, their expenditures i.e. fabrication costs and profits (plus taxes). The fabrication costs (KQP) are composed of expenditures on material inputs (QMP) (it is desirable to split them by industries and distinguish imports which typically calls for use of the I-O coefficients), depreciation of fixed capital (AP), wage compensation (ZP.N) and related expenditures, interest and other financial payments (BZRP):

$$KQP = QMP + AP + ZP.N \ (1 + \mu_1 + \ \mu_2) + BZRP \tag{18}$$

where $0 < \ \mu_1 < 1, \ \ 0 < \ \mu_2 < 1$ rates of social insurance and taxes, respectively.

Each of the above components must be separately explained. Total expenditures on material costs depend on technology, input prices and total output:

$$QMP = \sum_i \ (\ a_i \bullet P_i \ )Q \cong \ Q \ f \ (\ P_1 \ , \ldots , P_k \ ) \tag{19}$$

where a_i = technical coefficient showing the unit use of commodity i as an input, $P_1 \ , \ldots , P_k$) = prices of selected, important material inputs (including import prices).

The equation explaining depreciation will simply read:

$$AP = \delta KP \tag{20}$$

where KP - fixed capital (in current prices).

If the above cost decomposition is not explicitly reported (it happened in Poland in the 90s) the stochastic approximation of (18) must be used including material input prices, wage compensation, interest rate payments as explanatory variables.[15]

The difference between the total receipts of the firms and their fabrication costs (QP - KQP) covers a) turnover tax (TOP to be substituted by value added tax) being dependent on the firm's sales (given exogenous average tax rate), b) net subsidies - depending on the level of particular activities (given exogenously subsidy rates) and c) gross profits (ZYP). In the former CPE's under the weak financial constraint, the state-owned firms and the authorities tended to keep the gross-profits at levels ensuring "normal" average profitability. Hence, either net subsidies had to be treated as residual or it should affect the firms receipts via respective price adjustments. In the period of transition, where the hard financial constraints are observed - the gross profits become more and more characteristics of the firms' efficiency and (given the fabrication costs) of the impact of temporary gaps between demand and supply resulting in relevant price adjustments.

The gross profits must be further decomposed into (corporate) income tax (TDP) and other taxes (among them the antiinflationary tax on wage increases exceeding the assumed by the authorities portion of the consumers' price increase) and net profits (ZYNP). The latter are the source of financing autonomous investment, bonuses, social aid etc.

The specification of equations explaining the above distinguished components of the gross-income of firms is a new task as the majority of model builders treated them as exogenous in the past.

The modelling of firms funds and their allocation (stocks) including the problems of their liquidity needs elaboration. The data availability is very limited and its quality poor (the capital markets are just now developing).

The government budget receipts and expenditures were already in the last years the subject of macro-modelling exercises.[16] However, new types of taxes are being introduced (which poses the problem of continuity of time series), the role of subsidies is declining. Depending on the assumed concept of the public sector functions - other budget expenditures are allocated according to specific, changing principles (as, for instance, the transfers to households). The budget deficit slowly begins to play an active role as a macro policy instrument.

The receipts from taxes generated in other blocks (households, firms, foreign trade) will be added up, using additional bridge equations to overcome the differences in definitions (taxes due versus paid) and coverage. The receipts from sales of public property will probably be exogenous. Hence,

$$BYP = f (TYP + TOP + TDP + ...) \tag{21}$$

The budget expenditures cover a variety of components which need separate treatment: current expenditures (BCP) decomposed into: a) subsidies, b) payments to the social insurance funds, and other transfers, c) current expenditures of the public sector (wages and material expenditures) on education, health protection etc. and on general public administration (including military expenditures), d) debt service (domestic and foreign) and investment government expenditures (BJP, treated usually as a policy variable).

The attempts to explain the current expenditures of the public sector from the demand side are, in general, not very succesful. There are only a few characteristics of the relevant activities (number of students, teachers for educational sector) and - under recession - the requirements of these sectors are drastically cut down. Hence, an alternative approach will be advocated which defines the allocation equations allowing for inertia. The Polish experience suggests looking at the expenditure side as an iterative, adjustment process. Having been given initial income expectations and assumed budget deficit, we arrive at feasible initial expenditure level. The changes in budget receipts (for instance their decline) forces the government to cut off, in the first line, the current expenditures of the public sector (material uses and employment next). The budget deficit is also adjusted as result of a compromise. An iterative system following the above rules was developed by A. Welfe (1990) and it might prove useful to implement it into the new macro-models.

The money and capital flows pouring into the broadly understood banking system must be modelled to a much larger extent than in the past. In the former CPE's it was assumed that the money flows are following the real flows and banking operations were treated as accounting operations only (the interest rates stayed constant over years).

The modelling possibilities are, however, limited at the moment because of the unavailability of the data on major banking operations i.e. lending to the state-own firms.

The specification of equations explaining deposits of households and firms (OP - both in zlotys and foreign currencies), as well as of money 1 and money 2 aggregates, follows the theory of household savings and the institutional rules ensuring firms liquidity. Hence, the major explanatory variables are personal incomes or activity levels and interest rates on deposits (different for deposits in zlotys and dollars) compared with the rate of inflation.

Because of the above mentioned limitations on data, we had to abandon the idea of explaining banking loans (flows). For the past it seemed appropriate to explain the level of debt against banks (BZPP) in terms of stocks as a result of the allocation process of available deposits:

$$\text{BZPP} = \text{if} (\text{ OP, R} - \dot{\text{P}}\text{Q}) \tag{22}$$

where R - interest rate, $\dot{\text{P}}$Q - rate of growth of producers prices. This specification remains valid also for to-day.

In the not distant future, it seems feasible to construct also demand functions of flows - in the extreme case of banking loans (net of obligations paid back). It would partly enable to endogenize the interest rates (their nominal values are now much affected by the rates of inflation). The development of capital markets will call for further extensions of the models in the future, making use of the portfolio theory.

The balance of payments description requires serious additions to the so-far existing equations explaining mainly the receipts from exports and expenditures on imports. Transfers, in the current account, especially related to the interest payments, need explanation. The capital account must be incorporated into the model, changes in the foreign debt, being still its major item.[17]

VI. The Price System

The system of price equations plays an increasing role - being the basic tool in the analysis of the rate of inflation and ensuring full links between the real and financial flows. The core of the system is a subsystem of producers' prices. For Poland, it has been developed as soon as in the late 1970s (W. Welfe, 1992c). However, it put the emphasis on cost components in price formation, following the rules of calculating the administered prices. It, therefore, needs extensions to cover the new cost components and, first of all, to capture the impact of demand pressures.

In general, we can decompose the price of commodity j into unit costs of material inputs, depreciation, wage and related costs, financial costs, gross profit and turnover tax:

$$P_j = \sum_i a_{ij} P_i /Q_j + AP_j /Q_j + ZP_j \bullet N_j /Q_j (1 + \mu_j)$$

$$+ BZRP_j /Q_j + ZYP_j / Q_j + TOP_j /Q_j \tag{23}$$

where $\mu = \mu_1 + \mu_2$.
Gross profits can be substituted by a function of the rate of capacity utilization (UX_j) to allow for demand pull:

$$ZYP_j / Q_j = (ZYNP_j + POD_j) / Q_j = f (UX_j)   \tag{24}$$

Frequently, the equation (23) defined for the I-O system, is approximated by a regression equation to allow indirectly for (unknown) changes in I-O technical coefficients (a_{ij}) by means of selecting the major (important) material inputs, the prices of which are introduced as explanatory variables:

$$P_j = f(P_1, \ldots, P_k, PM, ZP_j N_j /Q_j \bullet(1+ \mu_j), BZRP_j /Q_j UX_j, \tau_j, P_{j,-1})   (25)$$

where $k \ne j$, PM = PM\$. WZ\$ $(1 + v_1 + v_2)$ and $\tau =$ rate of turnover tax.

This specification has the following advantages: a) it takes into account, major domestic interrelationships within the producers price system, b) links prices to the wage formation and, allowing for the inverse dependence of wages with respect to the rate of inflation (17), it defines the dynamic feedback known as inflationary spiral (loop), c) opens the domestic prices to the world prices shocks (imported inflation), d) shows the impact of fiscal policy (via changes in tax rates) and monetary policy (including changes in the interest rate and exchange rate), e) allows for market clearing represented by the changes in the rate of capacity utilization. Introduction of the two latter variables into the price equations poses several problems. Some of the institutional changes are new and the impact of demand pressures can be observed only in the last years after a change in the economic regime. Hence, there is the need for using extraneous information, partly from the parameter estimates of the quarterly model.

The prices of the commodity flows towards particular final users depend, first of all, on the producers prices as well on the imports prices.

In general, we can write for commodity i and user H:

$$PH_i = f (PQ_i , PM_i, \tau_i).   \tag{26}$$

We had to add the rate of indirect tax τ as in particular cases the taxes are superimposed on wholesale prices. In principle, these price equations are approximations of weighted averages of domestic and imports prices of relevant industries or commodity groups.

The major feedback characteristic for the financial sector and price system is the *inflationary spiral*. For the supply determined systems (see Figure 3.3), it was characterized by long lags. The wage (and other cost) increases in state-owned firms were in short-run "compensated" by subsidies. Only after several years and more recently - quarters, the administered prices were raised. The excess demand typically followed this price freeze. In the specific years, when price reforms were introduced labor compensations followed, but they were frequently lower than prices increases. Hence, the shortages declined or vanished for a short period of time, but the cost pressures remained or increased

FIGURE 3.3
Inflation - Supply Determined Models

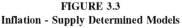

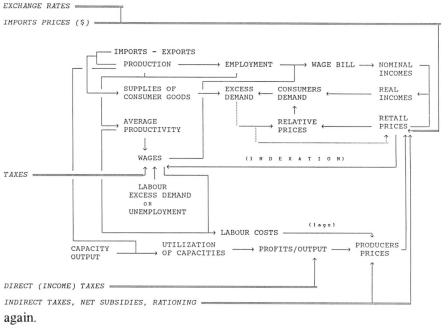

again.

In the market economies (see Figure 3.4) the prices become flexible, their adjustments only exceptionally being controlled by the authorities. Hence, the lags of adjustment are much shorter. However, if there are demand ceilings, the firms will not be able to raise prices, and the cost increase will lead to a decline in profits (being now residual). On the other hand, a hard fiscal control is imposed on wage increase (mainly in state-owned firms). In many countries the wage increase over a certain norm was highly progressively taxed, the norm being obtained by applying a centrally imposed portion of the rate of inflation. (This indexation scheme must be reflected in the wage equations as it has been shown above).

The development of money and capital markets and the more active role of the budgetary policies may lead to *further inflationary feedbacks.* Let us mention: - an increase in the interest rate on investment loans may lead to an increase in investment prices and to a decline in investment demand. The same may apply to the commercial loans, commodity prices and activity levels of the producers. The resulting increment of the rate of inflation may serve as an arguement for an increase in indirect tax rates may lead to an increase in commodity prices, and thus to a decline in real budget spending and in short-term to a further increase of tax rates, but in long-term rather to their decline (see Figure 3.4), - a too high increase in direct tax rates may deteriorate the profitability of firms and thus lead to a decline in their current activities and hence their tax base, as well as a decline in their investment activities, and thus undermine the future tax base. Unfortunately, it will be extremely difficult to introduce this feedback into the model empirically.

FIGURE 3.4
Financial Flows (Production Sector, Households, State Budget) and Inflation

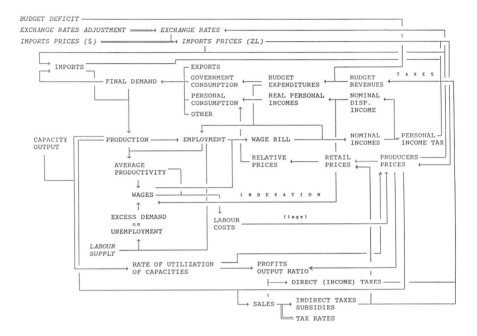

VII. Conclusions

The structure of the macromodels for the period of transition suggested in this paper is based on a comprehensive system of national accounts, which tries to bridge the "old" MPS and the "new" SNA. We are aware, that not all its blocks may be filled with sufficiently elaborated equations. The data constraint can be relaxed by looking into information of higher frequency or relying on geographical analogies. The specific behavioral characteristic of producers' responses cannot be neglected and the hypothesis about the changes in behaviour of commercialized and privatized public producers systematically tested. This applies both to the current adjustments and to the investment decisions of producers.

The public policy options must be broadly represented in the models.

Undoubtedly, the fiscal policy instruments i.e. changes in tax rates or subsidy rates (typically represented by parameter changes) and monetary policy measures (National Bank discount rate, exchange rates) will be explicitly introduced. Government budget deficit may not necessarily play this function; more frequently particular budget expenditures can be treated as policy variables (transfers or military, or investment expenditures). The model should, however, allow also for implementation of specific measures of industrial and agricultural policies, of antirecessionary policies aimed at promoting exports and investments. They need not necessarily consist only in fiscal (tax exemptions) or monetary (lower interest rates) measures.

These requirements are addressed, first of all, to those macro-models which are to be used in policy simulations and scenario-excercises. Perhaps, the macro-models to be used mainly for forecasting purposes may be less demanding with respect to the consistency of specifying the major economic interrelationships. This, however, lies beyond the scope of our study.

The progress depends, to some extent, on the inflow of new statistical information and the feasibility of the use the newly developed tests and estimation procedures. It is hoped that it will also stimulate further development of these techniques.

The above observations seem general enough to be identified with macro-model building activities in the majority of East and Central European economies in transition.

Władysław Welfe, Professor of Economics, Institute of Econometrics and Statistics, University of Łódź; 41, Rewolucji 1905 r., 90-214 Łódź, Poland .

NOTES

1 . Persistant continuation of tight stabilization policy over a period of two years has led to a long lasting recession (measured by the 33% decline in the industrial output within 1990-1991); first signs of recovery were observed only in 1992. See W. Welfe (1992a). .

2. Compare R. Portes (1991) and S. Gomulka (1992).

3. It follows with a serious lag the development in the industrial market economies. See L. R. Klein and E. Sojo (1989).

4. See A. Welfe (1989), and W. Welfe (1989).

5. See B. Salanie (1991).

6. See W. Welfe (1985, 1989, 1992b), G. Juszczak et al. (1991).

7. The authors wish to acknowledge the A.C.E. grant 90/999/040/001/031 under which the model was constructed.

8. See M. Kaźmierska, S. Narel (1992).

9. See G. Juszczak et al. (1992) and G. Gehrig, W. Welfe (1993).

10. See A. Welfe (1989).

11. No satisfactory solutions were found to this problem by the authors of W5 model. W. Charemza and M. Gronicki (1985) used the disequilibria indicators based on planned rates of growth.

12. See W. Welfe (1985).
13. See A. Welfe (1991).
14. The wages of employees paid from the government budget and the pensions were indexed on the the wage increase in productive sector.
15. See W. Welfe (1985) and G. Juszczak et al. (1991).
16. See N. Łapińska (1990), A. Welfe (1990).
17. See also B. Czyżewski (1987).

REFERENCES

Charemza, W. and Gronicki, M. (1985). *Ekonometryczna Analiza Nie równowagi Gospodarczej Polski* (An econometric analysis of disequilibria for the Polish economy). Warszawa: PWN.

Cuthbertson, K., Hall, S. G., Taylor, M. P. (1992). *Applied Econometric Techniques.* New York: Harrester, Weatsheaf.

Czyżewski, B. (1987). Symulacyjny model zadłużenia wolno-dewizowego Polski (Simulation model of the Polish hardcurrency debt). *Wiadomości Statystyczne*, no 4, 1-6.

Gehrig, G., Welfe, W. (1993). *A System of Models and Forecasts for Economics in Transition - Germany and Poland.* Heidelberg: Physica Verlag.

Gomulka, S. (1992). Polish economic reform, 1990-91: Principles, policies and outcomes. *Cambridge Journal of Ecoonomics*, *16*, 355-372.

Juszczak, G., Kaźmierska, M. M., Łapińska-Sobczak, N., Welfe, W. (1992). Quarterly model of the Polish economy in transition (with special emphasis on financial flows), paper presented at ESEM'92, Brussels, Materiały IEiS UŁ no 17/92, Łódź *Economic Modelling.* (1993). *10*, 127-149.

Juszczak, G., Kaźmierska, M. M., Welfe, W. (1991). *Ekonometryczny model gospodarki Larodowej Polski W5* (Wersja symulacyjna). (The econometric model of the Polish economy W5 Simulation Version), Prace IEiS UŁ, 93, UŁ, Łódź Wydawnictwo.

Kaźmierska, M. M., Narel, S. (1992). Unobservable macroaggregates in the quarterly econometric model of the Polish economy. Paper prepared for EEA '92 Congress, Dublin, Materiały IEiS UŁ, no 78/92, Discussion Papers, Łódź.

Klein L. R. (1983), *Lectures in Econometrics.* Amsterdam: North Holland.

Klein, L. R., Sojo, E. (1989). Combinations of high and low frequency data in macroeconometric models. In L. R. Klein and J. Marques (Eds.), *Economics in Theory and Practice: An Ecclectic Approach.* Dordrecht: Kluwer.

Łapińska-Sobczak, N. (1990). Założenia i struktura ekonometrycznego modelu przepływów finansowych w gospodarce polskiej (Assumptions and structure of an econometric model of financial flows in the Polish economy). *Wiadomości Statystyczne*, no. 9, 5-9.

Portes, R. (1991). Introduction. In *European Economy. The Path of Reform in Central and Eastern Europe, Special Edition no. 2, Commission of the European Communities*, Brussels-Luxemburg.

Salenie, B. (1991). Wage and price adjustment in a multimarket disequilibrium model. *Journal of Applied Econometrics.* 6, 1-15.

Welfe, A. (1989). Savings and consumption in the centrally planned economy: A disequilibrium approach. Iin C. Davis and W. Charemza (Eds.), *Models of Disequilibrium and Shortage in Centrally Planned Economies.* London, New York, Chapman and Hall.

Welfe, A. (1990). State budget and inflation processes. Estimates for Poland. *Journal of Public Economies.* 43, 161-180.

Welfe, A. (1991). Modeling wages in centrally planned economies: The case of Poland. *Economics of Planning*, 24, 47-58.

Welfe, W. (1983). Models of the socialist economy. In L.R. Klein (Ed.) *Lectures in Econometrics.* Amsterdam: North Holland.

Welfe, W. (1985), Econometric macromodels of unbalanced growth. *Prace IEiS* U$_K$, 52, Wydawnictwo Uł, łódź .

Welfe. W. (1989). Macroeconomic disequilibrium models in Poland. In C. Davis and W. Charemza (Eds.), *Models of Disequilibrium and Shortage in Centrally Planned Economies,* London: Chapman and Hall.

Welfe, W. (1991). Multisectoral econometric models of the centrally planned economies and the disequilibria. *Economics of Planning. 24,* 203-226.

Welfe, W. (1992a). The Outlook for the Polish Economy 1992-1996. The recession and policy options.Paper presented at the Project LINK Meeting, New York, March 4-7, 1992, Materiały IEiS U$_K$, 13/92, Discussion Papers. ₭ódź .

Welfe, W. (1992b). Modelling the Polish case. In W. D. Eberwein (Ed.), *Transformation Processes in Eastern Europe, Perspectives from the Modelling Laboratory.* Frankfurt am Main: Peter Lang.

Welfe, W. (1992c). *Ekonometryczne modele gospodarki nafodowej Polski* (Econometric Models of the Polish Economy). Warzawa: PWE.

An Analysis of the Effects of Economic Policies on Taiwan's Economic Growth and Stability

Tzong-shian Yu

Regardless of the political basis or administrative approach a country adopts, its government must play some role in stimulating economic activities in order to reduce recession and curb inflation. However, there are many policy measures from which to choose. As to which are most effective, no one can give an all-encompassing answer, even though policymakers are usually confident about what they are going to undertake. To determine why ``X'' policy is preferred to ``Y'' policy, past experience usually provides an important lesson, but this is not enough because economic conditions and social backgrounds change. Economic problems are quite different from natural science problems and it is extremely expensive to use the entire society or the whole economy to experiment with a given economic policy. To overcome this problem, a model-building approach has been herewith employed to look at the period following World War Two. The purpose of this paper is an attempt to examine the effects of economic policies on economic growth and price stability through a model-building approach to see which policy measures are most effective in achieving the goals set by the government.

The paper consists of five parts: (1) The selection of a model-building approach. We selected a quarterly econometric model since it is easy to manage under data constraints and can efficiently and quickly find the answers we need. (2) The nature of Taiwan's economy and building of the model. Taiwan is a trade-oriented economy and the government has played an important role in the process of economic development. The structure of the model has to take this into account. (3) The design of the quarterly econometric model includes theoretical considerations and an evaluation of the performance of the model. To assess the performance of the model, we make use of a turning point test and AAPE (average absolute percentage error) to evaluate the performance of its structural equations. The AAPE and the Theil inequality coefficient evaluate the performance of its simultaneous equations. (4) A comparison of the effects of policy measures on economic growth and price fluctuations. We simply use a one-period effect comparison, and the findings are interesting and significant. (5) Finally, there are some concluding remarks.

I. Selection of a Model-building Approach

Model-building approaches are used as a tool to test the validity of economic theories, predicting economic activities, and simulating the impact of an economic policy. Many models are employed for forecasting, but they are not appropriate tools for evaluating the impact of an economic policy. An example is time series models which are primarily based on historical patterns and do not relate the endogenous variables to a large number of exogenous and policy variables. They are thus unable to assist policymakers in knowing the direction of the relationships and the magnitudes involved. Compared with time series models, an econometric model can not only verify economic theories and predict the value of economic magnitudes, but it can also enable the policymaker to judge whether it is necessary to implement policy measures in order to influence the relevant economic variables, and, if it is, which policy measure is more effective.[1]

During the last two decades, additional model-building approaches have been developed and elaborated upon, such as the computational general equilibrium model (CGE Model), the INFORUM approach, and the vector autoregressive (VAR) models. The CGE model attempts to convert the Walrasian general-equilibrium structure from an abstract representation of an economy into a realistic model of an actual economy. This can then be used to evaluate policy options by specifying production and demand parameters and incorporating data reflective of real economies.[2] The shortcoming of this approach is that it ignores the existence of residuals and the influence of other factors. Furthermore, it also has the difficulty of choosing appropriate elasticities and parameter values. The INFORUM model is an internationally linkable dynamic, interindustry model, and relies heavily on regression-based econometrics for behavioral equations. In addition, input-output coefficients are projected to change.[3] The shortcoming of this approach is the data problem. For instance, an input-output table is usually prepared every five years and a detailed classification of trade data may not be available. Particularly for short-term economic policy analysis, or where rapid growth of industrial sectors is involved, this approach may not be very appropriate. The vector autoregressive (VAR) model lies at the unstructured extreme of the models assessed, so each variable is expressed as a function of its own lagged values and the lagged values of all the other variables. Though the model is easy to estimate and has built up a good track record in unconditional forecasting, its use in causality testing and policy analysis is still controversial.[4] After review of these four main approaches, we still prefer to use the econometric model to conduct policy analysis.

II. The Nature of Taiwan's Economy and Model-building

To see the effects of economic policies on economic growth and price stability, any attempt at model-building must be based on the characteristics of the economy in question. Taiwan, an island endowed with limited natural resources and a dense population, has achieved high economic growth and a relatively stable price level over the last forty years. The leading sector in

Taiwan's economic development has been its foreign trade. During this period of time, foreign trade underwent very rapid growth.[5] It can be seen that without the expansion of its foreign trade, it would have been impossible for Taiwan's economy to have made such a remarkable achievement. Since Taiwan's economy cannot be self-sufficient, imports are urgently required. To finance its imports, Taiwan has had to strengthen its exports. In turn, most of these imports are used for industrial and agricultural production, and this then comprises exports and domestic consumption.

However, during the same period of time, the Chinese government on Taiwan has adopted many policy measures in order to stimulate economic growth and reduce price fluctuation. Particularly, in the early stages of Taiwan's economic development, the government played a leading role in many economic activities not only because the majority of the people in Taiwan were poor and not well-educated, but because there were no large firms to support infrastructure and public utilities. The government thus made use of monetary policy, fiscal policy and other administrative measures to influence the private sector in achieving the desired goals.

Based on the above consideration, we built a quarterly econometric model which reflects the characteristics of Taiwan's economy and can help us explore the simulated effects of economic policies on economic growth and price stability. Due to the limitation of data, some of the functions in the model have had to be tailored so as to fit available data.

III. Design of the Quarterly Econometric Model

In designing the quarterly econometric model, we first took into account the theoretical framework for this model which includes domestic product, domestic demand, foreign trade, money and public finance, the labor force and unemployment, wages and prices, and definitions and identities. Except for definitions and identities, we constructed structural equations for each item and estimated their relationships. Second, we evaluated the performance of this model in order to see whether it meets the requirements statistically and theoretically. In this regard, we have used a turning point test[6] and AAPE to evaluate the performance of the structural equations, and further, the AAPE and Theil inequality coefficient[7] to evaluate the performance of the simultaneous equations.

A. Theoretical Framework of the Model

Domestic Product

To explain the production function, we do not follow the traditional approach emphasizing the effective combination of factors. Rather, we pay special attention to the importance of final demand: GDP = GDPAG + GDPIN + GDPSE, where gross domestic product (GDP) is composed of gross agricultural product (GDPAG), gross industrial product (GDPIN), and gross service product (GDPSE). GDPAG is assumed to be a function of private consumption expenditure on food and GDPAG lagged for one period. GDPIN is a function of domestic demand and exports, indicating industrial production not only for domestic demand but also for exports. Last, GDPSE is a residual of GDP. In estimating GDP, we follow the Keynesian approach: GDP = C + CG + I + J + X - M. GDP consists of private consumption expenditure (C), government consumption expenditure (CG), gross fixed capital formation (I), increase in stock (J), exports of goods and services (X), and imports of goods and services (M). Though the factors of production were not constraints before 1980, since then they have been subject to some constraints.[8] To display the possible influence of the constraints, we constructed a potential production function in which potential capital and potential labor are utilized as explanatory variables. In addition, technology, the exchange rate of the Japanese yen to the U.S. dollar, and oil prices are used as explanatory variables, of which the last two variables are used to stand for foreign imports on the production function because Taiwan's economy is heavily dependent on its exports and imports.

Domestic Demand

Domestic demand includes private consumption expenditure, government consumption expenditure, gross fixed capital formation, and increase in stock. Gross fixed capital formation can be divided into components of gross private, gross public, and gross government fixed capital formation. The last two are considered to be exogenous, while gross private fixed capital formation is mainly explained by gross industrial product, real market interest rate,[9] and gross private fixed capital formation lagged one period. Increase in stock is mainly determined by consumption expenditure (private and government) and gross fixed capital formation, the real interest rate, increase in stock lagged one period, and stock of inventory lagged one period. Private consumption expenditure is separated into two parts, namely, consumption expenditure on food, and consumption expenditure on nonfood items. The former is mainly explained by disposable personal income and private consumption expenditure on food lagged one period, while the latter is explained by liquid assets, average tariffs, and private consumption expenditure on nonfood items lagged one period. As for the stock of inventory, it is determined by domestic demand and the rate of change in the price deflator of GDP.

Foreign Trade

As mentioned above, foreign trade has been the leading sector in Taiwan's economic development, and it not only closely relates to domestic activities but also links to foreign markets. For the purpose of explanation, we divide the export of goods and services into the export of goods and the export of services. Similarly, we also divide the import of goods and services into the import of goods and the import of services. Since the import and export of goods in the customs data are different from those in the national income account, it is necessary to build a relationship between them, i.e., the import (or export) of goods in the national income account is a function of the import (or export) of goods in the customs data.

Here, the import of goods through customs is mainly explained by gross industrial product, price deflator of imports, and the import of goods through customs lagged one period. The import of services is explained by GNP and the import of services lagged one period. When dealing with exports, we explain them by importing countries which include the United States, Japan, Hong Kong and mainland China, and the rest of the world. Since trade with mainland China is mainly through Hong Kong and no direct trade is so far permitted, we have to combine these areas together. In explaining the export of goods to each country (or area), their GNP, export prices and lag variables are the main explanatory variables. However, when dealing with exports to the U.S., we add the wholesale price index as an explanatory variable. When dealing with exports to Japan, we add the exchange rate as an explanatory variable. Finally, when dealing with exports to Hong Kong and mainland China, we add two more variables as explanatory variables, i.e., total imports of goods from mainland China and the exchange rate. The export of goods to the rest of the world are explained by the world trade volume index. The export of services is explained by the GNP of the U.S. and Japan and the export of services lagged one period.

The adoption of a lag variable expresses the continuation of transactions in a short period of time, such as a quarter, and dummy variables are used for seasonal adjustment.

Money and Public Finance

In order to see the effects of economic policies such as monetary and fiscal policy measures, we take into consideration the sector of money and public finance. In this sector, the demand for money and the supply of money are equal, but these should be explained. The demand for money is determined by GNP and the market interest rate, while the supply of money (M_{1B}) is equal to the amount of currency in circulation and demand deposits. Usually, currency is issued by the Central Bank and should be completely controlled by the Central Bank, but currency in circulation cannot be totally controlled by the Central Bank. Accordingly, the amount of currency in circulation is assumed to be explained by foreign exchange reserves of all banks and gross domestic fixed capital formation and demand deposits which in turn is explained by the market interest rate, savings, and demand deposits lagged one period.

In order to estimate liquid assets, we need to add YDD and quasi-money together; quasi-money is explained by GNP and the market interest rate which in turn is mainly explained by savings and the rediscount rate of the Central Bank lagged one period. Foreign exchange reserves are primarily explained by the balance of trade and foreign exchange reserves lagged one period. The foreign exchange rate, i.e., the amount of N.T. dollars for which one U.S. dollar can be exchanged is explained by the amount of the previous year's balance of trade and the foreign exchange rate lagged one period.

Tax revenue, either through direct taxes or indirect taxes, is closely related to GNP. In this regard, we can explain net indirect taxes as a function of GNP and net indirect taxes lagged one period. Net indirect tax is the difference between indirect tax and subsidies. Total tax and monopoly revenue of public enterprises are explained by GNP and GNP lagged one period. Income from government property and enterprises is also a source of government revenue and is explained by domestic demand lagged one period.

Labor Force and Unemployment

For this paper, labor force is explained using population as the only variable, while the unemployment rate is assumed to be related to gross industrial product and the unemployment rate lagged one period. With regard to the unemployment rate, only data in the industrial sector is reliable.

Wages and Prices

In estimating the wage rate, we consider only the average wage index in manufacturing because of the availability of data and assume the wage rate is a function of labor productivity, the consumer price index lagged one period, and the difference between labor force and the level of employment, which implies the level of unemployment.

In the determination of prices, we assume the wholesale price index is a function of import prices, the ratio of the wage rate to potential labor productivity which indicates the net effect of the wage cost, and the wholesale price index lagged one period. The consumer price index is related to the wholesale price index and the ratio of money supply to gross domestic product lagged one period, which indicates the net effect of the money supply lagged one period and the wholesale price index lagged one period. With regard to import prices, we have two terms: one is the price deflator of imports; and the other is the unit value index of imports. The former is explained by the latter, while the latter is explained by the weighted average price index of exports of the U.S. and Japan, oil prices, and the foreign exchange rate. Similarly, export prices also have two different definitions: the price deflator of exports is explained by import prices since Taiwan's exports have been greatly influenced by its imports of raw materials and machinery equipment; and the unit value of exports is explained by the wholesale price index and foreign exchange rate. In this model, either export prices or import prices are substantially affected by the change in the foreign exchange rate. As for the price deflators of other factors of GDP, either they are related to the consumer price index or to the wholesale price index.

Others

Only the fixed capital consumption allowance is left here for explanation. This variable is assumed to depend on GDP which reflects economic activities.

B. Evaluation of the Performance of the Model

Though good performance of a model cannot absolutely guarantee high accuracy in forecasts, a model which performs well is likely to be much more accurate than one which has poor performance. In order to evaluate the performance of the model, usually two approaches are taken: one is to test the performance of the structural equations by means of the turning point test and the average absolute percentage error (AAPE), and the other is to test the performance of the simultaneous equations by means of comparison of the AAPE and the inequality coefficient as mentioned above.

First, we evaluate the performance of the structural equations. From Table 1.A, it can be found that among the 31 important variables of the model, the structural equations for 15 variables reach 50 percent of their turning points. Those for gross agricultural products, private consumption expenditure on nonfood items, exports of goods, exports of goods to Hong Kong and mainland China, imports of goods, gross private fixed capital formation, increases in stock, and the demand for money have the best performance since the percentage of their turning points hit is higher than 70 percent. The structural equations for exchange rate, foreign exchange reserves of all banks, market interest rate, potential GDP, exports of services, exports of goods to the rest of the world, and fixed capital consumption allowance each have a poor performance since the percentage of their turning points hit is lower than 30 percent.

It should be noted that when we compare the AAPEs, most of these give us different pictures of their performance. The structural equations which have a smaller AAPE are those for the labor force, wholesale price index, exchange rate, exports of goods, consumer price index, private consumption expenditure on food, market interest rate, gross industrial product and imports of goods with AAPEs of less than 2 percent, indicating that these have the best performance. The structural equations for the increase in stock, gross private fixed capital formation, total tax and monopoly revenue of public enterprises, unemployment rate, net indirect tax, imports of services, net amount of currency in circulation, exports of goods to Hong Kong and mainland China, and exports of services have larger AAPEs, indicating their performance is poor.

One may wonder why the two measures provide such different and inconsistent results, and which one is more reasonable. From the standpoint of forecasting, while both determining the direction of change and producing smaller estimate errors are of concern, the direction of change is more important to policymaking. The turning point test provides information as to how many turning points have been reached, while the AAPE only tells us the magnitude of the error of estimate. However, it should be mentioned that the structural equations for the financial sector are poor in terms of the turning points reached. For instance, the structural equations for the market interest rate, exchange rate, and foreign exchange reserves of all banks also have poor performance. One of the reasons for this is that all three structural equations have no apparent increasing or decreasing trend during the period of observation. Even so, their AAPEs are rather small.

Next, we evaluate the performance of the simultaneous equations by the two measures mentioned above. Using each measure, there are two conditions: a static condition, and a dynamic condition. When utilizing the AAPE, the estimates of the simultaneous equations have a larger AAPE in the dynamic condition than in the static condition. This is because we use actual values for endogenous variables in the static condition, while we only use estimated values for endogenous variables which are determined by the model in the dynamic condition. Roughly speaking, all 32 equations (except for the one regarding an increase in stock which has a high AAPE) performed very well under both conditions. In the estimation of structural equations, the increase in stock has the highest AAPE. Both the structural equation and the simultaneous equation for the labor force have the smallest AAPEs. When utilizing the Theil inequality coefficient, we also found that the simultaneous equation for the increase in stock has the largest coefficient under both conditions indicating the poorest performance, which is consistent with the measure of AAPE. Similarly, the Theil inequality coefficient under static conditions is smaller than that under dynamic conditions for the same reason mentioned above. (See Table 1.B.)

Obviously, the results of the two measures of the AAPE and the Theil inequality coefficient are not precisely consistent with each other. However, most simultaneous equations which have small (or large) AAPEs still have small (or large) Theil inequality coefficients. If we compare the performance of structural equations for a variable with the performance of its simultaneous equation by the measure of an AAPE, we do not reach the conclusion that the structural equation necessarily has a smaller AAPE than the simultaneous equation.

Table 1A: Evaluation of Model Perfomances
Evaluation of the Estimates of Structural Equations

ITEM	Turning Point Test			Rank	AAPE (%) (1980.1- 1990.4)	RANK
	# of Actual Turning Pts.	# of Turning Pts. Hit	% of Turning Pts. Hit			
	(1)	(2)	(3)	(4)	(5)	(6)
Domestic Product						
1. Potential GDP	59	7	11.86	27	4.73	20
2. Gross industrial product	43	22	51.16	16	1.62	8
3. Gross agricultural product	93	93	100.00	1	3.37	12
Domestic Demand						
4. Prvt. consumption exp. on food	56	34	60.71	12	1.52	6
5. Prvt. consumption exp. on nonfood	96	87	90.63	2	4.08	16
6. Gross prvt. fixed capital formation		40	76.92	6	11.92	31
7. Increase in stock	69	53	76.81	7	126.77	32
Foreign Trade						
8. Imports of goods	60	47	78.33	5	1.89	9
9. Imports of goods through customs	63	27	42.86	19	5.32	23
10.Imports of services	51	18	35.29	21	7.30	27
11.Exports of goods	51	45	88.24	3	0.97	4
12.Exports of goods to the U.S.	46	25	53.90	15	5.26	22
13.Exports of goods to Japan	57	27	47.37	17	4.97	21
14.Exports of goods to Hong Kong & PRC	76	61	80.26	4	5.67	25
15.Exports of goods to the rest of the world	47	13	27.66	25	4.33	19
16.Exports of services	62	9	14.52	26	5.45	24
Money and Public Finance						
17.Demand for money (MIB)	70	51	72.86	8	4.11	12
18.Demand for quasi- money	18	11	61.11	11	3.21	11
19.Demand deposits	48	28	58.33	14	3.59	14
20.Net amount of currency in circulation	42	13	30.95	23	5.83	26
21.Market interest rate	27	3	11.11	28	1.59	7
22.Exchange rate	16	1	6.25	30	0.96	3
23.Foreign exchange reserves of all banks	30	3	10.00	29	3.46	13
24.Net indirect tax	87	59	67.82	9	9.81	28
25.Tot. tax & monopoly rev. of pub. enterp	55	34	61.82	10	11.24	30
Labor Force and Unemployment						
26.Labor force	62	22	35.48	20	0.66	1
27.Unemployment rate	57	19	33.33	22	9.96	29
Wages and Prices						
28.Wholesale price index	30	13	43.33	18	0.78	2
29.Consumer price index	36	12	33.33	22	1.02	5
30.Average wage index in manufacturing	45	27	60.00	13	4.25	18
31.Fixed capital consumption allowance	50	14	28.00	24	1.91	10

So far as the performance of the model is concerned, it is sound and can be used as a tool to measure the effect of policy changes on economic activities. Needless to say, some individual equations which have poor performance should be further refined.

Table 1B. Evaluation of Model Performance
Evaluation of the Estimates of the Simultaneous Equations
Test Period: 1988.1-1990.4

ITEM	AAPE(%)				Theil inequality coefficient			
	Static		Dynamic		Static		Dynamic	
	value	rank	value	rank	value	rank	value	rank
Domestic Product								
1.Potential GDP	1.23	3	1.40	3	0.60	3	0.66	8
2. Gross industrial product	3.36	12	4.67	8	0.94	11	0.82	12
3. Gross agricultural product	3.88	14	4.27	7	0.20	1	0.18	1
Domestic Demand								
4. Priv. consumption expenditure on food	1.27	4	1.13	2	1.01	14	0.63	6
5. Priv. consumption expenditure on nonfood	5.66	21	9.33	20	0.63	4	0.58	5
6. Gross private fixed capital formation	16.25	31	16.67	29	0.65	5	0.65	7
7. Increase in stock	145.70	32	145.37	31	999.99	28	999.99	25
Foreign Trade								
8. Imports of goods	3.92	15	6.90	15	1.27	22	0.96	18
9. Imports of goods through customs	3.63	13	6.93	17	1.36	24	0.99	19
10.Imports of services	6.50	25	6.81	13	0.97	13	0.68	9
11.Exports of goods	2.76	8	2.99	5	0.79	7	0.54	3
12.Exports of goods to the U.S.	5.93	15	9.66	22	0.94	11	0.68	9
13.Exports of goods to Japan	5.31	20	7.48	18	1.25	21	0.80	11
14.Exports of goods to Hong Kong & PRC	5.98	23	11.05	24	0.58	2	0.49	2
15.Exports of goods to the rest of the world	2.93	9	6.86	14	0.95	12	0.56	4
16.Exports of services	5.09	19	5.35	10	1.07	16	0.90	16
Money and Public Finance								
17.Demand for money (M_{1B})	6.22	24	9.36	21	1.30	23	0.95	17
18.Demand for quasi-money	4.03	16	11.94	25	1.49	25	1.02	20
19.Demand deposits	4.71	18	10.52	23	1.05	15	0.87	14
20.Net amount of currency in circulation	4.12	17	5.90	11	1.18	19	1.34	23
21.Market interest rate	2.95	10	5.00	9	1.67	27	0.89	15
22.Exchange rate ($NT/$US)	1.37	5	2.21	4	1.01	14	0.96	18
23.Foreign exchange reserves of all banks	2.68	7	4.25	6	1.15	18	1.05	21
24.Net indirect tax	6.61	27	7.62	19	0.76	6	0.66	8
25.Tot. tax and monopoly rev. of pub. enter.	10.34	29	14.91	27	0.80	8	0.72	10
Labor Force and Unemployment								
26.Labor force	0.55	1	0.62	1	0.85	9	0.65	7
27.Unemployment rate	11.28	30	21.11	30	1.08	17	0.83	13
Wages and Prices								
28.Wholesale price index	1.08	2	6.23	12	1.54	26	1.49	24
29.Consumer price index	1.51	6	9.36	21	1.19	20	1.24	22
30.Average wage index in manufacturing	6.83	28	6.92	16	0.86	10	0.65	7
31.Fixed capital consumption allowance	3.35	11	13.23	26	1.05	15	0.96	18

IV. Comparison of the Effects of Policy Measures

In the realm of economics, experimentation in policy changes can have large impacts on society. Therefore, the best way to test the effect of a policy is through a model simulation designed for the purpose of testing the effect of a policy. According to the evaluation of the model performance, we found the above model can be used for this purpose. In order to compare the effects of policy measures on economic growth and price stability, we take the first quarter of 1993 as the simulation period.

The Effect on Economic Growth

First of all, we observe the effect of fiscal policy on economic growth. Table 2 provides a clear picture. If government consumption expenditure increases by NT$10 billion, gross national product will increase by NT$12.711 billion; gross agricultural product will increase by NT$0.026 billion; gross industrial product will increase by NT$2.306 billion; and gross service product will increase by NT$9.777 billion. If government fixed capital formation and fixed capital formation of public enterprises each increase by NT$10 billion, they will have a greater effect on economic growth than government consumption expenditure. The sequence of their effects on economic growth is as follows:

The effect of government consumption expenditure	<	The effect of government fixed capital formation	<	The effect of fixed capital formation of public enterprises

In the initial period, the change in government expenditures on foreign trade has different effects, namely, it has no effect on exports in the same period, but it has a positive effect on imports because the government needs goods for consumption, equipment for infrastructure, and materials and machinery for public enterprise production.[10] As a result, this causes a trade deficit.

If the tax level decreases by NT$10 billion, it has a positive effect on economic growth, which means that gross national product increases by NT$0.736 billion, gross agricultural product increases by NT$0.020 billion, gross industrial product increases by NT$0.142 billion, and gross services product increases by NT$0.590 billion. The effect of a change in taxes is less than that of a change in government expenditures.

The effect of monetary policy, such as an increase in the rediscount rate, a fall in the exchange rate (i.e., the New Taiwan dollar appreciates against the U.S. dollar), or the rise of the deposit reserve ratio are unfavorable for economic growth. If we compare the effects of a change in the rediscount rate with the effect of a change in the deposit reserve ratio on economic growth, we find that the former has a larger effect than the latter. However, the effect of a change in the rediscount rate on gross industrial production is less than the effect of a change in the deposit reserve ratio on gross industrial production. If the N.T. dollar either appreciates or depreciates against the U.S. dollar, this has a substantial effect on economic growth.

In reality, exports cannot be controlled. If exports increase by NT$10 billion, this has a larger effect on economic growth than government expenditure. Since Taiwan does not produce oil but depends heavily on its import, an increase in oil price also has a significant impact on economic growth.

Table 2A: Multiplier Effects of Policy Changes

On the following variables (NT$) Million; at 1986 prices)	The Multiplier Effects of			
	Increase in gov't consumption expenditure by NT$10 billion	Increase in gov't fixed capital form. by NT$10 billion	Increase in fixed capital formation of public enterprises by NT$10 billion	Increase in rediscount rate by 10%
1. Gross national product	12,711	12,790	13,512	-881
2. Gross agricultural product	26	26	27	-1
3. Gross industrial product	2,306	2,321	2,485	-167
4. Gross service product	9,777	9,840	10,423	-73
5. Private consumption expenditure on food	185	186	192	-9
6. Private consumption expenditure on nonfood	1,944	1,956	2,065	-128
7. Gross private fixed capital formation	538	539	553	-34.75
8. Increase in stock	167	163	128	-756
9. Imports of goods	359	361	382	-25
10.Imports of services	387	390	412	-27
11.Exports of goods	0	0	0	0
12.Exports of services	0	0	0	0
13.Balance of trade (US$M)	-27	-27	-28	-1.37
14.Demand for money (M_{1B})	14,510	14,596	15,394	-993
15.Demand for quasi-money	32,614	32,833	34,734	-2.167
16.Exchange rate (NT$/US$)	0.002	0	0	0
17.Total taxes and monopoly revenue of public enterprises	814	814	816	-36
18.Labor force (1,000 persons)	0	0	0	0
19.Unemployment rate(%)	-0.006	-0.006	-0.006	0
20.Wholesale price index (%)	0.017	0.017	0.018	-0.001
21.Consumer price index (%)	0.051	0.051	0.049	0.003
22.Average wage in manufacturing (%)	2.505	2.521	2.663	-0.171
23.Fixed capital consumption allowance	-58	-141	-3	-61

Table 2.B. Multiplier Effects of Policy Changes

The Multiplier Effects of

On the following variables (NT$ Million; at 1986 prices)	Fall of exchange rate by 10%(NT$ appreciates)	Rise of deposit reserve ratio by 10%	Decrease in tax by NT$10 B	Increase in exports of goods by NT$10 B	Increase in oil price by 10%
1. Gross national product	-4,449	-73	736	12,903	-1221
2. Gross agricultural product	0.309	-0.11	20	25	-1
3. Gross industrial product	-3,201	-13.62	142	5,216	-356
4. Gross service product	-2,262	-58	590	7,059	-992
5. Private consumption expenditure on food	-2	-0.75	145	183	-8.12
6. Private consumption expenditure on nonfood	-608	-10.75	598	1,959	-243.5
7. Gross private fixed capital formation	-459	-2.87	38	1,126	-116
8. Increase in stock	492	-62.90	8	118	-160
9. Imports of goods	203	-2.03	20	794	-217
10.Imports of services	-135	-2.25	22	393	-37
11.Exports of goods	-7,993	0	0	1,000	-245
12.Exports of services	4.3	0	0	0	0.14
13.Balance of trade (US$M)	92	0.12	-1	261	-118
14.Demand for money (M1B)	-4,679	-82	1,136	-14,677	-1,464
15.Demand for quasi-money	-11,681	-180	-2,116	33,091	-3,668
16.Exchange rate (NT$/US$)	-2.95	0	0	0	0
17.Total taxes and monopoly revenue of public enterprises	244	-2.93	-997	741	-145
18.Labor force (1,000 persons)	0	0	0	0	0
19.Unemployment rate(%)	0.008	0	0	-.0130	0.001
20.Wholesale price index (%)	-1.031	0	0.001	0.017	0.263
21.Consumer price index (%)	-0.309	0	-0.005	0.050	0.142
22.Average wage in manufacturing (%)	-0.878	-0.014	0.145	-2.535	0.232
23.Fixed capital consumption allowance	-318	-5.13	51	903	-84

The Effects on Price Fluctuations

Price fluctuations are always of great concern and a situation of hyperinflation not only worsens income distribution but also deteriorates the investment climate. When the government undertakes various policy measures, the consumer price index or wholesale price index must be taken into consideration. The information that the model has provided is that an increase in government consumption expenditure, an increase in government fixed capital formation and an increase in fixed capital formation of public enterprises have rather similar effects on wholesale prices, on consumer prices and on the wage index, respectively. However, an increase in government consumption expenditure has a different effect on wholesale prices, consumer prices and the wage index, as does an increase in government fixed capital formation and an increase in fixed capital formation of public enterprises. Their increases have a greater effect on consumer prices than on wholesale prices, but they have the greatest effect on the wage index. A change in taxes has a smaller effect on price fluctuations than a change in government expenditures.

The effect of monetary policies is that an increase in the rediscount rate has a positive effect on consumer prices but a negative effect on wholesale prices and on the wage index. A rise of the deposit reserve ratio has no effect on wholesale prices and consumer prices but has a negative effect on the wage index. A fall in the exchange rate has a substantial negative effect on prices and wages, but it has a larger negative effect on wholesale prices than on consumer prices because wholesale prices are more closely related to import prices and the change in exchange rate has a direct and large impact on import prices.

V. Concluding Remarks

When building an econometric model for an intended purpose, theoretical considerations and data limitations must both be taken into account. In fact, so far no model-building approach is completely free from data limitations. In order to make the model more suitable for the data, usually we have to tailor the model to some extent. In Taiwan, as in many developing countries, quarterly data are not available for many variables; when they are available, they are often not reliable; and when they are reliable, they may be not suitable. Accordingly, the structure of the model has to be simplified in its scale.

In this paper, we tested only the effect of one policy in one short period and let all other policies remain unchanged. Actually, the government sometimes undertakes more than one policy in the same period of time. Furthermore, any policy is likely to have a successive effect and not a once-and-for-all effect. What we have done in this paper is to utilize only one period. Apparently, this is not enough for us to see the cumulative effect of the particular policy we observe.

From an evaluation of the model, it can be seen that the performance of the model is fairly good. Thus, the model can be used as a tool to test the effect of a policy. Also, by comparing the effects of various policies, we found that the effect of a change in government expenditure on economic growth is much larger than that from a change in taxes; the effect of a change in the rediscount rate on economic growth is larger than that from a change in the deposit reserve ratio; and the effect of a change in the rediscount rate on price fluctuations is larger than that from a change in the deposit reserve ratio.

It should be noted that when a country's economy becomes more liberalized, its government has several policy measures to choose from. As far as monetary policy is concerned, if the central bank changes the rediscount rate by some percentage in order to influence the market interest rate, but private banks do not follow, such a policy would lose its effectiveness. Comparatively speaking, it is easier to implement monetary policies than to adopt fiscal policies, and it is also easier to reduce taxes than to increase taxes in a very short period. When the government wants to increase its expenditure, how to increase its revenue must be taken into consideration.

Tzong-Shian Yu, President, Chung-Hua Institute for Economic Research, 75 Chang Hsing Street, Taipei 106, Taiwan, Republic of China.

ACKNOWLEDGEMENTS

I am very grateful to Chin-sheun Ho for his assistance in computation and valuable suggestions.

NOTES

1. See Koutsoyiannis (1977).
2. See Shoven and Whalley (1984).
3. See Almon (1991) and McCarthy (1991).
4. See Sims (1988) and Wallis (1988).
5. Among the components of GNP, the average growth rate of exports (1952-90) was 15.01 percent. This was the highest component.
6. For a detailed analysis, see Theil (1955).
7. A systematic measure of the accuracy of the forecasts obtained from an econometric model was suggested by H. Theil. The value of the inequality coefficient is assumed to lie between 0 and ∞, and the smaller the value of the inequality coefficient, the better is the forecasting performance of the model. See Theil (1966).
8. Labor is no longer an abundant factor, and the shortage of labor has been found in several sectors.
9. In Taiwan, there are the official interest rate and the market interest rate. The former is controlled by the Central Bank of China and the latter is determined by market forces. Actually, the official interest rate has had a dominant influence on the market interest rate.
10. Another reason is that we only used one quarter as the experimental period. The effects of many policies are lagged and do not appear in the same period in which the policies are taken.

REFERENCES

Almon, Clopper. (1991). The INFORUM approach to interindustry modeling. *Economic Systems Research. 3*(1).

Bryant, Ralph C., Henderson, Dale W., Holtham, Gerald, Hooper, Peter, and Symansky, Steven A. (Eds.). (1988). *Empirical Macroeconomics for Interdependent Economics.* The Brookings Institution.

Klein, Lawrence R. (1979). Use of econometric models in the policy process. In P. Ormerod (Ed.). *Economic Modelling.* (pp. 309-29). London: Heinemann.

Koutsoyiannis, A. (1977). *Theory of Econometrics.* 2nd ed., Macmillan Press, pp.8-9.

Lo, Joan C., Li, Hung-Yi and Yu, Tzong-shian. (1992). A trade-oriented macroeconometric model of Taiwan. *Academia Economic Papers. 20*(2), Part II, September 1992, 625-677.

McCarthy, Margaret Buckler. (1991). LIFT: INFORUM's model of the U.S. economy. *Economic*

Systems Research. 3(1).

Prachowny, Martin F. J. (1984). *Macroeconomic Analysis for Small Open Economies.* Oxford: Clarendon Press.

Shoven, John B. and Whalley, John. (1984). Applied general-equilibrium models of taxation and international trade: An introduction and survey. *Journal of Economic Literature.* *XXII*(September), 1007-1051.

Sims, Christopher A. (1988). Identifying policy effects. In Ralph C. Bryant, et al. (Eds.) *Empirical Macroeconomics for Interdependent Economics.*

Tarkka, Juha and Willman, Alpo. (Eds.). (1985). The BOF & Quarterly Model of the Finnish Economy. Bank of Finland.

Theil, Henri. (1955). Who Forecasts Best. *International Economic Papers*, 1955, 5, 194-199.
_____. (1966). *Applied Economic Forecasting.* North-Holland, pp.26-36.

Toida, Mitsu and Liang, Youcai (Eds.). (1990). *Econometric Link Model of China and Japan.* Institute of Developing Economies, 1990.

Wallis, Kenneth F. (1988). Empirical models and macroeconomic policy analysis. In Ralph C. Bryant, et al. (Eds.). *Empirical Macroeconomics for Interdepentent Economics.* (pp.225-237).

Yu, Tzong-shian and Ho, Chin-Sheun. (Forthcoming). The effect of changes in the foreign exchange rate on Taiwan's foreign trade. To appear in a book in celebration of Professor S. C. Tsiang's 70th birthday.

APPENDIX

This is a nonlinear model including 113 equations, of which there are 46 behavioral equations and 67 definitions and identities. The estimate of the structural equations is made by means of the OLS method and the solution to the simultaneous equations for the estimation is obtained through an approximation of the Gauss-Seidel method. The figures in parentheses are the t-value, where R^2 stands for the corrected coefficient of determination and RHO is the serial correlation coefficient.

DOMESTIC PRODUCT

1. LNQF/KF = 11.6651 + 1.0708 LNNF/KF + 0.0126 LNETIME - 0.0969 LNEJA - 0.0047 LNPO
    (12.31)   (15.41)   (22.86)   (-3.04)   (-0.62)

    $\bar{R}^2$: 0.990   SEE: 0.36474E-01   D-W: 1.90   F(4,99): 2,454.005   Period of Fit: 1966.1 - 1991.4

2. GDPIN = 14,463.6094 + 0.1794 (C+CG+I+J) + 0.4590X+ 3,419.0068 Q1 - 1,643.4727 Q2 + 6,346.4102 Q3
              (0.95)   (6.51)   (13.76)   (2.67)   (-1.27)   (5.46)

    $\bar{R}^2$: 0.998   SEE: 6,318.6   D-W: 2.32   RHO(1): 0.948   F(5,97): 9,044.562   Period of fit: 1966.1 - 1991.4

3. GDPAG = 34,183.9297 + 0.1410 CF - 0.1328 GDPAG$_1$- 7,931.4883 Q1 - 4,361.0664 Q2 - 11,465.6211 Q3
            (17.59)   (10.32)   (-1.38)   (-6.07)   (-6.49)   (-12.32)

    $\bar{R}^2$: 0.899   SEE: 2,106.7   D-W: 1.74   F(5,98): 185.190   Period of fit: 1966.1 - 1991.4

DOMESTIC DEMAND

4. CF = 2,642.6760 + 0.0132 YDD + 0.9275 CF$_{-1}$ + 2,507.1531 Q1 -3,531.8115 Q2 + 1,045.5571 Q3
           (2.11)   (2.07)   (22.59)   (4.66)   (-5.81)   (2.00)

    $\bar{R}^2$: 0.997   SEE: 1,882.9   D-W: 2.43   F(5,98): 6,355.609   Period of fit: 1966.1 - 1991.4

5.   CO $=$ 90,790.7500 + 0.0452 LA + 0.2517 CO_{-1} - 5,183.1523 TAF+ 35,455.2227 Q1 - 4,413.2695 Q2
             (5.85)   (7.36)   (2.59)   (-5.51)   (8.72)   (-0.98)
       + 18,992.9570 Q3
        (4.48)

    $\bar{R}^2$: 0.982   SEE: 14,486   D-W: 2.53   F(6,97): 911.771   Period of fit: 1966.1 - 1991.4

6.   IBF $=$ 19,866.0391 + 0.2168 GDPIN - 264.3958 IRRB + 0.6662 IBF_{-1}- 28,359.9687 Q1 - 19,455.6680 Q2
            (4.65)   (9.32)   (1.50)   (1.87)   (-8.49)   (-5.20)
      - 6,245.7181 Q3
       (-1.73)

    $\bar{R}^2$: 0.903   SEE: 11,695   D-W: 1.71   F(6,96): 159.073   Period of fit: 1966.2 - 1991.4

7.   J $=$ 36,522.5000 + 0.0108 (TD-J) + 0.0988 J_{-1} - 0.0197 KJ_{-1} -718.9329 (IR-PGDP*) - 8,455.5859 Q1
          (1.71)   (1.33)   (0.98)   (-1.25)   (-3.82)   (-2.72)
      + 2,243.5435 Q2 -4,981.0039 Q3
       (0.73)   (-1.59)

    $\bar{R}^2$: 0.218   SEE: 11,041   D-W: 1.97   F(7,96): 5.106   Period of fit: 1966.1 - 1991.4

FOREIGN TRADE

8.   MG $=$ 3034,5642 + 0.9860 TVM
       (2.74)   (182.83)

   $\bar{R}^2$: 0.997   SEE: 6,909.1   D-W: 2.03   F(1,102): 33,428.098   Period of fit: 1966.1 - 1991.4

9.   TVM $=$ 7,167.7695 + 0.1516 GDPIN + 0.9036 TVM_{-1} - 214.2607 PM - 8,259.8984 Q1+ 7,332.8984 Q2
           (2.08)   (4.04)   (26.44)   (-3.74)   (-2.12)   (2.32)
      - 7,876.0430 Q3
       (-2.03)

    $\bar{R}^2$: 0.991   SEE: 11,769   D-W: 2.09   RHO(1): -0.271   F(6,95): 1,933.869   Period of fit: 1966.2 - 1991.4

10.   MS $= -$ 7,382.2227 + 0.0305 GNP + 0.6538 MS_{-1} + 1112.3967 Q1 + 4946.4180 Q2 + 5202.2148 Q3
          (-6.30)   (4.97)   (9.01)   (1.18)   (5.06)   (5.56)

    $\bar{R}^2$:   0.982   SEE: 3335.9   D-W: 1.94   F(5,98): 1,105.436   Period of fit: 1966.1 - 1991.4

11.   XG $= -$ 647.2305 + 1.0314 TVX
       (-1.04) (444.08)

    $\bar{R}^2$:   0.999   SEE: 4052.4   D-W: 1.62   F(1,102): 197,231.437   Period of fit: 1966.1 - 1991.4

12.   TVXUS $= -$ 52,878.1523 + 781.5330 GNPUS - 21,762.5234 PXROC\$ + 489.4080 WPIUS + 0.8567 $TVXUS_{-1}$
             (-3.13)   (2.61)   (-3.46)   (3.89)   (17.87)
      - 3,946.1938 Q1 + 14,840.5586 Q2 + 8,609.8320 Q3
       (-1.80)   (6.62)   (3.97)

    $\bar{R}^2$:   0.986   SEE: 7,784.8   D-W: 1.92   F(7,95): 1019.527   Period of fit: 1966.2 - 1991.4

13.   TVXJA $=$ 11,999.0820 + 94.1967 GNPJA - 2,952.6589 PXROC\$ - 36.5692 EJA + 0.8851 $TVXJA_{-1}$
           (2.25)   (1.82)   (-2.35)   (-2.99)   (22.76)
      - 2811.9661 Q1 + 2733.6467 Q2 - 1,590.1829 Q3
       (-3.52)   (3.45)   (-2.01)

    $\bar{R}^2$:   0.983 SEE: 2845.0 D-W: 2.66 RHO(1): 0.830 F(7,95): 840.977 Period of fit: 1966.2 - 1991.4

14. TVXHKM = 7,382.9297 + 1.9097 GNPHK - 2,522.9453 PXROC$ + 0.5050 TVXHKM$_{-1}$
            (2.46)   (6.77)   (-4.05)   (7.05)
    + 133.7929 TMMC$$ - 688.9944 EHK - 1,437.3132 Q1 + 1,442.8296 Q2 - 973.4329 Q3
     (0.66)   (-1.32)   (-1.93)   (2.11)   (-1.42)

    $\bar{R}^2$: 0.988   SEE: 2399.0   D-W: 1.60   F(8,94): 1070.556   Period of fit: 1966.2 - 1991.4

15. TVXOT = - 8,122.5312 + 82.3420 TW + 0.9861 TVXOT$_{-1}$- 2,370.6951 Q1 + 8,684.8320 Q2 + 6,512.5352 Q3
            (-1.81)   (1.45)   (43.01)   (-1.24)   (4.65)   (3.43)

    $\bar{R}^2$: 0.990   SEE: 6,715.9   D-W: 2.10   F(5,97): 2,057.001   Period of fit: 1966.2 - 1991.4

16. XS = - 7,608.5312 + 170.5690 GNPIUSJA + 0.6753 XS$_{-1}$ - 1,232.1162 Q1 - 414.7576 Q2 - 654.6948 Q3
      (-3.90)   (4.60)   (9.62)   (-2.64)   (-0.89)   (-1.41)

    $\bar{R}^2$: 0.975   SEE: 1674.2   D-W: 2.01   RHO(1): -0.099 F(5,98): 796.621   Period of fit: 1966.1 - 1991.4

MONEY AND FINANCE

17. MOND = 494,862.937 + 1.1084 GNP - 10,554.6406 IRB -19668.7383 Q1 - 11263.9453 Q2 - 32656.8242 Q3
           (1.20) (3.40)   (-1.73)   (-1.68)   (-0.95)   (-3.28

    $\bar{R}^2$: 0.991   SEE: 57441   D-W: 2.36   RHO(1): 0.990 F(5,97): 2135.184   Period of fit: 1966.1 - 1991.4

18. MQM = 521581.187 + 2.6963 GNP - 6276.3906 IRB+ 6223.8789 Q1 + 19671.7500 Q2 + 465.3557 Q3
         (1.03)   (6.73)   (-0.83)   (0.45)   (1.35)   (0.04

    $\bar{R}^2$: 0.997   SEE: 70,737   D-W: 2.04   RHO(1): 0.990 F(5,97): 6005.289   Period of fit: 1966.1 - 1991.4

19. MDD = 53084.6328 - 2743.7419 IRB + 0.4406 (YDD$ -C$) + 0.9508 MDD$_{-1}$ + 43,318.8164 Q1 - 12515.7383 Q2
        (1.64)   (-2.16)   (4.88)   (61.91)   (4.19)   (-1.24)
    - 6213.7109 Q3
     (0.62)

    $\bar{R}^2$: 0.996   SEE: 35,565   D-W: 1.88   F(6,96): 3851.946

20. CUR = - 2,709.2395 + 0.0757 AFR$ + 0.7932 I$ + 14,408.1445 Q1 - 3741.5945 Q2 + 2555.0320 Q3
        (-0.96)   (24.04)   (24.47)   (4.55)   (-1.21)   (0.82)

    $\bar{R}^2$: 0.991   SEE: 11,113   D-W: 1.68   F(5,98): 2,193.038   Period of fit: 1966.1 - 1991.4

21. IRB = 19.7388 - 0.00001 (YDD-C$) + 0.5512 IR$_{-1}$- 0.1520 Q1 + 0.1089 Q2 + 0.0324 Q3
     (10.30) (-0.82)   (4.08)   (-0.79)   (0.57)   (0.20)

    $\bar{R}^2$: 0.916   SEE: 0.87251   D-W: 2.14   RHO(1): 0.945   F(5,97): 224.626   Period of fit: 1966.1 - 1991.4

22. E = 1.3714 - 0.00001 $\sum\limits_{i=-1}^{4}$ BOT$ + 0.9659 E$_{-1}$

    (1.60) (-3.95)   (43.57)

    $\bar{R}^2$: 0.991   SEE: 0.42691   D-W: 1.69   RHO(1): 0.496 F(2,96): 5501.281   Period of fit: 1967.1 - 1991.4

23. AFR$ = 17,763.6680 + 1.0326 BOT$ + 0.9693 AFR$_{-1}$ - 13,887.0664 Q1 - 29,645.3125 Q2 - 3,349.7246 Q3
         (2.54)   (8.71)   (135.99)   (-1.51)   (-3.23)   (-0.36)

    $\bar{R}^2$: 0.998   SEE: 33089   D-W: 1.83   F(5,98): 11170.156   Period of fit: 1966.1 - 1991.4

24. (TI-SUB)\$ = 3,704.7998 - 0.0320 GNP\$ + 0.0694 GNP\$$_{-1}$+ 0.6495 (TI-SUB)\$$_{-1}$ - 3,176.4990 Q1
            (2.08)   (-0.48)   (1.04)   (7.53)   (-1.16)
    + 8731.8633 Q2 - 7959.2773 Q3
       (4.45)   (-2.74)

    $\overline{R}^2$: 0.958   SEE: 7551.4   D-W: 2.00 RHO(1): -0.380 F(6,96): 388.490 Period of fit: 1966.1 - 1991.4

25. TAXTT\$ = - 9,314.2305 + 0.0499 GNP\$ + 0.1291 GNP\$$_{-1}$+ 8368.6562 Q1 + 21671.8711 Q2 + 7180.8867 Q3
            (-2.83)   (0.47)   (1.18)   (2.08)   (5.40)   (1.74)

    $\overline{R}^2$: 0.951   SEE: 14.448   D-W: 1.34 RHO(1): 0.317 F(5,98): 397.147 Period of fit: 1966.1 - 1991.4

26. GOVPROA\$ = - 7,459.5664 + 0.0321 (C+CG+I+J)$_{-1}$ + 8675.9883 Q1+ 24167.8086 Q2 - 1209.1252 Q3
            (-2.81) (8.60)   (2.68)   (7.46)   (-0.38)

    $\overline{R}^2$: 0.595   SEE: 11,561   D-W: 2.06 RHO(1): 0.317   F(4,97): 38.102 Period of fit: 1966.3 - 1991.4

LABOR FORCE AND UNEMPLOYMENT

27. NF = - 4.7707 + 0.6527 N - 0.0407 Q1 - 0.0827 Q2 + 0.0842 Q3
        (-20.73)   (48.92)   (-3.18)   (-5.67)   (6.66

    $\overline{R}^2$: 0.998 SEE: 0.67084E-01 D-W: 2.10 RHO(1): 0.799 F(4,98): 12,266.680 Period of fit: 1966.1 - 1991.4

28. U = 0.5615 - 0.0121 GDPIN\$ + 0.7422 U$_{-1}$ - 0.1339 Q1 - 0.0632 Q2 + 0.3206 Q3
       (4.17)   (-2.58)   (13.93)   (-1.37)   (-0.65)   (3.25)

    $\overline{R}^2$: 0.663   SEE: 0.35100 D-W: 2.53   F(5,98): 41.446 Period of fit: 1966.1 - 1991.4

WAGES AND PRICES

29. WPI = 3.3461 + 0.3314 PM + 874.8000 PWM/PDT + 0.6120 WPI$_{-1}$ + 0.7933 Q1 + 0.5702 Q2 + 0.3946 Q3
        (2.80)   (6.04)   (0.57)   (9.06)   (2.38)   (1.67)   (1.33)

    $\overline{R}^2$: 0.997   SEE: 1.4163   D-W: 1.81   RHO(1): 0.634 F(6,96): 6,238.172 Period of fit: 1966.1 - 1991.4

30. CPI = 18.3262 + 0.4866 WPI + 2.1767 (MON/GDP)$_{-1}$+ 0.2532 WPI$_{-1}$ - 0.8699 Q1 - 0.6804 Q2 - 0.1849 Q3
        (2.29)   (7.00)   (1.33)   (3.61)   (-3.35)   (-2.93)   (-0.87)

    $\overline{R}^2$: 0.999 SEE: 1.1517 D-W: 0.90 RHO(1): 0.990 F(6,95): 12,885.535 Period of fit: 1966.2 - 1991.4

31. PWM = - 66.0569 + 0.0017 PGT + 0.1338 CPI$_{-1}$ - 11.6856 (NF-NE)+ 11.7311 Q1 - 0.1896 Q2 + 2.5217 Q3
         (-12.65)   (9.58)   (0.88)   (-0.51)   (5.56)   (-0.08)   (1.25)

    $\overline{R}^2$: 0.968   SEE: 9.0455   D-W: 2.05 RHO(1): 0.491 F(6,96): 513.167 Period of fit: 1966.1 - 1991.4

32. PM = 7.1942 + 90.3326 UVIM + 1.2079 UVIM$_{-1}$ - 0.0297 Q1+ 0.0429 Q2 + 0.0065 Q3
       (3.08)   (62.73)   (0.83)   (-0.40)   (0.57)   (0.10)

    $\overline{R}^2$: 1.000   SEE: 0.36982 D-W: 1.96 RHO(1): 0.990 F(5,96): 134,591.125 Period of fit: 1966.2 - 1991.4

33. UVIM = 0.0933 + 0.0012 PXIUSJA + 0.0045 PO + 0.0118 E- 0.0149 Q1 - 0.0085 Q2 - 0.0066 Q3
         (0.36)   (2.13)   (3.02)   (2.30)   (-2.77)   (-1.38)   (-1.24)

    $\overline{R}^2$: 0.991   SEE: 0.28635E-01 D-W: 1.34 RHO(1): 0.990 F(6,87): 1651.696 Period of fit: 1968.2 - 1991.4

34. PX = 9.0780 + 0.3232 PM + 0.5219 PM$_{-1}$
       (3.65)   (5.13)   (8.32)

    $\overline{R}^2$: 0.995   SEE: 1.8055   D-W: 2.17   RHO(1): 0.831 F(2,100): 10,997.086 Period of fit: 1966.1 - 1991.4

35. UVIX = - 0.1144 + 0.0099 WPI + 0.0026 E
               (-1.99)   (43.87)   (1.96)

$\overline{R}^2$: 0.993   SEE: 0.22132 E-01   D-W: 1.71   RHO(1): 0.615   F(2,100): 7482.328   Period of fit: 1966.1 - 1991.4

36. PCF = - 0.9701 + 1.3728 CPI - 0.3399 PCF_{-1}
               (-0.34)   (21.43)   (-5.97)

$\overline{R}^2$: 0.999   SEE: 1.1592 D-W: 1.68   RHO(1): 0.930   F(2,100): 39,545.051   Period of fit: 1966.1 - 1991.4

37. PCO = 0.3009 + 0.5274 CPI + 0.4670 PCO_{-1}
              (0.36)   (8.23)   (7.18)

$\overline{R}^2$:0.999   SEE: 0.86884   D-W: 2.20   RHO(1): 0.774   F(2,100): 68311.187   Period of fit: 1966.1 - 1991.4

38. PCG = - 11.3906 + 1.1016 CPI + 6.3383 Q1 + 0.4331 Q2 + 0.1828 Q3
               (-6.95)   (56.90)   (5.83)   (0.36)   (0.17)

$\overline{R}^2$: 0.984   SEE: 4.4709   D-W: 2.09   RHO(1): 0.126   F(4,98): 1,590.430   Period of fit: 1966.1 - 1991.4

39. PIBF = 17.1396 + 0.9269 WPI - 0.1248 WPI_{-1}
               (1.97)   (12.00)   (-1.61)

$\overline{R}^2$: 0.998   SEE: 1.2962 D-W: 1.89   RHO(1): 0.990   F(2,100): 25,399.328   Period of fit: 1966.1 - 1991.4

40. PIPC = 19.9183 + 0.8799 WPI - 0.1185 WPI_{-1}
               (2.02)   (10.03)   (-1.35)

$\overline{R}^2$: 0.997   SEE: 1.4718 D-W: 2.32   RHO(1): 0.990   F(2,100): 17,101.457   Period of fit: 1966.1 - 1991.4

41. PIG = 16.0337 + 1.0274 WPI - 0.1857 WPI_{-1}
              (1.37)   (9.85)   (-1.78)

$\overline{R}^2$: 0.997   SEE: 1.7504   D-W: 1.91   RHO(1): 0.990   F(2,100): 16,759.488   Period of fit: 1966.1 - 1991.4

42. PJ = 2.3530 + 0.9502 WPI
             (0.97)   (31.01)

$\overline{R}^2$: 0.924   SEE: 7.4233   D-W: 1.99   RHO(1): 0.119   F(1,101): 1234.274   Period of fit: 1966.1 - 1991.4

43. PFIA = 18.2633 + 0.8152 CPI
               (9.84)   (34.09)

$\overline{R}^2$: 0.998   SEE: 1.2791   D-W: 2.49   RHO(1): 0.859   F(1,101): 43,958.871   Period of fit: 1966.1 - 1991.4

44. PD = 15.7139 + 0.8253 WPI
             (1.77)   (13.34)

$\overline{R}^2$: 0.998   SEE: 1.3576   D-W: 1.93   RHO(1): 0.990   F(1,101): 46,143.707   Period of fit: 1966.1 - 1991.4

45. PTISUB = 34.4534 + 0.3939 PM + 0.0351 PWM + 2.2867 Q1+ 1.5481 Q2 + 0.8440 Q3
                  (2.42)   (4.59)   (0.91)   (4.84)   (3.28)   (2.03)

$\overline{R}^2$: 0.994   SEE: 2.2243   D-W: 2.15   RHO(1): 0.990   F(5,88): 3,063.229   Period of fit: 1966.1 - 1989.3

OTHERS

46. D = 6,534.0586 + 0.0701 GDP + 553.9236 Q1 + 476.9839 Q2 - 89.2071 Q3
            (0.92   (11.31)   (2.56)   (2.17)   (-0.48)

$\overline{R}^2$: 0.998   SEE: 1.079.5   D-W: 2.16   RHO(1): 0.990   F(4,98): 14,926.277   Period of fit: 1966.1 - 1991.4

DEFINITIONS AND IDENTITIES

47.	Private consumption expenditure at 1986 prices	$C = CF + CO$
48.	Government consumption expenditure at 1986 prices	$CG = 100 \times CG\$/PCG$
49.	Gross government fixed capital formation at 1986 prices	$IG = 100 \times IG\$/PIG$
50.	Gross fixed capital formation of public enterprises at 1986 prices	$IPC = 100 \times IPC\$/PIPC$
51.	Gross domestic fixed capital formation at 1986 prices	$I = IBF + IG + IPC$
52.	Exports of goods through customs	$TVX = TVXUS + TVXJA + TVXHKM + TVXOT$
53.	Exports of goods and services at 1986 prices	$X = XG + XS$
54.	Imports of goods and services at 1986 prices	$M = MG + MS$
55.	Gross domestic product at 1986 prices	$GDP = C + CG + I + J + X - M$
56.	Total demand at 1986 prices	$TD = C + CG + I + J + X$
57.	Domestic demand at current prices	$TD\$ = C\$ + CG\$ + I\$ + J\$ + X\$$
58.	Net indirect tax	$TI - SUB = 100 \times (TI - SUB)\$/PTISUB$
59.	Number of employment	$NE = NF \times (1-0.01U)$
60.	Domestic fixed capital consumption allowance at current prices	$D\$ = 0.01 \ PD \times D$
61.	Capital stock at 1986 prices	$K = K_1 + I - D$
62.	Stock of inventory at 1986 prices	$KJ = KJ_{-1} + J$
63.	Capital stock being utilized	$KE = KF \times (1-0.01U)$
64.	Private consumption expenditure on food at current prices	$CF\$ = 0.01 \ PCF \times CF$
65.	Private consumption expenditure on nonfood at current prices	$CO\$ = 0.01 \ PCO \times CO$
66.	Private consumption expenditure at current prices	$C\$ = CF\$ + CO\$$
67.	Price deflator of private consumption expenditure	$PC = 100 \ C\$/C$
68.	Gross private fixed capital formation at current prices	$IBF\$ = 0.01 \ PIBF \times IBF$
69.	Gross domestic fixed capital formation at current prices	$I\$ = IBF\$ + IG\$ + IPC\$$
70.	Price deflator of gross domestic fixed capital formation	$PI = 100 \ I\$/I$
71.	Increase in stock at current prices	$J\$ = 0.01 \ PJ \times J$
72.	Imports of goods and services at current prices	$M\$ = 0.01 \ PM \times M$
73.	Exports of goods and services at current prices	$X\$ = 0.01 \ PX \times X$
74.	Balance of trade in terms of NT$ at current prices	$BOT\$ = X\$ - M\$$
75.	Balance of trade in terms of US$ at current prices	$BOT\$\$ = BOT\$/E$
76.	Gross domestic product at current prices	$GDP\$ = C\$ + CG\$ + I\$ + J\$ + X\$ - M\$$
77.	Price deflator of GDP	$PGDP = 100 \ GDP\$/GDP$
78.	Rate of change in GDP price deflator	$PGDP^\star = 100(PGDP-PGDP_{-4})/PGDP_{-4}$
79.	Real rediscount rate	$IRR = IR - PGDP^\star$
80.	Real market interest rate	$IRRB = IRB - PGDP^\star$
81.	Potential labor productivity	$PDT = QF/NE$
82.	Labor productivity	$PGT = GDP/NE$
83.	Net factor income from abroad at 1986 prices	$FIA = 100 \ FIA\$/PFIA$
84.	Gross national product at 1986 prices	$GNP = GDP + FIA$
85.	Gross national product at current prices	$GNP\$ = GDP\$ + FIA\$$
86.	Price deflator of gross national product	$PGNP = 100 \ GNP\$/GNP$
87.	Rate of change in price deflator of gross national product	$PGNP^\star = 100 \ (PGNP-PGNP_{-4})/PGNP_{-4}$
88.	Personal disposable income (approximate value) at 1986 prices	$YDD = 100 \ YDD\$/PGDP$
89.	Personal disposable income (approximate value) at current prices	$YDD\$ = GNP\$ - TAXTT\$ -D\$$
90.	Liquid assets at 1986 prices	$LA = YDD + MQM$
91.	Direct tax	$TAXDD\$ = TAXTT\$ - (TI-SUB)\$ + SUB\$$
92.	Private savings at current prices	$PS\$ = YDD\$ - C\$$
93.	Supply of money at current prices	$MON\$ = CUR + MDD$
94.	Demand for money equals supply money	$MOND\$ = MON\$$
95.	Supply of money at 1986 prices	$MON = MON\$/(0.01 \times PGDP)$
96.	Demand for quasi-money at current prices	$MQM\$ = 0.01 \ MQM \times PGDP$
97.	Rate of change in GDP at 1986 prices	$GDP^\star = 100 \ (GDP-GDP_{-4})/GDP_{-4}$
98.	Rate of change in GNP at 1986 prices	$GNP^\star = 100 \ (GNP-GNP_{-4})/GNP_{-4}$
99.	Rate of change in GDPIN at 1986 prices	$GDPIN^\star = 100 \ (GDPIN-GDPIN_{-4})/GDPIN_{-4}$
100.	Prices of exports of goods of Japan in terms of US dollars	$PXJA\$ = PXJA/EJA$
101.	Prices of exports of goods of Hong Kong in terms of US dollars	$PXHK\$ = CPIHK \ / \ EHK$
102.	Prices of exports of goods of the U.S. in terms of NT dollars	$PXUSNT\$ = PXUS\$ \ X \ E$
103.	Prices of exports of goods of Japan in terms of NT dollars	$PXJANT\$ = PXJA\$ \ x \ E$
104.	Prices of exports of goods of Hong Kong in terms of NT dollars	$PXHKNT\$ = PXHK\$ \ x \ E$
105.		$PXUS/WPI = PXUSNT\$/WPI$
106.	Prices of exports of goods and services	$PXROC = PX$

107. Prices of exports of goods and services in terms of US dollars   PXROC$ = PX / E
108. Total exports of goods to the U.S. and Japan   TVXUSJA = TVXUS + TVXJA
109. The weighted average export price index of the US and Japan   PXIUSJA$ = RXUS x PXUS$ + RXJA x PXJA$
110. The ratio of exports of goods to the US to total exports of goods to the US and Japan   RXUS = TVXUS/TVXUSJA
111. The ratio of exports of goods to Japan to total exports of goods to the US and Japan   RXJA = TVXJA/TVXUSJA
112. The weighted average GNP index of the US and Japan   GNPIUSJA = RXUS x GNPIUS + RXJA x GNPIJA
113. Gross service product at 1986 prices   GDPSE = GDP - GDPIN - GDPAG

Value unit: NT$ million   Definitions of Variables
      US$ million

AFR$ = Foreign exchange reserves of all banks; BOT$ = Balance of trade in terms of the NT dollar at current prices;
BOT$$ = Balance of trade in terms of the U.S. dollar; C = Private consumption expenditure at 1986 prices;
C$ = Private consumption expenditure at current prices;
CF = Private consumption expenditure on food at 1986 prices;
CF$ = Private consumption expenditure on food at current prices;
CG = Government consumption expenditure at 1986 prices;
CG$ = Government consumption expenditure at current prices;
CO = Private consumption expenditure on nonfood at 1986 prices;
CO$ = Private consumption expenditure on nonfood at current prices; CPI = Consumer price index;
CPIHK = Consumer price index of Hong Kong; CUR = Net amount of currency in circulation;
D = Fixed capital consumption allowance at 1986 prices;
D$ = Fixed capital consumption allowance at current prices; E = Exchange rate: New Taiwan dollar/U.S. dollar;
EHK = Exchange rate: Hong Kong dollar/U.S. dollar; EJA = Exchange rate: Japanese yen/U.S. dollar;
FIA = Net factor income abroad at 1986 prices; FIA$ = Net factor income abroad at current prices;
GDP = Gross domestic product at 1986 prices; GDP$ = Gross domestic product at current prices;
GDPAG = Gross agricultural product at 1986 prices; GDPIN = Gross industrial product at 1986 prices;
GDPIN* = Gross industrial product at current prices; GNP = Gross national product at 1986 prices;
GNP$ = Gross national product at current prices; GNP* = Rate of change in GNP;
GNPHK = Real GNP of Hong Kong; GNPJA = Real GNP of Japan; GNPUS = Real GNP of the U.S.;
GNPIUSJA = Weighted average GNP index of Japan and the U.S.;
GOVPROA$ = Income from government property and enterprises;
I = Gross domestic fixed capital formation at 1986 prices;
I$ = Gross domestic fixed capital formation at current prices;
IBF = Gross private fixed capital formation at 1986 prices;
IBF$ = Gross private fixed capital formation at current prices;
IG = Gross government fixed capital formation at 1986 prices;
IG$ = Gross government fixed capital formation at current prices;
IPC = Gross public enterprise fixed capital formation at 1986 prices;
IPC$ = Gross public enterprise fixed capital formation at current prices;
IR = Rediscount rate of the Central Bank of China; IRB = Market interest rate; IRR = Real rediscount rate;
IRRB = Real market interest rate; J = Increase in stock at 1986 prices; J$ = Increase in stock at current prices;
K = Capital stock at 1986 prices; KE = Capital stock being utilized at 1986 prices;
KF = Potential capital stock at 1986 prices; KJ = Stock of inventory at 1986 prices;
LA = Liquid assets at 1986 prices; M = Imports of goods and services at 1986 prices;
M$ = Imports of goods and services at current prices; MDD = Demand deposits;
MG = Imports of goods at 1986 prices; MON = Supply of money at 1986 prices;
MON$ = Supply of money; MOND = Demand for money at 1986 prices; MOND$ = Demand for money;
MQM = Demand for quasi-money at 1986 prices; MQM$ = Demand for quasi-money;
MS = Imports of services at 1986 prices; N = Population in Taiwan;
NE = Number of employment (millions); NF = Labor force (millions);
PC = Price deflator of private consumption expenditure;
PCF = Price deflator of private consumption expenditure on food;
PCG = Price deflator of government consumption expenditure;
PCO = Price deflator of private consumption expenditure on nonfood;
PD = Price deflator of fixed capital consumption allowance at 1986 prices;
PDT = Potential labor productivity; PFIA = Price deflator of net income abroad; PGDP = Price deflator of GDP;
PGDP* = Rate of change in the price deflator of GDP; PGNP = Price deflator of GNP;
PGNP* = Rate of change in the price deflator of GNP; PGT = Labor productivity;
PHA = Price deflator of net factor income from abroad; PI = Price deflator of gross domestic fixed capital formation;
PIBF = Price deflator of gross private fixed capital formation;

PIG = Price deflator of gross government fixed capital formation;

PIPC = Price deflator of gross public enterprise fixed capital formation; PJ = Price deflator of the increase in stock;

PM = Price deflator of imports of goods and services; PO = Oil price (US$/per barrel) of Saudi Arabia;

PS$ = Private savings at current prices; PTISUB = Price deflator of net indirect taxes;

PX = Price deflator of exports of goods and services;

PXHK$ = Price of exports of goods of Hong Kong in terms of the U.S. dollar;

PXHKNT$ = Price of exports of goods of Hong Kong in terms of the N.T. dollar;

PXIUSJA = Weighted average price index of exports of the U.S. and Japan; PXJA = Price of exports of Japan;

PXJA$ = Price of exports of goods of Japan in terms of the U.S. dollar;

PXJANT$ = Price of exports of goods of Japan in terms of the N.T. dollar;

PXROC = Price of Taiwan's exports of goods and services in terms of the N.T. dollar;

PXROC$ = Price of Taiwan's exports of goods and services in terms of the U.S. dollar;

PXUS = Price of exports of the U.S.; PXUSNT$ = Price of exports of the U.S. in terms of the N.T. dollar;

PWM = Average wage index in manufacturing;

Q1 = Dummy Variable, 1 for the first quarter and 0 for other quarters;

Q2 = Dummy Variable, 1 for the second quarter and 0 for other quarters;

Q3 = Dummy Variable, 1 for the third quarter and 0 for other quarters; QF = Potential GDP;

RR = Reserve ratio for deposits;

RXJA = Exports of goods to Japan/Taiwan's total exports of goods to the U.S. and Japan;

RXUS = Exports of goods to the U.S./Taiwan's total exports of goods to the U.S.and Japan;

SUB = Government subsidies at 1986 prices; SUB$ = Government subsidies at current prices;

TAF = Average tariff rate; TAXDD$ = Direct taxes at current prices;

TAXTT$ = Total tax and monopoly revenue of public enterprises; TD = Domestic demand at 1986 prices;

TD$ = Domestic demand at current prices; (TI-SUB) = Net indirect taxes at 1986 prices;

(TI-SUB)$ = Net indirect taxes at current prices;

TMMC$$ = Total imports of goods of mainland China in terms of U.S.dollar;

TVM = Imports of goods through customs at 1986 prices; TVX = Exports of goods through customs at 1986 prices;

TVXHKM = Exports of goods to Hong Kong and mainland China at 1986 prices;

TVXJA = Exports of goods to Japan at 1986 prices;

TVXOT = Exports of goods to the rest of the world other than the U.S., Japan and Hong;

TVXUS = Exports of goods to the U.S. at 1986 prices;

TVXUSJA = Total exports of goods to the U.S. and Japan at 1986 prices;

TW = World trade volume index; U = Unemployment rate; UVIM = Unit value index of imports;

UVIX = Unit value index of exports; WPI = Wholesale price index;

WPIUS = Wholesale price index of the U.S.; X = Exports of goods and services at 1986 prices;

X$ = Exports of goods and services at current prices; XG = Exports of goods at 1986 prices;

XS = Exports of services at 1986 prices; YDD = Disposable personal income at 1986 prices;

YDD$ = Disposable personal income at current prices

China's Macroeconometric Model
for Project Link

Wu Jiapei, Liang Youcai, Zhang Yaxiong

The Chinese Macroeconometric Model built by the Chinese LINK modeling group at the State Information Center of China was linked in the world econometric model system of project LINK in 1986. With other country and region models, it has performed well and has contributed to the LINK forecasts for the world economy and various policy analyses. Each year, we added new annual data and did some improvements to get the updated version of our Chinese model. This year, in order to describe new characteristics of the China's economy, we revised some functions and did structure changes in some blocks (Price index block and finance block); added a GNP block and other related functions such as imports and exports of service; we also re-estimated all the stochastic equations using the expanded annual data to 1990 and did a medium-term economic forecast for the period of 1991 to 1997.

I. The General Structure of the Model

Since 1978, the policy of reform and opening to the outside world has been adopted in China, the economic system of China has changed greatly. Consequently, the economic development after 1949 can be divided into two different periods, before and after 1978. During the first period, a traditional centrally planned economic system was dominant, during the second period the old system coexists with the new one. As a result of the difference between these two economic systems, the economic achievements are quite different. So we are faced with much more difficulties in building China's macro-econometric model, because the model has to be built on the statistical data of both these different periods.

The difficulty of building the model and forecasting also lies in the fact that China's economic fluctuation is becoming more and more distinct in recent years. For example, the growth rate of gross value of investment in fixed assets of society was 24.1 percent from 1983 to 1988 annually, and was 7 percent only during the three-year rectification from 1989 to 1991. At the same time, the gross value of industrial production and national income was 16.5, 11.1 percent and 10.3, 5.4 percent respectively. In the model, we take the function of investment in fixed assets as below:

GDIC = f { (TEXP - EXAD - FED + LOAN - LOAN (-1),
      FAG (-1), GNPC },

where,

> GDIC: Investment in fixed assets of society;
> TEXP: Total financial expenditure;
> EXAD: Financial expenditure on culture, education, science, public health, national defense and administration. etc.;
> FED: Financial expenditure on debt;
> LOAN: Total value of loan (at the end of year);
> FAG: Original value of fixed assets of enterprises;
> GNPC: GNP(nominal).

The national income production is defined as the sum of the values added in five material production sectors, which are agriculture, industry, construction, transportation/communication and commerce. Agricultural production covers farming, forestry, husbandry, fishery and sideline industries. So we can get the function:

$$GVA = (GG, GVSP)$$

where,

> GVA: Agricultural output value;
> GG: Grain output;
> GVSP: Agricultural output value except farming.

Since the rural economic system reform, farmers have been encouraged by the government to engage in agriculture production except farming. The GVSP grows faster year by year. It is mainly regulated by the labor and funds available to the farmers. So we can have:

$$GVSP = g (LFA, RISD(-1) / PIRSSC),$$

where,

> LFA: Labour force in agriculture;
> RISD: Farmer's saving deposits;
> PIRSSC: Price index of retail sales of social commodities.

Combining the two functions, we have:

$$GVA = f (GG, g (LFA, RISD(-1) / PIRSSC))$$
$$= h (GG, LFA, RISD(-1) / PIRSSC)$$

The economic system of China is going from a centrally planned-supply decision system to a planned market economic system. In the past, the force of growth came from the planned supply. During the current years, growth is pulled by social demand and restricted by investment and supply of energy and raw materials. Because the values of import and export commodities increase very fast in last 10 years, foreign trade also has important influences on industrial production. So we have:

GVLI = f ((CONS – CFSS) / PIRSSC, PIRSSC, XX59*EXRA);
GVHIS = f (GDI, (GVE + GVP), (XX + MM)*EXRA).

where,

    GVLI: Gross output value of light industry;
    GVHIS: Gross output value of heavy industry except petroleum, coal
         and power industry;
    GVP: Gross output value of power industry;
    GVE: Gross output value of petroleum and coal industry;
    CONS: Total value of consumption;
    CFSS: Farmers' self-supply consumption;
    GDI: Investment in fixed assets of society;
    XX59: Export volume of SITC5-9;
    XX: Total export volume;
    MM: Total import volume;
    EXRA: Exchange rate of Chinese yuan to U.S. dollar.

The output value of transportation is mainly determined by the development of the agriculture, industry and commodities imports. So we have:

GVTC = f ((GVIN + GVA + MM*EXRA));

where,

    GVTC: Gross   output   value of transportation and
         communication;
    GVIN: Gross output value of industry.

The sum of the gross output value of five material production sectors multiplied by their corresponding net-output ratios respectively is the national income production. The net-output ratios are exogenous in the model.

According to the MPS, national income available is distributed into accumulation and consumption, consumption is divided into private and social consumption. According to the SNA, GNP is the sum of investment, consumption and net exports. So we calculate:

CONST = CIT + CST
CIT = CI + CIS

CST = CS + CSS

where,

        CIT:Private consumption (GNP);
        CI: Private consumption (NI);
        CST:Public consumption (GNP);
        CS: Social consumption (NI);
        CONST:Total consumption (GNP);

CIS and CSS are the other private and public consumption of services which are estimated.

There are seven price indexes (excluding price indexes of exports and imports) in the model: the national income deflator, GNP deflator, the retail price index of social commodities, the price index of industrial output value, the price index of agricultural output value, the index of retail price of industrial products in rural areas and the index of cost of living for staff and workers.

The retail price index of social commodities is determined by both supply and demand. On the supply side, social commodities are dominantly provided by agriculture and industry (more than 80 percent of the national income); on the demand side, gross retail sales of social commodities are determined by the monetary income of rural and urban inhabitants. Accompanied by the imports of commodities enlarged, the price of imports goods played an important role in the domestic prices. So, we get:

PIRSSC = f (NI / TP, MII / TP, PMM59)

where,

        PIRSSC: Retail price index of social commodities;
        TP: Population;
        MII: Monetary income of inhabitants;
        PMM59: Price index of imports of SITC5-9.

The national income deflator is also determined by supply-demand relationship. By the definition, it is the ratio of national income available (nominal) plus net exports to national income production (real). Such as:

LTP = (NIC + BFT*EXRA) / NI*100

where,

        LTP: NI deflator;
        NIC: National income available (nominal);
        BFT: Trade balance (current U.S. dollar).

So, we can get the GNP deflator function as follow:

PGNP = f (LTP, PSER)

where,

PGNP: GNP deflator;
PSER: Price index of service.

Since 1979, budget system reform has been carried out in China. The state collects the taxes from the enterprises instead of sharing profits with them as before. As a result, the composition of the revenue by source has changed greatly. In 1990, 85.2 percent of revenue was dervied from taxes, and in 1978 it was only 46.3 percent. The main parts of taxes are: agricultural & animal husbandry tax, industrial & commercial tax and customs duties.

We have not worked out the equation of budget expenditure because budget revenue and expenditures could generally keep balance with an insignificant deficit which is usually controlled by the government.

The functions mentioned above in the model are:

DFR = f (TOT);
TA   = f (NVA*MFAP, TA (-1));
TIB = f ((NI-NVA)*LTP);
TCO = f (VMM*EXRA, TCO (-1)).

where,

DFR: Domestic financial revenue;
TOT: Total value of taxes;
TA: Agricultural and animal husbandry tax;
TIB: Industrial and commercial tax;
TCO: Customs duties.

Interest rates are very low in China (the real interest rate was negative in the past few years) and can not adjust the value of loan effectively, so we only use bank credit receipts to explain the total loan.

The values of import and export commodities are from China's customs statistical data. Four classes of imports and exports are given which are SITC 0-1, SITC2+4, SITC3 and SITC5-9 in line with the LINK system. The factors determining the import goods are domestic demand, foreign exchange available and the ratio of international market price at domestic currency to corresponding domestic price.

The export goods are influenced by the domestic supply, the ratio of China's export price to the world market price and the world export volume.

China holds only a small weight in world trade, which is about 2 percent in 1990. So the foreign trade prices of China are basically determined by world market prices. They are also influenced by the domestic price at the same time.

We have built the import and export of service equations either. Because of the shortage of statistical materials, data on the import and export of services are estimated. We estimate the function as:

VMMS = f (VMMS(-1), VMM)
VXXS = f (RWGDP(-1), EXRA / PSER)

where,

        VMMS: Total value of imports of service;
        VXXS: Total value of exports of service;
        RWGDP: world GDP growth rate.

II. The Model

The revised version of the model consists of 114 equations, of which 59 are stochastic and 55 identified. The stochastic equations are estimated using annual observations for the period of 1970 to 1990, most of them are using the OLS estimation method. The model contains 152 variables (exclude dummy variables), of which 114 are endogenous and 38 exogenous. The model consists of nine blocks. They are allocation of investment, allocation of labor, gross output value of society, national income production, national income distribution, GNP, price indexes, finance, and foreign trade foreign exchange.

The relations among these nine blocks are as follows:
(1)   Production is determined by both supply and demand sides. On the supply side, there are mainly the factors of energy production, import, labor supply and etc. On the demand side, there are mainly the factors of investment, consumption and export;
(2)   We get net output value and national income produced by summing up every sector's net values;
(3)   In the field of national income distribution, private consumption plus public consumption equals total consumption. Consumption and accumulation equals available (or distributed) national income . The national income deflator is the ratio of national income available subtracted by net imports to national income produced; public consumption is exogenous, private consumption is mainly determined by private income and price index while private income is determined by production and prices. Accumulation is determined by investment. Investment is determined mainly by state budget outlays, bank loans and enterprises' self-collected funds;
(4)   Bank loan and state budget are separately determined by bank's credit

receipts (mainly deposits) and state budget revenue. Total value of deposits is determined by monetary income, price and interest rates. State budget revenue is determined by production level at current prices;

(5)   Exports are restrained by domestic supply capacity, international market demand and the ratio of China's export prices to world market prices. The export prices are determined by world market prices, exchange rate and domestic prices; Import is determined by domestic demand, the ratio of import prices at domestic currency to domestic price, and available foreign currency. Import prices are mainly determined by world market prices. Imports and exports determine foreign exchange receipts and expenditures.

Equations

I. Allocation of Investment

1. GDIC = 0.5562 (TEXP − EXAD − FED + LOAN − LOAN(− 1))   − 0.7354FAG(− 1) + 0.9596GNPC
          (3.01)   (− 5.12)   (6.52)

     + 1081.81D82 + 665.539D83 + 1037.02
        (3.49)   (2.61)   (4.15)

              RSQ = 0.9894 SER = 184.91 DW = 1.81   RANGE:1976 TO 1990

2. GDI = GDIC / MFIP*100

3. AIPI = DPC*GDI

4. AIEI = DEC*GDI

II. Allocation of Labor Force

5. UP = TP − AP

6. TL = 1.0198TL(− 1) + 0.2062(TP(− 17) − TP(− 18))
        (270.02)   (2.07)

              RSQ = 1.0 SER = 0.028 DW = 1.43   RANGE: 1975 TO 1990

7. LFIN = 0.5879LFIN(− 1) + 0.0000498GDI + 0.0871D78 + 0.1644
            (6.81)   (3.85)   (4.78)   (5.78)

              RSQ = 0.9944 SER = 0.0166 DW = 1.62   RANGE:1973 TO 1990

8. LFNPS = 0.7371LFENPS(− 1) + 0.013TWL/PUHC
            (3.77)   (1.69)

              RSQ = 0.9982 SER = 0.017 DW = 1.41   RANGE:1971 TO 1990 PE

9. LFPI = 0.9112LFPI(− 1) + 9.09(E − 5)(0.0109AIPI + 0.0183AIPI(− 1)
          (14.48)   (2.41)
            + 0.0219AIPI(− 2) + 0.0219AIPI(− 3) + 0.0183AIPI(− 4) + 0.0109AIPI(− 5)) + 0.0014
                                                                                              (2.44)

RSQ = 0.9960 SER = 0.0003 DW = 1.89   RANGE: 1976 TO 1990

10. LFEI = 0.7257LFEI(− 1) + 0.0001AIEI(− 1) + 0.0174D85 + 0.0097
        (6.40)   (3.31)   (4.40)   (2.35)

        RSQ = 0.9892 SER = 0.0037 DW = 2.42   RANGE: 1970 TO 1990

11. LFCS = 0.317LFCS(− 1) + 0.0649LFA(− 1) + 0.0162D85 + 0.000038GDI - 0.1532
        (2.19)   (2.80)   (2.29)   (4.05)   (− 2.40)

        RSQ = 0.9934 SER = 0.0060 DW = 1.88   RANGE:1971 TO 1990

12. LFTC = 5.98(E − 6)DEL(1:GVS) + 0.912LFTC(− 1) + 0.007
        (3.08)   (19.68)   (2.36)

        RSQ = 0.9894 SER = 0.0037 DW = 1.29   RANGE:1973 TO 1990

13. LFCM = 0.4264LFCM(− 1) + 0.000031TRSSCC/PIRSSC*100 + 0.0252
        (13.34)   (4.81)   (3.54)

        RSQ = 0.9834 SER = 0.009 DW = 1.76   RANGE:1971 TO 1990

14. LFA = TL − LFIN − LFNPS − LFCS − LFCM − LFTC

III. Gross Output Value of Society

15. GPC = GG/(0.5*(TP + TP(− 1))

16. GVA = 0.0429GG + 0.6265RISD(− 1)/PIRSSC*100 + 2207.88LFA − 308.912D82 − 364.45D83 − 5943.03
        (3.49)   (1.41)   (6.38)   (− 3.25)   (− 3.79)   (− 6.76)

        RSQ = 0.9877 SER = 91.26 DW = 1.46   RANGE:1971 TO 1990 PE

17. GVLI = 0.9766((CONS − CFSS)/PIRSSC*100 + 36.8961PIRSSC + 0.3112XX59*EXRA + 364.447D88 − 3137.04
        (27.38)   (12.78)   (2.38)   (4.03)   (− 12.64)

        RSQ = 0.9991 SER = 81.84 DW = 1.38   RANGE:1971 TO 1990

18. GVP = 0.0062(6.7161AIPI + 10.7457AIPI(− 1) + 12.089AIPI(− 2) + 10.7457AIPI(3) + 6.7161AIPI(− 4))
        (5.42)
                + 2976.88LFPI + 0.7523GVP(− 1)
                  (2.29)   (6.51)

        RSQ = 0.9996 SER = 5.53 DW = 2.67   RANGE:1975 T0 1990

19. GVE = 0.813GVE(− 1) + 1295.48LFEI
        (7.02)   (2.28)

        RSQ = 0.9982 SER = 19.95 DW = 2.17   RANGE:1971 TO 1990

20. GVHIS = 0.3405GDI + 1.6332(GVE + GVP) + 0.594(XX + MM)*EXRA + 631.339D89 − 391.26D86 + 402.183
        (2.55)   (3.71)   (6.30)   (3.65)   (− 2.45)   (2.72)

        RSQ = 0.9951 SER = 144.24 DW = 1.47   RANGE:1971 To 1990

21. GVHL = GVHIS + GVE + GVP

22. GVIN = GVLI + GVHL

23. GVCS = 0.4517GDI + 137.51D8082 + 163.754
        (33.59)   (3.03)   (6.40)

        RSQ = 0.9852 SER = 60.15 DW = 1.14   RANGE:1971 TO 1990

24. GVTC = 0.0285(GVIN + GVA + MM*EXRA) + 25.6934D8283 + 18.9834
    (58.59)   (2.72)   (3.32)

    RSQ = 0.9951 SER = 12.63 DW = 1.22   RANGE:1971 TO 1990 PE

25. GVCM = 0.2081TRSSCC/PIRSSC*100 – 15.6392
    (97.94)   (– 2.56)

    RSQ = 0.9982 SER = 11.08 DW = 2.0   RANGE:1972 TO 1990

26. GVS = GVA + GVIN + GVCS + GVTC + GVCM

IV. National Income Production

27. NVA = GVA*ANPC

28. NVIN = GVIN*INPC

29. NVCS = GVCS*CSNPC

30. NVTC = GVTC*TCNPC

31. NVCM = GVCM*CMNPC

32. NI = NVA + NVIN + NVCS + NVTC + NVCM

V. National Income Distribution

33. MIUI = 0.3317(NI – NVA – GVVIC/MFIP*100*INPC)*LTP/100 – 189.39D88 + 123.609
    (78.49)   (– 3.95)   (6.45)

    RSQ = 0.9981 SER = 42.88 DW = 2.39   RANGE:1975 TO 1990

34. MIRI = 0.5262GVVIC + 0.8877NVA*MFAP + 9.4238PIPFS – 149.551AP + 162.872D88 – 405.904D90
    (4.45)   (8.30)   (6.08)   (– 9.40)   (2.49)   (– 2.75)

    RSQ = 0.9996 SER = 55.41 DW = 2.21   RANGE:1971 TO 1990

35. MII = MIUI + MIRI

36. CFSS = 0.3214NVA*MFAP + 21.5889AP + 302.73D88
    (19.44)   (5.33)   (4.79)

    RSQ = 0.9961 SER = 57.49 DW = 2.0   RANGE:1971 TO 1990

37. CFNSS = 0.7023CFNSS(– 1) + 0.1975MIRI + 89.0444AP + 188.859D85-614.608
    (5.27)   (2.81)   (2.27)   (3.58)   (– 2.12)

    RSQ = 0.9982 SER = 46.53 DW = 1.71   RANGE:1971 TO 1990

38. CRI = CFSS + CFNSS

39. CUI = 0.8049MIUI + 6.2926PIRSSC + 215.247D88 – 497.773
    (14.69)   (4.09)   (6.38)   (– 4.65)

    RSQ = 0.9991 SER = 30.66 DW = 2.20   RANGE:1971 TO 1990

40. CI = CRI + CUI

41. CONS = CI + CS

42. ACC = 1.08464GDIC − 505.944D87 − 661.245D88 + [AR(1) = 0.6572]
    (35.60)   (− 4.41)   (− 5.56)

    RSQ = 0.9917 SER = 119.45 DW = 1.68   RANGE:1977 TO 1990

43. NIC = ACC + CONS

44. LTP = (NIC + BFT*EXRA)/NI*100

45. ARATE = ACC/NIC*100

46. CRIPE = CRI/(0.5*(AP + AP(− 1))

47. CUIPE = CUI/(0.5*(UP + UP(− 1))

48. CIPE = CI/(0.5*(TP + TP(− 1))

49. TRSSCC = 1.0761(CONS − CFSS) + 374.335D88 − 391.953D90 + 60.2255 + [AR(1) + 0.4899]
    (188.59)   (14.73)   (− 12.01)   (2.78)

    RSQ = 0.9997 SER = 27.67 DW = 1.74   RANGE:1972 TO 1990 PE

VI. GNP.

50. CIS = 1.2906CIS(− 1) − 164.855PSER/PIRSSC + 237.495D80 + 371.482D89
    (6.97)   (− 1.46)   (2.72)   (4.54)

    RSQ = 0.9906 SER = 78.60 DW = 2.05   RANGE:1978 TO 1990

51. CIT = CI + CIS

52. CST = CS + CSS

53. CONST = CIT + CST

54. GNPC = KIIC + GDIC + CONST + (VXXT − VMMT)*EXRA

55. GNP = GNPC/PGNP*100

56. PGNP = 0.7295LTP + 0.1376PSER + 13.499
    (16.29)   (4.23)   (8.32)

    RSQ = 0.9992 SER = 0.7557 DW = 2.30   RANGE:1979 TO 1990

VII. Price Indexes

57. MFAP = 0.1856MFAP(− 1) + 0.7704PIPFS + 10.1002D89 + 6.2969
    (2.43)   (13.36)   (− 5.59)   (3.34)

    RSQ = 0.9990 SER = 1.53 DW = 1.21   RANGE;1972 TO 1990

58. MFIP = 0.4471MFIP(− 1) + 0.0122TWL/LFIN + 0.026PMM59 + 8.276D89 + 39.0293
    (4.09)   (5.67)   (1.14)   (4.78)   (4.44)

    RSQ = 0.9916 SER = 1.47 DW = 1.45   RANGE:1972 TO 1990

59. PIRSSC = 0.3056MII/TP − 0.2217NI/TP + 0.1306PMM59 + 110.697
    (19.23)   (− 10.30)   (3.00)   (24.73)

    RSQ = 0.9948 SER = 2.54 DW = 2.20   RANGE:1971 TO 1990

60. PRR = 0.3585PRR(- 1) + 1.1429MFIP - 4.6212D87 - 48.8618
         (4.95)   (13.42)   (- 2.80)   (- 16.94)

     RSQ = 0.9953 SER = 1.56 DW = 1.33   RANGE:1970 TO 1990

61. PUHC = 0.6273PIRSSC + 0.3911PSER + 9.6967D89 + 11.3031D88
          (3.15)   (2.02)   (2.76)   (3.20)

     RSQ = 0.9997 SER = 2.82 DW = 0.87   RANGE:1979 TO 1990

62. GVAC = GVA*MFAP/100

63. GVINC = GVIN*MFIP/100

VIII. <u>Finance</u>

64. DFR = 0.7926TOT - 382.078D85 + 541.04
        (35.77)   (- 4.29)   (18.16)

     RSQ = 0.9817 SER = 84.04 DW = 1.29   RANGE:1971 TO 1990

65. TEXP = DFR + FRD - BFB

66. RISD = 0.139MIRI + 6.1835(IRSD - (PIRSSC/PIRSSC(- 1) - 1)*100) + 55.8171D86 + 99.5411
        (19.53)   (2.67)   (1.64)   (4.68)

     RSQ = 0.9823 SER = 32.39 DW = 1.80   RANGE:1979 TO 1990

67. UISD = 1.3942UISD(- 1) - 238.382PIRSSC/PIRSSC(- 1)*100 - 0.3952D88 + 0.1306(MIUI - MIUI(- 1))
        (259.13)   (- 12.70)   (- 6.75)   (3.25)

     RSQ = 1.0 SER = 15.23 DW = 1.27   RANGE:1975 TO 1990

68. VKD = 1.2947UISD + 1.156EDCCD + 445.744
        (30.94)   (23.32)   (11.19)

     RSQ = 0.9991 SER = 97.71 DW = 2.04   RANGE:1970 TO 1990

69. AR = 1.1003VKD + 1.4312M0 + 251.503D87 + 2.94503 + [AR(1) = 0.5892]
        (22.02)   (6.98)   (2.73)   (3.27)

     RSQ = 0.9990 SER = 89.93 DW = 1.67   RANGE:1976 TO 1990 PE

70. LOAN = 0.904AR + [AR(1) = 0.651]
         (147.97)

     RSQ = 0.9992 SER = 84.60 DW = 1.63   RANGE:1972 TO 1990

71. TIB = 0.0018(NI - NVA)*LTP + 210.605D86 + 252.423D85 + 105.715
        (61.95)   (5.39)   (6.51)   (7.91)

     RSQ = 0.9962 SER = 37.61 DW = 2.47   RANGE:1971 TO 1990

72. TA = 0.0043NVA*MFAP + 0.906TA(- 1) + 14.958D88 - 2.2964
        (3.35)   (12.01)   (5.76)   (- 1.66)

     RSQ = 0.9880 SER = 2.31 DW = 2.80   RANGE:1971 TO 1990

73. TCO = 0.495TCO(- 1) + 0.0389VMM*EXRA + 46.5499D84 + 108.177D85
        (4.31)   (3.99)   (3.86)   (8.23)

     RSQ = 0.9882 SER = 11.85 DW = 1.55   RANGE:1971 TO 1990

74. TOT = TIB + TA + TCO + TEL

IX. Foreign Trade and Foreign Exchange

75. MM01 = - 0.0082GPC(- 1) + 1.1205MM(- 1) + 46.9206(MFAP/(PMM01*EXPA)) - 15.4459D83 + 18.9966D87
         (- 2.53)   (6.99)   (3.02)   (- 2.26)   (3.15)

         RSQ = 0.9779 SER = 5.67 DW = 1.62   RANGE:1975 TO 1990

76. MM24 = 0.0049GVLI + 0.011WX24 + 11.1619D88
         (6.21)   (4.33)   (2.05)

         RSQ = 0.9878 SER = 4.86 DW = 1.43   RANGE:1975 TO 1990

77. MM59 = 0.8866MM59(- 1) + 50.3FEA/PMM59 + 0.0264DEL(1:GVHIS)
         (10.13)   (3.18)   (1.41)
         - 43.0748PMM59*EXRA/MFIP - 40.764D82 + 100.7
         (- 1.93)   (- 2.39)   (4.42)

         RSQ = 0.9967 SER = 15.97 DW = 2.52   RANGE:1971 TO 1990

78. MM = MM01 + MM24 + MM3 + MM59 PE

79. PMM01 = 0.6185PWX01 + 0.3973PMM01(- 1) - 21.2092D87 + 21.4353D89
         (6.97)   (4.44)   (- 4.64)   (4.38)

         RSQ = 0.9982 SER = 4.41 DW = 2.21   RANGE:1975 TO 1990

80. PMM24 = 0.8454DEL(1:PWX24) + 0.9998PMM24(- 1) - 15.8882D88
         (6.80)   (99.06)   (- 4.79)

         RSQ = 0.9989 SER = 3.07 DW = 1.62   RANGE:1975 TO 1990

81. PMM3 = 2.87DEL(1:PO) + 0.9844PMM3(- 1) + 41.1462D86 - 17.3032D90
         (9.36)   (60.88)   (5.96)   (- 3.56)

         RSQ = 0.9977 SER - 4.57 DW ▪ 1.53   RANGE:1975 TO 1990

82. PMM59 = 0.8837PWX59 - 6.6704D87 + 12.4637
         (20.28)   (- 1.94)   (3.05)

         RSQ = 0.9696 SER = 3.27 DW = 1.34   RANGE:1975 TO 1990

83. VMM01 = MM01*PMM01/100

84. VMM24 = MM24*PMM24/100

85. VMM3 = MM3*PMM3/100

86. VMM59 = MM59*PMM59/100

87. VMM = VMM01 + VMM24 + VMM3 + VMM59

88. PMM = VMM/MM*100

89. XX01 = 0.6377XX01(- 1) + 0.00096GG(- 1) - 16.7525PXX01/PWX01
         (4.69)   (2.82)   (- 2.19)

         RSQ = 0.9949 SER = 3.29 DW = 2.16   RANGE:1975 TO 1990

90. XX24 = 1.2591XX24(- 1) - 18.8989PXX24/PWX24 - 0.012PWX24 + 32.8797
     (6.71)   (- 2.74)   (- 1.89)   (3.38)

    RSQ = 0.9444 SER = 2.59 DW = 2.30   RANGE:1975 TO 1990 PE

91. XX3 = 3.14GVE*GVE/GVS - 2.848PXX3/PO + 28.2112D85 - 18.3926D89
     (12.61)   (- 3.67)   (3.35)   (- 2.16)

    RSQ = 0.9692 SER = 8.0 DW = 1.49   RANGE:1971 TO 1990

92. XX59 = 0.5496XX59(- 1) + 0.03477GVLI - 30.459PXX59/PWX59 - 57.897D85 - 42.9196D89
     (6.42)   (8.0)   (- 6.56)   (- 8.0)   (- 4.83)

    RSQ = 0.9995 SER = 6.95 DW = 2.12   RANGE:1975 TO 1990

93. XX = XX01 + XX24 + XX3 + XX59

94. PXX01 = 0.3225PXX01(- 1) + 0.8958PWX01 + 9.5317D89 - 15.7806
      (5.39)   (13.46)   (4.01)   (- 3.72)

    RSQ = 0.9860 SER = 2.13 DW = 2.59   RANGE:1975 TO 1990

95. PXX24 = 0.3924PXX24(- 1) + 0.6944PWX24 + 7.7243D89 + [AR(1) = 0.6029]
      (2.54)   (4.16)   (2.18)

    RSQ = 0.9921 SER = 4.09 DW = 1.59   RANGE:1976 TO 1990

96. PXX3 = 1.0146PXX3(- 1) + 3.2689(PO - PO(- 1)) + 44.35D86 - 9.2465D88
     (69.83)   (11.83)   (7.07)   (- 2.05)

    RSQ = 0.9985 SER = 4.0 DW = 2.48   RANGE:1975 TO 1990

97. PXX59 = 0.4153PWX59 + 0.972MFIP/EXRA - 15.1671D82 - 11.6757D83
      (15.37)   (17.06)   (- 3.89)   (- 3.03)

    RSQ = 0.9984 SER = 3.72 DW = 1.79   RANGE:1975 TO 1990

98. VXX01 = XX01*PXX01/100

99. VXX24 = XX24*PXX24/100

100. VXX3 = XX3*PXX3/100

101. VXX59 = XX59*PXX59/100

102. VXX = VXX01 + VXX24 + VXX3 + VXX59

103. PXX = VXX/XX*100

104. VMMS = 0.7053VMMS(- 1) + 0.042VMM + 54.2057D80 + 42.2412D81
       (6.38)   (2.48)   (4.22)   (3.16)

    RSQ = 0.9752 SER = 12.49 DW = 1.96   RANGE;1979 TO 1990

105. VXXS = 1.913RWGDP(- 1) + 524.365EXPA/PSER + 27.1708D83
       (5.80)   (0.44)   (2.38)
       - 19.9419D86 - 130.716
       (- 1.60)   (- 4.15)

    RSQ = 0.8842 SER = 10.65 DW = 2.73   RANGE:1979 TO 1990

106. VMMT = VMM + VMMS

107. VXXT = VXX + VXXS

108.BFT = VXX − VMM

109. FERT = 0.831VXX − 30.9658D90 + 10.0373
          (50.85)   (− 2.81)   (2.25)

          RSQ = 0.9964 SER = 8.74 DW = 2.05   RANGE:1975 TO 1990

110. FEET = 0.7983VMM + 51.95D84 + 71.2464D90 + 22.1082
          (35.55)   (3.42)   (4.37)   (3.08)

          RSQ = 0.9926 SER = 14.67 DW = 1.74   RANGE:1975 TO 1990

111. FER = FERT + FERNT + FERO

112. FEE = FEET + FEENT + FEEO

113. FES = FES(− 1) + FER − FEE

114. FEA = FES(− 1) + FER PE

List of Variables

I. ENDOGENOUS

ACC   Total value of accumulation;   AIEI   Investment in fixed assets in petroleum and coal   industry;
AIPI   Investment in fixed assets in power industry;   AR   All receipts of bank credit;
ARATE Accumulation rate;   BFT   Balance of foreign trade;   CFNSS Farmers' consumption on commodities;
CFSS   Farmers' self − supply consumption;   CI Consumption of inhabitants;
CIPE   Per capita consumption of inhabitants;   CIS   Other Service consumption of inhabitants;
CIT   Total consumption of inhabitants; CONS   Total consumption (NI)   CONST Total consumption (GNP)
CRI   Consumption of rural inhabitants; CRIPE Per capita consumption of rural inhabitants;
CST   Total social (public) consumption; CUI   Consumption of urban inhabitants;
CUIPE Per capita consumption of urban inhabitants; DFR   Domestic financial revenue;
FEA   Foreign exchange available;   FEE   Total expenditure of foreign exchange;
FEET   Foreign exchange expenditure on trade;   FER   Foreign exchange revenue;
FERT   Foreign exchange revenue from trade; FES   Foreign exchange reserve at the end of year;
GDI   Investment in fixed assets of society (1980 price);   GDIC   Investment in fixed assets of society;
GNP   Real GNP; GNPC Nominal GNP; GPC   Per capita grain output; GVA Gross output value of agriculture;
GVAC   Gross output value of agriculture at current price;   GVCM   Gross output value of commerce;
GVCS   Gross output value of construction;   GVE   Gross output value of petroleum and coal industry;
GVHIS Gross output value of heavy industry except petroleum, coal and power industry;
GVHL   Gross output value of heavy industry; GVIN   Gross output value of industry;
GVINC Gross output valve of industry at current price; GVLI Gross output value of light industry;
GVP   Gross output value of power industry; GVS   Total gross output value of society;
GVTC   Gross output value of transport and communication;   LFA   Number of labor force in agriculture;
LFCM   Number of labor force in commerce; LFCS   Number of labor force in construction;
LFEI   Number of labor force in petroleum and coal industries;   LFIN   Number of labor force in industry;
LFNPS Number of labor force in non − material product sectors;   LFPI   Number of labor force in power industry;
LFTC   Number of labor force in transportation and communication; LOAN   Total value of loan (at the end of year);
LTP   National income deflator; MFAP   Price modification factor of agriculture;
MFIP   Price modification factor of industry; MII   Monetary income of inhabitants;
MIUI   Monetary income of rural inhabitants; MIRI   Monetary income of urban inhabitants;
MM   Total import volume (1980 price, same below);   MM01   Import volume of SITC 0+1;
MM24   Import volume of SITC 2+4; MM59   Import volume of SITC 5+9;   NI National income at 1980 constant price;
 NIC   National income distributed (current price); NVA   Net output value of agriculture;
NVCM   Net output value of commerce; NVCS   Net output value of construction; NVIN   Net output value of industry;
 NVTC   Net output value of transport and communication; PGNP   GNP deflator
PIRSSC Price index of retail sales of social commodities; PMM   Price index of import commodities;

PMM01 Price index of import of SITC 0+l;   PMM24 Price index of import of SITC 2+4;

PMM3   Price index of import of SITC 3;   PMM59 Price index of import of SITC 5 TO 9;

PRR   Index of retail price of industrial products in rural area;   PUHC Index of cost of living price of staff and workers;

PXX   Price index of export commodities;   PXX01 Price index of SITC 0+l;   PXX24 Price index of SITC 2+4;

PXX3   Price index of SITC 3;   PXX59 Price index of 5 TO 9;

RISD   savings deposits of rural residents (end – year figure);   TA Agriculture and animal husbandry tax;

TCO   Customs duties;   TEXP Total financial expenditure;   TIB   Industrial and commercial tax;

TL   Total number of labor force;   TOT   Total value of taxes;

TRSSCC Total retail sales of social commodities   at current price;

UISD Savings deposits of urban residents (end – year figure);   UP Urban population;   VKD   Total value of deposits;

VMM Imports of commodity;   VMMS Imports of service;   VMMT Imports of commodity and service;

VMM01 Imports of SITC0+1;   VMM24 Imports of SITC2+4;   VMM3 Imports of SITC3;

VMM59 Imports of SITC5 – 9;   VXX   Exports of commodity;   VXXS Exports of service;

VXXT Exports of commodity and service;   VXX01 Exports of SITC0+1;   VXX24 Exports of SITC2+4;

VXX3 Exports of SITC3;   VXX59 Exports of SITC5 – 9;   XX Total volume of exports;

XX01   Export volume of SITC 0+1;   XX24 Export volume of SITC 2+4;   XX3   Export volume of SITC 3;

XX59   Export volume of SITC 5 TO 9;

II. EXOGENOUS

ANPC   Ratio of net output value to gross output value in agriculture;   AP Agricultural population;

BFB   Balance of financial budget;   CMNPC Ratio of net output value to gross output   value in commerce;

CS   Social (public) consumption;   CSNPC Ration of net output value to gross output   value of construction;

CSS   Other social (public) consumption of service;

DEC   Ratio of investment of fixed assets in petroleum and coal industry to total investment of society;

DPC   Ratio of investment of fixed assets in power industry to total investment of society;

D##   Dummy Variables (19## = 1 others 0): D##?? Dummy Variables (19## and 19?? = 1(others o);

EDCCD Enterprises and capital construction deposits;

EXAD   Financial expenditure on culture, education, science, public health,national defense and   state administration;

EXRA   Exchange rate of RMB Yuan to US\$;   FAG Original value of fixed assets of enterprises;

FED   Financial expenditure on debt;   FEENT Foreign exchange expenditure on non – trade (service);

FEEO   Foreign exchange expenditure on other purposes;   FERNT Foreign exchange revenue from service;

FERO   Foreign exchange revenue from other sources;   FRD Financial revenue from debt and borrowing;

GG   Grain output (unit:10 thousand ton);   GVVIC Gross output value of village industry (nominal);

INPC   Ratio of net output value to gross output value of industry;   IRSD Annual interest rate of savings deposits;

KIIC   Investment in circulating assets of society;   MM3 Import volume of SITC3;   M0 Currency in circulation;

PIPFS   Price index of purchases of farm and sideline products;   PO World market crude oil price;

PSER   Price index of service;   PWX01 Price index of world export of SITC 0+1;

PWX24 Price index of world export of SITC 2+4; PWX59 Price index of world export of SITC 5 – 9;

RWGDP Growth rate of world GDP;

TCNPC Ratio of net output value to gross output value of transportation and communication;   TEL Other taxes;

TP   Total population;   TWL Total wage of staff and workers;   WX24 Total world export of SITC 2+4.

III. China's Economic Prospects Based on the Model

Since Mr. Deng Xiaoping made his important speech in early 1992, the Chinese economy has entered a high growth phase. China ended 1992 with an overall performance even more impressive than before, recording a real GNP growth rate of 12.8 percent, with low inflation. Exports turned in another remarkable year's performance, growing 18.2 percent to reach US\$ 85 billion. Imports grew even faster, surging 26.2 percent to a total of U.S.\$ 80.6 billion. In addition, the economy attracted a remarkable level of foreign direct investment.

Entering 1993, this high growth pace of development has continued unabated. GNP grew at 13.9 percent in the first half of 1993, and the rapid pace of growth of domestic demand is beginning to affect the external sector. Import demand remains very high, growing at about 25 percent, but export growth has fallen to only 4.4 percent, reflecting diversion of export supply to the domestic market. The continued rapid growth was facilitated by the relaxed monetary policy and money supply.

Some signs did indicate that the economy was being stretched to its limits. Transportation bottlenecks, and shortages in certain key industrial raw materials and in energy suggest that the Chinese economy is growing at an unsustainable rate. Another very important issue is the threat of more broad based inflation in light of the government's relaxed monetary policy stance and over - expansion of investment.

In June, 1993, the State Council decided to restore financial order, enforce financial discipline and strengthen macro - economic control and promulgated a "16 points policy". The aim of the "16 points policy" is to rectify financial order and deepen financial reforms, and maintain sustained, steady and fast economic growth.

Released statistics from the State Statistical Bureau show that 1993 summer grain output achieved a harvest, recording a new level of 107.98 million tons, an increase of 4.7 million tons over 1992. Husbandry and fishery had a steady growth. In the first half of 1993, the output of pork, beef and mutton reached 14.4 million tons, growing by 9 percent over the same period of last year. Thanks to the people's hard work, the total output of crops will keep the same level as 1992 despite the decrease in sown area and bigger floods in most regions. The output of husbandry and fishery will increase more quickly than 1992. The gross output value of agriculture will increase by about 3.5 percent based on our forecast using the Chinese LINK model.

Since July 1992, the monthly growth rates of industrial production have kept more than 20 percent for 14 months in succession. According to the preliminary statistics, gross output of industry (excluding village industry) in the first eight months reached 2237.41 billion RMB Yuan, an increase of 24. 8 percent over the same period in 1992, calculated in terms of comparable prices, of which light industry accounted for 1043.2 billion RMB Yuan, up 22.5 percent, and heavy industry for 1194.21 billion RMB Yuan, up 26.9 percent. Output from collectives, especially township and village enterprises, and joint ventures grew the fastest. As a result, the non - state sector now accounts for over 50 percent of gross industrial output. Expansion is not evenly balanced throughout the country, and the coastal provinces contribute over 60 percent of the growth in the industrial output.

The rapid growth in industrial output is driven by a sharp rise in fixed investment. Fixed investment grew by more than 60 percent in the first half of 1993, of which state - owned sectors increased by 70.7 percent. After the central bank began to tighten their control over the country's money market, the speed of the industrial production began to fall. According to the business cycle, China's economy has tided over the peak, gradually reaching a rational level. The annual growth rate of industrial output will be 21.1 percent in 1993, based on our forecasting.

According to the Customs statistics, China's total trade value reached US$ 110.558 billion in the first eight months of 1993, an increase of 15.1 percent over the same period of 1992. Exports accounted for US$ 52.414 billion, only up 4.0 percent, while imports accounted for US$ 58.114 billion, up 27.3 percent.

There are two factors responsible for the slowing down of exports. First, with domestic demand rising, a lot of products that would otherwise have been exported are finding profitable outlets on the local market. Second, it appears that a number of implicit subsidies that were previously reserved for exports are no longer being applied, making export activity less attractive than before.

Import growth, as in the past, is fuelled by the strong rise in investment demand and industrial production. Along with the cooling off of the economy, export growth rate will be up, while import growth rate will be down. Exports and imports in fiscal 1993 are estimated to increase by 6.2 and 20.8 percent respectively over 1992. The trade balance will be US$7.14 deficit from the surplus of US$4.39 in 1992.

In 1992, China announced the extension of its open door policies to 28 cities along the Yangtze river and to 13 border cities in the North East, South West regions. New sectors such as transport, retails, and real estate were thrown open to foreign investors. Partly as a result, FDI flows show a strong growth trend. FDI flows in 1992 reached US$ 11.2 billion, an increase of more than 200 percent over 1991, and equivalent to almost a third of cumulative FDI flows into China since 1979. This year's FDI flows into China are expected to reach US$ 15 - 20 billion.

The rapid growth of investment and consumption demand will lead to an increase of about 13.1 percent of GNP in 1993.

Although money supply in 1992 showed strong growth over 1991, the general retail price index recorded only a moderate rise of about 5.4 percent over 1991, and about half of this can be attributed to price reform rather than price pressure. However, price rises have accelerated over the first seven months of 1993. Price indices have been rising faster every month. In the first six months, the price index of living costs of resident rose 12.5 percent over the same period of 1992, of which urban areas up 13 percent (for the 35 large and medium cities up 17.4 percent) , rural areas up 11.5 percent. The general retail price index rose 10.5 percent. In July, general retail price index increased by 14.9 percent, the price index of living costs of residents rose 13 percent, for urban areas up 14.1 percent (for the 35 cities up 18.2 percent), for rural areas up 12.2 percent. At the same time, the first half of 1993 saw further tightening in the markets for production materials. As a result, price inflation of production materials has picked up sharply with the price index of the "means of production" rising more than 40 percent by end – June. Obviously, the inflation rate showed a strong growth trend. The rapid increase in the loan and money supply, the depreciations of Chinese RMB against U.S. dollar and the sharp expansion of investment and consumption demand will continuously push domestic prices upward for a while, despite tightening the money supply and raising interest rates, because the measures of bring inflation rate down take effect in certain time. For the whole year of 1993, GNP deflator will rise about 15.5 percent, the general retail price index will rise 12.5 percent.

China's economic prospects for the remaining years in the decade, to a high degree, will be affected greatly by the policies and world economic situations. China made tremendous achievements in the 1980s, mainly because of the economic reform policies of the and opening up to the outside world. And we shall continue to carry out the successful policies to promote development of the economy.

Agriculture

Agriculture, particularly grain production,has played a very important role in stabilizing economy and society, because China has a population of more than 1.1 billion people. According to the Eighth Five – Year Plan and the tentative ideas of Ninth Five – Year Plan, the state will continue to adopt a series of measures, such as increasing investment in agriculture and deepening reform, to maintain a stable growth in output of grain, cotton and other key farm products. In order to adjust unreasonable agricultural structure, the state will pay great attention to develop husbandry, the fishing and sideline industries. Husbandry, fishing and sideline production will develop more quickly than before. Therefore, agriculture will develop steadily in the next five years. The average annual growth rate of agriculture will remain around 4.2 percent.

Industry

In order to maintain a rational speed of GDP, China will accelerate economic reforms so as to quickly establish a socialist market economy system. The state will change the way in which state – owned enterprises operate and increase their vitality and efficiency. To accelerate the changes in the functions of the government is another measure to further aid economic reform. The state will separate the functions of the government from those of enterprises. Governments at all levels should refrain from intervening in the business of enterprises, so as to increase the vitality of enterprises. All these measures will promote the development of industry in the next five years.

With a growing number of enterprises financed by foreign investment becoming operational, the strong growth of exports will continue. The strong investment demand will also keep for a long time. The increase of investment and exports will lead to a high growth of industry output. The annual growth rate of industrial output in the next five years will reamin about 14.5 percent or so. Inside industry, light industry will develop a little faster than heavy industry.

The External Sector

In the past few years, China made a series of reforms in the external sectors. A new foreign trade system has been established. Under the new system, foreign trade firms are responsible for their own profits and losses. The new rules also abolished all export subsidies, as a giant step toward free trade. In order to expand foreign trade and hard currency earnings, the state has set to empower more enterprises to deal directly in foreign trade. More and more outward oriented enterprises financed by foreign investment become operational. The exports of foreign funded enterprises will account for a bigger proportion in China's total exports. The fast growth in border trade will continue. All these will promote expansion of exports. For the imports, along with the development of domestic production and construction. China will appropriately increase the imports of advanced technology and key equipment, and some key raw materials. Exports and imports are expected to increase by an average annual growth rate of 14 and 16 percent respectively in the next five years.

In the 1990s, China's exchanges and economic cooperation with other countries (regions) will achieve greater achievements. FDI flows into China will keep a strong growth trend. By our forecast, in the next five years, China's GNP will grow at an average annual rate of 8 – 9 percent, GNP deflator will be about 8.1 percent.

Table 1: Model Forecasts
Aggreagate Demands **(Hundred Millions of Yuan, Current Prices)**

		1992	1993	1994	1995	1996	1997	1998
Priv.consumption	CIT	12154.33	15064.38	17196.84	19298.09	21933.88	24282.73	27034.95
			23.94%	14.16%	12.22%	13.66%	10.71%	11.33%
Publ.consumption	CST	2369.70	2821.00	3312.00	3843.00	4460.00	5157.00	5940.00
			19.04%	17.41%	16.03%	16.06%	15.63%	15.18%
Tot. consumption	CONST	14524.04	17885.38	20508.84	23141.09	26393.88	29439.73	32974.95
			23.14%	14.67%	12.83%	14.06%	11.54%	12.01%
Gross dom. inv.	GDIC	7460.00	10053.18	12406.32	14706.97	17336.53	20039.31	23073.19
			34.76%	23.41%	18.54%	17.88%	15.59%	15.14%

Balance of Payments (Hundred Millions of US Dollars, Current Prices)

		1992	1993	1994	1995	1996	1997	1998
Exports	VXX	849.99	902.28	1059.98	1221.98	1385.40	1579.69	1789.61
			6.15%	17.48%	15.28%	13.37%	14.02%	13.29%
Imports	VMM	806.10	973.66	1112.55	1263.09	1424.19	1604.56	1791.89
			20.79%	14.26%	13.35%	12.75%	12.66%	11.67%
FOB trade bal	BFT	43.89	- 71.38	- 52.57	- 41.12	- 38.80	- 24.86	- 2.28
			- 26.35%	- 21.78%	- 5.64%	- 35.93%		
Ex. rate (LOC/$)	EXRA	5.52	5.98	6.48	6.58	7.58	8.08	8.58
			8.33%	8.36%	1.54%	15.20%	6.60%	6.19%

Gross Value Production (Hundred Millions of Yuan, 1990 Prices)

		1992	1993	1994	1995	1996	1997	1998
Agriculture	GVA	8454.01	8749.90	9073.65	9475.61	9836.63	10230.10	10673.06
			3.50%	3.70%	4.43%	3.81%	4.00%	4.33%
Industry	GVIN	33329.80	40369.48	47363.91	54152.42	61860.17	70570.11	79621.70
			21.12%	17.33%	14.33%	14.23%	14.08%	12.83%
Industry Light	GVLI	16252.75	19894.53	23487.64	27081.33	30990.10	35539.80	40188.66
			22.41%	18.06%	15.30%	14.43%	14.68%	13.08%
Industry Heavy	GVHL	17077.04	20474.94	23876.28	27071.08	30869.89	35030.98	39435.29
			19.91%	16.61%	13.38%	14.03%	13.48%	12.57%
Gross Na. Prod.	GNP	21497.24	24306.34	26734.79	29218.33	31785.46	34604.59	37441.84
			13.07%	9.99%	9.29%	8.79%	8.87%	8.20%
Grain output *	GG	44266.00	44266.00	44624.55	44722.78	44821.12	44919.73	45018.55
			0.00%	0.81%	0.22%	0.22%	0.22%	0.22%

- - - - - - - -

* - - Ten thousand tons

Prices (base year 1990)

		1992	1993	1994	1995	1996	1997	1998
Price Index	PIRSSC	106.35	119.67	129.34	139.48	148.49	158.51	168.20
			12.52%	8.08%	7.84%	6.46%	6.75%	6.11%
GNP deflator	PGNP	108.03	124.82	136.28	148.51	159.27	171.64	184.02
			15.55%	9.17%	8.98%	7.24%	7.77%	7.21%

Population (Ten thousands)

		1992	1993	1994	1995	1996	1997	1998
Mid - year rural	AP	84799.00	85519.00	86246.00	86980.00	87710.00	88447.00	89181.00
			0.85%	0.85%	0.85%	0.84%	0.84%	0.83%
Mid - year urban	UP	38687.00	32372.00	33424.00	34477.00	35528.00	36580.00	37633.00
			3.25%	3.15%	3.05%	2.96%	2.88%	2.80%

IV. Conclusion

In general, the gross domestic product is determined by the capacity of production in China. Besides, the shortage of energy and raw materials has been a bottle – neck of economic development which restricts the industrial growth, and demand is becoming a main factor pulling up the growth of industry. In the most cases, aggregate demand is greater than aggregate supply. On the one hand, the demand expansion promotes production development. On the other hand, the excessive demand leads to severe inflation. For instance, in the past few years, the growth in consumption demand promoted the light industrial development directly and the changes in investment of industrial production. The weak demand of recent years before 1992 causes a very slow growth of industry.

Just because the current Chinese economy is jointly determined by both supply and demand, our model is supply and demand – oriented.

China's Macroeconometric Model for Project Link was built under MPS accounting system as other Chinese econometric models because the data which can be generated are those with the meanings of MPS. Now, we are doing the research work on building a macroeconometric model based on the SNA accounting system.

Jiapei Wu, Youcai Liang and Yaxiong Zhang, State Information Center, No. 58 Sanlihe Road, Beijing 100045, People's Republic of China.

REFERENCES

Klein, Lawrence R. (1983). *Lectures in Econometrics.* Amsterdam: North-Holland.
Klein, Lawrence R. & Young, R. M. (1980). *An Introduction to Econometric Forecasting and Forecasting Models.* Lexington Books.
Wu, Jiapei et al. (1983). *Quantitative Economic Theory: Models and Forecasting.* Energy Publisher.
Wu, Jiapei et al. (1986). *Studies on China Macro-Economic Modeling.* An Hui People's Publisher.
Wu, Jiapei et al. (1986). *Application of Econometric Methods in China.* China's Outlook Publisher.

Author Index

Subject Index

578